Nanomedicine for Inflammatory Diseases

Nanomedicine for Inflammatory Diseases

Edited by
Lara Scheherazade Milane
Mansoor M. Amiji

CRC Press
Taylor & Francis Group
Boca Raton London New York

CRC Press is an imprint of the
Taylor & Francis Group, an **informa** business

CRC Press
Taylor & Francis Group
6000 Broken Sound Parkway NW, Suite 300
Boca Raton, FL 33487-2742

First issued in paperback 2020

© 2017 by Taylor & Francis Group, LLC
CRC Press is an imprint of Taylor & Francis Group, an Informa business

No claim to original U.S. Government works

ISBN-13: 978-1-4987-4978-7 (hbk)
ISBN-13: 978-0-367-65783-3 (pbk)

Library of Congress Cataloging-in-Publication Data

Names: Milane, Lara, editor. | Amiji, Mansoor M., editor.
Title: Nanomedicine for inflammatory diseases / [edited by] Lara Milane and Mansoor M. Amiji.
Description: Boca Raton, FL : CRC Press/ Taylor & Francis Group, 2017. | Includes bibliographical references.
Identifiers: LCCN 2016053679 | ISBN 9781498749800 (hardback : alk. paper)
Subjects: | MESH: Autoimmune Diseases--therapy | Inflammation--therapy | Nanomedicine--methods
Classification: LCC RB131 | NLM WD 305 | DDC 616/.0473--dc23
LC record available at https://lccn.loc.gov/2016053679

Visit the Taylor & Francis Web site at
http://www.taylorandfrancis.com

and the CRC Press Web site at
http://www.crcpress.com

I dedicate this work to my family—my lovely wife and our three wonderful daughters. I also dedicate this work to past and present postdoctoral associates and graduate students who have contributed so much to the research success of my group.

MANSOOR M. AMIJI

I dedicate this work in loving memory to my twin sister, Samantha Tari Jabr; thank you for being my soulmate and for your unwavering love that keeps me going. I also dedicate this work to my daughter, Mirabella; you are the light of my life and I thank you for your eternal brilliance.

LARA SCHEHERAZADE MILANE

CONTENTS

PREFACE

Nanomedicine for Inflammatory Diseases is a critical resource for clinicians seeking advancements in the standard of care for inflammatory disease, for educators seeking a textbook for graduate-level courses in nanomedicine, and for both clinicians and scientists working at the intersection of inflammatory disease, nanomedicine, and translational science. *Nanomedicine for Inflammatory Diseases* unites the expertise of remarkable clinicians treating patients with inflammatory disease and high-caliber nanomedicine scientists working to develop new therapies for treating these diseases with the insight of translational medicine specialists, bridging the gap between the laboratory benchtop and the clinical bedside.

The effective treatment of inflammatory disease is a persistent clinical challenge, and managing inflammatory disease impacts the quality of life of many patients; asthma and multiple sclerosis are illustrative of these challenges. The inflammatory response and chronic inflammation is widespread in common disease. Prevalent diseases such as neurodegenerative disease, cancer, and diabetes are now being evaluated and understood in the context of inflammatory disease. Recent advances in immunology and immunotherapies have provided new insight into the molecular biology of the inflammatory response and inflammatory disease. New nanomedicine therapies have been developed to address the deficit of effective treatments for inflammatory disease and exploit the biology of these diseases. Nanomedicine offers many unique advantages for treating inflammatory disease, such as improved pharmacokinetics and decreased toxicity. Yet, the majority of these nanomedicine therapies have not transitioned into clinical application. The objective of this book is to promote the understanding and action of translation of nanomedicine for inflammatory disease by offering well-needed discussions of the challenges and details. The book is divided into three sections to address the fundamentals, primary inflammatory disease, and secondary inflammatory disease.

Part 1 covers the fundamentals. Chapter 1, "Fundamentals of Immunology and Inflammation," introduces the details of the inflammatory response, explains how these details can go awry and lead to chronic inflammation, and discusses exciting new discoveries, such as the formation of neutrophil extracellular traps. Neutrophil extracellular traps occur when neutrophils essentially sacrifice themselves to capture pathogens by unraveling their DNA and using DNA as a "net" to trap pathogens. Chapter 2, "Principles of Nanomedicine," answers some important questions, such as, what can nanomedicine really do, and what are the best nanomedicine formulations for particular applications? How are common nanomedicines made, and what is the fate of nanomedicine in the body? Chapter 3 addresses the important topic of nanotoxicity: What are the unique safety concerns that must be considered for the clinical use of nanomedicine? What are the main toxicity concerns, and how are they evaluated? Chapter 4, "Translational

Nanomedicine," discusses the history and progress in nanomedicine translation and highlights a crowning precedent for nanomedicine translation: the National Cancer Institute's Nanotechnology Characterization Laboratory (NCL). The NCL is developing and establishing standardized protocols with the National Institute of Standards and Technology and successfully outlining the process for nanomedicine translation for cancer. Although this is just for cancer, this is a powerful step for translational nanomedicine, as there is now a clear path to follow. This chapter also discusses the challenges of nanomedicine translation and the need for deliberate translational design with a schema for this design process.

Part 2 focuses on primary inflammatory disease, disease with established inflammatory etiology. The section foreword discusses rheumatoid arthritis as establishing a precedent for nanomedicine in primary inflammatory disease, as there are current clinical trials evaluating glucocorticoid liposomes for the treatment of rheumatoid arthritis. This section then goes into three disease-focused chapters for which nanomedicine translation is imperative: inflammatory bowel disease, multiple sclerosis, and asthma. Each chapter is divided into three sections:

- Section 1: Focuses on the biology of the disease and the current standard of care for the clinical treatment of the disease. The etiology and epidemiology of the disease are discussed, as are the specific concerns, challenges, and deficits for treatment.

- Section 2: Focuses on the nanomedicine in development for treating the disease. Nanomedicine and formulation design for the disease is contextualized and discussed. The current status of the disease-specific therapeutics that are being researched and evaluated in nanomedicine formulations is portrayed.

- Section 3: Focuses on the issues and challenges of bridging the gap between the bench (the nanomedicine research discussed in Section 2) and the clinic (the standard of care discussed in Section 1). A perspective of the current status of nanomedicine translation for the disease is detailed.

By dividing the chapters in Section 2 into these three parts, three distinct needs are addressed: (1) the need for a current assessment of inflammatory disease biology and the current standard of care of these diseases, (2) the need for a comprehensive analysis of nanotherapeutics that have been developed for these diseases, and (3) the need to understand the pathway for the clinical translation of these nanomedicine therapies as new treatments for inflammatory diseases. Comprehension of these three specific needs is essential for enabling successful nanomedicine translation for inflammatory disease.

Part 3, "The Emerging Role of Inflammation in Common Diseases," is the last section. In recent years, research into immune function and dysfunction in prominent disease has revealed an inflammatory component to many diseases that were not previously associated with an inflammatory etiology. These diseases are referred to as secondary inflammatory diseases. The disease-focused chapters of this section cover neurodegenerative disease, cancer, and diabetes. Each chapter discusses the disease in the context of inflammation and translational nanomedicine. Treating these secondary inflammatory diseases with nanomedicine is a promising approach, as demonstrated by current nanomedicine therapies for cancer. The pathways of nanomedicine translation for primary and secondary inflammatory disease intersect, and the National Cancer Institute's NCL offers a model for success.

Nanomedicine for Inflammatory Diseases is a translational medicine book that strives to push the field forward by offering insightful perspectives and interweaving the fundamentals of inflammation, nanomedicine, nanotoxicity, and translation; the biology and clinical treatment of inflammatory bowel disease, multiple sclerosis, and asthma; the nanomedicine therapies in development for these diseases; the pathway for translation of these therapies; the role of inflammation in neurodegenerative disease, cancer, and diabetes; and the current status of nanomedicine translation for these diseases. *Nanomedicine for Inflammatory Diseases* seeks to bridge the gaps between inflammation, nanomedicine, and translation by offering a foundational resource for the present and the future.

EDITORS

 **Lara Scheherazade Milane** recently joined Burrell College of Osteopathic Medicine (Las Cruces, New Mexico) as founding faculty in the Biomedical Sciences Department and is the director of online programing. Dr. Milane received her training as a National Cancer Institute/ National Science Foundation nanomedicine fellow at Northeastern University, Boston. She has a PhD in pharmaceutical science with specializations in nanomedicine and drug delivery systems (Northeastern University). She also earned her MS in biology and BS in neuroscience from Northeastern University.

Dr. Milane's research interests are in cancer biology, mitochondrial medicine, and translational nanomedicine. She is interested in developing a library of clinically translatable targeted nanomedicine therapies for cancer treatment. She teaches in the medical program and in the post-baccalaureate program. Dr. Milane is an advocate for women in the sciences and is a pioneer for outreach. She has published 18 peer-reviewed journal articles, 3 book chapters, and 3 white papers.

 **Mansoor M. Amiji** is currently the university distinguished professor in the Department of Pharmaceutical Sciences and codirector of the Northeastern University Nanomedicine Education and Research Consortium at Northeastern University in Boston. The consortium oversees a doctoral training program in nanomedicine science and technology that was cofunded by the National Institutes of Health and the National Science Foundation. Dr. Amiji earned his BS in pharmacy from Northeastern University in 1988 and a PhD in pharmaceutical sciences from Purdue University in 1992.

His research is focused on the development of biocompatible materials from natural and synthetic polymers, target-specific drug and gene delivery systems for cancer and infectious diseases, and nanotechnology applications for medical diagnosis, imaging, and therapy. His research has received more than $18 million in sustained funding from the National Institutes of Health, the National Science Foundation, private foundations, and the pharmaceutical/biotech industries.

Dr. Amiji teaches in the professional pharmacy program and in the graduate programs of pharmaceutical science, biotechnology, and nanomedicine. He has published six books and more than 200 book chapters, peer-reviewed articles, and conference proceedings. He has received a number of honors and awards, including the Nano Science and Technology Institute's Award for Outstanding Contributions toward the Advancement of Nanotechnology, Microtechnology, and Biotechnology; the American Association of Pharmaceutical Scientists Meritorious Manuscript Award; the Controlled Release Society's Nagai Award; and American Association of Pharmaceutical Scientists and Controlled Release Society fellowships.

CONTRIBUTORS

MANSOOR M. AMIJI
Department of Pharmaceutical Sciences
School of Pharmacy
Northeastern University
Boston, Massachusetts

P. ARNAUD
Unité de Technologies Chimiques et Biologiques
    pour la Santé (UTCBS)
Faculté des Sciences Pharmaceutiques
    et Biologiques
Paris, France

CHARUL AVACHAT
Department of Pharmaceutical Sciences
School of Pharmacy
Northeastern University
Boston, Massachusetts

BENJAMIN BLEIER
Department of Otolaryngology
Massachusetts Eye and Ear Infirmary
Harvard Medical School
Boston, Massachusetts

YONGHAO CAO
Departments of Neurology and Immunobiology
Yale School of Medicine
New Haven, Connecticut

BOBBY J. CHERAYIL
Mucosal Immunology and Biology Research
    Center
Department of Pediatrics
Massachusetts General Hospital
Boston, Massachusetts

GAIA CILLONI
Faculty of Pharmacy
University of Coimbra
Azinhaga de Santa Comba
Coimbra, Portugal

FERNANDA FERREIRA CRUZ
Laboratory of Pulmonary Investigation
Carlos Chagas Filho Institute of Biophysics
Federal University of Rio de Janeiro
Rio de Janeiro, Brazil

JON R. FELT
Carman and Ann Adams Department of Pediatrics
Wayne State University
Children's Hospital of Michigan
Detroit, Michigan

LEORAH FREEMAN
UTHealth
McGovern Medical School
Department of Neurology
Houston, Texas

SUSAN HUA
School of Biomedical Sciences and Pharmacy
University of Newcastle
Newcastle, New South Wales, Australia

and

Hunter Medical Research Institute
New Lambton Heights, New South Wales,
    Australia

JELENA M. JANJIC
Graduate School of Pharmaceutical Sciences
Mylan School of Pharmacy
Duquesne University
Pittsburgh, Pennsylvania

RIMA KANDIL
Department of Pharmacy, Pharmaceutical
    Technology and Biopharmaceutics
Ludwig-Maximilians-Universität München
Munich, Germany

MAHSA KHAYAT-KHOEI
UTHealth
McGovern Medical School
Department of Neurology
Houston, Texas

JOHN LINCOLN
UTHealth
McGovern Medical School
Department of Neurology
Houston, Texas

MARLENE LOPES
Faculty of Pharmacy
University of Coimbra
Azinhaga de Santa Comba
and
CNC—Center for Neurosciences and Cell Biology
University of Coimbra
Coimbra, Portugal

ADRIANA LOPES DA SILVA
Laboratory of Pulmonary Investigation
Carlos Chagas Filho Institute of Biophysics,
Federal University of Rio de Janeiro
Rio de Janeiro, Brazil

PRASHANT MAHAJAN
Carman and Ann Adams Department of Pediatrics
Wayne State University
Children's Hospital of Michigan
Detroit, Michigan

WILSON S. MENG
Graduate School of Pharmaceutical Sciences
Mylan School of Pharmacy
Duquesne University
Pittsburgh, Pennsylvania

OLIVIA M. MERKEL
Department of Pharmacy, Pharmaceutical
    Technology and Biopharmaceutics
Ludwig-Maximilians-Universität München
Munich, Germany

and

Department of Pharmaceutical Sciences
Eugene Applebaum College of Pharmacy and
    Health Sciences
and
Department of Oncology
Karmanos Cancer Institute
Wayne State University
Detroit, Michigan

DIDIER MERLIN
Institute for Biomedical Sciences
Center for Diagnostics and Therapeutics
Georgia State University
Atlanta, Georgia

and

Atlanta Veterans Affairs Medical Center
Decatur, Georgia

MARK MESSINA
Department of Neurology
Hofstra Northwell School of Medicine
Hempstead, New York

and

Department of Autoimmunity
Feinstein Institute for Medical Research
Manhasset, New York

MARCEL MENON MIYAKE
Department of Otolaryngology
Massachusetts Eye and Ear Infirmary
Harvard Medical School
Boston, Massachusetts

LARA SCHEHERAZADE MILANE
Department of Biomedical Sciences
Burrell College of Osteopathic Medicine
Las Cruces, New Mexico

RAQUEL MONTEIRO
Faculty of Pharmacy
University of Coimbra
Azinhaga de Santa Comba
Coimbra, Portugal

CHRISTOPHER J. MORAN
Mucosal Immunology and Biology Research
    Center
Department of Pediatrics
Massachusetts General Hospital
Boston, Massachusetts

ANGIE S. MORRIS
Department of Pharmaceutical Sciences and
    Experimental Therapeutics
College of Pharmacy
University of Iowa
Iowa City, Iowa

SOUHEL NAJJAR
Department of Neurology
Lenox Hill Hospital
New York, New York

and

Department of Neurology
Hofstra Northwell School of Medicine
Hempstead, New York

JOYCE J. PAN
Departments of Neurology and Immunobiology
Yale School of Medicine
New Haven, Connecticut

NEHA N. PARAYATH
Department of Pharmaceutical Sciences
School of Pharmacy
Northeastern University
Boston, Massachusetts

GRISHMA PAWAR
Department of Pharmaceutical Sciences
School of Pharmacy
Northeastern University
Boston, Massachusetts

ANTONIO J. RIBEIRO
Group Genetics of Cognitive Dysfunction
I3S—Instituto de Investigação e Inovação
    em Saúde
and
IBMC—Instituto de Biologia Molecular e Celular
Universidade do Porto
Porto, Portugal

and

Faculty of Pharmacy
University of Coimbra
Azinhaga de Santa Comba
Coimbra, Portugal

and

Unité de Technologies Chimiques et Biologiques
    pour la Santé (UTCBS)
Faculté des Sciences Pharmaceutiques
    et Biologiques
Paris, France

PATRICIA RIEKEN MACEDO ROCCO
Laboratory of Pulmonary Investigation
Carlos Chagas Filho Institute of Biophysics
Federal University of Rio de Janeiro
Rio de Janeiro, Brazil

ALIASGER K. SALEM
Department of Pharmaceutical Sciences and
    Experimental Therapeutics
College of Pharmacy
University of Iowa
Iowa City, Iowa

MAYA SHABBIR
Department of Autoimmunity
Feinstein Institute for Medical Research
Manhasset, New York

JOEL N. H. STERN
Department of Neurobiology and Behavior
Rockefeller University
and
Department of Neurology
Lenox Hill Hospital
New York, New York

and

Department of Neurology
Hofstra Northwell School of Medicine
Hempstead, New York

and

Department of Autoimmunity
Feinstein Institute for Medical Research
Manhasset, New York

INNA TABANSKY
Department of Neurobiology and Behavior
Rockefeller University
New York, New York

FRANCISCO VEIGA
Faculty of Pharmacy
University of Coimbra
Azinhaga de Santa Comba
and
CNC—Center for Neurosciences and Cell Biology
University of Coimbra
Coimbra, Portugal

MICHAEL E. WOODS
Department of Physiology & Pathology
Burrell College of Osteopathic Medicine
Las Cruces, New Mexico

PAUL WRIGHT
Department of Neurology
Hofstra Northwell School of Medicine
Hempstead, New York

BO XIAO
Institute for Clean Energy and Advanced Materials
Faculty of Materials and Energy
Southwest University
Chongqing, People's Republic of China

and

Institute for Biomedical Sciences
Center for Diagnostics and Therapeutics
Georgia State University
Atlanta, Georgia

YURAN XIE
Department of Pharmaceutical Sciences
Eugene Applebaum College of Pharmacy and
   Health Sciences
Wayne State University
Detroit, Michigan

PART ONE

Introduction

INTRODUCTION TO INFLAMMATORY DISEASE, NANOMEDICINE, AND TRANSLATIONAL NANOMEDICINE

Part 1 covers important foundational concepts in inflammation, nanomedicine, and translation. The inflammatory response is an important protective response; however, it is also central to primary inflammatory disease associated with chronic inflammation and secondary inflammatory disease, such as cancer. Why is inflammation associated with so many diseases? The inflammatory response is a very scripted process; understanding the normal physiology and transduction that occurs is helpful to understanding inflammatory dysfunction associated with disease etiologies and pathologies.

Understanding the benefits of nanomedicine is essential for understanding the need for translation. What does nanomedicine have to offer? How is it superior to traditional formulations? How are the desired properties of a nanomedicine formulation achieved through design? Foundational knowledge of the different nanomedicine platforms aids in understanding this important field of medicine. Being aware of nanotoxicity is also imperative. What are the risks of nanomedicine, and how are the safety concerns addressed? Are the risks of using nanomedicine worth the benefits? Being able to answer this question for individual therapies is important before translation begins.

Translational medicine has emerged as a distinct area of therapeutics. What is bionanotechnology, and what is the real "nanoappeal" for translational medicine? Translation has progressed from the Critical Path Initiative to the great model of the Nanotechnology Characterization Laboratory. How can this model be used to overcome the challenges of translation? What is the future of translational nanomedicine? These questions are discussed and contextualized to inflammatory disease.

The core concepts in inflammatory disease, nanomedicine, nanotoxicity, and translational nanomedicine are discussed and interconnected to establish foundational knowledge of nanomedicine translation for inflammatory disease. This section even offers a novel schema for translational design workflow. These concepts are the framework for the disease-focused discussions in Part 2 (primary inflammatory disease) and Part 3 (secondary inflammatory disease).

CHAPTER ONE

Fundamentals of Immunology and Inflammation

Michael E. Woods

CONTENTS

1.1 INTRODUCTION: INFLAMMATION IS THE BODY'S NATURAL RESPONSE TO INSULT AND INJURY

The immune system comprises a complex network of cells, tissues, and signaling molecules that detect, respond, adapt, and ultimately protect us from invading pathogens and tissue injury. It is a classic homeostatic system that is constantly sensing and responding to ever-changing environmental conditions. We classically divide the immune system into two major components: innate and adaptive immune responses. The non-specific innate defenses function to blunt the spread of invading pathogens early in the infection process (i.e., within minutes to hours) and return the tissue to normal as quickly as possible. Adaptive defenses, on the other hand, require days to weeks to develop and specifically target invading pathogens marking them for destruction and removal from the body. The reality, however, is that the innate and adaptive immune responses are intricately linked.

Acute inflammation is an early, almost immediate, nonspecific physiological response to tissue injury that is generally beneficial to the host and aims to remove the offending factors and restore tissue structure and function. Acute inflammation is the first line of defense against an injury or infection. It is characterized by four cardinal signs, as first described by the Roman physician Celsus almost 2000 years ago: calor (heat), rubor (redness), tumor (swelling), and dolor (pain). We now attribute these signs to increased blood flow to the site as a result of vasodilation (heat and redness), swelling due to the accumulation of fluid as a result of microvascular changes, and stimulation of nerve endings by secreted factors (pain). Rudolf Virchow later added a fifth sign, functio laesa (loss of function), in the nineteenth century, which denotes the restricted function of inflamed tissues (Heidland et al. 2006).

The mechanisms of infection-induced inflammation are understood much better than those of other inflammatory processes in response to tissue injury, stress, and malfunction, although many of the same processes apply. Invading microbes usually trigger an inflammatory response first through the interaction of microbial components and innate immune system receptors. Toll-like receptors (TLRs) and nucleotide-binding oligomerization domain protein (NOD)–like receptors (NLRs) recognize microbial components, such as bacterial lipopolysaccharide (LPS), double-stranded viral ribonucleic acid (RNA), or pepti-doglycan. Found in immune and nonimmune cells such as macrophages, dendritic cells (DCs), mast cells, and epithelium, these receptors trigger the production of several inflammatory mediators, including cytokines, chemokines, vasoactive amines, eicosanoids, prostaglandins, and other products. These mediators elicit an initial localized response whereby neutrophils and certain plasma proteins are allowed access through post-capillary venules to extravascular sites of injury, as illustrated in Figure 1.1. Here, the inflammatory response attempts to disable and destroy an invading pathogen through the action of activated neutrophils. Upon contact with a microbe, neutrophils release their granule contents, which includes reactive oxygen species (ROS) and nitrogen species and serine proteases, which nonspecifically damage the microbe. If the initial inflammatory response is successful and the microbe is destroyed, the body will recruit macrophages to the response site as part of the resolution and repair process. Lipid and nonlipid mediators, including lipoxins, resolvins, protec-tins, and transforming growth factor-β (TGF-β), initiate the transition from an acute inflammatory state to an anti-inflammatory state (Serhan 2010). During the resolution phase, neutrophil recruitment is inhibited and activated neutrophils undergo controlled cell death, and macrophages infiltrate the site to remove dead cell debris and initiate tissue remodeling.

If the acute inflammatory response continues unabated due to a defect in the system or subversion by microbial virulence factors, the inflammatory response may develop into a chronic, nonresolving state. This typically involves an increased presence of adaptive responses dominated by macrophages and T cells, as well as an overabundance of innate immune cell activity, primarily neutrophils, and progressive positive feedback loops that allow the inflammation to continue unabated. This eventually results in host tissue destruction due to excessive protease activity, as illustrated in Figure 1.2. These processes are also characteristic of many inflammatory diseases, which will be discussed in greater detail in the chapters to follow.

The following sections lay out the principal components of inflammation and immunity.

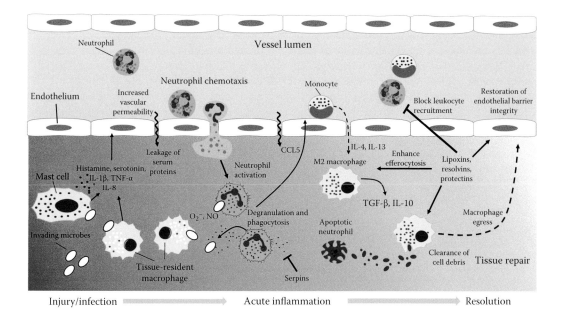

Figure 1.1. Acute inflammation is marked by the recruitment, infiltration, and activation of neutrophils into a site of injury or infection. This response, if successful, induces a series of counterbalancing responses to limit and resolve the inflammatory response in order to avoid extraneous host tissue damage. Alternatively activated macrophages play a role in removing apoptotic neutrophils and cell debris from the site and producing anti-inflammatory cytokines. SPMs, such as lipoxins, resolvins, and protectins, help orchestrate the resolution phase of inflammation. (Copyright © motifolio.com.)

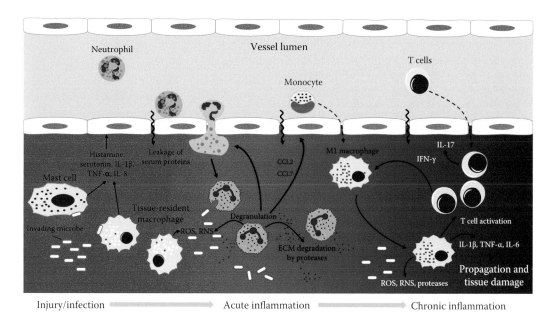

Figure 1.2. Progression of acute inflammation to chronic inflammation is dependent on excessive neutrophil and macrophage activity and can be propagated by aberrant lymphocyte activity. This process is dependent on unregulated inflammatory responses, including excessive protease and ROS production as a result of neutrophil and classically activated macrophage activity. Additionally, the presence of T lymphocytes can further propagate these responses through the induction of additional pro-inflammatory cytokines. (Copyright © motifolio.com.)

We discuss the general cell types involved in initiating, effecting, and regulating inflammation, followed by the primary soluble mediators involved in transmitting inflammatory signals between cells and coordinating the activation and infiltration of immune cells into the site of inflammation.

1.2 CELLS OF THE IMMUNE SYSTEM

The immune system is comprised of an army of cells and cell types with unique roles and responsibilities in inflammation and immunity. The cells of the immune system are generally divided into innate immune system cells and adaptive immune cells. Innate immune cells, including granulocytes, mononuclear phagocytes, and natural killer (NK) cells, generally respond to invading microbes in a nonspecific manner; that is, they recognize molecular patterns common to most microbes, or tumors in the case of NK cells, and respond using mechanisms capable of damaging both microbes and host tissues. This response is fast, often occurring within minutes to hours after injury or infection. Cells of the adaptive immune response target invading pathogens using mechanisms designed to specifically target unique features of the invading microbe; therefore, this response often requires days to weeks to develop and to effectively clear the pathogen. Most lymphocytes fall into this category. Table 1.1 lists the cellular components of the immune system, the primary role of each cell type, the unique cell surface for each cell type, and the main secretory compounds produced by each. Here, we present a broad overview of the general cell types; in reality, most cell types are comprised of diverse subsets with distinct roles in immunity.

1.2.1 Granulocytes

1.2.1.1 Mast Cells

Mast cells are a key component of the innate immune system with a role as first responders to many microbial infections and as key contributors to allergic reactions; however, it is now clear that mast cells are also intimately involved in many autoimmune and inflammatory diseases. Mast cells are critical to recruit neutrophils to sites of infection and inflammation, and they facilitate neutrophil recruitment by promoting localized increases in vascular permeability and the entry of inflammatory cells into the tissue. Mast cells mediate traditional immunoglobulin E (IgE)–mediated allergic responses, as well as diseases, such as multiple sclerosis and rheumatoid arthritis (Costanza et al. 2012; Kritas et al. 2013).

Mast cells are considered frontline defenders against infection due to their prevalence in tissues normally exposed to environmental insults, such as the skin, and intestinal, respiratory, and urinary tracts. Mast cells are also found in close association with blood and lymphatic vessels, where they contribute to angiogenesis, inflammation, and wound healing. $CD34^+$ hematopoietic precursor cells in the bone marrow produce immature mast cells, which circulate in the blood. Only after the immature mast cells establish residency in a particular tissue do they fully differentiate and mature (Okayama and Kawakami 2006).

The key role of mast cells is to initiate the early stages of inflammation by increasing local vascular permeability and recruiting neutrophils, resulting in escalation of host defenses. Mast cells primarily accomplish this by releasing the contents of granules or through selective release of certain pro-inflammatory cytokines. Upon activation, mast cells synthesize and/or secrete a wide array of vasoactive and pro-inflammatory compounds, which are listed in Table 1.2. These include histamine, serotonin, and proteases stored in secretory granules. Activated mast cells synthesize a number of lipid mediators (leukotrienes, prostaglandins, and platelet-activating factor [PAF]) from arachidonic acid, and numerous pro- and anti-inflammatory cytokines, including interleukin-1β (IL-1β), IL-6, IL-8, IL-13, and tumor necrosis factor-α (TNF-α) (Theoharides et al. 2012).

One of the best-understood mechanisms of mast cell activation is through IgE receptor cross-linking by antigen-bound IgE antibodies. This is the classic mechanism of allergic inflammatory responses, which results in mast cell degranulation and release of vasoactive peptides (Blank and Rivera 2004). Mast cells express a high-affinity receptor for IgE, FcεRI, and cross-linking of the receptor by its ligand induces granule translocation to the surface of the mast cell and calcium-dependent exocytosis of the granule contents. This process involves microRNA-221-promoted activation of the PI3K/Akt/PLCγ/Ca^{2+} signaling pathway (Xu et al. 2016). Activation of this pathway depletes Ca^{2+} from endoplasmic reticulum stores, which elicits oscillatory cytosolic Ca^{2+}

TABLE 1.1

Major cell types involved in inflammation and immunity and their primary function, identifying surface markers, and main secretory products.

Cell type	Primary function	Surface markers	Main secretory compounds	References
Mast cells	Initiation of inflammation	CD117, CD203c, FcεR1α	Histamine, heparin, thromboxane, PGD$_2$, LTC$_4$	Theoharides et al. 2012
Neutrophils	Phagocytosis	CD15, CD16, CD66b	Elastase, proteinase-3, cathepsin G, MMP-9	Beyrau et al. 2012
Basophils	IgE-mediated allergy	CD123, CD203c, Bsp-1	IL-4, histamine, LTC$_4$	Hennersdorf et al. 2005
Eosinophils	IgE-mediated allergy, parasitic infection	CD11b, CD193, EMR1	IL-4, IL-5, IL-6, IL-13, MBP	Long et al. 2016
Monocytes	Immune surveillance, differentiation into macrophages and DCs	CD14, CD16, CD33, CD64	IL-6, TNF-α	Ziegler-Heitbrock 2015
Macrophages	Phagocytosis, tissue repair	CD11b, CD14, CD33, CD68, CD163	TNF-α, IL-1β, IL-12, IL-23; TGF-β, PDGF	Murray and Wynn 2011
Dendritic cells	T cell activation; antigen presentation	CD1c, CD83, CD141, CD209, MHC II	IL-1β, IL-6, IL-23, TGF-β	Segura and Amigorena 2013
NK cells	Nonspecific cell killing of virally infected cells; antitumor immunity	CD11b, CD56, NKp46	IFN-γ, perforin, granzyme B	Fuchs 2016
T$_H$1 cells	Control of intracellular pathogens	CD3, CD4, IL-12R, CXCR3, CCR5	IFN-γ, IL-2	Raphael et al. 2015
T$_H$2 cells	Extracellular pathogens	CD3, CD4	IL-4, IL-5, IL-10, IL-13	Raphael et al. 2015
T$_H$17 cells	Pro-inflammatory	CD3, CD4, CD161	IL-17, IL-21	Korn et al. 2009
Treg cells	Suppression of effector T cell responses	CD3, CD4, CD25, FoxP3	IL-10, TGF-β	Vignali et al. 2008
γδ T cells	Local immunosurveillance	CD3, γδ TCR	IFN-γ, IL-4, IL-17	Ribot et al. 2009; Paul et al. 2015
B cells/plasma cells	Antibody production	CD19, CD20	IL-10, TGF-β1, IL-2, IL-4, TNF-α, IL-6	Lund 2008

NOTE: MHC, major histocompatibility complex.

elevations (Di Capite and Parekh 2009), which, in conjunction with activated protein kinase C, cause granule exocytosis (Ma and Beaven 2009). Furthermore, the pattern of calcium waves in cell protrusions during antigen stimulation correlates spatially with exocytosis, and likely involves TRPC1 channels for Ca^{2+} mobilization (Cohen et al. 2012).

Some triggers, such as LPS, parasites, and viruses, stimulate selective release of certain mediators without degranulation through TLR-mediated signaling. For example, LPS binding to TLR-4 induces TNF-α, IL-5, IL-10, and IL-13 secretion by mast cells without inducing degranulation (Okayama 2005). Binding of peptidoglycan to TLR-2 induces histamine release, as well as IL-4, IL-6, and IL-13 (Supajatura et al. 2002). *In vitro* studies have also demonstrated activation of mast cells through TLR-3, resulting in interferon (IFN) production (Kulka et al. 2004; Lappalainen et al. 2013), and TLR-9, resulting in IL-33 production (Tung et al. 2014). In turn, IL-33 induces Fcε receptor

TABLE 1.2

Key mast cell mediators involved in inflammation.

Mediators	Main physiological effects
Preformed in granules	
Histamine	Vasodilation, angiogenesis, pain
Serotonin (5-hydroxytryptamine [5-HT])	Vasoconstriction
IL-8, MCP-1, RANTES	Chemoattraction of leukocytes
Phospholipases	Arachidonic acid generation
Matrix metalloproteinases	ECM remodeling, modification of cytokines/ chemokines
Synthesized *de novo*	
Pro-inflammatory cytokines (IL-1, IL-4, IL-5, IL-6, IL-8, IL-13, IL-33, IFN-γ, TNF-α, MIP-1α, MCP-1)	Leukocyte activation and migration
Anti-inflammatory cytokines (IL-10, TGF-β)	Suppression of leukocyte activity
Nitric oxide	Vasodilation
Leukotriene B$_4$	Leukocyte adhesion and activation
Leukotriene C$_4$	Vasoconstriction, pain
Prostaglandin D$_2$	Leukocyte recruitment, vasodilation

1 (FcεR1)–independent production of IL-6, IL-8, and IL-13 in naïve human mast cells and enhances production of these cytokines in IgE- or anti-IgE-stimulated mast cells without inducing release of prostaglandin D$_2$ (PGD$_2$) or histamine (Iikura et al. 2007). This occurs through activation of mitogen-activated protein kinases (MAPKs), ERK, p38, JNK, and nuclear factor-κB (NF-κB) (Tung et al. 2014). Mast cell–derived IL-33 also plays a key role in T helper type 17 (Th17) cell maturation, indicating a role in autoimmune disorders and allergic asthma (Cho et al. 2012). Mast cells counteract regulatory T (Treg) cell inhibition of effector T cells in the presence of IL-6 and TGF-β, which establishes a Th17-mediated inflammatory response (Piconese et al. 2009). Additionally, mast cell–derived TNF-α is required for Th17-mediated neutrophilic airway hyperreactivity in the lungs of ovalbumin-challenged OTII transgenic mice, indicating that mast cells and IL-17 can contribute to antigen-dependent airway neutrophilia (Nakae et al. 2007).

Mast cells interact directly with a number of different cell types, which partly explains their role in certain autoimmune conditions. Mast cells bind directly to Treg cells via the OX40–OX40L axis. Mast cells constitutively express OX40L, which binds to OX40 constitutively expressed on Treg cells. This binding appears to result in

downregulation of FcεR1 expression and inhibition of FcεR1-dependent mast cell degranulation (Gri et al. 2008). However, this interaction also appears to cause a reversal of Treg suppression of T effector cells and a reduction in T effector cell susceptibility to Treg suppression by driving Th17 cell differentiation (Piconese et al. 2009). Under certain conditions, mast cells can express all the cytokines that drive Treg skewing to a Th17 phenotype, including IL-6, IL-21, IL-23, and TGF-β. These effects have been observed in some forms of cancer where mast cell IL-6 contributes to a pro-inflammatory Th17-dominated environment, leading to autoimmunity (Tripodo et al. 2010).

There is also a connection between mast cells and B cells, as evidenced by the mast cell expression of certain B cell–modulating molecules, and the importance of Ig receptor binding to antibodies produced by B cells. Mast cells exposed to monomeric IgE in the absence of antigen exhibit increased survival and priming (Kawakami and Galli 2002). Furthermore, mast cell–derived IL-6 and the expression of CD40–CD40L on B cells and mast cells, respectively, promote the differentiation of B cells into IgA-secreting CD138$^+$ plasma cells (Merluzzi et al. 2010). Mast cells also express IgG receptors FcγRII and FcγRIII, which induce degranulation in response to IgG–antigen complex–mediated cross-linking. These receptors are known to play important roles in numerous diseases associated with types II, III, and IV hypersensitivity reactions, such as rheumatoid arthritis, systemic lupus erythematosus (SLE), and experimental autoimmune encephalomyelitis (EAE) (Sayed et al. 2008).

1.2.1.2 Neutrophils

Neutrophils are arguably the most important mediator of the acute inflammatory response and are indispensable for protection against microbial pathogens. Neutrophils are the most abundant immune cell in circulation, comprising 50%–70% of circulating leukocytes in humans. The primary functions of neutrophils are to infiltrate the site of invasion, and engulf and then kill the microbes through both intracellular and extracellular defenses. Neutrophils mature in the bone marrow, where they acquire the ability to sense chemotactic gradients (Boner et al. 1982). They can then be mobilized and released to traffic to sites of injury. Once at the site of inflammation, neutrophils release a series of proteases from their granules to kill microbes directly or inactivate microbial toxins (Pham 2006); however, this response is normally counterbalanced by endogenous serine protease inhibitors, serpins, which serve to protect the body from excessive harmful proteolytic activity. As acute inflammation begins to resolve, neutrophils undergo apoptosis and are cleared by infiltrating macrophages through a process called efferocytosis (Figure 1.1).

Neutrophils are derived from myeloid precursor in the bone marrow, where they are continuously generated at a daily rate of up to 2×10^{11} cells. The process of neutrophil maturation is controlled by granulocyte colony-stimulating factor (G-CSF); however, the specific pathways responsible for G-CSF-induced neutrophil maturation are not completely understood. $\gamma\delta$ T cells and NK T-like cells are one source of G-CSF; these cells respond to IL-23 produced by tissue-resident macrophages and DCs and IL-17A to produce G-CSF (Ley et al. 2006; Smith et al. 2007). G-CSF exerts its effects by binding to a single homodimer receptor, G-CSFR; however, G-CSF does not interact directly with hematopoietic progenitor cells, instead exerting its effects indirectly, possibly through CD68+ monocytes (Christopher et al. 2011). Mature neutrophils are retained in the bone marrow by the balance of CXCR4/CXCL12, which retains mature neutrophils, and CXCR2/IL-8 signaling, which controls release into the peripheral circulation.

Neutrophils are mobilized and released to traffic to sites of inflammation through signaling mediated primarily by the chemokines IL-8 (CXCL8), macrophage inflammatory protein-2 (MIP-2) (CXCL2), and KC (CXCL1), which can bind two receptors, CXCR1 and CXCR2. Both receptors are members of the G protein–coupled receptor family that transduces a signal through a G protein–activated second messenger system, predominantly through the G$\beta\gamma$ subunit. Activation of these signaling pathways leads to cell polarization, which allows for directional migration of neutrophils, or chemotaxis (Mócsai et al. 2015). When exposed to a prototypical neutrophil chemoattractant, f-Met-Leu-Phe (fMLF), neutrophils will, within a 2- to 3-minute timeframe, rearrange their cytoskeleton to induce distinct subcellular arrangements. The leading edge, or pseudopod, is characterized by lamellipodia consisting primarily of F-actin bundles under the control of PI3Kγ, which drives the cell forward. The trailing edge, or uropod, contains myosin light chain under the control of Rho GTPase, which facilitates contraction of the rear of the cell.

Neutrophils must also regulate direct cell-to-cell contacts during recruitment as the cell tethers, rolls, adheres, and spreads from the vasculature, across the inflamed endothelium into tissue where the cell must interact with components of the extracellular matrix (ECM), such as fibrin. This process is primarily mediated by adhesion molecules of the β_2-integrin family (CD11/CD18), which are heterodimeric noncovalently linked transmembrane glycoproteins composed of one α and one β subunit. Neutrophils express three different β_2-integrins: LFA-1 (CD11a/CD18), Mac-1 (CD11b/CD18), and gp150/95 (CD11c/CD18). The most abundant molecule, Mac-1, is upregulated in response to neutrophil activation and is derived from intracellular stores, whereas LFA-1 and gp150/95 are constitutively expressed. These integrins all bind the same ligand, endothelial intercellular adhesion molecule-1 (ICAM-1); however, whereas LFA-1 is primarily responsible for slow rolling and induction of firm adhesion, Mac-1's predominant role is in intraluminal crawling (Smith et al. 1989; Phillipson et al. 2006). Ultimately, both molecules contribute to efficient emigration out of the vasculature. Mac-1 also mediates phagocytosis of complement-opsonized bacteria, which also induces the generation of ROS, indicating the importance of integrin-mediated cross talk with intracellular signaling pathways, in addition to adhesive properties (Abram and Lowell 2009).

Once at the site of inflammation, neutrophils are activated and will try to destroy invading microbes using an arsenal of antimicrobial defenses. One of the most important neutrophil effector functions is phagocytosis, which Ilya Metchnikoff first described in the late nineteenth century and later received the Nobel Prize in Physiology or Medicine for in 1908. Phagocytosis is an intricate process mediated by the complex interaction between membrane lipids, intracellular signaling cascades, and cytoskeletal rearrangement. Phagocytosis occurs in neutrophils within minutes of an opsonized microbe or particle binding to cell surface receptors, which include Fcγ receptors (Nimmerjahn and Ravetch 2006), complement receptors (Gordon et al. 1989), and other membrane-bound pathogen recognition receptors (PRRs), such as the C-type lectins (Kerrigan and Brown 2009). While relatively little is known about the precise mechanisms used by the C-type lectins to induce phagocytosis, Fcγ- and complement-mediated phagocytosis has been studied intensely. The process of phagocytosis begins upon receptor engagement, which in the case of the Fcγ receptor involves the recruitment of Syk and the activation of Rac, Cdc42, and PI3K (Cougoule et al. 2006). This results in actin reorganization to extend a membrane protrusion known as a pseudopod, which envelopes the microbe and draws it into the cell, forming a phagosome.

Once a phagocyte has ingested a microbe, it deploys a series of degradative processes to disable and kill the microbe. One of the most well-described pathways is by oxygen-dependent killing of the microbe, known as the "respiratory burst." The primary effectors of this killing mechanism are ROS—superoxide, hydroxide, hydrochlorous acid, and ozone—that occur downstream of O_2^- formed by the nicotinamide adenine dinucleotide phosphate (NADPH) oxidase complex. The NADPH oxidase complex assembles itself on the phagosomal membrane and is comprised of at least seven cytosolic and membrane-bound components (Segal 2008). The active complex pumps electrons from cytosolic NADPH across the membrane to the electron acceptor, molecular oxygen, generating superoxide anion in the vacuole. Conventionally, it is believed that O_2^- partitions into H_2O_2, hydroxide radical, and singlet oxygen, which are directly microbicidal; however, some have questioned the validity of this concept. Segal proposes an alternative concept whereby the main function of the NADPH oxidase–mediated electron transport is to optimize the pH inside the vacuole to facilitate proper function of the granule proteases (pH 8.5–9.5) (Segal 2008; Levine et al. 2015). This is dependent on K^+ influx in response to NADPH oxidase activation, which releases the granule proteases (Reeves et al. 2002). However, this hypothesis has been challenged by the observation that serine protease–deficient neutrophils show no impairment in killing bacteria (Sørensen et al. 2014). It is likely that both mechanisms play a role in intracellular killing of microbes.

The importance of NADPH oxidase–mediated ROS production in proper functioning of phagocytes is well accepted, and it is abundantly clear that elevated ROS production also contributes to tissue damage in a number of inflammatory conditions. For example, ROS production is responsible for loss of endothelial barrier integrity and subsequent vascular leakage (Fox et al. 2013). Oxidative stress causes downregulated expression of occludin (Krizbai et al. 2005), a component of endothelial tight junctions, as well as increased tyrosine phosphorylation of occludin-ZO-1 and E-cadherin–β-catenin complexes (Rao et al. 2002), resulting in dissociation of the junctional complexes from the cytoskeleton and subsequent endothelial barrier disruption.

Another oxygen-dependent effector function of phagocytic cells is the production of nitric oxide (NO). NO is a soluble gas produced from arginine by nitric oxide synthase (NOS). Three forms of NOS exist: endothelial (eNOS), neuronal (nNOS), and inducible (iNOS). eNOS and nNOS are constitutively expressed, and the NO they produce plays important roles in homeostasis as a regulator of vascular tone and as a neurotransmitter, respectively. iNOS is expressed primarily by neutrophils and macrophages and is upregulated in response to cytokine stimulation or microbial components. iNOS produces much higher amounts of NO than the other two isoforms and is important for microbial killing. NO reacts with O_2^- to form a highly reactive free radical peroxynitrite (ONOO$^-$). Peroxynitrite functions in a manner similar to that of ROS, whereby it damages the lipids, proteins, and nucleic acids of microbial and host cells.

Neutrophils also possess nonoxidative mechanisms for killing microbes composed of the serine proteases elastase, proteinase-3, cathepsin G, and azurocidin. They are components of azurophilic granules and possess both antimicrobial and pro-inflammatory activity. Activated neutrophils release their granule contents intracellularly into the phagosome, as well as extracellularly as a component of neutrophil extracellular traps (NETs) (Pham 2006). Structurally related to chymotrypsin, these enzymes have broad substrate specificity and cleave microbial and host proteins. For example, neutrophil elastase exerts antimicrobial activity by cleaving the outer membrane protein A of *Escherichia coli* (Belaaouaj et al. 2000). Neutrophil elastin and cathepsin G have also been shown to cleave the pro-inflammatory bacterial virulence factor flagellin (López-Boado et al. 2004). In terms of host proteins, proteinase-3 cleaves the proforms of TNF-α and IL-1β (Coeshott et al. 1999), and all three proteases cleave IL-8, liberating active cytokine (Padrines et al. 1994). Neutrophil serine proteases can also activate certain cellular receptors, such as the protease-activated receptors (PARs) found primarily on platelets and endothelial cells. Cathepsin G targets PAR-4, leading to platelet aggregation (Sambrano et al. 2000). All three proteases cleave PAR-1, inhibiting its activation by thrombin (Renesto et al. 1997).

Neutrophil-derived proteases play an important role in remodeling the ECM as part of the normal inflammatory response; however, improper regulation of this response can lead to chronic inflammation and tissue damage. Neutrophils release both serine proteases and matrix metalloproteases (MMPs) that cleave components of the ECM (Lu et al. 2011). While reorganization of the ECM is a normal, often beneficial process, if left unchecked, it can have devastating consequences on the host. For example, neutrophils expressing MMP-9 are required for revascularization of transplant tissue in a mouse islet cell transplant model via a mechanism whereby MMP-9 liberates vascular endothelial growth factor-A (VEGF-A) and other latent growth factors (Heissig et al. 2010; Christoffersson et al. 2012). The tripeptide N-acetyl proline–glycine–proline (PGP) is generated by the concerted action of MMP-8, MMP-9, and prolyl endopeptidase (PE) and is generated from cleavage of ECM components by these proteases. PCP acts as a chemokine mimetic and recruits neutrophils to the site through the engagement of CXCR1 and CXCR2 (Weathington et al. 2006), fostering a positive feedback loop of tissue destruction and chronic inflammation. Neutrophil-derived MMP-8 has also been shown to contribute to tissue destruction in the pulmonary cavities of chronic tuberculosis patients (Ong et al. 2015).

Serine proteases released by neutrophils can also play a role in regulating immune responses by cleaving essential receptors on the surface of immune cells. For example, neutrophil elastase cleaves CXCR1, the receptor for IL-8, on the surface of neutrophils. The consequence of CXCR1 cleavage is that it actually disables the bacterial killing capacity of neutrophils (Hartl et al. 2007). This is significant because neutrophils will downregulate CXCR2 during acute inflammation (Lee et al. 2015), which is the only other receptor for IL-8. Together, the cleavage of CXCR1 and downregulation of CXCR2 result in a severely impaired neutrophil response. Neutrophil serine proteases also cleave a number of complement receptors, including CR1 and C5aR on neutrophils, which normally regulate chemotaxis, degranulation, and phagocytosis. Other immune cell targets include CD14 on monocytes and the IL-2 and IL-6 receptors on T lymphocytes (Bank et al. 1999; Le-Barillec et al. 1999).

In order to prevent excessive tissue destruction, the immune system uses a class of molecules called serpins to regulate the effects of neutrophil serine proteases. The most abundant serpin is α-1 antitrypsin (AAT), which is synthesized primarily in the liver and released into the circulation as a component of the acute phase response. AAT's mechanism of action is to act as a pseudosubstrate for neutrophil elastase and proteinase-3, leading to a covalently bound complex that prevents further protease activity. AAT binds a number of other nonserine proteases and host proteins, the significance of which is only partly understood (Ehlers 2014). AAT inhibits the release of TNF-α and IL-1β, and enhances the release of anti-inflammatory IL-10, from LPS-stimulated monocytes through an unknown mechanism (Janciauskiene et al. 2004). AAT has also demonstrated the ability to prevent TNF-α-induced apoptosis through direct inhibition of caspase-3 (Zhang et al. 2007). Taken together, these data demonstrate the critical importance of serpins such as AAT in regulating inflammatory responses, so it should come as no surprise that numerous alleles associated with AAT deficiency are linked to lung and liver disease (Janciauskiene et al. 2011).

1.2.1.2.1 NEUTROPHIL EXTRACELLULAR TRAPS

NETs are a recently described mechanism whereby neutrophils release a web of extracellular DNA, composed of condensed chromatin, histones, and granule proteins, to ensnare, immobilize, and destroy invading microbes. NETs were first shown in 2004 to be capable of killing bacteria (Brinkmann et al. 2004), and a number of other studies have been conducted with bacteria, fungi, protozoa, and viruses. The process of NET formation involves the movement of elastase from granules to the nucleus, where the protease cleaves histones causing chromatin condensation. The neutrophil then actively releases the DNA, forming fibrils that ensnare free pathogens, exposing them to the degradative enzymes, as illustrated in Figure 1.3. During this process, generally referred to as NETosis, the neutrophil often dies; however, nonlytic NETosis can occur. NETosis occurs by NADPH oxidase–dependent and –independent pathways. Neutrophils treated with NADPH oxidase inhibitors are unable to release NETs (Fuchs et al. 2007), as are neutrophils from patients with chronic granulomatous disease (CGD), which is a disease characterized by a deficiency in this enzyme (Bianchi et al.

2009). The Raf–MEK–ERK pathway is involved in NET formation through activation of NADPH oxidase and also upregulates antiapoptotic proteins (Hakkim et al. 2011). In addition to ROS, other pathways for NET formation exist, including uric acid, a well-known ROS scavenger, which inhibits NET formation at low concentrations but surprisingly induces NET formation at high concentrations (Arai et al. 2014).

Although NET formation has initially been described as an antimicrobial defense mechanism, available evidence suggests that NET formation is equally or perhaps more important in autoimmune and inflammatory diseases. NETs have been linked to small-vessel vasculitis, which is an autoinflammatory condition linked to antineutrophil cytoplasm autoantibodies (ANCAs) that target NET components and autoantigens proteinase-3 and myeloperoxidase (Kessenbrock et al. 2009). The formation of NETs in pancreatic ducts is driven by pancreatic juice components, such as bicarbonate ions and calcium carbonate crystals, and is associated with ductal occlusion and pancreatitis (Leppkes et al. 2016). NETs have also been shown to contribute to SLE via the pro-inflammatory action of oxidized mitochondrial DNA, which stimulates type I IFN signaling (Wang et al. 2015; Lood et al. 2016). Furthermore, MMP-9 contained in NETs has been shown to activate endothelial MMP-2, leading to endothelial dysfunction in the form of impaired aortic endothelium-dependent vasorelaxation

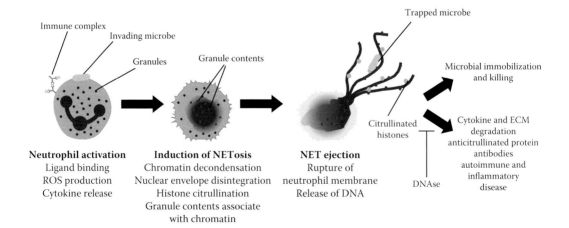

Figure 1.3. Process of NETosis. Activated neutrophils upregulate NADPH oxidase in response to pathogens or immune complexes, initiating the process of NETosis. Neutrophil granule contents mix with neutrophil DNA, which is then ejected from the cell, creating a NET. NETs are believed to contribute to several autoimmune and inflammatory conditions, as illustrated by the utility of DNAse in treating diseases such as cystic fibrosis.

NANOMEDICINE FOR INFLAMMATORY DISEASES

and increased endothelial cell apoptosis in murine endothelial cells, which can contribute to premature cardiovascular disease. Anti-MMP-9 autoantibodies are also present in human SLE sera and have been shown to induce NETosis and enhance MMP-9 activity (Carmona-Rivera et al. 2015). NETs also appear to play a role in cystic fibrosis, where extracellular DNA contributes to an increase in sputum viscosity. Cystic fibrosis is treated on a palliative basis with DNase, a nuclease that cleaves DNA (Papayannopoulos et al. 2011). NETs appear to contribute to several other diseases, including transfusion-related acute lung injury (Thomas et al. 2012), atherosclerosis (Döring et al. 2012), and rheumatoid arthritis (Khandpur et al. 2013; Pratesi et al. 2014).

1.2.1.3 Basophils

Basophils, which represent <1% of all blood leukocytes, are similar to mast cells but with a much less well-understood role in inflammatory disease. Basophils have long been suspected of playing a role in IgE-dependent allergic inflammation, similar to mast cells; they rapidly release histamine, heparin, and other inflammatory mediators upon cross-linking of the FcεR1 by IgE–allergen complexes. Basophil-derived IL-4 plays a major role in many Th2 inflammatory conditions, such as helminth infection, asthma, and atopic dermatitis (Wakahara et al. 2013; Kim et al. 2014). This appears to be mediated by direct cell-to-cell contact with CD4+ T cells (Sullivan et al. 2011; van Panhuys et al. 2011). Omalizumab, a monoclonal antibody to IgE that prevents IgE binding to the FcεR1, reduces the number of circulating basophils and decreases IL-4, IL-13, and IL-8 production by basophils in asthma patients receiving the antibody (Oliver et al. 2010; Hill et al. 2014). Basophils also appear to play a role in several non-allergic inflammatory diseases, such as irritable bowel syndrome. Patients with Crohn's disease have increased numbers of circulating basophils that drive a memory Th17/Th1 response in a contact-independent manner possibly reliant on histamine (Wakahara et al. 2012; Chapuy et al. 2014). Finally, recent data indicate that basophils may also possess immunoregulatory activity; basophil-derived IL-4 has been shown to attenuate skin inflammation by mediating monocyte differentiation into M2 macrophages (Egawa et al. 2013).

1.2.1.4 Eosinophils

Like basophils, eosinophils are nonprofessional phagocytic granulocytes traditionally associated with allergic diseases and parasite infection. Eosinophils represent 1%–6% of all blood lymphocytes, and are easily distinguished from other granulocytes in hematoxylin and eosin stain due to the intense red staining of the abundant intracellular granules as a result of the uptake of the acidophilic dye eosin. Like mast cells and basophils, eosinophils release an arsenal of toxic granule proteins and pro-inflammatory mediators that contribute to allergic inflammation and host defense against parasitic infection, as well as factors that promote tissue remodeling in response to damage. Table 1.3 lists the major physiological and pathological effects of eosinophil granules and the individual granule components associated with each. In addition to the individual granule components, eosinophils also produce ROS and inflammatory lipid mediators, including leukotrienes, prostaglandins, 5-hydroxyeicosatetraenoic (5-HETE), and PAF, as well as a number of proteinases (e.g., MMP-9). Eosinophils are common residents at sites of allergic inflammatory disease and are recruited mainly by IL-5 and eotaxin-1, both of which are released from activated eosinophils to recruit additional cells in an autocrine-like fashion. While the mechanisms triggering eosinophil degranulation during allergic inflammation are poorly understood, this likely involves epithelial cell–derived thymic stromal lymphopoietin (TSLP), which is produced by epithelial cells in response to certain allergens and environmental stimuli (Cook et al. 2012). While inducing

TABLE 1.3
Key eosinophil granule–derived mediators involved in inflammation.

Granule component	Main physiological effect
Proteases: MBP, EDN, ECP, EPO	Antimicrobial; tissue damage
Interleukins (IL-4, IL-6, IL-13, IL-24)	Inflammation, Th2 immunity
IL-5, GM-CSF, eotaxin-1	Eosinophil maturation and chemotaxis
TGF-β	Tissue remodeling and fibrosis

degranulation, TSLP has also been shown to stimulate the formation of eosinophil extracellular traps, which have a role in antibacterial responses (Morshed et al. 2012).

Eosinophils are well-known effectors in the destruction of antibody- or complement-opsonized parasites, as well as certain fungal, bacterial, and viral pathogens that are recognized through PRRs on the surface of the eosinophil (Kvarnhammar and Cardell 2012). These antimicrobial activities can be directed against both extracellular and phagocytosed organisms and are primarily mediated by the toxic granule proteins major basic protein (MBP), eosinophil peroxidase (EPO), eosinophil cationic protein (ECP), and eosinophil-derived neurotoxin (EDN). Eosinophils are stimulated by sIgA and IgA- and IgG-coated helminthes to release cytotoxic granule components to kill the parasites, but they also present parasite antigens to T cells to drive Th2 immunity (Shamri et al. 2011). Eosinophils also bind to some fungal pathogens and release MBP and EDN to kill the pathogen. Eosinophils can ingest bacteria, releasing MBP and ECP directly into the phagosome, or can mediate extracellular killing via eosinophil extracellular traps and oxygen-dependent mechanisms such as superoxide and EPO (Yousefi et al. 2008). Eosinophils also possess certain antiviral properties and can present viral peptides to T cells, promoting adaptive immune responses, especially for respiratory viruses that exacerbate asthma symptoms, as eosinophils are the predominant leukocyte in the airway of asthmatics (Drake et al. 2016).

Finally, eosinophils have a role in responding to tissue damage by recognizing necrotic cell debris through PRRs that bind to damage-associated molecular patterns (DAMPs), which enhance eosinophil survival and induce chemotactic migration to areas of tissue necrosis. While this can induce certain effects beneficial to wound healing, in diseases associated with eosinophilia, this contributes to chronic inflammation. Eosinophil-derived mediators such as TGF-β, IL-4, IL-13, MMPs, and granule proteins MBP and EDN can promote tissue fibrosis (Aceves and Ackerman 2009). This has been shown to contribute to diseases such as severe asthma (Aceves and Broide 2008), cardiac fibrosis during hypereosinophilic syndromes (Ogbogu et al. 2007), and eosinophilic esophagitis (Rawson et al. 2016).

1.2.2 Mononuclear Phagocyte System: Monocytes, Macrophages, and Dendritic Cells

1.2.2.1 Monocytes

Monocytes are a subset of leukocytes that normally circulate in the blood, bone marrow, and spleen, but during inflammation leave the bloodstream and migrate into tissues where they differentiate into macrophage or DC populations. The ability of monocytes to mobilize and traffic to a site where they are needed is central to their role in immune defenses; however, aberrant activation of monocytes is implicated in many inflammatory diseases, including atherosclerosis (Woollard and Geissmann 2010). Monocytes are short-lived and do not proliferate in the blood. Their role during homeostatic conditions is poorly understood but may involve the clearance of dead cells and toxins from the circulation (Auffray et al. 2009b).

Monocytes consist of at least two distinct subsets that originate in the bone marrow and have separate roles in inflammation and infection. $CD14^{++}CD16^-$ monocytes, or classical monocytes, are the most prevalent subset in the blood and also express CCR2, the primary chemokine receptor involved in monocyte recruitment. These cells differentiate into classically activated macrophages and are involved in microbial clearance. The $CD16^+$ monocytes are further subdivided into two subsets, $CD14^+CD16^{++}$, or nonclassical monocytes, and $CD14^{++}CD16^+$, or intermediate monocytes. Nonclassical monocytes are primarily involved in in vivo patrolling along the vascular lumen surface and differentiate into alternatively activated macrophages with a role in noninflammatory wound repair and tissue remodeling.

Classical monocytes (i.e., $CD16^-$) are recruited from the bone marrow and into the circulation through the activity of the chemokine receptor CCR2, which is the receptor for the two primary monocyte chemoattractants, CCL2 and CCL7, also known as monocyte chemoattractant protein-1 (MCP-1) and MCP-3, respectively. While very few cell types express the receptor, CCR2, most cells express CCL2 or CCL7 in response to proinflammatory cytokines or innate immune receptors. Many infections induce CCL2 and/or CCL7 expression, which results in high levels of these chemokines in serum and within inflamed tissues, which helps guide monocytes to the site. Deletion of either ccl2 or ccl7 in mice reduces the recruitment of monocytes by 40%–50% in response to *Listeria* infection, suggesting an additive response of these two chemokines (Jia et al. 2008).

On the other hand, nonclassical or intermediate monocytes (i.e., CD16[+]) respond to CX_3C–chemokine ligand 1 (CX_3CL1, or fractalkine), which is a membrane-tethered chemokine expressed in tissues that bind to CX_3CR1. This interaction is important in the patrolling activity of this monocyte subset (Auffray et al. 2007), and may also play a role in early recruitment of classical monocytes to the spleen (Auffray et al. 2009a). Mice deficient in either CX_3CR1 or CX_3CL1 show a significant reduction in the number of nonclassical monocytes in the circulation under both steady-state and inflammatory conditions, indicating that the CX_3CL1–CX_3CR1 axis also provides an essential survival signal to CD16[+] monocytes (Landsman et al. 2009).

Monocytes also express the chemokine receptors CCR1 and CCR5, which bind to a variety of shared ligands, including CCL3 (also known as MIP-1α) and CCL5. *In vitro* data indicate that CCR1 facilitates the arrest of monocytes along the vascular wall, CCR5 contributes to monocyte spreading, and both receptors assist in transendothelial migration toward CCL5 gradients (Weber et al. 2001). CCR1 and CCR5 have been implicated in several inflammatory diseases, including atherosclerosis, multiple sclerosis, and rheumatoid arthritis (Qidwai 2016); however, dissecting the specific roles of these two receptors in monocyte recruitment has been complicated by the fact that many cell types express both receptors. Therefore, any defect in monocyte recruitment may be secondary to defects in the recruitment of other cell types.

1.2.2.2 Macrophages

Macrophages are key mediators of inflammation and its resolution. Macrophages comprise diverse subpopulations of professional phagocytes, each with a unique anatomical location and function. Macrophage precursors are released into the circulation in the form of monocytes, which migrate into almost all tissues of the body and seed the tissue with mature, tissue-specific macrophages. Specialized tissue-resident macrophages include Kupffer cells in the liver, alveolar macrophages in the lung, and microglia in the brain. The main role of these tissue-specific macrophage populations is to ingest foreign material during homeostasis,

and to recruit additional macrophages from the circulation during periods of infection or injury. Characterization of macrophage subpopulations is complicated by a high degree of surface marker expression overlap, and few markers can definitively distinguish macrophages from monocytes and DCs because these cell types originate from a common myeloid progenitor. Nonetheless, several markers have been used in research to identify macrophage populations, including CD11b, CD11c, F4/80, CD68, LY6G, and LY6C (Murray and Wynn 2011). The most useful method for characterizing macrophage subpopulations is based on quantitative analysis of specific gene expression profiles following cytokine or microbial stimulation. This has revealed two broad categories of macrophages, namely, classically activated macrophages (M1 macrophages) and alternatively activated macrophages (M2 macrophages).

Classically activated macrophages, or M1 macrophages, mediate antimicrobial defenses and antitumor immunity. Following tissue injury or infection, these cells infiltrate the inflamed tissue and secrete pro-inflammatory mediators, such as TNF-α and IL-1β, and activate endogenous iNOS, producing NO, and NADPH oxidase during phagocytosis. The combination of NO, ROS, and derivatives such as peroxynitrite is harmful to microorganisms, as well as surrounding tissues, which can lead to inflammatory disease (Nathan and Ding 2010). Furthermore, M1 macrophages secrete a number of proteases, including MMP-9 that degrades components of the ECM (Roma-Lavisse et al. 2015), leading to tissue destruction and remodeling. Therefore, M1 macrophage responses must be tightly regulated in order to prevent excessive damage to the host.

Alternatively activated macrophages, or M2 macrophages, mediate anti-inflammatory responses and regulate wound healing through a process called efferocytosis. M2 macrophage differentiation is induced by anti-inflammatory cytokines IL-4 and IL-13 released from adaptive immune cells such as mast cells, Th2 cells, and basophils. These two cytokines induce the expression and/or activity of nuclear receptors, PPARγ and PPARδ, which respond to activating ligands such as 13-hydroxyoctadecadienoic acid (13-HODE) and 15-HETE. IL-4 also induces production of the two ligands through 15-lipoxygenase (15-LOX) activity (Huang et al. 1999). Together, these signaling cascades induce the process of efferocytosis, whereby M2 macrophages engulf apoptotic cells through a process more closely resembling micropinocytosis than phagocytosis, and initiate restoration of tissue structure and function. M2 macrophages exhibit downregulated expression of pro-inflammatory cytokines, ROS and reactive nitrogen species (RNS) production. M2 macrophages promote wound healing and fibrosis through the production of MMPs and growth factors, including TGF-β1 and platelet-derived growth factor (PDGF). Macrophage-derived TGF-β1 stimulates tissue repair by promoting fibroblast differentiation into myofibroblasts, by enhancing the expression of tissue inhibitors of metalloproteinases (TIMPs) that block the degradation of the ECM, and by stimulating the synthesis of ECM components by myofibroblasts (Roberts et al. 1986). TGF-β may also enhance PPARγ expression (Freire-de-Lima et al. 2006), leading to further efferocytic capacity.

M2 macrophages recognize apoptotic cells through a series of receptors that augment efferocytosis, creating a positive feedback loop of M2 macrophage activation and suppression of inflammatory responses. Efferocytic receptors, including stabilin-1 and stabilin-2, primarily recognize the phospholipid phosphatidylserine or its oxidized forms on the inner leaflet of apoptotic cells. Binding to phosphatidylserine induces autocrine secretion of IL-4, which further propagates signaling through PPARγ and PPARδ, leading to increased anti-inflammatory responses by the macrophage, including increased expression of efferocytic surface receptors and secretion of the bridge molecule, which are responsible for coupling apoptotic cells to the macrophage receptors (Park et al. 2009).

A number of enhancers augment the capacity of macrophages to engulf apoptotic cells. For example, lysophosphatidylserine (lyso-PS) is produced in dying neutrophils through an NADPH oxidase–dependent pathway and localizes to the neutrophil surface. Lyso-PS binds the macrophage G2A receptor, which stimulates prostaglandin E_2 (PGE_2) production, leading to a cAMP- and PKA-dependent increase in Rac1 activity (Frasch et al. 2008). Macrophages also release lipoxins, which are pro-resolving eicosanoids derived from arachidonic acid, during the resolution of inflammatory responses. Lipoxin A_4 enhances efferocytosis by binding to the macrophage ALX receptor and the annexin A1 bridge molecule

(Maderna et al. 2010), stimulating macrophages to engulf apoptotic neutrophils (Godson et al. 2000). Lipoxins and other pro-resolving mediators, such as resolvins, protectins, and maresins, have been observed to decrease production of pro-inflammatory cytokines. Importantly, production of these pro-resolving mediators occurs through PGE_2 signaling (Mancini and Di Battista 2011).

In addition to classically activated M1 macrophages and alternatively activated M2 macrophages, a number of other macrophage subsets have been described. So-called "regulatory" macrophages express high levels of IL-10 in response to Fcγ receptor ligation. This mechanism helps explain the immunosuppressive effects of intravenous Ig used to treat autoimmune and inflammatory disorders (Kozicky et al. 2015). Tumor-associated macrophages (TAMs) infiltrate the tumor microenvironment and contribute to cancer-related inflammation by promoting angiogenesis, immunosuppression, and tissue remodeling, which accelerates tumor progression (Belgiovine et al. 2016). Finally, myeloid-derived suppressor cells (MDSCs) are immature cells closely related to TAMs that exist mainly in the blood and lymphoid organs and suppress T cell functions. Although distinct from M2 macrophages, the regulatory macrophages, TAMs, and MDSCs all exhibit immune suppressive activity.

1.2.2.3 Dendritic Cells

DCs are a heterogeneous population of antigen-presenting cells with an important role in initiating adaptive immune responses; however, recent studies have identified a unique subpopulation of DCs that are present in many inflammatory conditions, including infections (De Trez et al. 2009), allergic asthma (Hammad et al. 2010), and rheumatoid arthritis (Reynolds et al. 2016). These so-called inflammatory DCs develop from monocytes and invade tissues during inflammation, and while similar, they are unique from inflammatory macrophages. The most important distinction between DCs and macrophages is that DCs, including inflammatory DCs, migrate from tissues to lymph nodes draining sites of infection. Additionally, inflammatory DCs are able to activate antigen-specific T cells in peripheral tissues (Wakim et al. 2008), which macrophages cannot. Unfortunately, there are limited data on the functional properties of inflammatory DCs in human disease. Inflammatory DCs from rheumatoid arthritis synovial fluid induce Th17 polarization *ex vivo*, and the same cells from tumor ascites secrete Th17-polarizing cytokines (Segura et al. 2013).

1.2.3 Lymphocytes

Lymphocytes are key effectors of adaptive immunity to provide defenses against pathogens; however, they can also play a major role in propagating chronic inflammation. Some of the strongest chronic inflammatory reactions are dependent on the presence of lymphocytes, and lymphocytes may be the dominant cell type in some types of chronic inflammatory states. The following section will focus on the role of lymphocytes in propagating chronic inflammation.

1.2.3.1 Natural Killer Cells

Conventional NK cells are a type of innate lymphoid cell (ILC) that, unlike T cells and B cells, lacks antigen-specific receptors. The primary role of NK cells is to modulate immune responses prior to the development of an adaptive response, primarily by secreting IFN-γ, and to target and destroy infected or malignant cells. NK cells are activated through the engagement of cell surface receptors that recognize infected cells and through the binding of activating cytokines. Activated NK cells not only secrete IFN-γ but also exhibit strong cytotoxic activity through the release of perforin and granzymes. Recently, it has been demonstrated that NK cells generate a pool of memory cells following expansion and contraction during a microbial infection, the mechanisms of which are still incompletely understood (O'Sullivan et al. 2015). These memory NK cells then possess enhanced cytokine and cytolytic responses following secondary exposure. Conventional NK cells primarily exist in the circulation or in secondary lymphoid tissue, such as the lymph nodes and spleen. A set of similar ILCs, ILC-1 cells or unconventional NK cells, reside in a variety of nonlymphoid tissues and exhibit many of the same properties as conventional NK cells, including IFN-γ production, but are derived from unique progenitors and appear to replenish themselves through local self-renewal. These cells play a greater role in localized inflammatory responses than do conventional NK cells, but also contribute to chronic inflammation (Fuchs 2016).

1.2.3.2 T Helper 1 Cells

Th1 cells are the classic immune cells involved in cell-mediated inflammatory reactions and play a critical role in defense against intracellular pathogens. Th1 cells primarily produce IL-2 and IFN-γ, but may also produce TNF-α and granulocyte–macrophage colony-stimulating factor (GM-CSF). IFN-γ plays a number of important roles in inflammatory reactions, including increasing expression of TLRs on innate immune cells and inducing the secretion of chemokines, leading to macrophage activation and increased phagocytosis (Bosisio et al. 2002; Schroder et al. 2004). There has been some debate about whether IFN-γ-secreting Th1 cells can play a pathogenic role in certain autoimmune conditions, or whether they primarily serve a protective or anti-inflammatory role in immune responses. Indeed, IFN-γ has been shown to exacerbate disease in models of SLE; IFN-γ-deficient and IFN-γ receptor–deficient mice experience less severe disease during murine lupus (Balomenos et al. 1998; Schwarting et al. 1998). However, IFN-γ-deficient mice experience more severe disease in EAE (Ferber et al. 1996), as well as asthma (Flaishon et al. 2002). One protective benefit of IFN-γ appears to be that it suppresses T cell differentiation toward more pathogenic Th subsets, such as Th17 cells. In fact, the increased disease severity during EAE in IFN-γ-deficient mice correlates with an increase in IL-17-producing Th cells (Komiyama et al. 2006), and IFN-γ-deficient mice also have a higher number of IL-17-producing Th cells during mycobacterial infection (Cruz et al. 2006).

1.2.3.3 T Helper 2 Cells

Th2 cells are primarily involved in host immune responses to multicellular parasites, as well as allergic and atopic diseases. Th2 cells primarily produce IL-4, IL-5, and IL-13, but are also important sources of IL-10. Th2 cells are often attributed to an anti-inflammatory role due to their ability to suppress Th1 cell responses, primarily through the activity of IL-4. IL-4 strongly suppresses Th1-activated macrophages by inhibiting pro-inflammatory cytokine secretion, including IL-1β and TNF-α, and the production of ROS and

RNS (Hwang et al. 2015). During parasite-induced inflammation, Th2 cell–derived IL-4 signaling promotes tissue repair by inducing M2 macrophage activity, and also downregulates pathogenic IL-17 expression while upregulating IL-10 expression (Chen et al. 2012).

1.2.3.4 T Helper 17 Cells

Th17 cells are a heterogeneous subset of Th lymphocytes that play a role in defense against bacteria and in the pathogenesis of many autoimmune diseases. These conflicting roles of Th17 cells in both protective responses and pathologic conditions have now been shown to be the result of diverse subsets of Th17 cells; Th17 cells can differentiate into pathogenic subsets in response to specific cytokine signals. Th17 cells differentiated in the presence of TGF-β1 and IL-6 produce IL-17 and IL-10 and do not induce inflammation (McGeachy et al. 2007). However, in the presence of IL-23 stimulation, these cells acquire the ability to induce pathogenic tissue inflammation. IL-17-producing T cells found in the lamina propria of the small intestines appear to be essential to maintain intestinal homeostasis by inducing the production of antimicrobial factors and secretory IgA (Cao et al. 2012), yet these cells do not induce any pathologic inflammation (Atarashi et al. 2008). In contrast, IL-23 induces pathogenic Th17 cell differentiation, which produces IL-17, IL-17F, IL-6, and TNF-α, but not IFN-γ and IL-4 (Aggarwal et al. 2003), contributing to inflammatory bowel disease in animal models (Yen et al. 2006).

1.2.3.5 Regulatory T Cells

Treg cells are a subset of CD4+ T cells with a role in suppressing immune responses and maintaining self-tolerance. Treg cells can be differentiated from conventional T cells by the expression of CD25 (i.e., CD4+CD25+). Treg cells can suppress proliferation and cytokine production by conventional CD4+CD25− T cells, and can also suppress the antigen-presenting capacity of antigen-presenting cells, thereby indirectly suppressing CD4+CD25− T cells. Treg cells interact directly with CD4+CD25− T cells by inhibiting IL-2 secretion via T cell receptor (TCR) engagement (Thornton and Shevach 1998), as well as transfer of cAMP through gap junctions (Bopp et al. 2007). Treg cells indirectly suppress conventional T cells by downregulating

costimulatory molecules on antigen-presenting cells via cytotoxic T lymphocyte–associated antigen-4 (CTLA-4) (Takahashi et al. 2000). Treg cells can also secrete the immunosuppressive cytokines IL-10 and TGF-β, which play a role in reducing inflammation and autoimmunity, respectively. Treg-derived IL-10 not only plays a role in the resolution of acute inflammation, but also promotes the maturation of memory CD8+ T cells during the resolution of infection (Laidlaw et al. 2015). Treg cells express both soluble and membrane-bound forms of TGF-β that are critical to suppressing CD4+CD25− T cells in models of intestinal inflammation (Nakamura et al. 2004). Taken together, Treg cells are critical to controlling immune responses primarily through the suppression of conventional T cell responses.

1.2.3.6 γδ T Cells

γδ T cells are specialized T cells comprising <5% of peripheral lymphocytes that are released from the thymus as mature cells that do not require TCR signaling for their differentiation and function. γδ T cells mobilize very early during immune responses and produce pro-inflammatory cytokines IFN-γ and TNF-α, as well as the anti-inflammatory cytokine IL-10 (Tsukaguchi et al. 1999). γδ T cells are also the major initial producers of IL-17 during acute infections (Lockhart et al. 2006). In contrast to Th17 cells, which require antigen-specific priming and an inflammatory environment to develop, γδ T cells respond without prior antigen exposure in an IL-23-dependent manner and possibly involving PRRs and/or inflammatory cytokine receptors (Martin et al. 2009). This initial burst of IL-17 helps recruit neutrophils to the site of infection long before Th17 cells are able to respond. Importantly, γδ T cells play an important role in regulating autoimmune disease, such as rheumatoid arthritis and SLE (Su et al. 2013), as well as facilitating cancer metastasis (Coffelt et al. 2015).

1.2.3.7 B Cells

B cells are a component of the adaptive immune response and are the source of Igs; therefore, B cells can play a role as indirect initiators of inflammation in response to pathogens or allergens. However, under certain conditions, B cells develop properties associated with immunosuppression. These cells are referred to as regulatory B (Breg) cells, and they support immunological tolerance and play a major role in regulating chronic inflammation. Breg cells produce the anti-inflammatory cytokines IL-10 and TGF-β, which suppress the expansion of pathogenic T cells (Carter et al. 2011, 2012; Rosser and Mauri 2015). Breg cell–derived IL-10 inhibits TNF-α-secreting monocytes, IL-12-producing DCs, IFN-γ-producing Th1 cells, and CD8+ cytotoxic T cells, and also activates Treg cells, which produce additional IL-10. TGF-β produced by LPS-activated Breg cells induces apoptosis of CD4+ T cells and activation-induced nonresponsiveness, or anergy, in CD8+ T cells (Tian et al. 2001; Parekh et al. 2003). Breg cells have been shown to be essential to controlling a number of autoimmune and inflammatory disorders, including arthritis, EAE, and colitis (Fillatreau et al. 2002; Mizoguchi et al. 2002; Mauri et al. 2003).

1.3 CYTOKINES ARE THE MESSENGERS OF THE IMMUNE SYSTEM

Many cytokines and chemokines play overlapping roles in inflammation and inflammatory disorders. Table 1.4 lists the major cytokines and chemokines involved in inflammatory immune responses; however, even this list is incomplete. The cytokines IL-1, IL-6, and TNF-α are elevated in most, if not all, inflammatory states and are recognized as targets for therapeutic intervention. For this reason, we focus our discussion on these three cytokines, with additional discussion around IL-17 as a now-recognized critical mediator of inflammatory disease.

1.3.1 Interleukin-1

The IL-1 family includes at least 11 members with both pro-inflammatory and anti-inflammatory effects, and they are expressed by multiple cell types. Of these, IL-1β is perhaps the most well-characterized cytokine in the family. IL-1β is a potent pyrogen and induces Th cell differentiation. IL-1β signaling is a tightly regulated process involving control processes at the expression, activation, and receptor-binding and signaling steps. Secretion of active IL-1β is a two-step process. First, IL-1β is synthesized in a pro-form lacking a secretory sequence. Then, following further signal input possibly involving a calcium- or calmodulin-dependent mechanism, the IL-1-converting enzyme (ICE) or caspase-1 cleaves IL-1β into its mature form, which is then secreted by the cell (Ainscough et al. 2015). Proteolytic maturation and secretion of IL-1β occurs through the action of a multiprotein complex called an inflammasome. The inflammasome forms in response to triggers binding to sensor molecules. Most inflammasomes contain a NLR sensor molecule, such as NLRP1 (NOD, leucine rich repeat [LRR], and pyrin domain-containing 1) or NLRP3 (Latz et al. 2013).

IL-1β binds a heterodimeric receptor complex on the surface of target cells. IL-1β first binds to IL-1 receptor 1 (IL-1R1), which then recruits the IL-1 receptor accessory protein (IL-1RAcP), forming a trimolecular signaling complex (Weber et al. 2010). IL-1R belongs to the IL-1R/TLR superfamily that contains a Toll/IL-1R (TIR) intracellular domain that interacts with MyD88 after ligand binding (O'Neill 2008). Signaling through MyD88 activates MAPKs and NF-κB, which results in inflammatory gene expression. IL-1β induces a multitude of physiological changes related to inflammation and immunity. First and foremost, IL-1β is the classic pyogen; it stimulates fever through the activation of cyclooxygenase-2 (COX-2) in the endothelium of the hypothalamus (Evans et al. 2015). COX-2 produces PGE_2, which then induces the release of noradrenaline from neurons, leading to elevated body temperature. IL-1β is also a critical regulator of the acute phase response by inducing IL-6 and CRP synthesis in hepatocytes (Kramer et al. 2008).

There are several mechanisms for regulating IL-1R-based signaling in the presence of active IL-1β. For example, IL-1β can bind a second receptor, IL-1R2, which lacks a functional intracellular domain and therefore serves as a decoy receptor (Garlanda et al. 2013). IL-1R2 is also expressed in a soluble form, which can bind and sequester IL-1β, thereby exerting an anti-inflammatory effect. Different isoforms of IL-1R2 can exert this effect at different locations. Extracellular, soluble IL-1R2 binds secreted IL-1β, whereas cytoplasmic isoforms bind pro-IL-1β with high affinity, thereby blocking interaction with caspase-1 (Smith et al. 2003). Finally, IL-1 receptor antagonist (IL-1Ra) plays an important role in regulating inflammation by limiting the activity of IL-1 signaling. IL-1Ra is a competitive inhibitor of IL-1 binding to cell surface receptors, and binding to IL-1R1 fails to recruit IL-1RAcP, thereby preventing

TABLE 1.4
Key cytokines and chemokines involved in inflammation.

Cytokine/chemokine	Primary sources	Main effect	References
IL-1β	Monocytes, macrophages, DCs, some epithelial cells	Pyrogenic, pro-inflammatory	Garlanda et al. 2013
IL-4	T cells, mast cells, basophils, eosinophils	Th cell differentiation, alternative macrophage activation; stimulates IgG and IgE production	Luzina et al. 2012
IL-5	T cells	Eosinophil maturation and differentiation; stimulates antibody production by activated B cells in mice	Kouro and Takatsu 2009
IL-6	T cells, macrophages, fibroblasts, endothelial cells	Neutrophil and monocyte chemotaxis, leukocyte transmigration, Th2 and Th17 differentiation	Scheller et al. 2011
IL-8 (CXCL8)	Macrophages, endothelial cells	Neutrophil chemotaxis	Russo et al. 2014
IL-10	Almost all immune cells	Anti-inflammatory; inhibits activity of Th1 cells, NK cells, and macrophages	Couper et al. 2008
IL-12	DCs, macrophages, B cells	Th1 differentiation; increased production of IFN-γ; activates NK cells	Vignali and Kuchroo 2012
IL-13	CD4+ T cells, NKT cells, mast cells, basophils, eosinophils	Downregulates macrophage activity, inhibiting inflammation	Rael and Lockey 2011
IL-17	Th17 cells, γδ T cells	Pro-inflammatory responses; recruitment of neutrophils and monocytes	Jin and Dong 2013
IL-23	Activated DCs and macrophages	Influences Th17 response	Vignali and Kuchroo 2012
IL-33	Epithelial, endothelial, and smooth muscle cells; released from necrotic cells	Th2 immunity and inflammation	Garlanda et al. 2013; Saluja et al. 2015
IFN-γ	NK cells, T cells, NKT cells	Macrophage activation, Th17 inhibition, Th1 development	Schroder et al. 2004
TNF-α	Macrophages, mast cells, endothelial cells, fibroblasts, T lymphocytes	Broad pro-inflammatory actions; induction of apoptosis	Wajant et al. 2003; Bradley 2008
G-CSF	Endothelial cells, macrophages, epithelial cells, fibroblasts	Neutrophil maturation	Bendall and Bradstock 2014
GM-CSF	T and B cells, monocytes/macrophages, endothelial cells, fibroblasts	Differentiation and proliferation of granulocytes and macrophages	Shiomi and Usui 2015
TGF-β	Many parenchymal cell types, lymphocytes, monocytes/macrophages, platelets	Induction of peripheral tolerance; T cell differentiation and innate immune cell suppression; wound healing and tissue repair	Sanjabi et al. 2009
MCP-1 (CCL2)	Monocytes/macrophages	Recruits monocytes, memory T cells, and NK cells	Deshmane et al. 2009

(Continued)

TABLE 1.4 (CONTINUED)

Cytokine/ chemokine	Primary sources	Main effect	References
MIP-1α (CCL3)	Macrophages	Recruits monocytes, eosinophils, basophils, and lymphocytes	Menten et al. 2002
Eotaxin (CCL11)	Epithelial cells, smooth muscle cells, endothelial cells, macrophages, eosinophils	Recruitment and activation of eosinophils	Adar et al. 2014
RANTES (CCL5)	T cells, macrophages, platelets, fibroblasts, epithelium	Recruits monocytes, eosinophils, basophils, and T cells; induces basophil histamine release; activates eosinophils; activation and proliferation of NK cells	Appay and Rowland-Jones 2001

downstream signaling. IL-1Ra is expressed as one of four isoforms as a result of alternative splicing: three cytosolic forms and one soluble secreted form (Arend and Guthridge 2000). The cytosolic isoforms likely serve as a reservoir of IL-1Ra and are released upon cell death, thereby regulating inflammation at sites of tissue damage. A recombinant, nonglycosylated form of IL-1Ra, Anakinra, has shown promise for treating certain conditions, such as rheumatoid arthritis and type 2 diabetes (Ruscitti et al. 2015).

1.3.2 Interleukin-6

IL-6 is a potent activator of the immune system with a broad range of activities, including a pivotal role during the transition from innate to acquired immunity. Early in the inflammatory process, IL-6 produced by endothelial cells facilitates, in conjunction with TNF-α and IL-1β, the infiltration of neutrophils into the site of inflammation. However, proteolytic processing of the IL-6 receptor (IL-6R) by neutrophil-derived proteases activates IL-6 *trans-signaling* in resident tissue cells, which then drives a transition to monocyte recruitment through suppression of neutrophil-attracting chemokines (e.g., IL-8) and enhancement of monocyte-attracting chemokines (e.g., MCP-1) (Kaplanski et al. 2003). Whereas IL-6 *classic signaling* is facilitated by conventional binding of IL-6 to cell surface–bound IL-6R plus gp130, IL-6R is only expressed on a subset of cells, including neutrophils, macrophages, some T cells, and hepatocytes. On the other hand, gp130

is expressed by a multitude of cell types. In IL-6 *trans-signaling*, soluble IL-6R released from the surface of neutrophils by the activity of serine proteases binds to IL-6, and the IL-6/sIL-6R complex then binds to membrane-bound gp130, mediating gp130 activation in an autocrine or paracrine manner on an expanded range of cell types (Rabe et al. 2008). IL-6 *trans-signaling* upregulates ICAM-1, vascular cell adhesion molecule-1 (VCAM-1), and E-selectin on endothelial cells, as well as L-selectin on lymphocytes, thereby enhancing lymphocyte transmigration (Chen et al. 2006).

IL-6 also plays a pivotal role in lymphocyte recruitment and differentiation. IL-6 *trans-signaling* activates the release of T cell–attracting chemokines (Suematsu et al. 1989), such as CCL4, CCL5, CCL17, and CXCL10 (McLoughlin et al. 2005), and prevents T cells from entering apoptosis (Curnow et al. 2004). IL-6 also enhances B cell function; IL-6-deficient mice have impaired IgG production after immunization with T cell–dependent antigens, and IL-6 has been shown to promote B cell antibody production by enhancing CD4[+] T cell production of IL-21 (Dienz et al. 2009; Eddahri et al. 2009). Finally, IL-6 has been shown to have an important role in skewing T cell differentiation toward Th2 and Th17 responses through STAT3 activation. Whereas TGF-β typically drives the development of Treg cells, which inhibit autoimmunity and prevent tissue damage, during inflammation or infection, the combination of TGF-β and IL-6 *trans-signaling* instead suppresses Treg development and favors the differentiation of effector Th17 cells (Bettelli et al. 2006;

Mangan et al. 2006; Dominitzki et al. 2007). In the absence of TGF-β, IL-6-mediated STAT3 activation instead favors an IL-4 autocrine loop in naïve T cells while blocking IFN-γ signaling, promoting Th2 differentiation (Sofi et al. 2009).

1.3.3 Tumor Necrosis Factor-α

TNF-α is perhaps the most studied pro-inflammatory cytokine and is attributed to multiple inflammatory diseases. TNF-α activity was first observed in the 1960s due to its ability to induce regression of tumors in mice; however, it was not until 1984–1985 when scientists first cloned TNF-α and the structurally related cytokine TNF-β (Aggarwal et al. 1984, 1985). TNF-α is now known to play a major role in a wide range of inflammatory and infectious conditions. Elevated serum and tissue TNF-α is known to correlate with the severity of numerous infections (Waage et al. 1987; Iwasaki et al. 2010), and anti-TNF antibodies or administration of soluble TNF receptors (TNFRs) is efficacious at controlling rheumatoid arthritis and other inflammatory conditions (Hernández et al. 2016). However, TNF-α is clearly essential to normal, protective immune responses, as indicated by the increased susceptibility to infection among rheumatoid arthritis patients receiving anti-TNF therapies (Crawford and Curtis 2008). It is the inappropriate or excessive production of TNF-α that is harmful.

Many of the pro-inflammatory effects of TNF-α signaling can be explained by its effects on the vascular endothelium and the impact this has on endothelial–leukocyte interactions. For example, TNF-α upregulates endothelial expression of certain cell adhesion molecules, including E-selectin, ICAM-1, and VCAM-1, which, in combination with chemokines, such as IL-8 and MCP-1, recruit leukocytes to the site of inflammation and facilitate their migration out of the blood and into the tissues (Munro et al. 1989). TNF-α induces the expression of COX-2 in endothelial cells, leading to the production of prostacyclin (PGI_2) and associated vasodilation (Mark et al. 2001). This is responsible for several of the cardinal signs of inflammation, including "rubor" and "calor," due to increased local blood flow. TNF-α also induces "tumor" as a result of increased vascular permeability, allowing the passage of fluid and macromolecules into the tissues, resulting in edema (Rochfort et al. 2016).

TNF signals through two receptors, TNFR1 (CD120a) and TNFR2 (CD120b), that utilize different signaling mechanisms, induce distinct biological responses, and are differentially regulated in various cell types in normal and diseased tissues. The pro-inflammatory and apoptotic signals attributed to TNF-α, as well as those associated with tissue injury, are largely mediated through TNFR1. In contrast, TNFR2 signaling promotes tissue repair and angiogenesis. Under some conditions, TNFR2 may contribute to TNFR1 responses under low concentrations of TNF-α through a mechanism known as "ligand passing" (Slowik et al. 1993), whereby TNFR2 catches TNF-α and passes it to nearby TNFR1 (Tartaglia et al. 1993). Certain pro- and anti-inflammatory signals are capable of regulating the balance between TNFR1 and TNFR2 expression. Binding of TNF, IL-1, and IL-10 is known to increase transcription of TNFR2, whereas these same signals often downregulate TNFR1 (Kalthoff et al. 1993; Winzen et al. 1993). Another pathway for regulating TNFR signaling is through soluble forms of the extracellular domain of TNFR1. NO and hydrogen peroxide, by-products of inflammatory cell activation, have been implicated in the activation of a metalloproteinase involved in shedding of TNFR1 (Hino et al. 1999). Soluble TNFR1 is then free to bind free TNF-α, preventing it from binding to the membrane-bound receptor and thereby limiting TNF signaling.

1.3.4 Interleukin-17

IL-17A, commonly referred to as IL-17, is one of six members of the IL-17 cytokine family, and in recent years has been the focus of intense investigation as a regulator of inflammation. The primary sources of IL-17 are Th17 cells and γδ T cells. The primary function of IL-17 is to promote the production of pro-inflammatory cytokines and chemokines that recruit neutrophils and macrophages to the site of inflammation. In response to IL-17, fibroblasts and epithelial cells upregulate the expression of CXCL1, CCL2, CCL7, CCL20, and MMP-3 and -13 (Park et al. 2005). This activity has been shown in mice to be important for clearing extracellular bacterial infections, including *Staphylococcus aureus* and *Klebsiella pneumoniae* (Ye et al. 2001; Chan et al. 2015). However, dysregulated IL-17 production can result in excessive pro-inflammatory cytokine expression, leading to tissue damage, autoimmunity, and inflammatory disease. IL-17 family cytokines have been linked to SLE (Li et al. 2015), rheumatoid arthritis (Kugyelka et al. 2016), and inflammation-induced malignancy (Kimura et al. 2016), and anti-IL-17 antibodies have been approved to treat patients with psoriasis (Shirley and Scott 2016).

1.4 LIPID MEDIATORS OF INFLAMMATION

1.4.1 Prostaglandins and Leukotrienes: Classic Inflammatory Mediators

Prostaglandins are lipid autocoids derived from arachidonic acid by the action of COX enzymes. Prostaglandins play a key role in inflammatory responses because they are a major contributor to the cardinal signs of acute inflammation. The four primary prostaglandins synthesized *in vivo* are PGE_2, PGI_2, PGD_2, and $PGF_2\alpha$. Prostaglandins are produced immediately following an injury, prior to the influx of leukocytes and other immune cells, due to the release of arachidonic acid from lipid membranes. COX-1 is constitutively expressed in most tissues, whereas COX-2 is induced by inflammatory stimuli, hormones, and growth factors. Nonetheless, both enzymes play a role in normal homeostatic maintenance of prostanoids, as well as during inflammatory responses (Rouzer and Marnett 2009). The clinical efficacy of nonsteroidal anti-inflammatory drugs (NSAIDs), such as aspirin, illustrates the importance of prostanoids as mediators of pain, fever, and inflammation; NSAIDs bind to and inactivate COX-1 and COX-2, blocking prostanoid synthesis (Vane 1971).

Prostaglandins bind to a subfamily of G protein–coupled receptors that exert their effect via a range of intracellular signaling pathways, with many ultimately regulating cAMP levels. PGE_2, the most abundant and widely characterized prostaglandin, binds to one of four cognate receptors, EP1 to EP4, and each EP receptor subtype shows a unique cellular distribution in tissues. PGE_2 signaling contributes to all the classic signs of inflammation. PGE_2 binding to the EP3 receptor augments arterial dilation and increased microvascular permeability through the activation of mast cells, leading to redness and edema (Morimoto et al. 2014). PGE_2 binding to the EP1 receptor acts on peripheral sensory neurons and on central sites in the spinal cord and brain to induce pain (Moriyama

et al. 2005). PGE_2 signaling can also modulate the function of macrophages, DCs, and lymphocytes. For example, PGE_2 binding to EP4 facilitates Th1 and IL-23-dependent Th17 differentiation, which promotes inflammation (Yao et al. 2009). PGE_2 signaling through EP2 and EP4 also promotes the migration and maturation of DCs (Legler et al. 2006). In contrast, PGE_2 has also been shown to exert anti-inflammatory action on neutrophils, monocytes, and NK cells (Harris et al. 2002). The other three main prostaglandins exhibit similar properties and induce many of the same physiological responses as PGE_2 (Ricciotti and FitzGerald 2011).

Leukotrienes are another important family of lipid meditators with a significant role in inflammatory immune responses. Leukotrienes are synthesized in leukocytes from arachidonic acid via the activity of 5-LOX, which first produces leukotriene A_4 (LTA_4). LTA_4 is then hydrolyzed by LTA_4 hydrolase into LTB_4, or conjugated with reduced glutathione by LTC_4 synthase to produce LTC_4. LTB_4 is perhaps the best characterized of the leukotrienes, and like the other leukotrienes, it interacts with a G protein–coupled receptor. Neutrophils, macrophages, and DCs are the primary sources of LTB_4, and the molecule induces numerous pro-inflammatory and antimicrobial effects in an autocrine fashion. LTB_4 induces neutrophil and lymphocyte migration and trafficking (Ford-Hutchinson et al. 1980; Tager et al. 2003; Lv et al. 2015), and increases leukocyte survival by inhibiting apoptosis (Hébert et al. 1996). LTB_4 also enhances phagocytic activity in macrophages and neutrophils (Okamoto et al. 2010), stimulates the release of lysosomal and antimicrobial enzymes in neutrophils (Flamand et al. 2004), and enhances NADPH oxidase–dependent killing of bacteria (Soares et al. 2013). Finally, LTB_4 enhances the production of a number of inflammatory cytokines, including TNF-α, IL-6, MCP-1, and IL-8 (Rola-Pleszczynski and Stanková 1992; McCain et al. 1994; Huang et al. 2004), which act to further propagate localized inflammatory reactions.

1.4.2 Pro-Resolving Lipid Mediators

Ideally, tissue injury or microbial invasion should induce an acute inflammatory response that is protective and self-limiting. Ungoverned acute inflammation, characterized by continuous polymorphonuclear (PMN) activation, can lead to irreversible tissue injury, chronic inflammation, scarring, and fibrosis. For decades, scientists believed that the resolution of acute inflammation was a passive process, whereby acute inflammation resolved itself due to such actions as dilution of cytokines and lymph drainage; it was believed that it simply "fizzled out." However, we now know that resolution is a highly orchestrated biosynthetically active response involving specialized pro-resolving mediators (SPMs), including lipoxins derived from arachidonic acid and the resolvins, protectins, and maresins derived from polyunsaturated fatty acids.

SPMs mediate the five general stages in the resolution of acute inflammation: (1) termination of PMN activation and infiltration, (2) a return to normal of vascular permeability and edema at the site of injury, (3) clearance of neutrophils via apoptosis, (4) nonphlogistic infiltration of monocytes and macrophages (i.e., without eliciting inflammatory cytokine secretion), and (5) removal of apoptotic PMNs, microbes, and cell debris by macrophages (Serhan 2010). SPMs are distinct from classical anti-inflammatory compounds such as IL-10 and TGF-β. The key actions of SPMs are to stop PMN infiltration and chemotaxis, block prostaglandin and leukotriene synthesis, reduce pro-inflammatory cytokine release, recruit monocytes without stimulating release of pro-inflammatory cytokines, and induce phagocytic removal of apoptotic PMNs (Tjonahen et al. 2006; Campbell et al. 2007; Ramon et al. 2016).

The lipoxins were the first mediators to be discovered with dual anti-inflammatory and pro-resolution activities. Lipoxins are synthesized by leukocytes, platelets, vascular endothelium, and epithelium via the action of LOXs, a group of non–heme iron–containing enzymes that catalyze the addition of oxygen to arachidonic acid. In addition to lipoxins, LOXs are responsible for producing leukotrienes and hydroperoxyeicosatetraenoic (HPETE) acids (Kim et al. 2008). Lipoxins modulate inflammation by decreasing a number of responses in neutrophils, including chemotaxis, adherence and transmigration, ROS production, and inflammatory cytokine release. Lipoxins also stimulate monocyte chemotaxis and upregulate the nonphlogistic phagocytosis of apoptotic neutrophils by macrophages (Mitchell et al. 2002). Some pathogens have exploited this response. For example, *Pseudomonas aeruginosa*

possesses a LOX that converts host arachidonic acid to 15-HETE, leading to local lipoxin production (Vance et al. 2004). *Toxoplasma gondii* also encodes its own LOX (Bannenberg et al. 2004). In both cases, microbial-induced production of lipoxins likely generates local anti-inflammatory mediators, subverting the host immune response.

The resolvins and protectins are synthesized from omega-3 polyunsaturated fatty acids and also play a major role in reducing neutrophil influx into a site of inflammation and stimulating macrophages to engulf apoptotic neutrophils (Schwab et al. 2007). Resolvin E1 accelerates phagocytosis-induced neutrophil apoptosis by binding to the LTB$_4$ receptor, BLT1, which causes an increase in NADPH oxidase–derived ROS generation and subsequent activation of caspases important in cell death (El Kebir et al. 2012). This interaction also suppresses certain apoptosis-suppressing signals. In addition, resolvin E1 and protectin D1 upregulate the expression of CCR5 on dying neutrophils, which binds CCL3 and CCL5 and sequesters them from inducing further activity (Ariel et al. 2006). Resolvins E1 and D1 inhibit LPS-induced TNF-α, IL-6, and IL-1β gene expression in microglia cells *in vitro* through unique pathways; resolvin E1 modulates NF-κB signaling by suppressing IκB-α phosphorylation, whereas resolvin D1 modulates the expression of miRNAs involved in inflammatory responses (Rey et al. 2016). Protectin D1, produced by Th2-skewed monocytes, inhibits TNF-α and IFN-γ secretion and induces apoptosis in T cells (Ariel et al. 2005).

1.5 SUMMARY

In conclusion, the inflammatory response is a complex process composed of both cellular and humoral components that are critical to maintaining homeostasis; however, the inflammatory response is often dysregulated in many disease states. Not only is inflammation critical to controlling infection, but it also plays a role in sterile injury and tissue repair. The primary components of acute inflammation are neutrophils, tissue-resident macrophages, and infiltrating monocytes, as well as a multitude of soluble factors, including pro-inflammatory cytokines and lipid-derived mediators. A successful inflammatory response is both effective and self-limiting; it accomplishes the task of containing and removing the offending agent while not causing excessive harm to the surrounding host tissues. The pro-inflammatory activities of many cellular and cytokine mediators are tempered by the activity of counterregulating signals, including the activities of IL-1 and TGF-β. Finally, the resolution phase of acute inflammation is a metabolically active process, rather than a passive process, that is coordinated by a series of lipid-derived mediators, namely, the lipoxins, resolvins, and protectins. These mediators act to suppress the activities of neutrophils, and to facilitate the removal of neutrophils by alternatively activated macrophages. In many infections and disease states, the primary acute inflammatory response is either insufficient or deficient in clearing the injurious agent; this often leads to a nonresolving or chronic inflammatory state characterized by a greater presence of components of the adaptive immune response, including lymphocytes. In chronic inflammation, the progressive accumulation of pro-inflammatory signals feeds positive feedback loops, eventually leading to excessive host tissue damage, which is the primary complication associated with most chronic inflammatory diseases. The chapters that follow will address in greater detail several of these conditions.

GLOSSARY

IL-1β: a pro-inflammatory cytokine produced by numerous cell types with an important role in modulating antimicrobial responses and fever.

IL-6: a pro-inflammatory cytokine with an important role in the transition from innate to acquired immunity.

IL-10: an anti-inflammatory cytokine that inhibits the activity of Th1 cells, NK cells, and macrophages.

IL-17: one member of a larger group of pro-inflammatory cytokines that is important in recruiting neutrophils and macrophages to the site of inflammation.

leukotriene: a family of lipid mediators derived from arachidonic acid that enhance the activity of neutrophils and macrophages.

lipoxin: a family of lipid mediators derived from arachidonic acid that inhibit the activity of neutrophils and promote nonphlogistic recruitment of macrophages, promoting resolution of inflammation.

macrophage: a professional phagocytic cell that is important in clearing microbial invaders and in clearing apoptotic neutrophils after an inflammatory response.

NETosis: the process by which neutrophils extrude strands of DNA to ensnare and digest extracellular bacteria.

neutrophil: an innate immune cell that is critically important in inflammatory responses and the response to infection; kills invading pathogens via an arsenal of serine proteases and oxygen-dependent killing mechanisms.

prostaglandin: a family of lipid mediators derived from arachidonic acid that are responsible for many of the cardinal signs of inflammation, including heat, redness, pain, and swelling.

TGF-β: an anti-inflammatory cytokine or growth factor that is important in suppressing innate immune cell responses during inflammation.

TNF-α: a pro-inflammatory cytokine with a broad range of activities, including the ability to induce apoptosis in cancer cells.

REFERENCES

Abram, C. L., and C. A. Lowell. 2009. The ins and outs of leukocyte integrin signaling. *Annu Rev Immunol* 27:339–62.

Aceves, S. S., and S. J. Ackerman. 2009. Relationships between eosinophilic inflammation, tissue remodeling, and fibrosis in eosinophilic esophagitis. *Immunol Allergy Clin North Am* 29 (1):197–211.

Aceves, S. S., and D. H. Broide. 2008. Airway fibrosis and angiogenesis due to eosinophil trafficking in chronic asthma. *Curr Mol Med* 8 (5):350–8.

Adar, T., S. Shteingart, A. Ben Ya'acov, A. Bar-Gil Shitrit, and E. Goldin. 2014. From airway inflammation to inflammatory bowel disease: Eotaxin-1, a key regulator of intestinal inflammation. *Clin Immunol* 153 (1):199–208.

Aggarwal, B. B., W. J. Henzel, B. Moffat, W. J. Kohr, and R. N. Harkins. 1985. Primary structure of human lymphotoxin derived from 1788 lymphoblastoid cell line. *J Biol Chem* 260 (4):2334–44.

Aggarwal, B. B., B. Moffat, and R. N. Harkins. 1984. Human lymphotoxin. Production by a lymphoblastoid cell line, purification, and initial characterization. *J Biol Chem* 259 (1):686–91.

Aggarwal, S., N. Ghilardi, M. H. Xie, F. J. de Sauvage, and A. L. Gurney. 2003. Interleukin-23 promotes a distinct CD4 T cell activation state characterized by the production of interleukin-17. J Biol Chem 278 (3):1910–4.

Ainscough, J. S., G. F. Gerberick, I. Kimber, and R. J. Dearman. 2015. Interleukin-1β processing is dependent on a calcium-mediated interaction with calmodulin. J Biol Chem 290 (52):31151–61.

Appay, V., and S. L. Rowland-Jones. 2001. RANTES: A versatile and controversial chemokine. Trends Immunol 22 (2):83–7.

Arai, Y., Y. Nishinaka, T. Arai, M. Morita, K. Mizugishi, S. Adachi, A. Takaori-Kondo, T. Watanabe, and K. Yamashita. 2014. Uric acid induces NADPH oxidase-independent neutrophil extracellular trap formation. Biochem Biophys Res Commun 443 (2):556–61.

Arend, W. P., and C. J. Guthridge. 2000. Biological role of interleukin 1 receptor antagonist isoforms. Ann Rheum Dis 59 (Suppl 1):i60–4.

Ariel, A., G. Fredman, Y. P. Sun, A. Kantarci, T. E. Van Dyke, A. D. Luster, and C. N. Serhan. 2006. Apoptotic neutrophils and T cells sequester chemokines during immune response resolution through modulation of CCR5 expression. Nat Immunol 7 (11):1209–16.

Ariel, A., P. L. Li, W. Wang, W. X. Tang, G. Fredman, S. Hong, K. H. Gotlinger, and C. N. Serhan. 2005. The docosatriene protectin D1 is produced by TH2 skewing and promotes human T cell apoptosis via lipid raft clustering. J Biol Chem 280 (52):43079–86.

Atarashi, K., J. Nishimura, T. Shima, Y. Umesaki, M. Yamamoto, M. Onoue, H. Yagita, N. Ishii, R. Evans, K. Honda, and K. Takeda. 2008. ATP drives lamina propria T(H)17 cell differentiation. Nature 455 (7214):808–12.

Auffray, C., D. Fogg, M. Garfa, G. Elain, O. Join-Lambert, S. Kayal, S. Sarnacki, A. Cumano, G. Lauvau, and F. Geissmann. 2007. Monitoring of blood vessels and tissues by a population of monocytes with patrolling behavior. Science 317 (5838):666–70.

Auffray, C., D. K. Fogg, E. Narni-Mancinelli, B. Senechal, C. Trouillet, N. Saederup, J. Leemput et al. 2009a. CX3CR1+ CD115+ CD135+ common macrophage/DC precursors and the role of CX3CR1 in their response to inflammation. J Exp Med 206 (3):595–606.

Auffray, C., M. H. Sieweke, and F. Geissmann. 2009b. Blood monocytes: Development, heterogeneity, and relationship with dendritic cells. Annu Rev Immunol 27:669–92.

Balomenos, D., R. Rumold, and A. N. Theofilopoulos. 1998. Interferon-gamma is required for lupus-like disease and lymphoaccumulation in MRL-lpr mice. J Clin Invest 101 (2):364–71.

Bank, U., D. Reinhold, C. Schneemilch, D. Kunz, H. J. Synowitz, and S. Ansorge. 1999. Selective proteolytic cleavage of IL-2 receptor and IL-6 receptor ligand binding chains by neutrophil-derived serine proteases at foci of inflammation. J Interferon Cytokine Res 19 (11):1277–87.

Bannenberg, G. L., J. Aliberti, S. Hong, A. Sher, and C. Serhan. 2004. Exogenous pathogen and plant 15-lipoxygenase initiate endogenous lipoxin A4 biosynthesis. J Exp Med 199 (4):515–23.

Belaaouaj, A., K. S. Kim, and S. D. Shapiro. 2000. Degradation of outer membrane protein A in Escherichia coli killing by neutrophil elastase. Science 289 (5482):1185–8.

Belgiovine, C., M. D'Incalci, P. Allavena, and R. Frapolli. 2016. Tumor-associated macrophages and anti-tumor therapies: Complex links. Cell Mol Life Sci 73 (13):2411–24.

Bendall, L. J., and K. F. Bradstock. 2014. G-CSF: From granulopoietic stimulant to bone marrow stem cell mobilizing agent. Cytokine Growth Factor Rev 25 (4):355–67.

Bettelli, E., Y. Carrier, W. Gao, T. Korn, T. B. Strom, M. Oukka, H. L. Weiner, and V. K. Kuchroo. 2006. Reciprocal developmental pathways for the generation of pathogenic effector TH17 and regulatory T cells. Nature 441 (7090):235–8.

Beyrau, M., J. V. Bodkin, and S. Nourshargh. 2012. Neutrophil heterogeneity in health and disease: A revitalized avenue in inflammation and immunity. Open Biol 2 (11):120134.

Bianchi, M., A. Hakkim, V. Brinkmann, U. Siler, R. A. Seger, A. Zychlinsky, and J. Reichenbach. 2009. Restoration of NET formation by gene therapy in CGD controls aspergillosis. Blood 114 (13):2619–22.

Blank, U., and J. Rivera. 2004. The ins and outs of IgE-dependent mast-cell exocytosis. Trends Immunol 25 (5):266–73.

Boner, A., B. J. Zeligs, and J. A. Bellanti. 1982. Chemotactic responses of various differentiational stages of neutrophils from human cord and adult blood. Infect Immun 35 (3):921–8.

Bopp, T., C. Becker, M. Klein, S. Klein-Hessling, A. Palmetshofer, E. Serfling, V. Heib et al. 2007. Cyclic adenosine monophosphate is a key component of regulatory T cell-mediated suppression. J Exp Med 204 (6):1303–10.

Bosisio, D., N. Polentarutti, M. Sironi, S. Bernasconi, K. Miyake, G. R. Webb, M. U. Martin, A. Mantovani, and M. Muzio. 2002. Stimulation of toll-like receptor 4 expression in human mononuclear phagocytes by interferon-gamma: A molecular basis for priming and synergism with bacterial lipopolysaccharide. *Blood* 99 (9):3427–31.

Bradley, J. R. 2008. TNF-mediated inflammatory disease. *J Pathol* 214 (2):149–60.

Brinkmann, V., U. Reichard, C. Goosmann, B. Fauler, Y. Uhlemann, D. S. Weiss, Y. Weinrauch, and A. Zychlinsky. 2004. Neutrophil extracellular traps kill bacteria. *Science* 303 (5663):1532–5.

Campbell, E. L., N. A. Louis, S. E. Tomassetti, G. O. Canny, M. Arita, C. N. Serhan, and S. P. Colgan. 2007. Resolvin E1 promotes mucosal surface clearance of neutrophils: A new paradigm for inflammatory resolution. *FASEB J* 21 (12):3162–70.

Cao, A. T., S. Yao, B. Gong, C. O. Elson, and Y. Cong. 2012. Th17 cells upregulate polymeric Ig receptor and intestinal IgA and contribute to intestinal homeostasis. *J Immunol* 189 (9):4666–73.

Carmona-Rivera, C., W. Zhao, S. Yalavarthi, and M. J. Kaplan. 2015. Neutrophil extracellular traps induce endothelial dysfunction in systemic lupus erythematosus through the activation of matrix metalloproteinase-2. *Ann Rheum Dis* 74 (7):1417–24.

Carter, N. A., E. C. Rosser, and C. Mauri. 2012. Interleukin-10 produced by B cells is crucial for the suppression of Th17/Th1 responses, induction of T regulatory type 1 cells and reduction of collagen-induced arthritis. *Arthritis Res Ther* 14 (1):R32.

Carter, N. A., R. Vasconcellos, E. C. Rosser, C. Tulone, A. Muñoz-Suano, M. Kamanaka, M. R. Ehrenstein, R. A. Flavell, and C. Mauri. 2011. Mice lacking endogenous IL-10-producing regulatory B cells develop exacerbated disease and present with an increased frequency of Th1/Th17 but a decrease in regulatory T cells. *J Immunol* 186 (10):5569–79.

Chan, L. C., S. Chaili, S. G. Filler, K. Barr, H. Wang, D. Kupferwasser, J. E. Edwards et al. 2015. Nonredundant roles of interleukin-17A (IL-17A) and IL-22 in murine host defense against cutaneous and hematogenous infection due to methicillin-resistant *Staphylococcus aureus*. *Infect Immun* 83 (11):4427–37.

Chapuy, L., M. Bsat, H. Mehta, M. Rubio, K. Wakahara, V. Q. Van, N. Baba et al. 2014. Basophils increase in Crohn disease and ulcerative colitis and favor mesenteric lymph node memory TH17/TH1 response. *J Allergy Clin Immunol* 134 (4):978–81.e1.

Chen, F., Z. Liu, W. Wu, C. Rozo, S. Bowdridge, A. Millman, N. Van Rooijen, J. F. Urban, T. A. Wynn, and W. C. Gause. 2012. An essential role for TH2-type responses in limiting acute tissue damage during experimental helminth infection. *Nat Med* 18 (2):260–6.

Chen, Q., D. T. Fisher, K. A. Clancy, J. M. Gauguet, W. C. Wang, E. Unger, S. Rose-John, U. H. von Andrian, H. Baumann, and S. S. Evans. 2006. Fever-range thermal stress promotes lymphocyte trafficking across high endothelial venules via an interleukin 6 trans-signaling mechanism. *Nat Immunol* 7 (12):1299–308.

Cho, K. A., J. W. Suh, J. H. Sohn, J. W. Park, H. Lee, J. L. Kang, S. Y. Woo, and Y. J. Cho. 2012. IL-33 induces Th17-mediated airway inflammation via mast cells in ovalbumin-challenged mice. *Am J Physiol Lung Cell Mol Physiol* 302 (4):L429–40.

Christoffersson, G., E. Vågesjö, J. Vandooren, M. Lidén, S. Massena, R. B. Reinert, M. Brissova, A. C. Powers, G. Opdenakker, and M. Phillipson. 2012. VEGF-A recruits a proangiogenic MMP-9-delivering neutrophil subset that induces angiogenesis in transplanted hypoxic tissue. *Blood* 120 (23):4653–62.

Christopher, M. J., M. Rao, F. Liu, J. R. Woloszynek, and D. C. Link. 2011. Expression of the G-CSF receptor in monocytic cells is sufficient to mediate hematopoietic progenitor mobilization by G-CSF in mice. *J Exp Med* 208 (2):251–60.

Coeshott, C., C. Ohnemus, A. Pilyavskaya, S. Ross, M. Wieczorek, H. Kroona, A. H. Leimer, and J. Cheronis. 1999. Converting enzyme-independent release of tumor necrosis factor alpha and IL-1beta from a stimulated human monocytic cell line in the presence of activated neutrophils or purified proteinase 3. *Proc Natl Acad Sci USA* 96 (11):6261–6.

Coffelt, S. B., K. Kersten, C. W. Doornebal, J. Weiden, K. Vrijland, C. S. Hau, N. J. Verstegen, M. Ciampricotti, L. J. Hawinkels, J. Jonkers, and K. E. de Visser. 2015. IL-17-producing γδ T cells and neutrophils conspire to promote breast cancer metastasis. *Nature* 522 (7556):345–8.

Cohen, R., K. Corwith, D. Holowka, and B. Baird. 2012. Spatiotemporal resolution of mast cell granule exocytosis reveals correlation with Ca2+ wave initiation. *J Cell Sci* 125 (Pt 12):2986–94.

Cook, E. B., J. L. Stahl, E. A. Schwantes, K. E. Fox, and S. K. Mathur. 2012. IL-3 and TNFα increase thymic stromal lymphopoietin receptor (TSLPR) expression on eosinophils and enhance TSLP-stimulated degranulation. *Clin Mol Allergy* 10 (1):8.

Costanza, M., M. P. Colombo, and R. Pedotti. 2012. Mast cells in the pathogenesis of multiple sclerosis and experimental autoimmune encephalomyelitis. *Int J Mol Sci* 13 (11):15107–25.

Cougoule, C., S. Hoshino, A. Dart, J. Lim, and E. Caron. 2006. Dissociation of recruitment and activation of the small G-protein Rac during Fcgamma receptor-mediated phagocytosis. *J Biol Chem* 281 (13):8756–64.

Couper, K. N., D. G. Blount, and E. M. Riley. 2008. IL-10: The master regulator of immunity to infection. *J Immunol* 180 (9):5771–7.

Crawford, M., and J. R. Curtis. 2008. Tumor necrosis factor inhibitors and infection complications. *Curr Rheumatol Rep* 10 (5):383–9.

Cruz, A., S. A. Khader, E. Torrado, A. Fraga, J. E. Pearl, J. Pedrosa, A. M. Cooper, and A. G. Castro. 2006. Cutting edge: IFN-gamma regulates the induction and expansion of IL-17-producing CD4 T cells during mycobacterial infection. *J Immunol* 177 (3):1416–20.

Curnow, S. J., D. Scheel-Toellner, W. Jenkinson, K. Raza, O. M. Durrani, J. M. Faint, S. Rauz et al. 2004. Inhibition of T cell apoptosis in the aqueous humor of patients with uveitis by IL-6/soluble IL-6 receptor trans-signaling. *J Immunol* 173 (8):5290–7.

Deshmane, S. L., S. Kremlev, S. Amini, and B. E. Sawaya. 2009. Monocyte chemoattractant protein-1 (MCP-1): An overview. *J Interferon Cytokine Res* 29 (6):313–26.

De Trez, C., S. Magez, S. Akira, B. Ryffel, Y. Carlier, and E. Muraille. 2009. iNOS-producing inflammatory dendritic cells constitute the major infected cell type during the chronic *Leishmania major* infection phase of C57BL/6 resistant mice. *PLoS Pathog* 5 (6):e1000494.

Di Capite, J., and A. B. Parekh. 2009. CRAC channels and Ca2+ signaling in mast cells. *Immunol Rev* 231 (1):45–58.

Dienz, O., S. M. Eaton, J. P. Bond, W. Neveu, D. Moquin, R. Noubade, E. M. Briso et al. 2009. The induction of antibody production by IL-6 is indirectly mediated by IL-21 produced by CD4+ T cells. *J Exp Med* 206 (1):69–78.

Dominitzki, S., M. C. Fantini, C. Neufert, A. Nikolaev, P. R. Galle, J. Scheller, G. Monteleone, S. Rose-John, M. F. Neurath, and C. Becker. 2007. Cutting edge: Trans-Signaling via the soluble IL-6R abrogates the induction of FoxP3 in naive CD4+CD25 T cells. *J Immunol* 179 (4):2041–5.

Drake, M. G., E. R. Bivins-Smith, B. J. Proskocil, Z. Nie, G. D. Scott, J. J. Lee, N. A. Lee, A. D. Fryer, and D. B. Jacoby. 2016. Human and mouse eosinophils have antiviral activity against parainfluenza virus. *Am J Respir Cell Mol Biol* 5 (3):387–94.

Döring, Y., H. D. Manthey, M. Drechsler, D. Lievens, R. T. Megens, O. Soehnlein, M. Busch et al. 2012. Auto-antigenic protein-DNA complexes stimulate plasmacytoid dendritic cells to promote atherosclerosis. *Circulation* 125 (13):1673–83.

Eddahri, F., S. Denanglaire, F. Bureau, R. Spolski, W. J. Leonard, O. Leo, and F. Andris. 2009. Interleukin-6/STAT3 signaling regulates the ability of naive T cells to acquire B-cell help capacities. *Blood* 113 (11):2426–33.

Egawa, M., K. Mukai, S. Yoshikawa, M. Iki, N. Mukaida, Y. Kawano, Y. Minegishi, and H. Karasuyama. 2013. Inflammatory monocytes recruited to allergic skin acquire an anti-inflammatory M2 phenotype via basophil-derived interleukin-4. *Immunity* 38 (3):570–80.

Ehlers, M. R. 2014. Immune-modulating effects of alpha-1 antitrypsin. *Biol Chem* 395 (10):1187–93.

El Kebir, D., P. Gjorstrup, and J. G. Filep. 2012. Resolvin E1 promotes phagocytosis-induced neutrophil apoptosis and accelerates resolution of pulmonary inflammation. *Proc Natl Acad Sci USA* 109 (37):14983–8.

Evans, S. S., E. A. Repasky, and D. T. Fisher. 2015. Fever and the thermal regulation of immunity: The immune system feels the heat. *Nat Rev Immunol* 15 (6):335–49.

Ferber, I. A., S. Brocke, C. Taylor-Edwards, W. Ridgway, C. Dinisco, L. Steinman, D. Dalton, and C. G. Fathman. 1996. Mice with a disrupted IFN-gamma gene are susceptible to the induction of experimental autoimmune encephalomyelitis (EAE). *J Immunol* 156 (1):5–7.

Fillatreau, S., C. H. Sweenie, M. J. McGeachy, D. Gray, and S. M. Anderton. 2002. B cells regulate autoimmunity by provision of IL-10. *Nat Immunol* 3 (10):944–50.

Flaishon, L., I. Topilski, D. Shoseyov, R. Hershkoviz, E. Fireman, Y. Levo, S. Marmor, and I. Shachar. 2002. Cutting edge: Anti-inflammatory properties of low levels of IFN-gamma. *J Immunol* 168 (8):3707–11.

Flamand, L., P. Borgeat, R. Lalonde, and J. Gosselin. 2004. Release of anti-HIV mediators after administration of leukotriene B4 to humans. *J Infect Dis* 189 (11):2001–9.

Ford-Hutchinson, A. W., M. A. Bray, M. V. Doig, M. E. Shipley, and M. J. Smith. 1980. Leukotriene B, a potent chemokinetic and aggregating substance released from polymorphonuclear leukocytes. *Nature* 286 (5770):264–5.

Fox, E. D., D. S. Heffernan, W. G. Cioffi, and J. S. Reichner. 2013. Neutrophils from critically ill septic patients mediate profound loss of endothelial barrier integrity. *Crit Care* 17 (5):R226.

Frasch, S. C., K. Z. Berry, R. Fernandez-Boyanapalli, H. S. Jin, C. Leslie, P. M. Henson, R. C. Murphy, and D. L. Bratton. 2008. NADPH oxidase-dependent generation of lysophosphatidylserine enhances clearance of activated and dying neutrophils via G2A. *J Biol Chem* 283 (48):33736–49.

Freire-de-Lima, C. G., Y. Q. Xiao, S. J. Gardai, D. L. Bratton, W. P. Schiemann, and P. M. Henson. 2006. Apoptotic cells, through transforming growth factor-beta, coordinately induce anti-inflammatory and suppress pro-inflammatory eicosanoid and NO synthesis in murine macrophages. *J Biol Chem* 281 (50):38376–84.

Fuchs, A. 2016. ILC1s in tissue inflammation and infection. *Front Immunol* 7:104.

Fuchs, T. A., U. Abed, C. Goosmann, R. Hurwitz, I. Schulze, V. Wahn, Y. Weinrauch, V. Brinkmann, and A. Zychlinsky. 2007. Novel cell death program leads to neutrophil extracellular traps. *J Cell Biol* 176 (2):231–41.

Garlanda, C., C. A. Dinarello, and A. Mantovani. 2013. The interleukin-1 family: Back to the future. *Immunity* 39 (6):1003–18.

Godson, C., S. Mitchell, K. Harvey, N. A. Petasis, N. Hogg, and H. R. Brady. 2000. Cutting edge: Lipoxins rapidly stimulate nonphlogistic phagocytosis of apoptotic neutrophils by monocyte-derived macrophages. *J Immunol* 164 (4):1663–7.

Gordon, D. L., J. L. Rice, and P. J. McDonald. 1989. Regulation of human neutrophil type 3 complement receptor (iC3b receptor) expression during phagocytosis of *Staphylococcus aureus* and *Escherichia coli*. *Immunology* 67 (4):460–5.

Gri, G., S. Piconese, B. Frossi, V. Manfroi, S. Merluzzi, C. Tripodo, A. Viola, S. Odom, J. Rivera, M. P. Colombo, and C. E. Pucillo. 2008. CD4+CD25+ regulatory T cells suppress mast cell degranulation and allergic responses through OX40-OX40L interaction. *Immunity* 29 (5):771–81.

Hakkim, A., T. A. Fuchs, N. E. Martinez, S. Hess, H. Prinz, A. Zychlinsky, and H. Waldmann. 2011. Activation of the Raf-MEK-ERK pathway is required for neutrophil extracellular trap formation. *Nat Chem Biol* 7 (2):75–7.

Hammad, H., M. Plantinga, K. Deswarte, P. Pouliot, M. A. Willart, M. Kool, F. Muskens, and B. N. Lambrecht. 2010. Inflammatory dendritic cells—not basophils—are necessary and sufficient for induction of Th2 immunity to inhaled house dust mite allergen. *J Exp Med* 207 (10):2097–111.

Harris, S. G., J. Padilla, L. Koumas, D. Ray, and R. P. Phipps. 2002. Prostaglandins as modulators of immunity. *Trends Immunol* 23 (3):144–50.

Hartl, D., P. Latzin, P. Hordijk, V. Marcos, C. Rudolph, M. Woischnik, S. Krauss-Etschmann et al. 2007. Cleavage of CXCR1 on neutrophils disables bacterial killing in cystic fibrosis lung disease. *Nat Med* 13 (12):1423–30.

Heidland, A., A. Klassen, P. Rutkowski, and U. Bahner. 2006. The contribution of Rudolf Virchow to the concept of inflammation: What is still of importance? *J Nephrol* 19 (Suppl 10):S102–9.

Heissig, B., C. Nishida, Y. Tashiro, Y. Sato, M. Ishihara, M. Ohki, I. Gritli, J. Rosenkvist, and K. Hattori. 2010. Role of neutrophil-derived matrix metalloproteinase-9 in tissue regeneration. *Histol Histopathol* 25 (6):765–70.

Hennersdorf, F., S. Florian, A. Jakob, K. Baumgärtner, K. Sonneck, A. Nordheim, T. Biedermann, P. Valent, and H. J. Bühring. 2005. Identification of CD13, CD107a, and CD164 as novel basophil-activation markers and dissection of two response patterns in time kinetics of IgE-dependent upregulation. *Cell Res* 15 (5):325–35.

Hernández, M. V., R. Sanmartí, and J. D. Cañete. 2016. The safety of tumor necrosis factor-alpha inhibitors in the treatment of rheumatoid arthritis. *Expert Opin Drug Saf* 15 (5):613–24.

Hill, D. A., M. C. Siracusa, K. R. Ruymann, E. D. Tait Wojno, D. Artis, and J. M. Spergel. 2014. Omalizumab therapy is associated with reduced circulating basophil populations in asthmatic children. *Allergy* 69 (5):674–7.

Hino, T., H. Nakamura, S. Abe, H. Saito, M. Inage, K. Terashita, S. Kato, and H. Tomoike. 1999. Hydrogen peroxide enhances shedding of type I soluble tumor necrosis factor receptor from pulmonary epithelial cells. *Am J Respir Cell Mol Biol* 20 (1):122–8.

Huang, J. T., J. S. Welch, M. Ricote, C. J. Binder, T. M. Willson, C. Kelly, J. L. Witztum, C. D. Funk, D. Conrad, and C. K. Glass. 1999. Interleukin-4-dependent production of PPAR-gamma ligands in macrophages by 12/15-lipoxygenase. *Nature* 400 (6742):378–82.

Huang, L., A. Zhao, F. Wong, J. M. Ayala, M. Struthers, F. Ujjainwalla, S. D. Wright, M. S. Springer, J. Evans, and J. Cui. 2004. Leukotriene B4 strongly increases monocyte chemoattractant protein-1 in human monocytes. *Arterioscler Thromb Vasc Biol* 24 (10):1783–8.

Hwang, I., J. Yang, S. Hong, E. Ju Lee, S. H. Lee, T. Fernandes-Alnemri, E. S. Alnemri, and J. W. Yu. 2015. Non-transcriptional regulation of NLRP3 inflammasome signaling by IL-4. *Immunol Cell Biol* 93 (6):591–9.

Hébert, M. J., T. Takano, H. Holthöfer, and H. R. Brady. 1996. Sequential morphologic events during apoptosis of human neutrophils. Modulation by lipoxygenase-derived eicosanoids. *J Immunol* 157 (7):3105–15.

Iikura, M., H. Suto, N. Kajiwara, K. Oboki, T. Ohno, Y. Okayama, H. Saito, S. J. Galli, and S. Nakae. 2007. IL-33 can promote survival, adhesion and cytokine production in human mast cells. *Lab Invest* 87 (10):971–8.

Iwasaki, H., J. Mizoguchi, N. Takada, K. Tai, S. Ikegaya, and T. Ueda. 2010. Correlation between the concentrations of tumor necrosis factor-alpha and the severity of disease in patients infected with *Orientia tsutsugamushi*. *Int J Infect Dis* 14 (4):e328–33.

Janciauskiene, S., S. Larsson, P. Larsson, R. Virtala, L. Jansson, and T. Stevens. 2004. Inhibition of lipopolysaccharide-mediated human monocyte activation, in vitro, by alpha1-antitrypsin. *Biochem Biophys Res Commun* 321 (3):592–600.

Janciauskiene, S. M., R. Bals, R. Koczulla, C. Vogelmeier, T. Köhnlein, and T. Welte. 2011. The discovery of α1-antitrypsin and its role in health and disease. *Respir Med* 105 (8):1129–39.

Jia, T., N. V. Serbina, K. Brandl, M. X. Zhong, I. M. Leiner, I. F. Charo, and E. G. Pamer. 2008. Additive roles for MCP-1 and MCP-3 in CCR2-mediated recruitment of inflammatory monocytes during *Listeria monocytogenes* infection. *J Immunol* 180 (10):6846–53.

Jin, W., and C. Dong. 2013. IL-17 cytokines in immunity and inflammation. *Emerg Microbes Infect* 2 (9):e60.

Kalthoff, H., C. Roeder, M. Brockhaus, H. G. Thiele, and W. Schmiegel. 1993. Tumor necrosis factor (TNF) up-regulates the expression of p75 but not p55 TNF receptors, and both receptors mediate, independently of each other, up-regulation of transforming growth factor alpha and epidermal growth factor receptor mRNA. *J Biol Chem* 268 (4):2762–6.

Kaplanski, G., V. Marin, F. Montero-Julian, A. Mantovani, and C. Farnarier. 2003. IL-6: A regulator of the transition from neutrophil to monocyte recruitment during inflammation. *Trends Immunol* 24 (1):25–9.

Kawakami, T., and S. J. Galli. 2002. Regulation of mast-cell and basophil function and survival by IgE. *Nat Rev Immunol* 2 (10):773–86.

Kerrigan, A. M., and G. D. Brown. 2009. C-type lectins and phagocytosis. *Immunobiology* 214 (7):562–75.

Kessenbrock, K., M. Krumbholz, U. Schönermarck, W. Back, W. L. Gross, Z. Werb, H. J. Gröne, V. Brinkmann, and D. E. Jenne. 2009. Netting neutrophils in autoimmune small-vessel vasculitis. *Nat Med* 15 (6):623–5.

Khandpur, R., C. Carmona-Rivera, A. Vivekanandan-Giri, A. Gizinski, S. Yalavarthi, J. S. Knight, S. Friday et al. 2013. NETs are a source of citrullinated autoantigens and stimulate inflammatory responses in rheumatoid arthritis. *Sci Transl Med* 5 (178):178ra40.

Kim, B. S., K. Wang, M. C. Siracusa, S. A. Saenz, J. R. Brestoff, L. A. Monticelli, M. Noti, E. D. Tait Wojno, T. C. Fung, M. Kubo, and D. Artis. 2014. Basophils promote innate lymphoid cell responses in inflamed skin. *J Immunol* 193 (7):3717–25.

Kim, C., J. Y. Kim, and J. H. Kim. 2008. Cytosolic phospholipase A(2), lipoxygenase metabolites, and reactive oxygen species. *BMB Rep* 41 (8):555–9.

Kimura, Y., N. Nagai, N. Tsunekawa, M. Sato-Matsushita, T. Yoshimoto, D. J. Cua, Y. Iwakura et al. 2016. IL-17A-producing CD30(+) Vδ1 T cells drive inflammation-induced cancer progression. *Cancer Sci* 107 (9):1206–14.

Komiyama, Y., S. Nakae, T. Matsuki, A. Nambu, H. Ishigame, S. Kakuta, K. Sudo, and Y. Iwakura. 2006. IL-17 plays an important role in the development of experimental autoimmune encephalomyelitis. *J Immunol* 177 (1):566–73.

Korn, T., E. Bettelli, M. Oukka, and V. K. Kuchroo. 2009. IL-17 and Th17 cells. *Annu Rev Immunol* 27:485–517.

Kouro, T., and K. Takatsu. 2009. IL-5- and eosinophil-mediated inflammation: From discovery to therapy. *Int Immunol* 21 (12):1303–9.

Kozicky, L. K., Z. Y. Zhao, S. C. Menzies, M. Fidanza, G. S. Reid, K. Wilhelmsen, J. Hellman, N. Hotte, K. L. Madsen, and L. M. Sly. 2015. Intravenous immunoglobulin skews macrophages to an anti-inflammatory, IL-10-producing activation state. *J Leukoc Biol* 98 (6):983–94.

Kramer, F., J. Torzewski, J. Kamenz, K. Veit, V. Hombach, J. Dedio, and Y. Ivashchenko. 2008. Interleukin-1beta stimulates acute phase response and C-reactive protein synthesis by inducing an NFkappaB- and C/EBPbeta-dependent autocrine interleukin-6 loop. *Mol Immunol* 45 (9):2678–89.

Kritas, S. K., A. Saggini, G. Varvara, G. Murmura, A. Caraffa, P. Antinolfi, E. Toniato et al. 2013. Mast cell involvement in rheumatoid arthritis. *J Biol Regul Homeost Agents* 27 (3):655–60.

Krizbai, I. A., H. Bauer, N. Bresgen, P. M. Eckl, A. Farkas, E. Szatmári, A. Traweger, K. Wejksza, and H. C. Bauer. 2005. Effect of oxidative stress on the junctional proteins of cultured cerebral endothelial cells. *Cell Mol Neurobiol* 25 (1):129–39.

Kugyelka, R., Z. Kohl, K. Olasz, K. Mikecz, T. A. Rauch, T. T. Glant, and F. Boldizsar. 2016. Enigma of IL-17 and Th17 cells in rheumatoid arthritis and in autoimmune animal models of arthritis. *Mediators Inflamm* 2016:6145810.

Kulka, M., L. Alexopoulou, R. A. Flavell, and D. D. Metcalfe. 2004. Activation of mast cells by double-stranded RNA: Evidence for activation through Toll-like receptor 3. *J Allergy Clin Immunol* 114 (1):174–82.

Kvarnhammar, A. M., and L. O. Cardell. 2012. Pattern-recognition receptors in human eosinophils. *Immunology* 136 (1):11–20.

Laidlaw, B. J., W. Cui, R. A. Amezquita, S. M. Gray, T. Guan, Y. Lu, Y. Kobayashi, R. A. Flavell, S. H. Kleinstein, J. Craft, and S. M. Kaech. 2015. Production of IL-10 by CD4(+) regulatory T cells during the resolution of infection promotes the maturation of memory CD8(+) T cells. *Nat Immunol* 16 (8):871–9.

Landsman, L., L. Bar-On, A. Zernecke, K. W. Kim, R. Krauthgamer, E. Shagdarsuren, S. A. Lira, I. L. Weissman, C. Weber, and S. Jung. 2009. CX3CR1 is required for monocyte homeostasis and atherogenesis by promoting cell survival. *Blood* 113 (4):963–72.

Lappalainen, J., J. Rintahaka, P. T. Kovanen, S. Matikainen, and K. K. Eklund. 2013. Intracellular RNA recognition pathway activates strong anti-viral response in human mast cells. *Clin Exp Immunol* 172 (1):121–8.

Latz, E., T. S. Xiao, and A. Stutz. 2013. Activation and regulation of the inflammasomes. *Nat Rev Immunol* 13 (6):397–411.

Le-Barillec, K., M. Si-Tahar, V. Balloy, and M. Chignard. 1999. Proteolysis of monocyte CD14 by human leukocyte elastase inhibits lipopolysaccharide-mediated cell activation. *J Clin Invest* 103 (7):1039–46.

Lee, S. K., S. D. Kim, M. Kook, H. Y. Lee, J. Ghim, Y. Choi, B. A. Zabel, S. H. Ryu, and Y. S. Bae. 2015. Phospholipase D2 drives mortality in sepsis by inhibiting neutrophil extracellular trap formation and down-regulating CXCR2. *J Exp Med* 212 (9):1381–90.

Legler, D. F., P. Krause, E. Scandella, E. Singer, and M. Groettrup. 2006. Prostaglandin E2 is generally required for human dendritic cell migration and exerts its effect via EP2 and EP4 receptors. *J Immunol* 176 (2):966–73.

Leppkes, M., C. Maueröder, S. Hirth, S. Nowecki, C. Günther, U. Billmeier, S. Paulus et al. 2016. Externalized decondensed neutrophil chromatin occludes pancreatic ducts and drives pancreatitis. *Nat Commun* 7:10973.

Levine, A. P., M. R. Duchen, S. de Villiers, P. R. Rich, and A. W. Segal. 2015. Alkalinity of neutrophil phagocytic vacuoles is modulated by HVCN1 and has consequences for myeloperoxidase activity. *PLoS One* 10 (4):e0125906.

Ley, K., E. Smith, and M. A. Stark. 2006. IL-17A-producing neutrophil-regulatory Tn lymphocytes. *Immunol Res* 34 (3):229–42.

Li, D., B. Guo, H. Wu, L. Tan, C. Chang, and Q. Lu. 2015. Interleukin-17 in systemic lupus erythematosus: A comprehensive review. *Autoimmunity* 48 (6):353–61.

Lockhart, E., A. M. Green, and J. L. Flynn. 2006. IL-17 production is dominated by gammadelta T cells rather than CD4 T cells during *Mycobacterium tuberculosis* infection. *J Immunol* 177 (7):4662–9.

Long, H., W. Liao, L. Wang, and Q. Lu. 2016. A player and coordinator: The versatile roles of eosinophils in the immune system. *Transfus Med Hemother* 43 (2):96–108.

Lood, C., L. P. Blanco, M. M. Purmalek, C. Carmona-Rivera, S. S. De Ravin, C. K. Smith, H. L. Malech, J. A. Ledbetter, K. B. Elkon, and M. J. Kaplan. 2016. Neutrophil extracellular traps enriched in oxidized mitochondrial DNA are interferogenic and contribute to lupus-like disease. *Nat Med* 22 (2):146–53.

López-Boado, Y. S., M. Espinola, S. Bahr, and A. Belaaouaj. 2004. Neutrophil serine proteinases cleave bacterial flagellin, abrogating its host response-inducing activity. *J Immunol* 172 (1):509–15.

Lu, P., K. Takai, V. M. Weaver, and Z. Werb. 2011. Extracellular matrix degradation and remodeling in development and disease. *Cold Spring Harb Perspect Biol* 3 (12): a005058.

Lund, F. E. 2008. Cytokine-producing B lymphocytes—Key regulators of immunity. *Curr Opin Immunol* 20 (3):332–8.

Luzina, I. G., A. D. Keegan, N. M. Heller, G. A. Rook, T. Shea-Donohue, and S. P. Atamas. 2012. Regulation of inflammation by interleukin-4: A review of "alternatives." *J Leukoc Biol* 92 (4):753–64.

Lv, J., L. Zou, L. Zhao, W. Yang, Y. Xiong, B. Li, and R. He. 2015. Leukotriene B_4 leukotriene B_4 receptor axis promotes oxazolone-induced contact dermatitis by directing skin homing of neutrophils and CD8[+] T cells. *Immunology* 146 (1):50–8.

Ma, H. T., and M. A. Beaven. 2009. Regulation of Ca2+ signaling with particular focus on mast cells. *Crit Rev Immunol* 29 (2):155–86.

Maderna, P., D. C. Cottell, T. Toivonen, N. Dufton, J. Dalli, M. Perretti, and C. Godson. 2010. FPR2/ALX receptor expression and internalization are critical for lipoxin A4 and annexin-derived peptide-stimulated phagocytosis. *FASEB J* 24 (11):4240–9.

Mancini, A. D., and J. A. Di Battista. 2011. The cardinal role of the phospholipase A(2)/cyclooxygenase-2/prostaglandin E synthase/prostaglandin E(2) (PCPP) axis in inflammostasis. *Inflamm Res* 60 (12):1083–92.

Mangan, P. R., L. E. Harrington, D. B. O'Quinn, W. S. Helms, D. C. Bullard, C. O. Elson, R. D. Hatton, S. M. Wahl, T. R. Schoeb, and C. T. Weaver. 2006. Transforming growth factor-beta induces development of the T(H)17 lineage. *Nature* 441 (7090):231–4.

Mark, K. S., W. J. Trickler, and D. W. Miller. 2001. Tumor necrosis factor-alpha induces cyclooxygenase-2 expression and prostaglandin release in brain microvessel endothelial cells. *J Pharmacol Exp Ther* 297 (3):1051–8.

Martin, B., K. Hirota, D. J. Cua, B. Stockinger, and M. Veldhoen. 2009. Interleukin-17-producing gammadelta T cells selectively expand in response to pathogen products and environmental signals. *Immunity* 31 (2):321–30.

Mauri, C., D. Gray, N. Mushtaq, and M. Londei. 2003. Prevention of arthritis by interleukin 10-producing B cells. *J Exp Med* 197 (4):489–501.

McCain, R. W., E. P. Holden, T. R. Blackwell, and J. W. Christman. 1994. Leukotriene B4 stimulates human polymorphonuclear leukocytes to synthesize and release interleukin-8 in vitro. *Am J Respir Cell Mol Biol* 10 (6):651–7.

McGeachy, M. J., K. S. Bak-Jensen, Y. Chen, C. M. Tato, W. Blumenschein, T. McClanahan, and D. J. Cua. 2007. TGF-beta and IL-6 drive the production of IL-17 and IL-10 by T cells and restrain T(H)-17 cell-mediated pathology. *Nat Immunol* 8 (12):1390–7.

McLoughlin, R. M., B. J. Jenkins, D. Grail, A. S. Williams, C. A. Fielding, C. R. Parker, M. Ernst, N. Topley, and S. A. Jones. 2005. IL-6 trans-signaling via STAT3 directs T cell infiltration in acute inflammation. *Proc Natl Acad Sci USA* 102 (27):9589–94.

Menten, P., A. Wuyts, and J. Van Damme. 2002. Macrophage inflammatory protein-1. *Cytokine Growth Factor Rev* 13 (6):455–81.

Merluzzi, S., B. Frossi, G. Gri, S. Parusso, C. Tripodo, and C. Pucillo. 2010. Mast cells enhance proliferation of B lymphocytes and drive their differentiation toward IgA-secreting plasma cells. *Blood* 115 (14):2810–7.

Mitchell, S., G. Thomas, K. Harvey, D. Cottell, K. Reville, G. Berlasconi, N. A. Petasis et al. 2002. Lipoxins, aspirin-triggered epi-lipoxins, lipoxin stable analogues, and the resolution of inflammation: Stimulation of macrophage phagocytosis of apoptotic neutrophils in vivo. *J Am Soc Nephrol* 13 (10):2497–507.

Mizoguchi, A., E. Mizoguchi, H. Takedatsu, R. S. Blumberg, and A. K. Bhan. 2002. Chronic intestinal inflammatory condition generates IL-10-producing regulatory B cell subset characterized by CD1d upregulation. *Immunity* 16 (2):219–30.

Mócsai, A., B. Walzog, and C. A. Lowell. 2015. Intracellular signalling during neutrophil recruitment. *Cardiovasc Res* 107 (3):373–85.

Morimoto, K., N. Shirata, Y. Taketomi, S. Tsuchiya, E. Segi-Nishida, T. Inazumi, K. Kabashima, S. Tanaka, M. Murakami, S. Narumiya, and Y. Sugimoto. 2014. Prostaglandin E2-EP3 signaling induces inflammatory swelling by mast cell activation. *J Immunol* 192 (3):1130–7.

Moriyama, T., T. Higashi, K. Togashi, T. Iida, E. Segi, Y. Sugimoto, T. Tominaga, S. Narumiya, and M. Tominaga. 2005. Sensitization of TRPV1 by EP1 and IP reveals peripheral nociceptive mechanism of prostaglandins. *Mol Pain* 1:3.

Morshed, M., S. Yousefi, C. Stöckle, H. U. Simon, and D. Simon. 2012. Thymic stromal lymphopoietin stimulates the formation of eosinophil extracellular traps. *Allergy* 67 (9):1127–37.

Munro, J. M., J. S. Pober, and R. S. Cotran. 1989. Tumor necrosis factor and interferon-gamma induce distinct patterns of endothelial activation and associated leukocyte accumulation in skin of *Papio anubis*. *Am J Pathol* 135 (1):121–33.

Murray, P. J., and T. A. Wynn. 2011. Protective and pathogenic functions of macrophage subsets. *Nat Rev Immunol* 11 (11):723–37.

Nakae, S., H. Suto, G. J. Berry, and S. J. Galli. 2007. Mast cell-derived TNF can promote Th17 cell-dependent neutrophil recruitment in ovalbumin-challenged OTII mice. *Blood* 109 (9):3640–8.

Nakamura, K., A. Kitani, I. Fuss, A. Pedersen, N. Harada, H. Nawata, and W. Strober. 2004. TGF-beta 1 plays an important role in the mechanism of CD4+CD25+ regulatory T cell activity in both humans and mice. *J Immunol* 172 (2):834–42.

Nathan, C., and A. Ding. 2010. Nonresolving inflammation. *Cell* 140 (6):871–82.

Nimmerjahn, F., and J. V. Ravetch. 2006. Fcgamma receptors: Old friends and new family members. *Immunity* 24 (1):19–28.

Ogbogu, P. U., D. R. Rosing, and M. K. Horne. 2007. Cardiovascular manifestations of hypereosinophilic syndromes. *Immunol Allergy Clin North Am* 27 (3):457–75.

Okamoto, F., K. Saeki, H. Sumimoto, S. Yamasaki, and T. Yokomizo. 2010. Leukotriene B4 augments and restores Fc gammaRs-dependent phagocytosis in macrophages. *J Biol Chem* 285 (52):41113–21.

Okayama, Y. 2005. Mast cell-derived cytokine expression induced via Fc receptors and Toll-like receptors. *Chem Immunol Allergy* 87:101–10.

Okayama, Y., and T. Kawakami. 2006. Development, migration, and survival of mast cells. *Immunol Res* 34 (2):97–115.

Oliver, J. M., C. A. Tarleton, L. Gilmartin, T. Archibeque, C. R. Qualls, L. Diehl, B. S. Wilson, and M. Schuyler. 2010. Reduced FcepsilonRI-mediated release of asthma-promoting cytokines and chemokines from human basophils during omalizumab therapy. *Int Arch Allergy Immunol* 151 (4):275–84.

O'Neill, L. A. 2008. The interleukin-1 receptor/Toll-like receptor superfamily: 10 years of progress. *Immunol Rev* 226:10–8.

Ong, C. W., P. T. Elkington, S. Brilha, C. Ugarte-Gil, M. T. Tome-Esteban, L. B. Tezera, P. J. Pabisiak et al. 2015. Neutrophil-derived MMP-8 drives AMPK-dependent matrix destruction in human pulmonary tuberculosis. *PLoS Pathog* 11 (5):e1004917.

O'Sullivan, T. E., J. C. Sun, and L. L. Lanier. 2015. Natural killer cell memory. *Immunity* 43 (4):634–45.

Padrines, M., M. Wolf, A. Walz, and M. Baggiolini. 1994. Interleukin-8 processing by neutrophil elastase, cathepsin G and proteinase-3. *FEBS Lett* 352 (2):231–5.

Papayannopoulos, V., D. Staab, and A. Zychlinsky. 2011. Neutrophil elastase enhances sputum solubilization in cystic fibrosis patients receiving DNase therapy. *PLoS One* 6 (12):e28526.

Parekh, V. V., D. V. Prasad, P. P. Banerjee, B. N. Joshi, A. Kumar, and G. C. Mishra. 2003. B cells activated by lipopolysaccharide, but not by anti-Ig and anti-CD40 antibody, induce anergy in CD8+ T cells: Role of TGF-beta 1. *J Immunol* 170 (12):5897–911.

Park, H., Z. Li, X. O. Yang, S. H. Chang, R. Nurieva, Y. H. Wang, Y. Wang, L. Hood, Z. Zhu, Q. Tian, and C. Dong. 2005. A distinct lineage of CD4 T cells regulates tissue inflammation by producing interleukin 17. *Nat Immunol* 6 (11):1133–41.

Park, S. Y., M. Y. Jung, S. J. Lee, K. B. Kang, A. Gratchev, V. Riabov, J. Kzhyshkowska, and I. S. Kim. 2009. Stabilin-1 mediates phosphatidylserine-dependent clearance of cell corpses in alternatively activated macrophages. *J Cell Sci* 122 (Pt 18):3365–73.

Paul, S., Shilpi, and G. Lal. 2015. Role of gamma-delta (γδ) T cells in autoimmunity. *J Leukoc Biol* 97 (2):259–71.

Pham, C. T. 2006. Neutrophil serine proteases: Specific regulators of inflammation. *Nat Rev Immunol* 6 (7):541–50.

Phillipson, M., B. Heit, P. Colarusso, L. Liu, C. M. Ballantyne, and P. Kubes. 2006. Intraluminal crawling of neutrophils to emigration sites: A molecularly distinct process from adhesion in the recruitment cascade. *J Exp Med* 203 (12):2569–75.

Piconese, S., G. Gri, C. Tripodo, S. Musio, A. Gorzanelli, B. Frossi, R. Pedotti, C. E. Pucillo, and M. P. Colombo. 2009. Mast cells counteract regulatory T-cell suppression through interleukin-6 and OX40/OX40L axis toward Th17-cell differentiation. *Blood* 114 (13):2639–48.

Pratesi, F., I. Dioni, C. Tommasi, M. C. Alcaro, I. Paolini, F. Barbetti, F. Boscaro, F. Panza, I. Puxeddu, P. Rovero, and P. Migliorini. 2014. Antibodies from patients with rheumatoid arthritis target citrullinated histone 4 contained in neutrophils extracellular traps. *Ann Rheum Dis* 73 (7):1414–22.

Qidwai, T. 2016. Chemokine genetic polymorphism in human health and disease. *Immunol Lett* 176:128–38.

Rabe, B., A. Chalaris, U. May, G. H. Waetzig, D. Seegert, A. S. Williams, S. A. Jones, S. Rose-John, and J. Scheller. 2008. Transgenic blockade of interleukin 6 transsignaling abrogates inflammation. *Blood* 111 (3):1021–8.

Rael, E. L., and R. F. Lockey. 2011. Interleukin-13 signaling and its role in asthma. *World Allergy Organ J* 4 (3):54–64.

Ramon, S., J. Dalli, J. M. Sanger, J. W. Winkler, M. Aursnes, T. V. Tungen, T. V. Hansen, and C. N. Serhan. 2016. The protectin PCTR1 is produced by human M2 macrophages and enhances resolution of infectious inflammation. *Am J Pathol* 186 (4):962–73.

Rao, R. K., S. Basuroy, V. U. Rao, K. J. Karnaky Jr., and A. Gupta. 2002. Tyrosine phosphorylation and dissociation of occludin-ZO-1 and E-cadherin-beta-catenin complexes from the cytoskeleton by oxidative stress. *Biochem J* 368 (Pt 2):471–81.

Raphael, I., S. Nalawade, T. N. Eagar, and T. G. Forsthuber. 2015. T cell subsets and their signature cytokines in autoimmune and inflammatory diseases. *Cytokine* 74 (1):5–17.

Rawson, R., T. Yang, R. O. Newbury, M. Aquino, A. Doshi, B. Bell, D. H. Broide, R. Dohil, R. Kurten, and S. S. Aceves. 2016. TGF-β1-induced PAI-1 contributes to a profibrotic network in patients with eosinophilic esophagitis. *J Allergy Clin Immunol* 138 (3):791–800.e4.

Reeves, E. P., H. Lu, H. L. Jacobs, C. G. Messina, S. Bolsover, G. Gabella, E. O. Potma, A. Warley, J. Roes, and A. W. Segal. 2002. Killing activity of neutrophils is mediated through activation of proteases by K+ flux. *Nature* 416 (6878):291–7.

Renesto, P., M. Si-Tahar, M. Moniatte, V. Balloy, A. Van Dorsselaer, D. Pidard, and M. Chignard. 1997. Specific inhibition of thrombin-induced cell activation by the neutrophil proteinases elastase, cathepsin G, and proteinase 3: Evidence for distinct cleavage sites within the aminoterminal domain of the thrombin receptor. *Blood* 89 (6):1944–53.

Rey, C., A. Nadjar, B. Buaud, C. Vaysse, A. Aubert, V. Pallet, S. Layé, and C. Joffre. 2016. Resolvin D1 and E1 promote resolution of inflammation in microglial cells in vitro. *Brain Behav Immun* 55:249–59.

Reynolds, G., J. R. Gibbon, A. G. Pratt, M. J. Wood, D. Coady, G. Raftery, A. R. Lorenzi et al. 2016. Synovial CD4+ T-cell-derived GM-CSF supports the differentiation of an inflammatory dendritic cell population in rheumatoid arthritis. *Ann Rheum Dis* 75 (5):899–907.

Ribot, J. C., A. deBarros, D. J. Pang, J. F. Neves, V. Peperzak, S. J. Roberts, M. Girardi, J. Borst, A. C. Hayday, D. J. Pennington, and B. Silva-Santos. 2009. CD27 is a thymic determinant of the balance between interferon-gamma- and interleukin 17-producing gammadelta T cell subsets. *Nat Immunol* 10 (4):427–36.

Ricciotti, E., and G. A. FitzGerald. 2011. Prostaglandins and inflammation. *Arterioscler Thromb Vasc Biol* 31 (5):986–1000.

Roberts, A. B., M. B. Sporn, R. K. Assoian, J. M. Smith, N. S. Roche, L. M. Wakefield, U. I. Heine, L. A. Liotta, V. Falanga, and J. H. Kehrl. 1986. Transforming growth factor type beta: Rapid induction of fibrosis and angiogenesi0s in vivo and stimulation of collagen formation in vitro. *Proc Natl Acad Sci USA* 83 (12):4167–71.

Rochfort, K. D., L. E. Collins, A. McLoughlin, and P. M. Cummins. 2016. Tumour necrosis factor-α-mediated disruption of cerebrovascular endothelial barrier integrity in vitro involves the production of proinflammatory interleukin-6. *J Neurochem* 136 (3):564–72.

Rola-Pleszczynski, M., and J. Stanková. 1992. Leukotriene B4 enhances interleukin-6 (IL-6) production and IL-6 messenger RNA accumulation in human monocytes in vitro: Transcriptional and posttranscriptional mechanisms. *Blood* 80 (4):1004–11.

Roma-Lavisse, C., M. Tagzirt, C. Zawadzki, R. Lorenzi, A. Vincentelli, S. Haulon, F. Juthier et al. 2015. M1 and M2 macrophage proteolytic and angiogenic profile analysis in atherosclerotic patients reveals a distinctive profile in type 2 diabetes. *Diab Vasc Dis Res* 12 (4):279–89.

Rosser, E. C., and C. Mauri. 2015. Regulatory B cells: Origin, phenotype, and function. *Immunity* 42 (4):607–12.

Rouzer, C. A., and L. J. Marnett. 2009. Cyclooxygenases: Structural and functional insights. *J Lipid Res* 50 (Suppl):S29–34.

Ruscitti, P., P. Cipriani, L. Cantarini, V. Liakouli, A. Vitale, F. Carubbi, O. Berardicurti, M. Galeazzi, M. Valenti, and R. Giacomelli. 2015. Efficacy of inhibition of IL-1 in patients with rheumatoid arthritis and type 2 diabetes mellitus: Two case reports and review of the literature. *J Med Case Rep* 9:123.

Russo, R. C., C. C. Garcia, M. M. Teixeira, and F. A. Amaral. 2014. The CXCL8/IL-8 chemokine family and its receptors in inflammatory diseases. *Expert Rev Clin Immunol* 10 (5):593–619.

Saluja, R., M. Khan, M. K. Church, and M. Maurer. 2015. The role of IL-33 and mast cells in allergy and inflammation. *Clin Transl Allergy* 5:33.

Sambrano, G. R., W. Huang, T. Faruqi, S. Mahrus, C. Craik, and S. R. Coughlin. 2000. Cathepsin G activates protease-activated receptor-4 in human platelets. *J Biol Chem* 275 (10):6819–23.

Sanjabi, S., L. A. Zenewicz, M. Kamanaka, and R. A. Flavell. 2009. Anti-inflammatory and proinflammatory roles of TGF-beta, IL-10, and IL-22 in immunity and autoimmunity. *Curr Opin Pharmacol* 9 (4):447–53.

Sayed, B. A., A. Christy, M. R. Quirion, and M. A. Brown. 2008. The master switch: The role of mast cells in autoimmunity and tolerance. *Annu Rev Immunol* 26:705–39.

Scheller, J., A. Chalaris, D. Schmidt-Arras, and S. Rose-John. 2011. The pro- and anti-inflammatory properties of the cytokine interleukin-6. *Biochim Biophys Acta* 1813 (5):878–88.

Schroder, K., P. J. Hertzog, T. Ravasi, and D. A. Hume. 2004. Interferon-gamma: An overview of signals, mechanisms and functions. *J Leukoc Biol* 75 (2):163–89.

Schwab, J. M., N. Chiang, M. Arita, and C. N. Serhan. 2007. Resolvin E1 and protectin D1 activate inflammation-resolution programmes. *Nature* 447 (7146):869–74.

Schwarting, A., T. Wada, K. Kinoshita, G. Tesch, and V. R. Kelley. 1998. IFN-gamma receptor signaling is essential for the initiation, acceleration, and destruction of autoimmune kidney disease in MRL-Fas(lpr) mice. J Immunol 161 (1):494–503.

Segal, A. W. 2008. The function of the NADPH oxidase of phagocytes and its relationship to other NOXs in plants, invertebrates, and mammals. Int J Biochem Cell Biol 40 (4):604–18.

Segura, E., and S. Amigorena. 2013. Inflammatory dendritic cells in mice and humans. Trends Immunol 34 (9):440–5.

Segura, E., M. Touzot, A. Bohineust, A. Cappuccio, G. Chiocchia, A. Hosmalin, M. Dalod, V. Soumelis, and S. Amigorena. 2013. Human inflammatory dendritic cells induce Th17 cell differentiation. Immunity 38 (2):336–48.

Serhan, C. N. 2010. Novel lipid mediators and resolution mechanisms in acute inflammation: To resolve or not? Am J Pathol 177 (4):1576–91.

Shamri, R., J. J. Xenakis, and L. A. Spencer. 2011. Eosinophils in innate immunity: An evolving story. Cell Tissue Res 343 (1):57–83.

Shiomi, A., and T. Usui. 2015. Pivotal roles of GM-CSF in autoimmunity and inflammation. Mediators Inflamm 2015:568543.

Shirley, M., and L. J. Scott. 2016. Secukinumab: A review in psoriatic arthritis. Drugs 76 (11):1135–45.

Slowik, M. R., L. G. De Luca, W. Fiers, and J. S. Pober. 1993. Tumor necrosis factor activates human endothelial cells through the p55 tumor necrosis factor receptor but the p75 receptor contributes to activation at low tumor necrosis factor concentration. Am J Pathol 143 (6):1724–30.

Smith, C. W., S. D. Marlin, R. Rothlein, C. Toman, and D. C. Anderson. 1989. Cooperative interactions of LFA-1 and Mac-1 with intercellular adhesion molecule-1 in facilitating adherence and transendothelial migration of human neutrophils in vitro. J Clin Invest 83 (6):2008–17.

Smith, D. E., R. Hanna, Della Friend, H. Moore, H. Chen, A. M. Farese, T. J. MacVittie, G. D. Virca, and J. E. Sims. 2003. The soluble form of IL-1 receptor accessory protein enhances the ability of soluble type II IL-1 receptor to inhibit IL-1 action. Immunity 18 (1):87–96.

Smith, E., A. Zarbock, M. A. Stark, T. L. Burcin, A. C. Bruce, P. Foley, and K. Ley. 2007. IL-23 is required for neutrophil homeostasis in normal and neutrophilic mice. J Immunol 179 (12):8274–9.

Soares, E. M., K. L. Mason, L. M. Rogers, C. H. Serezani, L. H. Faccioli, and D. M. Aronoff. 2013. Leukotriene B4 enhances innate immune defense against the puerperal sepsis agent Streptococcus pyogenes. J Immunol 190 (4):1614–22.

Sofi, M. H., W. Li, M. H. Kaplan, and C. H. Chang. 2009. Elevated IL-6 expression in CD4 T cells via PKCtheta and NF-kappaB induces Th2 cytokine production. Mol Immunol 46 (7):1443–50.

Sørensen, O. E., S. N. Clemmensen, S. L. Dahl, O. Østergaard, N. H. Heegaard, A. Glenthøj, F. C. Nielsen, and N. Borregaard. 2014. Papillon-Lefèvre syndrome patient reveals species-dependent requirements for neutrophil defenses. J Clin Invest 124 (10):4539–48.

Su, D., M. Shen, X. Li, and L. Sun. 2013. Roles of γδ T cells in the pathogenesis of autoimmune diseases. Clin Dev Immunol 2013:985753.

Suematsu, S., T. Matsuda, K. Aozasa, S. Akira, N. Nakano, S. Ohno, J. Miyazaki, K. Yamamura, T. Hirano, and T. Kishimoto. 1989. IgG1 plasmacytosis in interleukin 6 transgenic mice. Proc Natl Acad Sci USA 86 (19):7547–51.

Sullivan, B. M., H. E. Liang, J. K. Bando, D. Wu, L. E. Cheng, J. K. McKerrow, C. D. Allen, and R. M. Locksley. 2011. Genetic analysis of basophil function in vivo. Nat Immunol 12 (6):527–35.

Supajatura, V., H. Ushio, A. Nakao, S. Akira, K. Okumura, C. Ra, and H. Ogawa. 2002. Differential responses of mast cell Toll-like receptors 2 and 4 in allergy and innate immunity. J Clin Invest 109 (10):1351–9.

Tager, A. M., S. K. Bromley, B. D. Medoff, S. A. Islam, S. D. Bercury, E. B. Friedrich, A. D. Carafone, R. E. Gerszten, and A. D. Luster. 2003. Leukotriene B4 receptor BLT1 mediates early effector T cell recruitment. Nat Immunol 4 (10):982–90.

Takahashi, T., T. Tagami, S. Yamazaki, T. Uede, J. Shimizu, N. Sakaguchi, T. W. Mak, and S. Sakaguchi. 2000. Immunologic self-tolerance maintained by CD25(+)CD4(+) regulatory T cells constitutively expressing cytotoxic T lymphocyte-associated antigen 4. J Exp Med 192 (2):303–10.

Tartaglia, L. A., D. Pennica, and D. V. Goeddel. 1993. Ligand passing: The 75-kDa tumor necrosis factor (TNF) receptor recruits TNF for signaling by the 55-kDa TNF receptor. J Biol Chem 268 (25):18542–8.

Theoharides, T. C., K. D. Alysandratos, A. Angelidou, D. A. Delivanis, N. Sismanopoulos, B. Zhang, S. Asadi, M. Vasiadi, Z. Weng, A. Miniati, and D. Kalogeromitros. 2012. Mast cells and inflammation. Biochim Biophys Acta 1822 (1):21–33.

Thomas, G. M., C. Carbo, B. R. Curtis, K. Martinod, I. B. Mazo, D. Schatzberg, S. M. Cifuni et al. 2012. Extracellular DNA traps are associated with the pathogenesis of TRALI in humans and mice. *Blood* 119 (26):6335–43.

Thornton, A. M., and E. M. Shevach. 1998. CD4+CD25+ immunoregulatory T cells suppress polyclonal T cell activation in vitro by inhibiting interleukin 2 production. *J Exp Med* 188 (2):287–96.

Tian, J., D. Zekzer, L. Hanssen, Y. Lu, A. Olcott, and D. L. Kaufman. 2001. Lipopolysaccharide-activated B cells down-regulate Th1 immunity and prevent autoimmune diabetes in nonobese diabetic mice. *J Immunol* 167 (2):1081–9.

Tjonahen, E., S. F. Oh, J. Siegelman, S. Elangovan, K. B. Percarpio, S. Hong, M. Arita, and C. N. Serhan. 2006. Resolvin E2: Identification and anti-inflammatory actions: Pivotal role of human 5-lipoxygenase in resolvin E series biosynthesis. *Chem Biol* 13 (11):1193–202.

Tripodo, C., G. Gri, P. P. Piccaluga, B. Frossi, C. Guarnotta, S. Piconese, G. Franco et al. 2010. Mast cells and Th17 cells contribute to the lymphoma-associated pro-inflammatory microenvironment of angioimmunoblastic T-cell lymphoma. *Am J Pathol* 177 (2):792–802.

Tsukaguchi, K., B. de Lange, and W. H. Boom. 1999. Differential regulation of IFN-gamma, TNF-alpha, and IL-10 production by CD4(+) alphabetaTCR+ T cells and vdelta2(+) gammadelta T cells in response to monocytes infected with *Mycobacterium tuberculosis*-H37Ra. *Cell Immunol* 194 (1):12–20.

Tung, H. Y., B. Plunkett, S. K. Huang, and Y. Zhou. 2014. Murine mast cells secrete and respond to interleukin-33. *J Interferon Cytokine Res* 34 (3):141–7.

Vance, R. E., S. Hong, K. Gronert, C. N. Serhan, and J. J. Mekalanos. 2004. The opportunistic pathogen *Pseudomonas aeruginosa* carries a secretable arachidonate 15-lipoxygenase. *Proc Natl Acad Sci USA* 101 (7):2135–9.

Vane, J. R. 1971. Inhibition of prostaglandin synthesis as a mechanism of action for aspirin-like drugs. *Nat New Biol* 231 (25):232–5.

van Panhuys, N., M. Prout, E. Forbes, B. Min, W. E. Paul, and G. Le Gros. 2011. Basophils are the major producers of IL-4 during primary helminth infection. *J Immunol* 186 (5):2719–28.

Vignali, D. A., L. W. Collison, and C. J. Workman. 2008. How regulatory T cells work. *Nat Rev Immunol* 8 (7):523–32.

Vignali, D. A., and V. K. Kuchroo. 2012. IL-12 family cytokines: Immunological playmakers. *Nat Immunol* 13 (8):722–8.

Waage, A., A. Halstensen, and T. Espevik. 1987. Association between tumour necrosis factor in serum and fatal outcome in patients with meningococcal disease. *Lancet* 1 (8529):355–7.

Wajant, H., K. Pfizenmaier, and P. Scheurich. 2003. Tumor necrosis factor signaling. *Cell Death Differ* 10 (1):45–65.

Wakahara, K., N. Baba, V. Q. Van, P. Bégin, M. Rubio, P. Ferraro, B. Panzini et al. 2012. Human basophils interact with memory T cells to augment Th17 responses. *Blood* 120 (24):4761–71.

Wakahara, K., V. Q. Van, N. Baba, P. Bégin, M. Rubio, G. Delespesse, and M. Sarfati. 2013. Basophils are recruited to inflamed lungs and exacerbate memory Th2 responses in mice and humans. *Allergy* 68 (2):180–9.

Wakim, L. M., J. Waithman, N. van Rooijen, W. R. Heath, and F. R. Carbone. 2008. Dendritic cell-induced memory T cell activation in nonlymphoid tissues. *Science* 319 (5860):198–202.

Wang, H., T. Li, S. Chen, Y. Gu, and S. Ye. 2015. Neutrophil extracellular trap mitochondrial DNA and its autoantibody in systemic lupus erythematosus and a proof-of-concept trial of metformin. *Arthritis Rheumatol* 67 (12):3190–200.

Weathington, N. M., A. H. van Houwelingen, B. D. Noerager, P. L. Jackson, A. D. Kraneveld, F. S. Galin, G. Folkerts, F. P. Nijkamp, and J. E. Blalock. 2006. A novel peptide CXCR ligand derived from extracellular matrix degradation during airway inflammation. *Nat Med* 12 (3):317–23.

Weber, A., P. Wasiliew, and M. Kracht. 2010. Interleukin-1 (IL-1) pathway. *Sci Signal* 3 (105):cm1.

Weber, C., K. S. Weber, C. Klier, S. Gu, R. Wank, R. Horuk, and P. J. Nelson. 2001. Specialized roles of the chemokine receptors CCR1 and CCR5 in the recruitment of monocytes and T(H)1-like/CD45RO(+) T cells. *Blood* 97 (4):1144–6.

Winzen, R., D. Wallach, O. Kemper, K. Resch, and H. Holtmann. 1993. Selective up-regulation of the 75-kDa tumor necrosis factor (TNF) receptor and its mRNA by TNF and IL-1. *J Immunol* 150 (10):4346–53.

Woollard, K. J., and F. Geissmann. 2010. Monocytes in atherosclerosis: Subsets and functions. *Nat Rev Cardiol* 7 (2):77–86.

Xu, H., L. N. Gu, Q. Y. Yang, D. Y. Zhao, and F. Liu. 2016. MiR-221 promotes IgE-mediated activation of mast cells degranulation by PI3K/Akt/PLCγ/Ca(2+) pathway. *J Bioenerg Biomembr* 48 (3):293–9.

Yao, C., D. Sakata, Y. Esaki, Y. Li, T. Matsuoka, K. Kuroiwa, Y. Sugimoto, and S. Narumiya. 2009. Prostaglandin E2-EP4 signaling promotes immune inflammation through Th1 cell differentiation and Th17 cell expansion. *Nat Med* 15 (6):633–40.

Ye, P., P. B. Garvey, P. Zhang, S. Nelson, G. Bagby, W. R. Summer, P. Schwarzenberger, J. E. Shellito, and J. K. Kolls. 2001. Interleukin-17 and lung host defense against *Klebsiella pneumoniae* infection. *Am J Respir Cell Mol Biol* 25 (3):335–40.

Yen, D., J. Cheung, H. Scheerens, F. Poulet, T. McClanahan, B. McKenzie, M. A. Kleinschek et al. 2006. IL-23 is essential for T cell-mediated colitis and promotes inflammation via IL-17 and IL-6. *J Clin Invest* 116 (5):1310–6.

Yousefi, S., J. A. Gold, N. Andina, J. J. Lee, A. M. Kelly, E. Kozlowski, I. Schmid, A. Straumann, J. Reichenbach, G. J. Gleich, and H. U. Simon. 2008. Catapult-like release of mitochondrial DNA by eosinophils contributes to antibacterial defense. *Nat Med* 14 (9):949–53.

Zhang, B., Y. Lu, M. Campbell-Thompson, T. Spencer, C. Wasserfall, M. Atkinson, and S. Song. 2007. Alpha1-antitrypsin protects beta-cells from apoptosis. *Diabetes* 56 (5):1316–23.

Ziegler-Heitbrock, L. 2015. Blood monocytes and their subsets: Established features and open questions. *Front Immunol* 6:423.

CHAPTER TWO

Principles of Nanomedicine

Wilson S. Meng and Jelena M. Janjic

CONTENTS

2.1 INTRODUCTION

The concept of nanomedicine stemmed from a standing paradigm of clinical pharmacology: getting the appropriate drug to diseased tissues for the duration needed (Strebhardt and Ullrich 2008). It builds on the ability to engineer materials between the molecular and microscopic scale. More than 50 nanomedicine-based products are currently being tested in humans (Etheridge et al. 2013). These clinical trials include small molecules, recombinant proteins, antisense oligonucleotides, siRNA, and plasmid DNA as active pharmaceutical ingredients (APIs). Concurrent with the advances in nanoscience is the recognition that inflammation drives the underlying pathogenesis of many chronic ailments. While Doxil (Safra et al. 2000; Barenholz 2012) and Abraxane (Ibrahim et al. 2002; Gradishar et al. 2005) are examples of clinically validated nanomedicines, the field encompasses many platforms that have matured over the past two decades. While solid tumors continue to be important targets for developing new nanomedicines, chronic inflammatory diseases have been investigated actively in recent years. There are currently more than 80 products studied in humans that are indicated for cancer, approximately 20 indicated for infectious disease, and several products for hepatitis, cardiovascular, and autoimmune diseases at the commercial or investigational stage (Etheridge et al. 2013).

While the number of nano-based products moving into clinical phase is increasing, it is generally recognized that the clinical potential of nanomedicine has yet to be fully realized. The goal of this chapter is to put forth foundational principles of nanomedicine. The broad definition of nanomedicine connotes nanosized devices, scaffolds, and particulates designed for therapeutic or diagnostic purposes, or both. Herein, we address the current scope of nanotherapeutics (NTs), and their distinguishing features compared with traditional dosage forms. We examine the design parameters in NTs with respect to efficacy and safety in vivo. The impact of design on performance and clinical translation are analyzed. Strategies and challenges for pharmaceutical-grade manufacturing of nanomedicines and considerations of scale and quality control are discussed. Immune cells participating in inflammatory conditions are examined as therapeutic targets. Multifunctional systems, armed with targeting ligands and molecular imaging agents, are considered. The emphasis is placed on colloids: liposomes, polymer-based particulates, micelles, nanoemulsions, and microemulsions. PEGylated recombinant protein therapeutics, due to their distinct physical properties and manufacturing schemes, are not discussed here. The vast amount of data accumulated from preclinical and clinical studies allow defining converging concepts that should aid in the evaluation of future NTs. From these, we project emerging paradigms by which the complex pathogenesis in chronic inflammation may be dramatically altered by carefully designed NTs. As such, these inflammation-targeted NTs may lead to rationally designed new nanotechnology platforms that can be cross-utilized in multiple diseases.

Broadly speaking, inflammation is a shared pathology across many diseases. It is also a shared target for nanomedicine development. One of the standing arguments against NT development is the projected high cost of pharmaceutical-scale production impeding clinical translation. Many NTs designed for inflammation can be used as platforms delivering agents for a multitude of disease states; in the long run, nanomedicine has

the advantage in terms of economical and clinical values. Finally, implementing modern pharmaceutical manufacturing methodologies and quality by design (QbD) approaches to inflammation, nanomedicine can lead to consistent outcomes and personalized treatments. We offer arguments for nanomedicine development for inflammatory diseases while fully embracing the depth of challenges of nanomedicine as a whole, spanning preclinical efforts to supply chain management, pharmaceutical manufacturing, and quality control (Satalkar et al. 2016).

2.2 HISTORICAL PERSPECTIVE

The discovery of buckminsterfullerene propelled technological advances at the nanoscale (Kroto et al. 1985). Nanotechnology has been defined as processes used in manipulating matters with at least one dimension between 1 and 100 nm (Szebeni et al. 2003). One-dimensional materials measured at the nanoscale include films and coacervates. Two-dimensional nanomaterials include nanowire and nanotubes. Particulates, including liposomes and polymeric particles, are considered three-dimensional nanomaterials. Medical and pharmaceutical descriptions of NTs, however, include systems with dimensions between 1 and 1000 nm (Duncan and Gaspar 2011; Etheridge et al. 2013). A 2013 report showed that close to 80 commercial and investigational NTs are below 100 nm, with approximately 70 NTs near 200 nm (Etheridge et al. 2013). Fewer than 30 NT products fall between 200 and 1000 nm, with close to 20 APIs formulated as nanocrystals between 1000 and 2000 nm. Applied nanomedicine also extends to tissue engineering in codelivering protein factors and cells (Langer and Tirrell 2004; Dvir et al. 2011). Nanomedicine application to diagnostics has been reviewed elsewhere (Janib et al. 2010). Here, we discuss the inclusion of imaging moieties in nanomedicine formulations only in the context of imaging serving a supporting role to therapeutic delivery.

Beyond physical dimension, nanomaterials are characterized by interdependent entities from which the collective properties can be leveraged to enhance performance. The unique nanomedicine attributes center on the high ratio of surface area to volume and, in some cases, quantum effects in particles of sizes falling between 1 and 20 nm (Daniel and Astruc 2004). Unlike molecular species, nanoscale substances exhibit surface-dominant effects, with a high percentage of atoms on the outermost layer. Their colloidal properties promote interactions among the particulates, forming networks of correlated motion while remaining physically distinct entities. NTs include various common configurations and designs. Loading of APIs into the materials is accomplished through encapsulation, permeation, surface adsorption, chemical conjugation, or noncovalent affinity binding. Drug release occurs through diffusion from, or erosion of, the material matrix, or desorption from the particle surface. The kinetics of these processes are governed by the physiochemical properties of the nanocarriers and the APIs; the approaches taken for small organic molecules differ significantly from those for macromolecules, in recombinant proteins and oligomeric nucleic acids.

2.3 NANOMEDICINE RATIONALES

2.3.1 Rationale for Nanomedicine Therapeutic Strategies with Small Molecules

The rapid discovery of new chemical entities (NCEs) for chronic inflammatory diseases demands new strategies for inflammation-targeted NT design. Table 2.1 lists examples where NTs were applied to inflammatory diseases. A major hurdle for small-molecule anti-inflammatory APIs is unfavorable pharmacokinetics; poor drug accumulation at target tissues resulting from distribution and degradation is attributed to the drug's intrinsic physiochemical properties. Low-molecular-weight compounds (<55 kDa) are typically eliminated rapidly due to renal filtration (Mogensen 1968). Lipophilic molecules with poor aqueous solubility tend to distribute extensively in adipose tissues, often in off-target vital organs. These challenges can be overcome in part by developing novel formulations of approved drugs, as an alternative approach to advancing NCEs.

As drug delivery systems, NTs are designed to perform as a unit in carrying therapeutic cargos to the diseased tissues with (ideally) increased target selectivity and specificity over healthy tissues. The nanoscale is meant to mimic the physical attributes of endogenous components. With respect to size, polymer–drug conjugates and micelles would fall in the same scale as globular proteins. Liposomes, polymersomes, and polymeric particles have sizes similar to those of multimeric protein complexes, while microparticles are

TABLE 2.1

Application of nanomedicines in chronic and acute inflammatory diseases.

Inflammatory diseases	Therapeutic agents		Nanodelivery systems	
	Small molecule	Macromolecule	Organic	Inorganic
Chronic				
Cancer	Anthracyclines, taxanes, vinca alkaloids, platinum agents	Cytokines, mAbs, antigenic proteins and peptides, ASOs, siRNA, shRNA, plasmid DNA	Liposomes, polymeric particles, solid lipid nanoparticles, dendrimers, self-assembling proteins and peptides, PFC nanoemulsions	Iron oxide, gold, silver nanoparticles, quantum dots
Type I diabetes	ROS scavengers	ASOs, cytokines, mAbs, siRNA, plasmid DNA		
IBD/vascular	COX inhibitors	Cytokine		
Cutaneous ulcers	Antibiotics	Growth factors, siRNA		
Acute				
Infection	Amphotericin B	Liposomes		

NOTE: COX, cyclooxygenase; ROS, reactive oxygen species.

comparable to packed chromosomes in mammalian cells. The nanoscale is perceived to have the advantage of deep penetration into the interstitium, as well as carrying drug molecules en masse into tissues and cells. Nanoformulated APIs may exhibit altered half-lives and distribution to the extent that a large fraction of the drug molecules may avoid premature degradation and metabolism. In tissues, nanosized materials are taken into cells through endocytic and fusogenic pathways (Sahay et al. 2010). The proportional increase in surface area may enhance the efficiency of targeted delivery to specific cells via surface markers (Bergen et al. 2006; Farokhzad and Langer 2009). This is because higher densities of ligands can be grafted to exploit multivalent binding to membrane-bound receptors to trigger receptor-mediated endocytosis.

Recently developed nanoformulations have led to novel therapies for a range of ailments for which curative treatments are lacking, in particular, autoimmune diseases such as type I diabetes (TID), rheumatoid arthritis (RA), inflammatory bowel diseases (IBDs), and metastatic cancers (Table 2.1). These examples have paved the way for moving the field of nanomedicine into clinical translation. Due to the chronic nature of inflammatory diseases, long-term treatments are needed, and NT formulations are used to reduce the need for frequent dosing (Van Deventer et al. 1997; Schwab and Klotz 2001). Prolonged exposure to certain drugs may result in severe toxicities not seen in short-term therapies. To this end, nanosystems are engineered to impart selective distribution of APIs to affected tissues. The ability to concentrate drug molecules in cellular targets would reduce the dose needed, perhaps lowering it to the microgram range, as opposed to milligram doses. Decreasing the dose administered by an order of magnitude while retaining the same level of efficacy would accelerate the development of NCEs with narrow therapeutic windows.

Another motivation for formulating small molecules into nanoformulations is to overcome poor aqueous solubility of lipophilic compounds. An early example was modified cyclosporine (Neoral), which exhibits higher bioavailability than the first-generation, nonmodified formulation, Sandimmune (Choc 1997). In Neoral, the lipophilic API is emulsified with micelle-forming hydrogenated castor oil. A liposomal formulation of amphotericin B was developed to create a colloidal suspension for intravenous infusion. Parenteral formulations of rapamycin and paclitaxel have been developed using a mixture of ethanol and polyepoxylated castor oil (Cremophor EL) (Weiss et al. 1990). Because Cremophor is associated with high incidences of hypersensitivity reactions, a polymeric micelle formulation of paclitaxel (Genexol-PM) is marketed as a Cremophor-free alternative (Kim et al. 2004). Human serum albumin is used as an alternative to Cremophor to formulate paclitaxel in Abraxane (Ibrahim et al. 2002). These APIs share the common features of low solubility and high permeability (Yu et al. 2002).

Small molecules have been formulated into NTs in order to decrease drug clearance rates. Low-molecular-weight drugs, including small molecules and short peptides less than 55 kDa, tend to be removed from blood circulation through renal filtration (Mogensen 1968); encapsulation of these molecules is used to increase the overall size beyond the threshold. For example, the circulating half-life of vasoactive intestinal peptide is extended significantly when delivered within micelles (Önyüksel et al. 1999; Farokhzad and Langer 2009). Another motivation is to reduce toxicities by passive targeting. Doxil, PEGylated liposomes loaded with doxorubicin, was developed to steer the anthracycline away from the heart in order to mitigate dose-dependent irreversible cardiomyopathies (O'Brien et al. 2004). Studies using radiotracers in humans show that accumulation of the liposomes to heart tissues rapidly declines after 24 hours, while uptake by tumors continues to increase for 96 hours. The ability to penetrate through the extracellular matrix (ECM) is another rationale to construct drug carriers at the nanoscale. At the cellular level, NTs may overcome the P-glycoprotein (P-gp) efflux pump (Blanco et al. 2015). Rapid drug accumulation in the cytosol rendered by NTs may saturate the pump, sparing a fraction of the APIs for binding to molecular targets.

2.3.2 Rationale for Nanomedicine Therapeutic Strategies with Biologics

Early efforts in formulating biologics as nanomedicines include plasmid DNA and antisense oligodeoxynucleotides (ASOs). The discovery of disease-associated genes paved the way for using gene replacement or suppression as pharmacological strategies. It was recognized early on that the impact of these agents, upstream of protein synthesis, could not be realized unless the DNA was delivered inside target cells intact. Being highly negatively charged, short or long oligomers of nucleotides infused into blood circulation would be neutralized by serum proteins and digested by extracellular nucleases (Panyam and Labhasetwar 2003). ASOs with thiol-modified backbone exhibit improved chemical stability, rendering passage into the cytosol as the rate-limiting step. The plasma membrane, covered in glycocalyx, would bind proteins but repel nucleic acids. For plasmid DNA, the nuclear membrane is an additional barrier. Thus, the approach has been to protect cargo DNA from extracellular degradation and facilitate passage through cellular barriers in a single system. The efforts have generated important molecular pharmaceutics tools, including chemical moieties that promote the escape of the cargo from endosomes.

Protein drug delivery follows a somewhat different rationale (Lee 1990; Chen et al. 1995). Cytokines and growth factors need to be protected from premature degradation *in vivo*. A critical challenge of proteins as APIs is that, relative to DNA and small organic molecules, polypeptides are more liable to physiochemical stresses, even slight changes in pH and temperature. While biological functions of small peptides do not necessarily depend on adopting a stable conformation in solution, the activities of larger proteins are readily degraded through denaturation or enzymatic digestion. Native conformations of proteins are stabilized by relatively weak net free energies, estimated between 10 and 15 kcal/mol, in burying hydrophobic side chains. Protein aggregation occurs even with slight changes in the environmental pH. Noncovalent interactions with plasma components can alter the native conformation, potentially resulting in denaturation. Unless shielded from peptidases and acids, proteins are digested or denatured in biological fluids. Thus, the main focus is to protect cargo proteins from physical and chemical degradation to the extent that sufficient accumulation in diseased tissues occurs. In some cases, the NT system is designed to mimic the endogenous temporal cycles of growth factor expression in the body. A common strategy is to encapsulate proteins within polymer matrices to shield them from degradation.

Nanocarriers made with biodegradable matrices could be designed to control the drug release rate. A key challenge is to limit the damage to the protein cargo during fabrication of the NTs; size reduction using sonication and homogenization can cause protein denaturation. Another complication is that proteins may denature due to accumulation of acidic polymer oligomers, by-products generated from matrix degradation (Kang and Schwendeman 2002; Li and Schwendeman 2005). Poly(D,L-lactide-co-glycolide acid) (PLGA) is a case in point; encapsulated proteins are released by surface or bulk erosion of the polymer chains through hydrolysis. The latter may be mitigated by reducing the size of the particles to the nanoscale; the less tortuous pathways allow rapid escape of acidic fragments. However, size reduction would require higher input of energy, which in turn exerts energies that damage proteins during the fabrication process. Alternatively, proteins can be loaded onto the surface of nanocarriers. Antigens adsorbed onto nickel-decorated particles are effective in eliciting adaptive immune responses (Patel et al. 2007). Heparin adsorbed on polymeric particles can act as a surface substrate for loading growth factors.

2.3.3 Rationale for Multifunctional Nanomedicines: Theranostics

An important rationale for the development of nanomedicine is the possibility for engineering multiple functions into a single system. Inflammation is a highly variable process among individuals, but also in a given patient during the course of pathological progression and treatment. As the patient progresses through the disease and responds to treatment, whether favorably or unfavorably, the nature and intensity of inflammation is expected to change. These changes in inflammation severity in the spatial and temporal sense would ideally be addressed by nanomedicine delivery that correlates to these changes. One way to achieve this is to combine therapeutic entities with imaging agents that provide measurable signatures corresponding to these changes and allow imaging of inflammation in living subjects, whether humans or animals.

Integration of diagnosis with therapy (theranostics) in a single nanoparticle facilitates *in vivo* monitoring and imaging of nanoparticle biodistribution, drug delivery to target tissues, and measurement of treatment response (Kelkar and Reineke 2011). This ultimately enables assessment of the safety, toxicity, and efficacy of the nanomedicine in an individual patient over time, leading to personalized medicine. In a typical scenario, the nanosystem (e.g., liposome or nanoparticle) contains an API, which is encapsulated and infused intravenously, circulating in the bloodstream and eventually partitioning out and accumulating at target tissues. As the nanosystem carries a molecular imaging probe, we can monitor nanosystem distribution *in vivo*. However, this ability of nanosystem monitoring *in vivo* is only the first level of theranostic development. If the theranostic is to serve as a true diagnostic tool, it must provide information on the disease state and potential changes induced by the delivered drug on that state. In this view, inflammation lends itself as a perfect target for theranostic nanomedicine development, where imaging inflammation is achieved with the same nanoparticle that is delivering the anti-inflammatory drug. In the following narrative, we discuss converging principles derived from preclinical studies using different systems and models and reported clinical results.

2.4 NANOMEDICINE PLATFORMS

Nanocarriers can be classified into inorganic, or "hard," and organic, or "soft" materials (Table 2.1). Among the inorganic structures are quantum dots, fullerenes, carbon nanotubes, graphene, gold, silver, and iron oxide. Soft nanomaterials include systems consisting of lipids, dendrimers, polymers, colloids (nanoemulsions and microemulsions), micelles, and combinations thereof.

Lipid-based systems include liposomes, micelles, and solid lipid nanoparticles (SLNs) (Figure 2.1). There have been many designs reported in the literature; highlighted herein are platforms that have been tested extensively in *in vivo* models of inflammatory diseases, including cancer.

2.4.1 Liposomes

Formed by phospholipids into multilamellar structures with an aqueous interior, liposomes can be used for encapsulating hydrophilic and hydrophobic compounds. Typically, the vesicles are generated from rehydration of a solvent-evaporated lipid film, followed by extrusion through polycarbonate membranes. In many formulations, the sizes of liposomes range from 80 to 200 nm (Duncan and Gaspar 2011). Hydrophilic compounds are loaded into the interior aqueous core during the rehydration step, while hydrophobic molecules are partitioned into the lipid bilayer. The efficiencies of loading into liposomes in both types of APIs are typically above 50%. Delivery of the cargo occurs by binding of the liposomes with the cellular membrane. The bilayer configuration allows fusion of the vesicular lipids with the cellular plasma membrane, thereby emptying the content of the liposomes into the cytoplasm (Zhou and Huang 1994; Farhood et al. 1995). Early formulations of liposomes were found to be cleared rapidly from the plasma. The short half-lives can be attributed to opsonization, a mechanism mediated by adsorption of serum proteins and eventual internalization by cells in the mononuclear phagocyte system (MPS). Coating liposomes with polyethylene glycol (PEG), which slightly increases

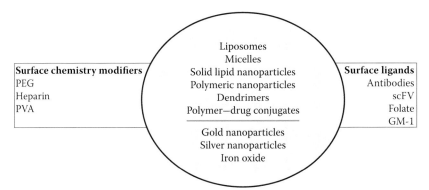

Figure 2.1. Nanomedicine platforms and designs, emphasizing particulate systems. Examples of surface modifications include targeting ligands and polymers used for steric stabilization. PVA, polyvinyl alcohol; scFV, single-chain variable fragment; GM-1, gangliosides.

vesicle size (but still below 200 nm), extends their life spans in plasma by reducing complement activation and sequestration in the spleen and liver (Bradley et al. 1998).

Doxorubicin, represented by the products Doxil and Myocet, is a well-known application of liposomes in cancer chemotherapy. A combination product, CPX-351, is a liposomal formulation in which cytarabine and danuorubicin are coencapsulated. Solid tumors have been targeted using immunoliposomes. The advent of antitumor antibodies provided new modalities in developing doxorubicin-encapsulated liposomes targeted to cancer markers, as in the case of anti–epidermal growth factor receptor (EGFR) liposomes (clinicaltrials.gov: NCT01702129). Arming the liposomes with the antibody transforms the passive targeting into active targeting, whereby anticancer efficacy can be enhanced while at the same time reducing the cardiac toxicities of doxorubicin. Despite these advantages, several challenges were noted in early liposomal formulations. Loaded drug molecules may leak from the vesicles in circulation prior to accumulating at diseased tissues. Hydrophobic APIs may destabilize vesicles by disrupting organization of the lipid bilayer. In addition, certain APIs may not remain stable in liposomes. Proteins loaded into the aqueous core may denature through interactions with the lipids over time. Manufacturing at large scale and preserving liposomes for extended periods are challenging.

Liposomes have been used to deliver therapeutic plasmid DNA and oligonucleotides (antisense and siRNA) *in vivo* (Gao and Huang 1991; Li and Huang 1997; Sorgi et al. 1997; Li et al. 1998, 2010). To overcome multiple barriers in a single system, several functions are engineered into modular designs. DNA is condensed and complexed with cationic lipids and polycations (Farhood et al. 1995). The inner core is stabilized by protamine (Sorgi et al. 1997) and calcium phosphate (Li et al. 2010). The ability to attain efficient gene transfection is accomplished through the helper lipid dioleoyl phosphatidylethanolamine (DOPE), which destabilizes the endosome membrane (Farhood et al. 1995). Targeting is introduced by grafting the surface with folate and other ligands (Lee and Huang 1996). *In vivo* delivery of a plasmid encoding the cystic fibrosis transmembrane conductance regulator (CFTR), a 189 kb gene, has been reported (Caplen et al. 1995), demonstrating the capacity of liposomes to deliver large genes. The ability to accumulate in tumors has been demonstrated (Sindrilaru et al. 2011), contributed by the vesicles' net positive charges, which enhances binding to endothelial cells, resulting in enhanced uptake in tumor vasculature (Thurston et al. 1998).

2.4.2 Nanoemulsions

Nanoemulsions are kinetically stable emulsions with a droplet size typically between 100 and 500 nm. The droplets have high oil content and are stabilized with low amounts of surfactants (McClements 2012). Nanoemulsions have wide applications in the pharmaceutical industry, where they are typically used to increase tissue penetration of poorly soluble drugs (transdermal), and improve drug solubility and bioavailability (injectable) (Sarker 2005; McClements and Rao 2011; Rajpoot et al. 2011; Kotta et al. 2012; Shakeel et al. 2012). They are also easily incorporated into other dosage forms, such as capsules and gels, and they can be manufactured on an industrial scale (Sarker 2005; Mitri et al. 2012; Müller et al. 2012). A particular type of nanoemulsion, perfluorocarbon (PFC) nanoemulsion, was extensively studied in preclinical models of inflammation. Although PFC formulations are erroneously referred to as "nanoparticles" in some literature, they are always prepared as nanoemulsions with PFC oil as an internal phase. Passive targeting to circulating inflammatory cells by PFC nanoemulsions has been extensively investigated for the purpose of ^{19}F MRI of inflammation (Weise et al. 2011; Stoll et al. 2012; Balducci et al. 2013). PFCs are chemically and biologically inert ^{19}F MRI agents that allow for quantitative and qualitative assessment of inflammation *in vivo* without apparent background. PFC nanoemulsions as inflammation-specific agents were used to image inflammation in breast cancer (Balducci et al. 2013), transplantation (Hitchens et al. 2011), abscess (Hertlein et al. 2011), and IBD (Kadayakkara et al. 2012). PFC nanoemulsions designed as dual-mode imaging agents for optical and photoacoustic imaging were reported by Akers et al. (2010) and were also investigated as ultrasound imaging contrast agents in preclinical models (Rapoport et al. 2011). As a theranostic platform, PFC nanoemulsions have attracted increasing interest over the past 10 years. In one example, Wickline and coworkers incorporated a cytolitic peptide (melittin) as an anticancer agent into the lipid layer of the PFC nanoemulsion droplet (Soman et al. 2008). They showed that melittin pharmacokinetics and tumor targeting were successfully achieved with a dramatic reduction of tumor size (Soman et al. 2009). At the same time, the nanoemulsion provided tumor-specific ^{19}F MRI in a mouse model. Further, paclitaxel was formulated in a block polymer–stabilized perfluoropentane nanoemulsion for image-guided delivery using ultrasound and MRI in a mouse tumor model (Rapoport et al. 2011).

Another recent study used PFC nanoemulsions to deliver antigen to dendritic cells (DCs) with the goal of boosting the immune response in DC-based vaccines (Dewitte et al. 2013). For active targeting strategies in nanoemulsion-based theranostics, the targeting agent is attached to the droplet surface by covalent conjugation. For example, PFC nanoemulsions targeted to $\alpha_v\beta_3$ integrins for atherosclerosis imaging have been developed (Caruthers et al. 2009). In an earlier study, Lanza et al. (2002) reported targeted delivery of antiproliferative drugs to smooth muscle cells using perfluorooctyl bromide (PFOB) nanoemulsions. We reported a new design of PFC nanoemulsions (O'Hanlon et al. 2012; Patel et al. 2013a,b) where hydrocarbon oil is combined with PFC in the internal phase, leading to a much enhanced capacity of PFC nanoemulsions to carry the lipophilic payload. These nanoemulsions are a triphasic colloidal system, where two immiscible oils, PFC and a hydrocarbon oil (synthetic or natural), are dispersed in water as small-sized droplets (110–180 nm is a typical range) and stabilized by nonionic surfactants.

2.4.3 Micelles

Typically with diameters between 15 and 200 nm, micelles are formed by self-assembly of amphiphilic block copolymers (Önyüksel et al. 1999; Krishnadas et al. 2003; Duncan and Gaspar 2011). The monodispersed vesicles consist of a hydrophobic core surrounded by a hydrophilic outer shell stabilized by surfactants. Hydrophobic drugs are loaded into the core, with the outer layer decorated by PEG. These sterically stabilized vesicles exhibit extended circulation time. Phospholipids with low critical micelle concentrations render stable micelles *in vivo*. These properties are leveraged in delivering poorly water-soluble

compounds to cancer patients. Several clinical trials are ongoing in testing paclitaxel-loaded polymeric micelles (clinicaltrials.gov: NCT00912639, NCT01023347, NCT01426126, and NCT00111904). Other studies of micellar formulations in humans include curcumin and vitamin D (NCT01644890, NCT01925287, NCT01925547, and NCT01807845). Another example was aminfostine (NCT02587442), formulated in micelles for protection against radiation. The product (RadProtect) is constructed with iron (ferrous)-linked PEG-b-PGA and remains in the blood circulation by binding to transferrin.

2.4.4 Dendrimers

Dendrimers are star-shaped polymers embedded with amines with mixed pK$_a$, resulting in a composite of ionizable potentials. These vesicles containing a hydrophobic core and hydrophilic edge exhibit micelle-like performance *in vivo*. These branched structures can be synthesized to generate particles of different sizes. APIs can be encapsulated within the polymeric core, or conjugated through covalent linkages to the end groups on the exterior. These include poly(glycerol–succinic acid), polyamidoamine (PAMAM), 2,2-bis(hydroxymethyl) propanoic acid–based dendrimer, star amphiphilic block copolymer containing poly(ε-caprolactone), and PEG. In some cases, hemolysis is noted in preclinical toxicological studies. Masking the peripheral cationic groups can mitigate this effect. The particle's surface may be grafted with PEG or modified with targeting ligands, for example, lactoferrin, arginine–glycine–asparate (RGD) motifs, and vascular endothelial growth factors (VEGFs). PAMAM has been used to carry the anti-inflammatory drug flurbiprofen. Entry into the cell cytoplasm is mediated by two properties; the cationic groups bind to the plasma membrane, allowing efficient endocytosis, while the less basic amines trigger endosome burst through the proton sponge effect. Freeman et al. (2013) have simulated the PAMAM-triggered endosomal burst mechanism by modeling the flux of protons and chloride across theoretical membranes via channels and ATPase pumps.

2.4.5 Solid Lipid Nanoparticles

SLNs are formed by high-pressure homogenization or sonication in dispersing drug molecules into an aqueous medium of highly purified triglycerides (Müller et al. 2002). Hydrophobic compounds, such as dexamethasone, can be incorporated into the matrix by mixing with lipids such as stearic acid. The resulting particles are stabilized by surfactants. The proposed advantages of SLNs, compared with liposomes, include improved physical stability and greater control of drug release rate. On the other hand, drug loading is limited by solubility of the compound of interest in the triglycerides. Oral delivery of insulin has been attempted with SLNs. Clinical studies of SLNs loaded with Myc-specific siRNA are ongoing (clinicaltrials. gov: NCT02110563 and NCT02314052). Evidence indicates that the lipid-based vesicles extend drug half-lives in plasma for as long as 10 hours (clinicaltrials.gov: NCT00495079, NCT00144963, and NCT00145041).

2.4.6 Polymersomes and Polymeric Nanosystems

Polymersomes are formed by high-molecular-weight, nonlipidic, amphiphilic molecules (Discher and Ahmed 2006). The large hydrophobic blocks add to the mechanical strength of the particles, resulting in relatively good physical stability in storage and *in vivo*. The hydrophilic and hydrophobic layers allow loading of APIs with diverse physiochemical properties; doxorubicin, paclitaxel, and cisplatin have been tested with polymersomes as carriers in mouse models of cancer (Levine et al. 2008). Various polymers have been used to prepare polymersomes, including polyphosphazene, PEO-b-PCL, PEO-block-poly(γ-methyl-ε-caprolactone), poly(γ-benzyl-L-glutamate)-block-hyaluronan, and poly(butadiene)-β-PEO. Polymersomes can be readily functionalized with targeting ligands to enhance efficacy *in vivo*. The cytotoxic action of tumor necrosis factor α (TNFα) is improved with polymersomes functionalized with integrins (α5β1) (Demirgöz et al. 2009).

Reducible polyethylenimine (PEI) incorporated with histidines and nuclear localization sequences have been developed as DNA vectors (Oupický et al. 2002; Manickam and Oupický 2006). Chitosan is used to manufacture nanocarriers to deliver anti-inflammatory compounds to the gastrointestinal (GI) tract (Hejazi and Amiji 2003). The mucoadhesive property and cationic nature make it attractive for the delivery of macromolecules to sites of inflammation. Prepared by ionic cross-linking and participation, complexes of chitosan and plasmid DNA are effective in transferring genes into intestinal cells. Successful mucosal vaccinations have been shown with increased uptake by M cell antigens complexed with chitosan (Illum et al. 2001). A critical parameter is the chitosan charge density, which is a function of the degree of deacetylation. Another design is constructed around β-cyclodextrin (β-CD) (Bartlett and Davis 2007) and has been used to deliver siRNA into tumor cells. The complex contains pH buffering (imidazole) and polycation (amidine) domains. There are also systems in which the polymers themselves exert therapeutic effects. Degradation of polymeric spermine in tumor cells generates an accumulation of ornithine decarboxylase inhibitors (Wetzler et al. 2000).

Gene therapy has been a major focus of NT development for organ transplantation. While adenoviruses, a widely used gene vector, generally confer a high degree of transfection efficiency, inflammatory reactions triggered by viral components may complicate induction of graft tolerance (Sen et al. 2001). Synthetic polymeric carriers have emerged as nonimmunogenic alternatives to viral gene vectors. In general, these systems serve to compact plasmid DNA molecules, protect DNA from enzymatic degradation in plasma and interstitial fluids, and facilitate entry of gene constructs through the plasma membrane (Niidome and Huang 2002). Ectopic expression of interleukin-10 (IL-10) in graft parenchyma has been attempted with liposomes (Batteux et al. 1999; Sen et al. 2001; Hong et al. 2002). These systems appear to afford superior kinetics: the IL-10 gene delivered by liposomes resulted in a higher overall expression and slower decline of the cytokine in animals with a transplanted heart than an adenoviral vector (Sen et al. 2001). It is well recognized, however, that a limiting factor of nonviral vectors is their relatively low level of transfection efficiency *in vivo*.

Nonlipidic polymers used in the fabrication of drug carriers include poly(anhydrides), poly(carpolactone), poly-D,L-lactide, and polyglycolide. For nanoparticles, PLGA, a diblock polymer consisting of lactic and glycolic acids, is often used (Panyam and Labhasetwar 2003). Poly(ε-caprolactone) nanoparticles have been formulated to deliver the anticancer compound tamoxifen (Chawla and Amiji 2002). Paclitaxel-embedded nanoparticles have been generated from poly(ethylene oxide)–modified poly(β-amino ester) as a pH-sensitive delivery system (Potineni et al. 2003). This is effective in targeted delivery of hydrophobic drugs into solid tumors (Shenoy et al. 2005; Devalapally et al. 2007). These matrix-dispersed drugs are released upon hydrolysis of the polymer chains. The rate of drug release can be tuned by the polymers' physiochemical properties. The relative bond stability determines the rate of degradation of the polymer backbone into fragments; drug release is coupled to dissolution of the matrix. The half-life of poly(anhydrides) is significantly less than that of poly(esters) such as PLGA. The hydrophobicity of the polymers regulates the rate of water penetration into the particle matrix. Studies of polyesters show that the degradation rate decreases with increasing hydrophobicity. The steric effect of polymer side chains could hinder hydrolytic reactions. Bulky side chains block nucleophilic attacks by water molecules, as exemplified by PLG degrading more rapidly than poly(lactic acid) (PLA). The nature of the hydrolytic products may also alter the rate of matrix degradation. In PLA, PLG, and PLGA, autocatalysis of hydrolysis can occur if the acidic oligomers are trapped inside. It has been shown that the intramatrix pH in PLGA microparticles can be as low as 1.8. Hydrolysis of poly(anhydrides) produces chains with both ends with acidic groups. This effect is especially relevant when drug release is a function of bulk erosion of the particle matrix.

2.4.6.1 PLGA Particle Fabrication Methods

Conventional fabrication methods entail embedding small-molecule drugs or bioactive macromolecules in the matrix of particulates. In this mode, entrapped plasmid molecules are protected from enzymatic digestion in biological fluids when delivered *in vivo*. Two drawbacks arise from this methodology. First, sonic or mechanical energies employed in the emulsification steps can damage the protein, leading to poor loading efficiency, with losses of bioactive proteins reaching up to 70% (Capan et al. 1999). Second, macromolecules entrapped in the particles can undergo chemical degradation as the polyester matrix erodes. This latter phenomenon has been documented with PLGA microparticles. Hydration of PLGA matrices results in microclimates in which a pH as low as 2 has been registered (Li and Schwendeman 2005).

Such a condition can lead to acid denaturation of proteins. Incorporation of base salts into PLGA microparticles reverses the acidification (Kang and Schwendeman 2002). Generally, changing polymer chemistry can modulate drug release rates. Matrices are tailored to the need with varying degrees of hydrophobicity, concentrations of reactive groups, and sensitivities to temperature, pH, and enzymes. Drug release from PLGA particles occurs by surface or bulk erosion. Surface erosion dominates if the rate of bond cleavage (t_c) is much greater than the rate of water penetration into the matrix (t_{diff}). Such surface-level degradation would generate zero-order drug release because the loss of polymer mass is linear over time. In cases where t_c is slower than t_{diff}, bulk erosion would dominate, resulting in drug release driven by diffusion out of the hollowing matrix. The implication for NTs is that the pathway for polymer oligomers to escape is less tortuous than that for microparticles, which would make the matrix less acidic, and more amenable to protein encapsulation.

PLGA particles are also used for delivering nucleic acids (Wen and Meng 2014). Capan et al. (1999) used poly-(L-lysine) in the formulation of plasmid DNA in PLGA microspheres. A low-molecular-weight polycation, ornithine-histidine peptide (O10H6), functions as a nucleic acid binding domain on the surface (Chamarthy et al. 2003) and in matrices (Kovacs et al. 2005) of PLGA nanoparticles. O10H6 was used to complex with oligonucleotides prior to emulsification (Zheng et al. 2006). Another strategy is to decouple matrix formation from loading, by adsorbing plasmid DNA on a PLGA surface coated with O10H6 (Kovacs et al. 2005). Ramsey and coworkers (Pouton et al. 1998; Ramsay et al. 2000; Ramsay and Gumbleton 2002) have shown that polymers consisting of ornithine, a nonnatural amino acid with the side chain $-(CH_2)_3-NH_2$ (protonated at physiological pH), are more effective transfection agents than poly-L-lysine. O10H6 is less cytotoxic than the similar lysine-based peptide at micromolar concentrations in primary DC cultures (Chamarthy et al. 2003). Complexation of the peptide with DNA results in stable condensates resistant to serum destabilization (Chamarthy et al. 2003). A fraction of DNA cointernalized with O10H6 escapes endosomal sequestration (Kovacs et al. 2005; Zheng et al. 2006), presumably aided by the buffering effect of imidazole moieties of the contiguous histidine tract (Midoux et al. 1998; Pichon et al. 2001). Imidazole, with a pK_a between 5 and 6 (depending on solvent accessibility), acts as a "proton sponge" to impair the acidification of the endosome lumen (Pack et al. 2000; Putnam et al. 2001). This causes vacuolar proton ATPase (V-ATPase) to import hydrogen ions in excess, leading to an influx of chloride and water and osmotic burst of endosomes.

2.5 PHARMACEUTICAL MANUFACTURING CONSIDERATIONS IN NANOMEDICINE DEVELOPMENT

One design, one disease has been a typical development strategy for the majority of therapies developed to date, with infrequent exceptions. High-grade specialization is omnipresent in the nanomedicine field as a whole. Targeting specificity and selectivity are considered of utmost importance. Nanomedicine development for cancer treatment has been primarily concerned with improving target specificity, so drug reaches only cancerous tissues and spares the healthy ones. It is easy to imagine how nanomedicines, by striving for this level of specialization, grew to become more and more complex in nature. What this complexity drove was obvious: manufacturing complexities leading to high cost and regulatory hurdles, complex processes needed for quality assurance and hence lowering clinical translation. This is where nanomedicine for inflammatory diseases may overcome development hurdles and provide a new paradigm for carrier drug design. We propose that if we simplify the overall nanomedicine design and chose a therapeutic target that is broadly applicable to many diseases, such as inflammation, clinical translation of nanomedicine would be much faster. To reach patients, nanomedicine must progress through the same development stages as any other therapeutic entity. However, nanomedicine development poses unique challenges for the industry and regulators. In a recent article, Kaur et al. (2015) listed a number of challenges nanomedicine development faces, such as scalability and quality control, reproducibility, and negative public impressions on nanotechnology, to name a few, which hinder clinical translation and commercialization. They go on to highlight the often undervalued factors, such as physicians and their views on nanotechnology, industry leaders, and regulators. Safety,

efficacy, and quality are the primary concerns of nanotechnology development.

We agree that nanotechnology is uniquely challenging for clinical development, but we strongly argue that its benefits outweigh the potential risks. However, the literature on QbD approaches applied to nanomedicine as a whole is rather limited. One reason may be that nanomedicine, broadly defined, is not always reported as such. Many of the nanomaterials developed for drug delivery are not reported as nanomedicines per se. One reason lies in terminology. A second reason is manufacturing on scale and process development. Nanomedicine preparation is typically reported on a very small lab scale. Further, as researchers are pressured to report innovative methodologies, this leads to often complex procedures, and with significant variability. Multifunctional nanomedicines, such as theranostics, present an even bigger challenge from a manufacturing perspective. There is a need for adequate balance between the deliverables, diagnostic (imaging agents), and therapeutic. Each functional component must maintain its properties throughout the manufacturing process, and quality control must concern both entities individually and in combination in the final nanomedicine product. Our group has recently reported a theranostic nanoemulsion scale-up development where we laid out some basic strategies for quality control (Liu et al. 2015). However, the overall scale achieved in this study was 1 L. Although this scale is rather large compared with typical amounts reported in the literature for theranostic nanoparticles (<1 g), the scale of 1 L is not fully representative of future manufacturing for clinical applications, which may require batch sizes that are orders of magnitude larger. Investigations on process development for nanomedicine scale-up can be cost-prohibitive, and access to high-quality raw materials may be limited. There are limited examples in the literature of large-scale manufacturing of nanomaterials for drug delivery and/or imaging.

Although QbD application to nanomedicines has not been explicitly reported, there have been systematic studies on QbD applied to nanosized drug delivery systems. Hence, it is our opinion that nanomedicine must fully engage pharmaceutical development principles such as QbD, regardless of the nanomedicine ultimate purpose, diagnostic or therapeutic, or both. However, only a few studies have been reported in this regard. For example, a QbD approach to an intranasal delivery system was recently reported (Pallagi et al. 2015). In this study, the quality target product profile (QTPP), critical quality attributes (CQAs), and critical process parameters (CPPs) were defined for the production of nanocrystals for the poorly soluble anti-inflammatory drug meloxicam distributed in a hydrogel matrix. QbD was also applied by Patil et al. (2015) to SLN production by hot-melt extrusion (HME) with continuous processing on a large scale. SLNs were produced with a size below 200 nm at a 60 mg/ml lipid solution and flow rate of 100 ml/min (Patil et al. 2015). Good manufacturing practice (GMP)–grade immune liposomes for EGFR-targeted doxorubicin delivery were produced with an exceptionally high level of quality control for first-in-human clinical trials (Wicki et al. 2015). Ten batches of nanomedicine were produced with robust process design and rigorous batch-to-batch quality assessments. This study demonstrates that large-scale GMP production of targeted nanoparticles is feasible, and that CQAs (particle size, polydispersity, chemical and colloidal stability, etc.) can remain within strictly defined values when production conditions are stringent and batch-to-batch quality control is maintained. Hence, this study highlights the profound impact QbD can have on nanomedicine.

2.6 FATE OF NANOMEDICINE *IN VIVO*

NTs are administered through parenteral, oral and topical routes. Parenteral routes include ophthalmic, nasal, and pulmonary delivery, while intravenous infusion is by far the most common route (Etheridge et al. 2013). A critical attribute of intravenous formulations is that the NTs must be hemorheologically compatible (Blanco et al. 2015). Particles that are larger than 1 micron or tend to adhere to endothelia would cause thrombosis or, worse, embolism. The physical properties presented to blood elements and substances therein dictate in part the accumulation of the nanosystem in target tissues. The fraction of dose accumulated in bystander, nontargeted tissues is also a function of the distribution of the NTs. That the APIs may dissociate from the carrier in circulation further complicates predictive modeling, in part because the free drug would be subjected to hepatic metabolism. The physical properties and surface chemistry of NTs strongly predict clearance from the circulation (Table 2.2). The ratio of

surface area to volume, surface hydrophobicity, deformability, and grafting of targeting ligands have been shown to affect plasma half-lives of NTs.

The MPS extracts particulates from blood circulation. The anatomical structures of the highly perfused liver and spleen resemble in-line blood filters. The MPS resident phagocytic cells, Kupffer cells, and macrophages remove particles as a function of physical and chemical attributes, in surface chemistry, size, charge, and shape. Extraction of liposomes by the liver and spleen is especially sensitive to the size of the vesicles, with significant effects with even 10 nm changes (Liu et al. 1992; Tenzer et al. 2011). Circulating NTs interact with each organ's unique vasculature. Particles below 5 nm are rapidly removed by renal filtration (Table 2.2). Nanoparticles larger than 150 nm are trapped in the lung capillaries. Positively charged vesicles have a higher propensity to accumulate in the lungs, liver, and spleen than negatively charged particles (Blanco et al. 2015). NTs with positive surface potential are also more likely to adhere to endothelium (margination), leading to higher extravasation (Thurston et al. 1998). Highly positively or negatively charged micellar nanoparticles have significant liver uptake (Xiao et al. 2011). In general, spherical NTs are less likely to deposit in these organs than disc-shaped carriers. The aspect ratio, or curvature, of NTs is related to uptake by phagocytosis (Champion and Mitragotri 2006, 2009) and margination in circulation (Gentile et al. 2008). While size, charge, and shape each contribute to clearance and distribution, tissue accumulation of NTs is an interplay of all three properties (Blanco et al. 2015).

The surface chemistry of NTs determines the propensity to adsorb proteins that elicit opsonization. In blood, protein corona (adsorption layer of proteins to the surface of NTs) is typically formed by immunoglobulins (Igs), fibrinogen,

TABLE 2.2
Influences of physical properties on NT clearance.

	Size	Charge	Shape
Renal filtration	<5 nm	–	–
MPS	>150 nm	Positive > negative	Disc > rod > sphere

NOTE: MPS, mononuclear phagocyte system; NT, nano-therapeutic.

NANOMEDICINE FOR INFLAMMATORY DISEASES

and complement proteins. The adsorption step occurs at varying rates, correlating with clearance of NTs from circulation. The corona is not static but undergoes dynamic changes (Li et al. 1999). The longer the half-life of an NT, the more likely the corona changes because of adsorption and desorption of plasma proteins over time. Consequently, the NTs may adopt a pattern of tissue distribution later in circulation distinct from the first few hours after infusion. Notably, surface protein corona decreases the targeting efficiency of transferrin-guided nanoparticles (Salvati et al. 2013).

Some nanocarriers infused into blood circulation can be internalized by blood monocytes, which tend to congregate in the spleen and inflammatory tissues. Hence, trafficking of the monocytes drives a second phase of drug distribution. Erythrocytes may sequester NTs through membrane adsorption. Their sheer abundance, at 4 million to 6 million cells per cubic milliliter of blood, makes them a potentially significant mechanism for trapping NTs in the spleen. The elimination of the free drug and NTs together determines the concentration available at the target tissues. At the target site, the NTs must penetrate into the tissues with the APIs remaining intact. The elevated interstitial fluid pressure in some solid tumor types might reduce exposure of the molecular target to the drug. For inflammation in the central nervous system (CNS), the blood–brain barrier (BBB) poses a rate-limiting step in the uptake of the NTs. When administered through intramuscular and subcutaneous routes, draining to lymph nodes from the site of injection represents a major route of elimination. The slow-moving lymphatic fluid brings particulates to phagocytic cells. In inflammatory tissues, the uptake of NTs is reduced, owing to loss of microvalves in lymphatic vessels, thereby allowing retrograde diffusion of entered particulates (Hirakawa et al. 2014).

Some APIs are formulated as NTs to increase cellular uptake (Sahay et al. 2010). Accumulation in target cells is crucial for nucleic acid drugs, including plasmid DNA, ASOs, and siRNA. Cellular pharmacokinetics of NTs have been investigated in detail (Mahato et al. 1997a). For nontargeted NTs, translocation across the plasma membrane occurs mainly through endocytosis or membrane fusion (Figure 2.2). The latter is mediated through lipid exchange with certain viruses and liposomes. Endocytosis occurs through pinocytosis (micropinocytosis) in most cells, or phagocytosis in certain leukocytes. Neutrophils, DCs, and macrophages are equipped with phagocytic capacity. Membrane-destabilizing peptides, including H5WYG and GALA, can be attached to help NTs translocate across the lipid bilayer. Cell penetration can be enhanced by appending receptor-specific ligands to the carriers. The rationale is that targeting ligands facilitate tissue localization, rendering prolonged resident time of the vesicle on the cell surface, thereby enhancing internalization via clathrin-mediated endocytosis (Liu and Huang 2002).

An early example was using peptides containing the integrin binding RGD motif to enhance gene transfer in tumor endothelia (Pierschbacher and Ruoslahti 1984; Lu et al. 1993; Hart et al. 1995, 1998; Harbottle et al. 1998; Schneider et al. 1998, 1999; Erbacher et al. 1999; Colin et al. 2000; Jost et al. 2001; Muller et al. 2001). Incorporating RGD motifs into NTs can increase gene expression up to 200-fold. Transferrin has also been used

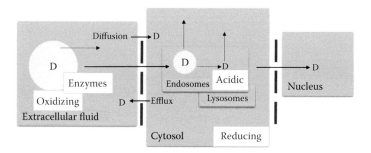

Figure 2.2. Cellular and intracellular barriers of delivery of nanomedicines. In some cases, the drug molecules need to be ushered into cellular compartments in which molecular targets are located. In cytosol, microtubules and nuclear receptors are targets of drugs. Plasmid DNA and antisense oligonucleotides need to accumulate in the nucleus to be effective.

to increase the expression of genes delivered via nanocarriers (Kircheis et al. 1997; Uike et al. 1998; Vinogradov et al. 1999; Wightman et al. 1999). Other common ligands include folate (Mislick et al. 1995), mannose (Mahato et al. 1997b), asialoglycoprotein (Choi et al. 1999), low-density lipoprotein (Yu et al. 2001), epidermal growth factor (Sosnowski et al. 1999; Blessing et al. 2001), antiplatelet endothelial cell adhesion molecule antibody (Li et al. 2000), and anti-CD3 antibody (O'Neill et al. 2001). A single-chain variable fragment is used to target siRNA and microRNA to tumors *in vivo* (Chen et al. 2010). The HIV TAT peptide facilitates intracellular delivery through micropinocytosis (Torchilin et al. 2001, 2003; Kaplan et al. 2005). Also known as peptide transduction domain (PTD), TAT-based systems have been tested in more than 25 clinical trials, with one agent in a current phase III trial (van den Berg and Dowdy 2011).

Endocytosed materials typically do not enter into the cytosol (Figure 2.2). Early and late endosomes are progressively acidified, and the vesicles eventually fuse with lysosomes within which exogenous substances are degraded or sequestrated. Escape from the endolysosomal pathway can occur through disruption of the vesicle membrane integrity by fusogenic hemagglutinin peptides (Lee et al. 2002), or through the proton sponge effect. The pH buffering effect of polymers containing multiple secondary and tertiary amines has been verified experimentally by Sonawane et al. (2003) and Midoux et al. (1998; Pichon et al. 2001). Imidazole, with a pK_a between 5 and 6 (depending on solvent accessibility), acts as a proton scavenger to impair the acidification of the endosome lumen (Pack et al. 2000; Putnam et al. 2001). This causes V-ATPase to import hydrogen ions in excess, leading to an influx of chloride and water and osmotic burst of the endosomes. Acidification is driven by transporters embedded in the endosomal membrane; the balanced movement of ions through the H+ ATPase pump, Na+ K+ ATPase pump, and chloride channel is shifted by the presence of a high concentration of amines. Computational models created by Rybak et al. (1997) report that the acidification is largely dependent on the number of H+ ATPase present and the diffusion of ions through the endosome membrane. Other studies indicate that the endosome membrane is also permeable to Cl− diffusion, and that Na+ K+ channels are also present

(Grabe and Oster 2001). Of note is that caveola-directed uptake may bypass the endolysosomal pathway, providing direct entry into the cytosol. Unlike endosomes, the pH of the interior of caveosomes is neutral. NTs exploiting the caveola pathway may be able to achieve increased intracellular drug delivery (Hillaireau and Couvreur 2009).

Molecules that escape the endosomes could still be cleared from the cytosol via P-gp. Many small-molecule drugs are substrates of the ATP-driven efflux pump. For nucleic-based agents, nuclear entry is another barrier. Viruses use nuclear localization signal peptides to open nuclear pores to deliver their genetic materials. For synthetic non-viral DNA carriers, dissolution of the nuclear membrane during mitosis is exploited in rapidly dividing cells. However, this strategy typically yields low efficiency in percent dose of plasmid DNA accumulation in the nucleus. Cellular tracking of transgene expression using polymerase chain reaction allowed modeling of the kinetics of plasmid DNA passages with PEI as the carrier (Zhou et al. 2007). Another cellular organelle targeted for therapeutic purposes is mitochondria. Mutations in mitochondrial DNA are associated with optical neuropathies and other diseases (Murphy and Smith 2000; Weissig et al. 2006).

2.7 NANOMEDICINE IN INFLAMMATORY DISEASES

In chronic inflammation, there appears to be an ongoing dynamic feedback between locoregional diseased sites and systemic vascular and clearance compartments (Morgan 2009). Tissues undergoing inflammation are targeted therapeutically based on two main physiological features: vasculature and biomarkers. Upon extravasation at the site of the intended organ, NTs would encounter several barriers before engaging target cells. Tissue accumulation of NTs could be considered from a compartmental perspective (Figure 2.3). The vasculature may be populated with immature vessels, and gap junctions between endothelial cells may be widened by vascular factors, including bradykinin, histamine, and leukotrienes. Such fenestrations in vasculature are characterized extensively in mouse models of solid tumors in which the enhanced permeability effect (EPR) has been demonstrated (Matsumura and Maeda 1986).

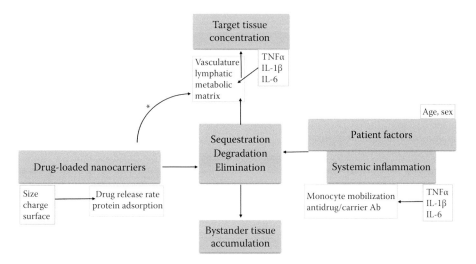

Figure 2.3. Factors contributing to nanomedicine delivery in chronic inflammation. Physiological factors in nanomedicine accumulation in tissues. Removal of nanomedicine in circulation depends on the particles' physicochemical properties. Cytokines can mobilize monocytes from bone marrow, thereby accelerating nanomedicine removal. Accumulation of drugs is a function of the local vascular, immune, and metabolic microenvironment.

It is generally recognized that therapeutic expectations of the EPR have not been met with consistent clinical benefits (Venditto and Szoka 2013). This is exemplified in the poor distribution of macromolecules in xenograft tumors (Baxter and Jain 1989, 1990, 1991), and the reduced permeability of macromolecules (Yuan et al. 1995). Jain and coworkers have studied the limiting effects of interstitial hypertension on the distribution of IgG molecules in solid tumors (Jain 1998). The elevated pressure resulting from fluids extruding (microvascular hypertension) from the center of the tumor overrides diffusive movement. The lymphatic outlets of solid tumors may be impeded, thereby exerting hydrostatic pressure to resist diffusion of particulates. This results in an outward convective current, moving NTs in interstitial fluid to the periphery of the tumor (Jain 1998). Consequently, NTs do not accumulate in tumors to the extent expected. It is generally recognized that the EPR effect is tumor type dependent. In mouse models, the mouse mammary tumor 4T1 exhibits relatively poor permeability, whereas lung carcinoma 3LL and colon cell–derived CT26 tumors exhibit relatively high permeability. A key differentiator is the density of collagen, regulated by tumor-derived MMP-9 and TIMP-1 activities. Jain and coworkers have used losartan to reduce collagen density in order to ease resistance of fluid flow in the tumor mass. VEGF inhibitors have been used to restore normalcy of the vasculature (Diop-Frimpong et al. 2011).

Similar to tumors, metabolic factors, including pH and redox gradients, may change the transport behaviors of NTs in noncancerous tissues undergoing inflammation (Figure 2.3). The composition of the ECM would hinder diffusional movement. These processes, crucial to steady-state drug concentrations, are affected by inflammatory mediators directly or indirectly. Acute inflammation is triggered by activation of innate immune cells. Often, reactions are caused by injuries to tissues or microbial infections. The initial waves of neutrophils, typically at peak on day 2, are followed by macrophages infiltrating in the subsequent 4–5 days to the damaged area and releasing TNFα, IL-1β, and IL-6. These cytokines regulate acute-phase protein synthesis to trigger downstream inflammatory mechanisms, including production of prostaglandins and leukotrienes, which may impact the systemic clearance of therapeutic agents (Morgan 2009). IL-6 has been shown to upregulate the cytochrome isozyme CYP3A4, which may accelerate the elimination of free drug released from NTs. This mechanism might be considered negligible if the APIs remained largely in the carrier with a limited rate of extraction into the liver over time. Alteration in drug transporters in inflammation may also affect the final drug concentration at target tissues.

Pro-coagulation cytokines may heighten the risk of embolism of NTs delivered by intravenous infusion. Systemic, chronic inflammation also alters the composition of plasma proteins. The compositions of protein corona might evolve as the nature and intensity of inflammation changes. Systemic inflammation also alters the synthesis of plasma proteins by the liver, resulting in dynamic changes in NT protein corona. Cytokines might also have an impact on NT interactions with blood cells. IL-6 has been implicated in the mobilization of monocytes (Ly6-high) from the bone marrow into blood, and from blood into tissues. The increased blood monocytes may enhance sequestration of NTs in circulation. Once extravasated, NTs in tissues may be removed by macrophages and DCs. In the inflammatory state, DCs exhibit altered trafficking, costimulation, and cross-priming functionalities. As immune sentinels, immature DCs constantly take up particulates in the surrounding environment through phagocytosis and endocytosis (Mellman and Steinman 2001). The maturing DCs could also raise the potential for inducing anti-NT antibodies.

RA is another chronic inflammatory condition for which nanomedicines have been used. The common mouse models of arthritis, induced by adjuvant or collagen, exhibit pathologies that mimic human RA in that TNFα, IL-1β, and IL-6 are the main drivers of inflammation. The fibrosis may inhibit the penetration of nanoparticles into the tissues. The vasculature in arthritic joints may resemble the tumor microenvironment because of the overlapping inflammatory mechanisms. The main targets are infiltrating leukocytes. The challenge is the rapid turnover of neutrophils and macrophages, in that anti-inflammatory drugs in the affected joints are interacting with moving targets. Consequently, drug may not accumulate in a sufficient number of cells because new cells are constantly infiltrating into the tissues. Neutrophil infiltration is an essential factor in inflammatory diseases. While neutrophils play an important role in sterilizing tissues and wounds, these innate immune cells also prolong recovery and lead to scar tissue through fibrosis. Neutrophils survive in tissues for only 1–2 days, a small window of time during which they may not encounter drugs infused systemically. In punctured cutaneous wounds, their turnover is rapid, registered at 1 million cells in 3 hours (Kim et al. 2008).

An opportunity for tailoring nanomedicine to the expanding knowledge in biology is found in wound healing. In diabetic wounds, the process is suspended at the inflammatory phase, sustained by a cascade of events characterized by apoptotic-resistant neutrophils and macrophages, and increased apoptosis of fibroblasts and keratinocytes (Alikhani et al. 2004; Liu et al. 2004). Consequently, clearance of cell debris is slowed, matrix synthesis is impoverished, and epithelial restitution is delayed. The unrelenting inflammation further attracts more neutrophils and macrophages by which cycles of tissue destruction ensue. The inability of the lesions to heal raises the risk of developing bacterial biofilms, a major cause of hospitalization among diabetic patients (Zhao et al. 2013). Thus, there appears to be an excessive neutrophilic response in diabetic wounds (Wetzler et al. 2000). Rather than peaking on day 1 or 2 postinjury, the neutrophils stay on and resist apoptosis. The ongoing tissue damage and ensuing inflammation may account for the slow healing process in ulcers of diabetic patients. In inflammatory environments, macrophages and DCs traffic to draining lymph nodes, and are constantly replenished by blood monocytes. Targeting these disease-associated cells would require optimized dosing regimens. But the kinetics of trafficking of these immune cells is likely unique at different disease stages and in each patient. The uncertainty in extrapolating the complex pathogenesis from animal to human also makes it difficult to use dosing alone as a solution. The neutrophil elastase is a protease that degrades components in the ECM. Locoregional inhibition of this enzyme may promote healing. Conventional topical formulations, however, are limited by poor local retention and drug bioavailability. Sustained release of elastase inhibitors may reverse the negative impact of neutrophils without adversely impacting the healing process. NTs can be used to deliver anti-inflammatory agents in a sustained manner, thereby increasing the fraction of cells exposed.

2.7.1 Evading Systemic Sequestration

The size and charge of nanocarriers are tunable properties in controlling plasma circulation half-lives. Those below 20 nm are rapidly filtered through the renal glomerulus (Blanco et al. 2015). Particles larger than 1 micron are trapped in the spleen. Clearance of particulates with dimensions between 100 and 500 nm depends on surface charge. Adsorption of serum proteins alters the surface chemistry and may enhance clearance by opsonization. Fibrinogen may aggregate on particles. Complement activation is triggered by adsorptive properties (Salvador-Morales and Sim 2013); C3b, a product of C3 lysis, is the first step of the alternative pathway and is used to measure the extent of the activation. Surface charge is a critical factor in protein adsorption. Huang and coworkers have developed lipid vesicles with an optimized charge ratio and with charges completely shielded on the exterior (Yang and Huang 1997; Li and Huang 2009).

Considered pharmacologically and immunologically inert, PEG is used to sterically stabilize NTs in vivo. PEG-coated liposomes exhibit increased plasma half-lives (Klibanov et al. 1990; Torchilin et al. 1994). Surfaces tethered with linear PEG (with a typical molecular weight between 2 and 5 kDa) are partially protected from protein adsorption through steric hindrance (Mori et al. 1991) and reduce protein adsorption to surfaces (Kingshott and Griesser 1999; Winblade et al. 2000), improving hemocompatibility (Amiji and Park 1993). The extended coils of the polymer chains screen van der Waals and electrostatic interactions with proteins, which behave as colloids at the material–tissue interface. PEG chains also render a thermodynamic barrier; hydrophilic but uncharged, the semimobile, partial helical ethylene glycols are well hydrated. The water cloud at the interface slows the diffusion of proteins in solution toward the materials (Unsworth et al. 2008; Walkey et al. 2012). PEGylation extends half-lives of liposomes from minutes to hours. The tight water hydration layer impedes protein adsorption. Nanocarriers grafted with PEG exhibit improved tumor targeting (van Vlerken et al. 2007b). PEG-grafted gelatin exhibits long plasma half-lives in vivo and at the same time is taken up by cells efficiently (Kaul and Amiji 2002). In immune liposomes, PEG is incorporated with antibodies for targeted delivery (Torchilin et al. 1992). A "sheddable" PEG design has been developed to negate the hindrance of the steric barrier on lipid fusion that leads to cell entry (Li and Huang 2010).

Coverage imperfections can still exist in PEG layers on surfaces (Gon et al. 2010; Toda et al. 2010; Gon and Santore 2011). Adsorption of albumin and other serum proteins in PEG-coated surfaces has been documented (Satulovsky et al. 2000; Swartzlander et al. 2015). Generally tethered via covalent conjugation or chemisorption, PEG grafting methods are still being improved. Coverage of covalent-tethered PEG is limited by configuration of the polymer conjugate. Chemisorption typically results in higher coverage, but adsorbed polymer chains can desorb; serum proteins and other endogenous macromolecules may displace PEG. Such erosion of coverage would accelerate in the inflammatory milieu concentrated with proteins and metabolites. The resulting flaws leave the materials patchy and adsorptive. Congruent with this paradigm is that the density of PEG needed for resisting protein adsorption in vivo is much higher than previously predicted in vitro (Walkey et al. 2012; Yang et al. 2014).

Active evasion of MPS has been investigated by attaching the phagocytosis inhibitor CD47 on NTs. In addition, particles cloaked with complement regulators CD55 and CD59 have been shown to prevent generation of C3b and subsequent opsonization. NTs might be transported in circulation as passengers in blood cells; erythrocytes have been used as vehicles, as have monocytes. Infused PFC nanoemulsions are captured by monocytes, which are home to inflammatory tissues (Wen et al. 2013; Janjic et al. 2014). Membrane isolates obtained from human leukocytes (Parodi et al. 2013) and platelets (Modery-Pawlowski et al. 2013) serve as biomimetic coatings of NTs, with the latter used to target tumor cells. Red blood cell membranes have been used to camouflage polymeric nanoparticles (Hu et al. 2011). Challenges of the approach include the purity and uniformity of the isolates from donated human plasma. Another application of the concept is found in NTs grafted with polysorbate-80. The surfactant captures endogenous apolipoproteins in circulation, resulting in enhanced passage of the particles through the BBB (Hu et al. 2011).

2.7.2 Local–Regional Delivery

Despite the myriad targeting approaches, the fraction accumulated in target tissues (e.g., tumors) as a function of total dose infused is low. An alternative to extending circulation time would be direct administration of NTs into diseased tissues (Figure 2.4). Typically, this mode of delivery is limited by externally accessible disease sites through injection or magnetic manipulation of drugs infused

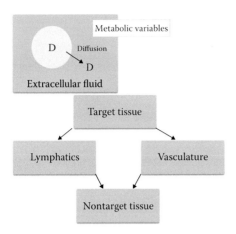

Figure 2.4. Local–regional delivery of nanomedicines to diseased tissues. Direct application to target organs can bypass systemic clearance mechanisms and vascular barriers. It is limited, however, to accessible tissues.

systemically. Alexiou et al. (2000) have reported using magnetic drug targeting to concentrate chemotherapeutics in squamous cell carcinoma in a rabbit model. Direct injection is common in preclinical studies. NTs injected into diseased tissues (e.g., tumors) or to the draining lymphatics should be evaluated using different sets of endpoints. Localized delivery can avoid serum proteins and bypass the vasculatures in the liver and the spleen. In the case of tumors, the inflammatory milieu provides an opportunity for immune modulation, locally or systemically. Tumor-associated antigens released by cancer cells as soluble factors or within exosomes are captured by macrophages and DCs. Strong antitumor responses have been shown by intratumoral injection of liposomes coencapsulating paclitaxel and a plasmid encoding IL-12, a cytokine that promotes cytotoxic T cell responses (Wang et al. 2006).

Local delivery is used to deliver immunosuppressants to circumvent systemic adverse effects. Nanocarriers loaded with IL-10, either plasmid encoding the protein or the recombinant protein itself, have been studied in mouse models of allogenic skin transplants and IBD. In the latter case, microspheres encapsulating IL-10 are delivered efficiently to the colonic region with strong evidence of suppression of local inflammation (Bhavsar and Amiji 2008). The BBB is a special case for which tailored systems have been designed. The intranasal route may allow NTs to gain entry into the CNS, bypassing the endothelial

barrier. Select targeting of regional clusters of lymph nodes can be achieved by carefully selecting injection sites. Another extravascular mode of delivery is to combine NTs with hydrogels to establish stable local drug depots. Such hybrid systems are designed to exploit the stabilizing effects of matrix scaffolding on the smaller nanosized particulates.

Stimuli-responsive fibrillar materials have been used to anchor drug-loaded nanoparticles. Polymers that undergo sol-gel phase transition upon exposure to changes in ionic strength or temperature can be coinjected with the particles. Self-assembling peptides remain dissolved in aqueous solution before injecting in vivo to initiate gelation. The resulting network of entangled fibrils composed of β-sheet-rich membrane-like scaffolds remaining stable in vivo (Wen et al. 2014) is well suited to facilitate particle adhesive moieties and can be incorporated into the scaffolds to enhance anchorage. The fibrillar networks allow diffusion of nutrients and release of metabolic waste, and can promote neovascularization. The injectable feature offers a versatile and robust translational potential. Such hybrids of gel and particle properties have been applied extensively in tissue engineering applications, including the regeneration of connective tissues. Seeding of mesenchymal stem cells in a fibrillar matrix requires the constant release of protein growth factors. Encapsulation of bioactive molecules in NTs allows an extended duration of action. Immobilizing NTs serve to restrict their clearance via diffusion and convection in extracellular fluid draining to the lymphatics. This could delay NT clearance by inhibiting phagocytosis.

2.8 TRANSLATION OF NANOMEDICINE TO THE CLINIC

Recent debates have raised questions on the lack of nano-based products on the market, despite decades of intense research (Venditto and Szoka 2013). A clear conceptual framework, based on converging data, would move the field forward to become mainstream. All checkpoints must be overcome or circumvented (Duncan and Gaspar 2011). To this end, novel multifunctional systems with optimized physiochemical and biological properties may prove a fast track to clinical applications (Giannoukakis and Meng 2015). In addition to size, charge, and steric stabilization, nanocarriers can be engineered with functionalities to inhibit or overcome multidrug resistance in cells (van Vlerken et al. 2007a; Batrakova and Kabanov 2008; Jabr-Milane et al. 2008; Sharma et al. 2008; Torchilin 2009; Abbasi et al. 2010; Xiong and Lavasanifar 2011). Biocompatibility should be qualified in the context of dose, model, and method of measurements. It should be defined in quantitative terms for each NT system, and not simply as a class. A substance judged "generally recognized as safe" (GRAS) should be specified with the route of administration and dosage for the particular application. Each component in the system must be evaluated in the context of the formulation and disease indications.

The drive toward evading sequestration must be balanced against safety criteria in that the nanomedicine should be biodegradable into metabolites that can be eliminated from the body. Considerations must be given to long-term toxicities, as nanocarriers may be cytotoxic, carcinogenic, mutagenic, or allergenic. Allergic reactions to NTs, including infusion reactions triggered by complement activation, are an important consideration when transitioning from preclinical to clinical testing. The level of endotoxin must be reduced to acceptable levels and be verified. Standard and effective methods of sterilization must be developed. To these ends, inflammatory potentials of NT materials have been characterized at the cellular and molecular levels. Bennewitz and Babensee (2005) have systemically characterized the phenotypic changes of DCs exposed to polymeric materials. Such in-depth knowledge is needed to drive the manufacturing of NT drugs designed for inducing immunological tolerance in autoimmune diseases.

There is a need for adequate pharmacokinetic modeling of NTs in animal studies. Preclinical studies should yield predictive models in half-lives and drug exposure (area under the curve [AUC]). Clinical performance is affected by route and frequency of administration, distribution, and metabolism. Ligand-directed localization should be optimized for the target and models to establish in vitro–in vivo correlations. Optimization may be enhanced by quantitative profiling of protein coronas on nanocarriers (Docter et al. 2014). These variables are especially important because of the advent of generic versions of nanomedicine for chronic inflammatory diseases such as RA and IBD. Indications for these diseases pose an additional tier of challenges in which multiple doses of NTs are given repeatedly. Azaya's ATI-0918 for Doxil and Sorrento's IG-001 for Abraxane must demonstrate bioequivalence to gain regulatory approvals. Metrics, including cumulative drug exposure (AUC), C_{max}, and T_{max}, would reflect the formulation parameters of the nanomedicine. Critical to the evaluation is the analytical capability to discriminate encapsulated and free fractions of the APIs. The total drug, accounting for both fractions, would be the basis for bioequivalence. Differentiating protein-bound and encapsulated APIs poses a challenge. Studies of a generic version of Caelyx marketed in Europe show high variability (coefficient of variation 30%–60%) of free doxorubicin in subjects. The European Medicines Agency (EMA) assessment report suggests that sufficient sample size (power) and robust fractionation methods are needed for accurate determination of bioequivalence.

2.9 CONCLUSION

The science and application of colloids formed the basis for advancing applications of particulate nanomedicines. The diverse chemical functionalization of the nanomaterial surface introduces the versatility of the approach. The next generation of nanomedicines would include hybrid systems with increasing complexities and functions, by engineering into a single system features of two or more traditional platforms. Systems will be specifically tailored to molecular definitions of inflammatory diseases, factoring in the effects of pro-inflammatory cytokines on the clearance and distribution of NTs because of altered plasma protein composition and monocyte mobilization. There would be increasing efforts in conforming nano-based formulations with regulatory specifications in the preclinical stage, with increased emphasis on content uniformity and purity, and other CQAs. We hope that modern pharmaceutical manufacturing approaches become a standard in nanomedicine production from the earliest stages of their development to clinical translation. Without this, we believe, nanomedicine will remain confined within preclinical walls and clinical examples will remain an exception rather than the expected outcome of nanomedicine development.

ACKNOWLEDGMENTS

This work was supported in part by NIH grants R21 AI113000 (W.S.M.) and R21 DA039621 (J.M.J.). We are grateful for the assistance provided by Wen Liu and Michele Herneisey in preparing this chapter.

REFERENCES

Abbasi, M., A. Lavasanifar, L. G. Berthiaume, M. Weinfeld, and H. Uludag. 2010. Cationic polymer-mediated small interfering RNA delivery for P-glycoprotein down-regulation in tumor cells. *Cancer* 116 (23):5544–54.

Akers, W. J., C. Kim, M. Berezin, K. Guo, R. Fuhrhop, G. M. Lanza, G. M. Fischer, E. Daltrozzo, A. Zumbusch, and X. Cai. 2010. Noninvasive photoacoustic and fluorescence sentinel lymph node identification using dye-loaded perfluorocarbon nanoparticles. *ACS Nano* 5 (1):173–82.

Alexiou, C., W. Arnold, R. J. Klein, F. G. Parak, P. Hulin, C. Bergemann, W. Erhardt, S. Wagenpfeil, and A. S. Lubbe. 2000. Locoregional cancer treatment with magnetic drug targeting. *Cancer Res* 60 (23):6641–8.

Alikhani, M., Z. Alikhani, M. Raptis, and D. T. Graves. 2004. TNF-α in vivo stimulates apoptosis in fibroblasts through caspase-8 activation and modulates the expression of pro-apoptotic genes. *J Cell Physiol* 201 (3):341–8.

Amiji, M., and K. Park. 1993. Surface modification of polymeric biomaterials with poly (ethylene oxide), albumin, and heparin for reduced thrombogenicity. *J Biomater Sci Polym Ed* 4 (3):217–34.

Balducci, A., Y. Wen, Y. Zhang, B. M. Helfer, T. K. Hitchens, W. S. Meng, A. K. Wesa, and J. M. Janjic. 2013. A novel probe for the non-invasive detection of tumor-associated inflammation. *Oncoimmunology* 2 (2):e23034.

Barenholz, Y. 2012. Doxil(R)—The first FDA-approved nano-drug: Lessons learned. *J Control Release* 160 (2):117–34.

Bartlett, D. W., and M. E. Davis. 2007. Physicochemical and biological characterization of targeted, nucleic acid-containing nanoparticles. *Bioconjug Chem* 18:456–68.

Batrakova, E. V., and A. V. Kabanov. 2008. Pluronic block copolymers: Evolution of drug delivery concept from inert nanocarriers to biological response modifiers. *J Control Release* 130 (2):98–106.

Batteux, F., H. Trebeden, J. Charreire, and G. Chiocchia. 1999. Curative treatment of experimental autoimmune thyroiditis by in vivo administration of plasmid DNA coding for interleukin-10. *Eur J Immunol* 29 (3):958–63.

Baxter, L. T., and R. K. Jain. 1989. Transport of fluid and macromolecules in tumors. I. Role of interstitial pressure and convection. *Microvasc Res* 37 (1):77–104.

Baxter, L. T., and R. K. Jain. 1990. Transport of fluid and macromolecules in tumors. II. Role of heterogeneous perfusion and lymphatics. *Microvasc Res* 40 (2):246–63.

Baxter, L. T., and R. K. Jain. 1991. Transport of fluid and macromolecules in tumors. III. Role of binding and metabolism. *Microvasc Res* 41 (1):5–23.

Bennewitz, N. L., and J. E. Babensee. 2005. The effect of the physical form of poly(lactic-co-glycolic acid) carriers on the humoral immune response to co-delivered antigen. *Biomaterials* 26 (16):2991–9.

Bergen, J. M., H. A. Von Recum, T. T. Goodman, A. P. Massey, and S. H. Pun. 2006. Gold nanoparticles as a versatile platform for optimizing physicochemical parameters for targeted drug delivery. *Macromol Biosci* 6 (7):506–16.

Bhavsar, M. D., and M. M. Amiji. 2008. Oral IL-10 gene delivery in a microsphere-based formulation for local transfection and therapeutic efficacy in inflammatory bowel disease. *Gene Ther* 15 (17):1200–9.

Blanco, E., H. Shen, and M. Ferrari. 2015. Principles of nanoparticle design for overcoming biological barriers to drug delivery. *Nat Biotechnol* 33 (9):941–51.

Blessing, T., M. Kursa, R. Holzhauser, R. Kircheis, and E. Wagner. 2001. Different strategies for formation of pegylated EGF-conjugated PEI/DNA complexes for targeted gene delivery. *Bioconjug Chem* 12 (4):529–37.

Bradley, A. J., D. V. Devine, S. M. Ansell, J. Janzen, and D. E. Brooks. 1998. Inhibition of liposome-induced complement activation by incorporated poly (ethylene glycol)-lipids. *Arch Biochem Biophys* 357 (2):185–94.

Capan, Y., B. H. Woo, S. Gebrekidan, S. Ahmed, and P. P. DeLuca. 1999. Preparation and characterization of poly (D,L-lactide-co-glycolide) microspheres for controlled release of poly(L-lysine) complexed plasmid DNA. *Pharm Res* 16 (4):509–13.

Caplen, N. J., E. W. F. W. Alton, P. G. Middleton, J. R. Dorin, B. J. Stevenson, X. Gao, S. R. Durham, P. K. Jeffery, M. E. Hodson, and C. Coutelle. 1995.

Liposome-mediated CFTR gene transfer to the nasal epithelium of patients with cystic fibrosis. *Nat Med* 1 (1):39–46.

Caruthers, S. D., T. Cyrus, P. M. Winter, S. A. Wickline, and G. M. Lanza. 2009. Anti-angiogenic perfluorocarbon nanoparticles for diagnosis and treatment of atherosclerosis. *Wiley Interdiscipl Rev Nanomed Nanobiotechnol* 1 (3):311–23.

Chamarthy, S. P., J. R. Kovacs, E. McClelland, D. Gattens, and W. S. Meng. 2003. A cationic peptide consists of ornithine and histidine repeats augments gene transfer in dendritic cells. *Mol Immunol* 40 (8):483–90.

Champion, J. A., and S. Mitragotri. 2006. Role of target geometry in phagocytosis. *Proc Natl Acad Sci USA* 103(13):4930–4.

Champion, J. A., and S. Mitragotri. 2009. Shape induced inhibition of phagocytosis of polymer particles. *Pharm Res* 26 (1):244–9.

Chawla, J. S., and M. M. Amiji. 2002. Biodegradable poly (ε-caprolactone) nanoparticles for tumor-targeted delivery of tamoxifen. *Int J Pharm* 249 (1):127–38.

Chen, J., S. Jo, and K. Park. 1995. Polysaccharide hydrogels for protein drug delivery. *Carbohydr Polym* 28 (1):69–76.

Chen, Y., X. Zhu, X. Zhang, B. Liu, and L. Huang. 2010. Nanoparticles modified with tumor-targeting scFv deliver siRNA and miRNA for cancer therapy. *Mol Ther* 18 (9):1650–6.

Choc, M. G. 1997. Bioavailability and pharmacokinetics of cyclosporine formulations: Neoral® vs Sandimmune®. *Int J Dermatol* 36 (Suppl 1):1–6.

Choi, Y. H., F. Liu, J. S. Choi, S. W. Kim, and J. S. Park. 1999. Characterization of a targeted gene carrier, lactose-polyethylene glycol-grafted poly-L-lysine and its complex with plasmid DNA. *Hum Gene Ther* 10 (16):2657–65.

Colin, M., M. Maurice, G. Trugnan, M. Kornprobst, R. P. Harbottle, A. Knight, R. G. Cooper, A. D. Miller, J. Capeau, C. Coutelle, and M. C. Brahimi-Horn. 2000. Cell delivery, intracellular trafficking and expression of an integrin-mediated gene transfer vector in tracheal epithelial cells. *Gene Ther* 7 (2):139–52.

Daniel, M.-C., and D. Astruc. 2004. Gold nanoparticles: Assembly, supramolecular chemistry, quantum-size-related properties, and applications toward biology, catalysis, and nanotechnology. *Chem Rev* 104 (1):293–346.

Demirgöz, D., T. O. Pangburn, K. P. Davis, S. Lee, F. S. Bates, and E. Kokkoli. 2009. PR_b-targeted delivery of tumor necrosis factor-α by polymersomes for the treatment of prostate cancer. *Soft Matter* 5 (10):2011–9.

Devalapally, H., D. Shenoy, S. Little, R. Langer, and M. Amiji. 2007. Poly (ethylene oxide)-modified poly (beta-amino ester) nanoparticles as a pH-sensitive system for tumor-targeted delivery of hydrophobic drugs. Part 3. Therapeutic efficacy and safety studies in ovarian cancer xenograft model. *Cancer Chem Pharmacol* 59 (4):477–84.

Dewitte, H., B. Geers, S. Liang, U. Himmelreich, J. Demeester, S. C. De Smedt, and I. Lentacker. 2013. Design and evaluation of theranostic perfluorocarbon particles for simultaneous antigen-loading and 19 F-MRI tracking of dendritic cells. *J Control Release* 169 (1):141–49.

Diop-Frimpong, B., V. P. Chauhan, S. Krane, Y. Boucher, and R. K. Jain. 2011. Losartan inhibits collagen I synthesis and improves the distribution and efficacy of nanotherapeutics in tumors. *Proc Natl Acad Sci USA* 108 (7):2909–14.

Discher, D. E., and F. Ahmed. 2006. Polymersomes. *Annu Rev Biomed Eng* 8:323–41.

Docter, D., U. Distler, W. Storck, J. Kuharev, D. Wunsch, A. Hahlbrock, S. K. Knauer, S. Tenzer, and R. H. Stauber. 2014. Quantitative profiling of the protein coronas that form around nanoparticles. *Nat Protoc* 9 (9):2030–44.

Duncan, R., and R. Gaspar. 2011. Nanomedicine(s) under the microscope. *Mol Pharm* 8 (6):2101–41.

Dvir, T., B. P. Timko, D. S. Kohane, and R. Langer. 2011. Nanotechnological strategies for engineering complex tissues. *Nat Nanotechnol* 6 (1):13–22.

Erbacher, P., J. S. Remy, and J. P. Behr. 1999. Gene transfer with synthetic virus-like particles via the integrin-mediated endocytosis pathway. *Gene Ther* 6 (1):138–45.

Etheridge, M. L., S. A. Campbell, A. G. Erdman, C. L. Haynes, S. M. Wolf, and J. McCullough. 2013. The big picture on nanomedicine: The state of investigational and approved nanomedicine products. *Nanomedicine* 9 (1):1–14.

Farhood, H., N. Serbina, and L. Huang. 1995. The role of dioleoyl phosphatidylethanolamine in cationic liposome mediated gene transfer. *Biochim Biophys Acta* 1235 (2):289–95.

Farokhzad, O. C., and R. Langer. 2009. Impact of nanotechnology on drug delivery. *ACS Nano* 3 (1):16–20.

Freeman, E. C., L. M. Wieland, and W. S. Meng. 2013. Modeling the proton sponge hypothesis: Examining proton sponge effectiveness for enhancing intracellular gene delivery through multiscale modeling. *J Biomater Sci Polym Ed* 24 (4):398–416.

Gao, X., and L. Huang. 1991. A novel cationic liposome reagent for efficient transfection of mammalian cells. *Biochem Biophys Res Commun* 179 (1):280–5.

Gentile, F., C. Chiappini, D. Fine, R. C. Bhavane, M. S. Peluccio, M. M. Cheng, X. Liu, M. Ferrari, and P. Decuzzi. 2008. The effect of shape on the margination dynamics of non-neutrally buoyant particles in two-dimensional shear flows. *J Biomech* 41 (10):2312–8.

Giannoukakis, N., and W. S. Meng. 2015. Nanotherapeutics for autoimmunity becomes mainstream. *Clin Immunol* 160 (1):1–2.

Gon, S., M. Bendersky, J. L. Ross, and M. M. Santore. 2010. Manipulating protein adsorption using a patchy protein-resistant brush. *Langmuir* 26 (14):12147–54.

Gon, S., and M. M. Santore. 2011. Single component and selective competitive protein adsorption in a patchy polymer brush: Opposition between steric repulsions and electrostatic attractions. *Langmuir* 27 (4):1487–93.

Grabe, M., and G. Oster. 2001. Regulation of organelle acidity. *J Gen Physiol* 117 (4):329–344.

Gradishar, W. J., S. Tjulandin, N. Davidson, H. Shaw, N. Desai, P. Bhar, M. Hawkins, and J. O'Shaughnessy. 2005. Phase III trial of nanoparticle albumin-bound paclitaxel compared with polyethylated castor oil-based paclitaxel in women with breast cancer. *J Clin Oncol* 23 (31):7794–803.

Harbottle, R. P., R. G. Cooper, S. L. Hart, A. Ladhoff, T. McKay, A. M. Knight, E. Wagner, A. D. Miller, and C. Coutelle. 1998. An RGD-oligolysine peptide: A prototype construct for integrin-mediated gene delivery. *Hum Gene Ther* 9 (7):1037–47.

Hart, S. L., C. V. Arancibia-Carcamo, M. A. Wolfert, C. Mailhos, N. J. O'Reilly, R. R. Ali, C. Coutelle et al. 1998. Lipid-mediated enhancement of transfection by a nonviral integrin-targeting vector. *Hum Gene Ther* 9 (4):575–85.

Hart, S. L., R. P. Harbottle, R. Cooper, A. Miller, R. Williamson, and C. Coutelle. 1995. Gene delivery and expression mediated by an integrin-binding peptide. *Gene Ther* 2 (8):552–4.

Hejazi, R., and M. Amiji. 2003. Chitosan-based gastrointestinal delivery systems. *J Control Release* 89 (2):151–65.

Hertlein, T., V. Sturm, S. Kircher, T. Basse-Lüsebrink, D. Haddad, K. Ohlsen, and P. Jakob. 2011. Visualization of abscess formation in a murine thigh infection model of *Staphylococcus aureus* by 19 F-magnetic resonance imaging (MRI). *PLoS One* 6 (3):e18246.

Hillaireau, H., and P. Couvreur. 2009. Nanocarriers' entry into the cell: Relevance to drug delivery. *Cell Mol Life Sci* 66 (17):2873–96.

Hirakawa, S., M. Detmar, and S. Karaman. 2014. Lymphatics in nanophysiology. *Adv Drug Deliv Rev* 74:12–8.

Hitchens, T. K., Q. Ye, D. F. Eytan, J. M. Janjic, E. T. Ahrens, and C. Ho. 2011. 19F MRI detection of acute allograft rejection with in vivo perfluorocarbon labeling of immune cells. *Magn Reson Med* 65 (4):1144–53.

Hong, Y. S., H. Laks, G. Cui, T. Chong, and L. Sen. 2002. Localized immunosuppression in the cardiac allograft induced by a new liposome-mediated IL-10 gene therapy. *J Heart Lung Transplant* 21 (11):1188–200.

Hu, C.-M. J., L. Zhang, S. Aryal, C. Cheung, R. H. Fang, and L. Zhang. 2011. Erythrocyte membrane-camouflaged polymeric nanoparticles as a biomimetic delivery platform. *Proc Natl Acad Sci USA* 108 (27):10980–5.

Ibrahim, N. K., N. Desai, S. Legha, P. Soon-Shiong, R. L. Theriault, E. Rivera, B. Esmaeli, S. E. Ring, A. Bedikian, G. N. Hortobagyi, and J. A. Ellerhorst. 2002. Phase I and pharmacokinetic study of ABI-007, a Cremophor-free, protein-stabilized, nanoparticle formulation of paclitaxel. *Clin Cancer Res* 8 (5):1038–44.

Illum, L., I. Jabbal-Gill, M. Hinchcliffe, A. N. Fisher, and S. S. Davis. 2001. Chitosan as a novel nasal delivery system for vaccines. *Adv Drug Deliv Rev* 51 (1):81–96.

Jabr-Milane, L. S., L. E. van Vlerken, S. Yadav, and M. M. Amiji. 2008. Multi-functional nanocarriers to overcome tumor drug resistance. *Cancer Treat Rev* 34 (7):592–602.

Jain, R. K. 1998. Delivery of molecular and cellular medicine to solid tumors. *J Control Release* 53 (1–3):49–67.

Janib, S. M., A. S. Moses, and J. A. MacKay. 2010. Imaging and drug delivery using theranostic nanoparticles. *Adv Drug Deliv Rev* 62 (11):1052–63.

Janjic, J. M., P. Shao, S. Zhang, X. Yang, S. K. Patel, and M. Bai. 2014. Perfluorocarbon nanoemulsions with fluorescent, colloidal and magnetic properties. *Biomaterials* 35 (18):4958–68.

Jost, P. J., R. P. Harbottle, A. Knight, A. D. Miller, C. Coutelle, and H. Schneider. 2001. A novel peptide, THALWHT, for the targeting of human airway epithelia. *FEBS Lett* 489 (2–3):263–9.

Kadayakkara, D. K., S. Ranganathan, W.-B. Young, and E. T. Ahrens. 2012. Assaying macrophage activity in a murine model of inflammatory bowel disease using fluorine-19 MRI. *Lab Invest* 92 (4):636–45.

Kang, J., and S. P. Schwendeman. 2002. Comparison of the effects of Mg(OH)2 and sucrose on the stability of bovine serum albumin encapsulated in injectable poly(D,L-lactide-co-glycolide) implants. *Biomaterials* 23 (1):239–45.

Kaplan, I. M., J. S. Wadia, and S. F. Dowdy. 2005. Cationic TAT peptide transduction domain enters cells by macropinocytosis. *J Control Release* 102 (1):247–53.

Kaul, G., and M. Amiji. 2002. Long-circulating poly (ethylene glycol)-modified gelatin nanoparticles for intracellular delivery. *Pharm Res* 19 (7):1061–7.

Kaur, R., T. Garg, U. D. Gupta, P. Gupta, G. Rath, and A. K. Goyal. 2015. Preparation and characterization of spray-dried inhalable powders containing nano-aggregates for pulmonary delivery of anti-tubercular drugs. *Artif Cells Nanomed Biotechnol* 44 (1):182–7.

Kelkar, S. S., and T. M. Reineke. 2011. Theranostics: Combining imaging and therapy. *Bioconjug Chem* 22 (10):1879–903.

Kim, M. H., W. Liu, D. L. Borjesson, F. R. Curry, L. S. Miller, A. L. Cheung, F. T. Liu, R. R. Isseroff, and S. I. Simon. 2008. Dynamics of neutrophil infiltration during cutaneous wound healing and infection using fluorescence imaging. *J Invest Dermatol* 128 (7):1812–20.

Kim, T.-Y., D.-W. Kim, J.-Y. Chung, S. G. Shin, S.-C. Kim, D. S. Heo, N. K. Kim, and Y.-J. Bang. 2004. Phase I and pharmacokinetic study of Genexol-PM, a cremophor-free, polymeric micelle-formulated paclitaxel, in patients with advanced malignancies. *Clin Cancer Res* 10 (11):3708–16.

Kingshott, P., and H. J. Griesser. 1999. Surfaces that resist bioadhesion. *Curr Opin Solid State Mater Sci* 4 (4):403–12.

Kircheis, R., A. Kichler, G. Wallner, M. Kursa, M. Ogris, T. Felzmann, M. Buchberger, and E. Wagner. 1997. Coupling of cell-binding ligands to polyethylenimine for targeted gene delivery. *Gene Ther* 4 (5):409–18.

Klibanov, A. L., K. Maruyama, V. P. Torchilin, and L. Huang. 1990. Amphipathic polyethyleneglycols effectively prolong the circulation time of liposomes. *FEBS Lett* 268 (1):235–7.

Kotta, S., A. W. Khan, K. Pramod, S. H. Ansari, R. K. Sharma, and J. Ali. 2012. Exploring oral nano-emulsions for bioavailability enhancement of poorly water-soluble drugs. *Expert Opin Drug Deliv* 9 (5):585–98.

Kovacs, J. R., Y. Zheng, H. Shen, and W. S. Meng. 2005. Polymeric microspheres as stabilizing anchors for oligonucleotide delivery to dendritic cells. *Biomaterials* 26 (33):6754–61.

Krishnadas, A., I. Rubinstein, and H. Önyüksel. 2003. Sterically stabilized phospholipid mixed micelles: In vitro evaluation as a novel carrier for water-insoluble drugs. *Pharm Res* 20 (2):297–302.

Kroto, H. W., J. R. Heath, S. C. O'Brien, R. F. Curl, and R. E. Smalley. 1985. C 60: Buckminsterfullerene. *Nature* 318 (6042):162–3.

Langer, R., and D. A. Tirrell. 2004. Designing materials for biology and medicine. *Nature* 428 (6982):487–92.

Lanza, G. M., X. Yu, P. M. Winter, D. R. Abendschein, K. K. Karukstis, M. J. Scott, L. K. Chinen, R. W. Fuhrhop, D. E. Scherrer, and S. A. Wickline. 2002. Targeted antiproliferative drug delivery to vascular smooth muscle cells with a magnetic resonance imaging nanoparticle contrast agent: Implications for rational therapy of restenosis. *Circulation* 106 (22):2842–7.

Lee, H., J. H. Jeong, and T. G. Park. 2002. PEG grafted polylysine with fusogenic peptide for gene delivery: High transfection efficiency with low cytotoxicity. *J Control Release* 79 (1):283–91.

Lee, R. J., and L. Huang. 1996. Folate-targeted, anionic liposome-entrapped polylysine-condensed DNA for tumor cell-specific gene transfer. *J Biol Chem* 271 (14):8481–7.

Lee, V. 1990. *Peptide and Protein Drug Delivery*. Vol. 4. Boca Raton, FL: CRC Press.

Levine, D. H., P. P. Ghoroghchian, J. Freudenberg, G. Zhang, M. J. Therien, M. I. Greene, D. A. Hammer, and R. Murali. 2008. Polymersomes: A new multifunctional tool for cancer diagnosis and therapy. *Methods* 46 (1):25–32.

Li, J., Y.-C. Chen, Y.-C. Tseng, S. Mozumdar, and L. Huang. 2010. Biodegradable calcium phosphate nanoparticle with lipid coating for systemic siRNA delivery. *J Control Release* 142 (3):416–21.

Li, L., and S. P. Schwendeman. 2005. Mapping neutral microclimate pH in PLGA microspheres. *J Control Release* 101 (1–3):163–73.

Li, S., W. C. Tseng, D. Beer Stolz, S. P. Wu, S. C. Watkins, and L. Huang. 1999. Dynamic changes in the characteristics of cationic lipidic vectors after exposure to mouse serum: Implications for intravenous lipofection. *Gene Ther* 6 (4):585–94.

Li, S., and L. Huang. 1997. In vivo gene transfer via intravenous administration of cationic lipid-protamine-DNA (LPD) complexes. *Gene Ther* 4 (9):891–900.

Li, S., Y. Tan, E. Viroonchatapan, B. R. Pitt, and L. Huang. 2000. Targeted gene delivery to pulmonary endothelium by anti-PECAM antibody. *Am J Physiol Lung Cell Mol Physiol* 278 (3):L504–11.

Li, S. D., and L. Huang. 2009. Nanoparticles evading the reticuloendothelial system: Role of the supported bilayer. *Biochim Biophys Acta* 1788 (10):2259–66.

Li, S.-D., and L. Huang. 2010. Stealth nanoparticles: High density but sheddable PEG is a key for tumor targeting. *J Control Release* 145 (3):178.

Li, S. M. A. S. L., M. A. Rizzo, S. Bhattacharya, and L. Huang. 1998. Characterization of cationic lipid-protamine-DNA (LPD) complexes for intravenous gene delivery. *Gene Ther* 5 (7):930–7.

Liu, D., A. Mori, and L. Huang. 1992. Role of liposome size and RES blockade in controlling biodistribution and tumor uptake of GM 1-containing liposomes. *Biochim Biophys Acta* 1104 (1):95–101.

Liu, F., and L. Huang. 2002. Development of non-viral vectors for systemic gene delivery. *J Control Release* 78 (1–3):259–66.

Liu, L., C. Bagia, and J. M. Janjic. 2015. The first scale-up production of theranostic nanoemulsions. *Biores Open Access* 4 (1):218–28.

Liu, R., T. Desta, H. He, and D. T. Graves. 2004. Diabetes alters the response to bacteria by enhancing fibroblast apoptosis. *Endocrinology* 145 (6):2997–3003.

Lu, X., J. J. Deadman, J. A. Williams, V. V. Kakkar, and S. Rahman. 1993. Synthetic RGD peptides derived from the adhesive domains of snake-venom proteins: Evaluation as inhibitors of platelet aggregation. *Biochem J* 296 (Pt 1):21–4.

Mahato, R. I., A. Rolland, and E. Tomlinson. 1997a. Cationic lipid-based gene delivery systems: Pharmaceutical perspectives. *Pharm Res* 14 (7):853–9.

Mahato, R. I., S. Takemura, K. Akamatsu, M. Nishikawa, Y. Takakura, and M. Hashida. 1997b. Physicochemical and disposition characteristics of antisense oligonucleotides complexed with glycosylated poly(L-lysine). *Biochem Pharmacol* 53 (6):887–95.

Manickam, D. S., and D. Oupický. 2006. Multiblock reducible copolypeptides containing histidine-rich and nuclear localization sequences for gene delivery. *Bioconjug Chem* 17 (6):1395–403.

Matsumura, Y., and H. Maeda. 1986. A new concept for macromolecular therapeutics in cancer chemotherapy: Mechanism of tumoritropic accumulation of proteins and the antitumor agent smancs. *Cancer Res* 46 (12 Pt 1):6387–92.

McClements, D. J. 2012. Nanoemulsions versus microemulsions: Terminology, differences, and similarities. *Soft Matter* 8 (6):1719–29.

McClements, D. J., and J. Rao. 2011. Food-grade nanoemulsions: Formulation, fabrication, properties, performance, biological fate, and potential toxicity. *Crit Rev Food Sci Nutr* 51 (4):285–330.

Mellman, I., and R. M. Steinman. 2001. Dendritic cells: Specialized and regulated antigen processing machines. *Cell* 106 (3):255–8.

Midoux, P., A. Kichler, V. Boutin, J. C. Maurizot, and M. Monsigny. 1998. Membrane permeabilization and efficient gene transfer by a peptide containing several histidines. *Bioconjug Chem* 9 (2):260–7.

Mislick, K. A., J. D. Baldeschwieler, J. F. Kayyem, and T. J. Meade. 1995. Transfection of folate-polylysine DNA complexes: Evidence for lysosomal delivery. *Bioconjug Chem* 6 (5):512–5.

Mitri, K., C. Vauthier, N. Huang, A. Menas, C. Ringard-Lefebvre, C. Anselmi, M. Stambouli, V. Rosilio, J.-J. Vachon, and K. Bouchemal. 2012. Scale-up of nanoemulsion produced by emulsification and solvent diffusion. *J Pharm Sci* 101 (11):4240–7.

Modery-Pawlowski, C. L., A. M. Master, V. Pan, G. P. Howard, and A. S. Gupta. 2013. A platelet-mimetic paradigm for metastasis-targeted nanomedicine platforms. *Biomacromolecules* 14 (3):910–9.

Mogensen, C. E. 1968. The glomerular permeability determined by dextran clearance using Sephadex gel filtration. *Scand J Clin Lab Invest* 21 (1):77–82.

Morgan, E. T. 2009. Impact of infectious and inflammatory disease on cytochrome P450–mediated drug metabolism and pharmacokinetics. *Clin Pharmacol Ther* 85 (4):434–8.

Mori, A., A. L. Klibanov, V. P. Torchilin, and L. Huang. 1991. Influence of the steric barrier activity of amphipathic poly (ethyleneglycol) and ganglioside GM1 on the circulation time of liposomes and on the target binding of immunoliposomes in vivo. *FEBS Lett* 284 (2):263–6.

Muller, K., T. Nahde, A. Fahr, R. Muller, and S. Brusselbach. 2001. Highly efficient transduction of endothelial cells by targeted artificial virus-like particles. *Cancer Gene Ther* 8 (2):107–17.

Müller, R. H., D. Harden, and C. M. Keck. 2012. Development of industrially feasible concentrated 30% and 40% nanoemulsions for intravenous drug delivery. *Drug Dev Ind Pharm* 38 (4):420–30.

Müller, R. H., M. Radtke, and S. A. Wissing. 2002. Solid lipid nanoparticles (SLN) and nanostructured lipid carriers (NLC) in cosmetic and dermatological preparations. *Adv Drug Deliv Rev* 54:S131–55.

Murphy, M. P., and R. A. J. Smith. 2000. Drug delivery to mitochondria: The key to mitochondrial medicine. *Adv Drug Deliv Rev* 41 (2):235–50.

Niidome, T., and L. Huang. 2002. Gene therapy progress and prospects: Nonviral vectors. *Gene Ther* 9 (24):1647–52.

O'Brien, M. E., N. Wigler, M. Inbar, R. Rosso, E. Grischke, A. Santoro, R. Catane et al. 2004. Reduced cardiotoxicity and comparable efficacy in a phase III trial of pegylated liposomal doxorubicin HCl (CAELYX/Doxil) versus conventional doxorubicin for first-line treatment of metastatic breast cancer. *Ann Oncol* 15 (3):440–9.

O'Hanlon, C. E., K. G. Amede, R. O. Meredith, and J. M. Janjic. 2012. NIR-labeled perfluoropolyether nanoemulsions for drug delivery and imaging. *J Fluorine Chem* 137:27–33.

O'Neill, M. M., C. A. Kennedy, R. W. Barton, and R. J. Tatake. 2001. Receptor-mediated gene delivery to human peripheral blood mononuclear cells using anti-CD3 antibody coupled to polyethylenimine. *Gene Ther* 8 (5):362–8.

Önyüksel, H., H. Ikezaki, M. Patel, X.-P. Gao, and I. Rubinstein. 1999. A novel formulation of VIP in sterically stabilized micelles amplifies vasodilation in vivo. *Pharm Res* 16 (1):155–60.

Oupický, D., A. L. Parker, and L. W. Seymour. 2002. Laterally stabilized complexes of DNA with linear reducible polycations: Strategy for triggered intracellular activation of DNA delivery vectors. *J Am Chem Soc* 124 (1):8–9.

Pack, D. W., D. Putnam, and R. Langer. 2000. Design of imidazole-containing endosomolytic biopolymers for gene delivery. *Biotechnol Bioeng* 67 (2):217–23.

Pallagi, E., R. Ambrus, P. Szabó-Révész, and I. Csóka. 2015. Adaptation of the quality by design concept in early pharmaceutical development of an intranasal nanosized formulation. *Int J Pharm* 491 (1):384–92.

Panyam, J., and V. Labhasetwar. 2003. Biodegradable nanoparticles for drug and gene delivery to cells and tissue. *Adv Drug Deliv Rev* 55 (3):329–47.

Parodi, A., N. Quattrocchi, A. L. Van De Ven, C. Chiappini, M. Evangelopoulos, J. O. Martinez, B. S. Brown, S. Z. Khaled, I. K. Yazdi, and M. V. Enzo. 2013. Synthetic nanoparticles functionalized with biomimetic leukocyte membranes possess cell-like functions. *Nature Nanotechnol* 8 (1):61–68.

Patel, J. D., R. O'Carra, J. Jones, J. G. Woodward, and R. J. Mumper. 2007. Preparation and characterization of nickel nanoparticles for binding to his-tag proteins and antigens. *Pharm Res* 24 (2):343–52.

Patel, S. K., J. Williams, and J. M. Janjic. 2013a. Cell Labeling for 19F MRI: New and improved approach to perfluorocarbon nanoemulsion design. *Biosensors* 3 (3):341–59.

Patel, S. K., Y. Zhang, J. A. Pollock, and J. M. Janjic. 2013b. Cyclooxygenase-2 inhibiting perfluoropoly (ethylene glycol) ether theranostic nanoemulsions—In vitro study. *PloS One* 8 (2):e55802.

Patil, H., X. Feng, X. Ye, S. Majumdar, and M. A. Repka. 2015. Continuous production of fenofibrate solid lipid nanoparticles by hot-melt extrusion technology: A systematic study based on a quality by design approach. *AAPS J* 17 (1):194–205.

Pichon, C., C. Goncalves, and P. Midoux. 2001. Histidine-rich peptides and polymers for nucleic acids delivery. *Adv Drug Deliv Rev* 53 (1):75–94.

Pierschbacher, M. D., and E. Ruoslahti. 1984. Cell attachment activity of fibronectin can be duplicated by small synthetic fragments of the molecule. *Nature* 309 (5963):30–3.

Potineni, A., D. M. Lynn, R. Langer, and M. M. Amiji. 2003. Poly (ethylene oxide)-modified poly (β-amino ester) nanoparticles as a pH-sensitive biodegradable system for paclitaxel delivery. *J Control Release* 86 (2):223–34.

Pouton, C. W., P. Lucas, B. J. Thomas, A. N. Uduehi, D. A. Milroy, and S. H. Moss. 1998. Polycation-DNA complexes for gene delivery: A comparison of the biopharmaceutical properties of cationic polypeptides and cationic lipids. *J Control Release* 53 (1–3):289–99.

Putnam, D., C. A. Gentry, D. W. Pack, and R. Langer. 2001. Polymer-based gene delivery with low cytotoxicity by a unique balance of side-chain termini. *Proc Natl Acad Sci USA* 98 (3):1200–5.

Rajpoot, P., K. Pathak, and V. Bali. 2011. Therapeutic applications of nanoemulsion based drug delivery systems: A review of patents in last two decades. *Recent Pat Drug Deliv Formul* 5 (2):163–72.

Ramsay, E., and M. Gumbleton. 2002. Polylysine and polyornithine gene transfer complexes: A study of complex stability and cellular uptake as a basis for their differential in-vitro transfection efficiency. *J Drug Target* 10 (1):1–9.

Ramsay, E., J. Hadgraft, J. Birchall, and M. Gumbleton. 2000. Examination of the biophysical interaction between plasmid DNA and the polycations, polylysine and polyornithine, as a basis for their differential gene transfection in-vitro. *Int J Pharm* 210 (1–2):97–107.

Rapoport, N., K.-H. Nam, R. Gupta, Z. Gao, P. Mohan, A. Payne, N. Todd, X. Liu, T. Kim, and J. Shea. 2011. Ultrasound-mediated tumor imaging and nanotherapy using drug loaded, block copolymer stabilized perfluorocarbon nanoemulsions. *J Control Release* 153 (1):4–15.

Rybak, S. L., F. Lanni, R. F. Murphy. 1997. Theoretical considerations on the role of membrane potential in the regulation of endosomal pH. *Biophys J* 73:674–87.

Safra, T., F. Muggia, S. Jeffers, D. D. Tsao-Wei, S. Groshen, O. Lyass, R. Henderson, G. Berry, and A. Gabizon. 2000. Pegylated liposomal doxorubicin (Doxil): Reduced clinical cardiotoxicity in patients reaching or exceeding cumulative doses of 500 mg/m². *Ann Oncol* 11 (8):1029–33.

Sahay, G., D. Y. Alakhova, and A. V. Kabanov. 2010. Endocytosis of nanomedicines. *J Control Release* 145 (3):182–95.

Salvador-Morales, C., and R. B. Sim. 2013. Complement activation. In *Handbook of Immunological Properties of Engineered Nanomaterials*, ed. M. A. Dobrovolskaia and S. E. McNeil, 357–84. Singapore: World Scientific Publishing.

Salvati, A., A. S. Pitek, M. P. Monopoli, K. Prapainop, F. B. Bombelli, D. R. Hristov, P. M. Kelly, C. Aberg, E. Mahon, and K. A. Dawson. 2013. Transferrin-functionalized nanoparticles lose their targeting capabilities when a biomolecule corona adsorbs on the surface. *Nat Nanotechnol* 8 (2):137–43.

Sarker, D. K. 2005. Engineering of nanoemulsions for drug delivery. *Curr Drug Deliv* 2 (4):297–310.

Satalkar, P., B. S. Elger, P. Hunziker, and D. Shaw. 2016. Challenges of clinical translation in nanomedicine: A qualitative study. *Nanomedicine* 12 (4):893–900.

Satulovsky, J., M. A. Carignano, and I. Szleifer. 2000. Kinetic and thermodynamic control of protein adsorption. *Proc Natl Acad Sci USA* 97 (16):9037–41.

Schneider, H., R. P. Harbottle, Y. Yokosaki, P. Jost, and C. Coutelle. 1999. Targeted gene delivery into alpha9beta1-integrin-displaying cells by a synthetic peptide. *FEBS Lett* 458 (3):329–32.

Schneider, H., R. P. Harbottle, Y. Yokosaki, J. Kunde, D. Sheppard, and C. Coutelle. 1998. A novel peptide, PLAEIDGIELTY, for the targeting of alpha9beta1-integrins. *FEBS Lett* 429 (3):269–73.

Schwab, M., and U. Klotz. 2001. Pharmacokinetic considerations in the treatment of inflammatory bowel disease. *Clin Pharmacokinet* 40 (10):723–51.

Sen, L., Y. S. Hong, H. Luo, G. Cui, and H. Laks. 2001. Efficiency, efficacy, and adverse effects of adenovirus vs. liposome-mediated gene therapy in cardiac allografts. *Am J Physiol Heart Circ Physiol* 281 (3):H1433–41.

Shakeel, F., S. Shafiq, N. Haq, F. K. Alanazi, and I. A. Alsarra. 2012. Nanoemulsions as potential vehicles for transdermal and dermal delivery of hydrophobic compounds: An overview. *Expert Opin Drug Deliv* 9 (8):953–74.

Sharma, A. K., L. Zhang, S. Li, D. L. Kelly, V. Y. Alakhov, E. V. Batrakova, and A. V. Kabanov. 2008. Prevention of MDR development in leukemia cells by micelle-forming polymeric surfactant. *J Control Release* 131 (3):220–7.

Shenoy, D., S. Little, R. Langer, and M. Amiji. 2005. Poly (ethylene oxide)-modified poly (β-amino ester) nanoparticles as a pH-sensitive system for tumor-targeted delivery of hydrophobic drugs. 1. In vitro evaluations. *Mol Pharm* 2 (5):357–66.

Sindrilaru, A., T. Peters, S. Wieschalka, C. Baican, A. Baican, H. Peter, A. Hainzl, S. Schatz, Y. Qi, and A. Schlecht. 2011. An unrestrained proinflammatory M1 macrophage population induced by iron impairs wound healing in humans and mice. *J Clin Invest* 121 (3):985–7.

Soman, N. R., S. L. Baldwin, G. Hu, J. N. Marsh, G. M. Lanza, J. E. Heuser, J. M. Arbeit, S. A. Wickline, and P. H. Schlesinger. 2009. Molecularly targeted nanocarriers deliver the cytolytic peptide melittin specifically to tumor cells in mice, reducing tumor growth. *J Clin Invest* 119 (9):2830–42.

Soman, N. R., G. M. Lanza, J. M. Heuser, P. H. Schlesinger, and S. A. Wickline. 2008. Synthesis and characterization of stable fluorocarbon nanostructures as drug delivery vehicles for cytolytic peptides. *Nano Lett* 8 (4):1131–6.

Sonawane, N. D., F. C. Szoka Jr., and A. S. Verkman. 2003. Chloride accumulation and swelling in endosomes enhances DNA transfer by polyamine-DNA polyplexes. *J Biol Chem* 278 (45):44826–31.

Sorgi, F. L., S. Bhattacharya, and L. Huang. 1997. Protamine sulfate enhances lipid-mediated gene transfer. *Gene Ther* 4 (9):961–8.

Sosnowski, B. A., D. L. Gu, M. D'Andrea, J. Doukas, and G. F. Pierce. 1999. FGF2-targeted adenoviral vectors for systemic and local disease. *Curr Opin Mol Ther* 1 (5):573–9.

Stoll, G., T. Basse-Lüsebrink, G. Weise, and P. Jakob. 2012. Visualization of inflammation using 19F-magnetic resonance imaging and perfluorocarbons. *Wiley Interdiscip Rev Nanomed Nanobiotechnol* 4 (4):438–47.

Strebhardt, K., and A. Ullrich. 2008. Paul Ehrlich's magic bullet concept: 100 years of progress. *Nat Rev Cancer* 8 (6):473–80.

Swartzlander, M. D., C. A. Barnes, A. K. Blakney, J. L. Kaar, T. R. Kyriakides, and S. J. Bryant. 2015. Linking the foreign body response and protein adsorption to PEG-based hydrogels using proteomics. *Biomaterials* 41:26–36.

Szebeni, J., L. Baranyi, S. Savay, J. Milosevits, M. Bodo, R. Bunger, and C. R. Alving. 2003. The interaction of liposomes with the complement system: In vitro and in vivo assays. *Methods Enzymol* 373:136–54.

Tenzer, S., D. Docter, S. Rosfa, A. Wlodarski, J. Kuharev, A. Rekik, S. K. Knauer et al. 2011. Nanoparticle size is a critical physicochemical determinant of the human blood plasma corona: A comprehensive quantitative proteomic analysis. *ACS Nano* 5 (9):7155–67.

Thurston, G., J. W. McLean, M. Rizen, P. Baluk, A. Haskell, T. J. Murphy, D. Hanahan, and D. M. McDonald. 1998. Cationic liposomes target angiogenic endothelial cells in tumors and chronic inflammation in mice. *J Clin Invest* 101 (7):1401–13.

Toda, M., Y. Arima, and H. Iwata. 2010. Complement activation on degraded polyethylene glycol-covered surface. *Acta Biomater* 6 (7):2642–9.

Torchilin, V. 2009. Multifunctional and stimuli-sensitive pharmaceutical nanocarriers. *Eur J Pharm Biopharm* 71 (3):431–44.

Torchilin, V. P., A. L. Klibanov, L. Huang, S. O'Donnell, N. D. Nossiff, and B. A. Khaw. 1992. Targeted accumulation of polyethylene glycol-coated immunoliposomes in infarcted rabbit myocardium. *FASEB J* 6 (9):2716–9.

Torchilin, V. P., T. S. Levchenko, R. Rammohan, N. Volodina, B. Papahadjopoulos-Sternberg, and G. G. M. D'Souza. 2003. Cell transfection in vitro and in vivo with nontoxic TAT peptide–liposome–DNA complexes. *Proc Natl Acad Sci USA* 100 (4):1972–7.

Torchilin, V. P., V. G. Omelyanenko, M. I. Papisov, A. A. Bogdanov, V. S. Trubetskoy, J. N. Herron, and C. A. Gentry. 1994. Poly (ethylene glycol) on the liposome surface: On the mechanism of polymer-coated liposome longevity. *Biochim Biophys Acta* 1195 (1):11–20.

Torchilin, V. P., R. Rammohan, V. Weissig, and T. S. Levchenko. 2001. TAT peptide on the surface of liposomes affords their efficient intracellular delivery even at low temperature and in the presence of metabolic inhibitors. *Proc Natl Acad Sci USA* 98 (15):8786–91.

Uike, H., R. Sakakibara, K. Iwanaga, M. Ide, and M. Ishiguro. 1998. Efficiency of targeted gene delivery of ligand-poly-L-lysine hybrids with different crosslinks. *Biosci Biotechnol Biochem* 62 (6):1247–8.

Unsworth, L. D., H. Sheardown, and J. L. Brash. 2008. Protein-resistant poly (ethylene oxide)-grafted surfaces: Chain density-dependent multiple mechanisms of action. *Langmuir* 24 (5):1924–9.

van den Berg, A., and S. F. Dowdy. 2011. Protein transduction domain delivery of therapeutic macromolecules. *Curr Opin Biotechnol* 22 (6):888–93.

Van Deventer, S. J., C. O. Elson, and R. N. Fedorak. 1997. Multiple doses of intravenous interleukin 10 in steroid-refractory Crohn's disease. Crohn's Disease Study Group. *Gastroenterology* 113 (2):383–9.

van Vlerken, L. E., Z. Duan, M. V. Seiden, and M. M. Amiji. 2007a. Modulation of intracellular ceramide using polymeric nanoparticles to overcome multidrug resistance in cancer. *Cancer Res* 67 (10):4843–50.

van Vlerken, L. E., T. K. Vyas, and M. M. Amiji. 2007b. Poly (ethylene glycol)-modified nanocarriers for tumor-targeted and intracellular delivery. *Pharm Res* 24 (8):1405–14.

Venditto, V. J., and F. C. Szoka Jr. 2013. Cancer nanomedicines: So many papers and so few drugs! *Adv Drug Deliv Rev* 65 (1):80–8.

Vinogradov, S., E. Batrakova, S. Li, and A. Kabanov. 1999. Polyion complex micelles with protein-modified corona for receptor-mediated delivery of oligonucleotides into cells. *Bioconjug Chem* 10 (5):851–60.

Walkey, C. D., J. B. Olsen, H. Guo, A. Emili, and W. C. W. Chan. 2012. Nanoparticle size and surface chemistry determine serum protein adsorption and macrophage uptake. *J Am Chem Soc* 134 (4):2139–47.

Wang, Y., S. Gao, W.-H. Ye, H. S. Yoon, and Y.-Y. Yang. 2006. Co-delivery of drugs and DNA from cationic core–shell nanoparticles self-assembled from a biodegradable copolymer. *Nat Mater* 5 (10):791–6.

Weise, G., T. C. Basse-Luesebrink, C. Wessig, P. M. Jakob, and G. Stoll. 2011. In vivo imaging of inflammation in the peripheral nervous system by 19 F MRI. *Exp Neurol* 229 (2):494–501.

Weiss, R. B., R. C. Donehower, P. H. Wiernik, T. Ohnuma, R. J. Gralla, D. L. Trump, J. R. Baker, D. A. Van Echo, D. D. Von Hoff, and B. Leyland-Jones. 1990. Hypersensitivity reactions from taxol. *J Clin Oncol* 8 (7):1263–8.

Weissig, V., S. V. Boddapati, S.-M. Cheng, and G. G. M. D'souza. 2006. Liposomes and liposome-like vesicles for drug and DNA delivery to mitochondria. *J Liposome Res* 16 (3):249–64.

Wen, Y., H. R. Kolonich, K. M. Kruszewski, N. Giannoukakis, E. S. Gawalt, and W. S. Meng. 2013. Retaining antibodies in tumors with a self-assembling injectable system. *Mol Pharm* 10 (3):1035–44.

Wen, Y., and W. S. Meng. 2014. Recent in vivo evidences of particle-based delivery of small-interfering RNA (siRNA) into solid tumors. *J Pharm Innov* 9 (2):158–73.

Wen, Y., S. L. Roudebush, G. A. Buckholtz, T. R. Goehring, N. Giannoukakis, E. S. Gawalt, and W. S. Meng. 2014. Coassembly of amphiphilic peptide EAK16-II with histidinylated analogues and implications for functionalization of beta-sheet fibrils in vivo. *Biomaterials* 35 (19):5196–205.

Wetzler, C., H. Kämpfer, B. Stallmeyer, J. Pfeilschifter, and S. Frank. 2000. Large and sustained induction of chemokines during impaired wound healing in the genetically diabetic mouse: Prolonged persistence of neutrophils and macrophages during the late phase of repair. *J Invest Dermatol* 115 (2):245–53.

Wicki, A., R. Ritschard, U. Loesch, S. Deuster, C. Rochlitz, and C. Mamot. 2015. Large-scale manufacturing of GMP-compliant anti-EGFR targeted nanocarriers: Production of doxorubicin-loaded anti-EGFR-immunoliposomes for a first-in-man clinical trial. *Int J Pharm* 484 (1):8–15.

Wightman, L., E. Patzelt, E. Wagner, and R. Kircheis. 1999. Development of transferrin-polycation/DNA based vectors for gene delivery to melanoma cells. *J Drug Target* 7 (4):293–303.

Winblade, N. D., I. D. Nikolic, A. S. Hoffman, and J. A. Hubbell. 2000. Blocking adhesion to cell and tissue surfaces by the chemisorption of a poly-L-lysine-graft-(poly(ethylene glycol); phenylboronic acid) copolymer. *Biomacromolecules* 1 (4):523–33.

Xiao, K., Y. Li, J. Luo, J. S. Lee, W. Xiao, A. M. Gonik, R. G. Agarwal, and K. S. Lam. 2011. The effect of surface charge on in vivo biodistribution of PEG-oligocholic acid based micellar nanoparticles. *Biomaterials* 32 (13):3435–46.

Xiong, X. B., and A. Lavasanifar. 2011. Traceable multifunctional micellar nanocarriers for cancer-targeted co-delivery of MDR-1 siRNA and doxorubicin. *ACS Nano* 5 (6):5202–13.

Yang, J. P., and L. Huang. 1997. Overcoming the inhibitory effect of serum on lipofection by increasing the charge ratio of cationic liposome to DNA. *Gene Ther* 4 (9):950–60.

Yang, Q., S. W. Jones, C. L. Parker, W. C. Zamboni, J. E. Bear, and S. K. Lai. 2014. Evading immune cell uptake and clearance requires PEG grafting at densities substantially exceeding the minimum for brush conformation. *Mol Pharm* 11 (4):1250–8.

Yu, K. C., W. Chen, and A. D. Cooper. 2001. LDL receptor-related protein mediates cell-surface clustering and hepatic sequestration of chylomicron remnants in LDLR-deficient mice. *J Clin Invest* 107 (11):1387–94.

Yu, L. X., G. L. Amidon, J. E. Polli, H. Zhao, M. U. Mehta, D. P. Conner, V. P. Shah, L. J. Lesko, M.-L. Chen, and V. H. L. Lee. 2002. Biopharmaceutics classification system: The scientific basis for biowaiver extensions. *Pharm Res* 19 (7):921–5.

Yuan, F., M. Dellian, D. Fukumura, M. Leunig, D. A. Berk, V. P. Torchilin, and R. K. Jain. 1995. Vascular permeability in a human tumor xenograft: Molecular size dependence and cutoff size. *Cancer Res* 55 (17):3752–6.

Zhao, G., M. L. Usui, S. I. Lippman, G. A. James, P. S. Stewart, P. Fleckman, and J. E. Olerud. 2013. Biofilms and inflammation in chronic wounds. *Adv Wound Care* 2 (7):389–99.

Zheng, Y., J. R. Kovacs, E. S. Gawalt, H. Shen, and W. S. Meng. 2006. Characterization of particles fabricated with poly(D,L-lactic-co-glycolic acid) and an ornithine-histidine peptide as carriers of oligodeoxynucleotide for delivery into primary dendritic cells. *J Biomater Sci Polym Ed* 17 (12):1389–403.

Zhou, J., J. W. Yockman, S. W. Kim, and S. E. Kern. 2007. Intracellular kinetics of non-viral gene delivery using polyethylenimine carriers. *Pharm Res* 24 (6):1079–87.

Zhou, X., and L. Huang. 1994. DNA transfection mediated by cationic liposomes containing lipopolylysine: Characterization and mechanism of action. *Biochim Biophys Acta* 1189 (2):195–203.

CHAPTER THREE

Nanotoxicity

Angie S. Morris and Aliasger K. Salem

CONTENTS

3.1 INTRODUCTION

Nanomaterials, having long been used throughout history, are recognized for possessing unique material characteristics and have inspired scientists to gain a better understanding of their properties and potential applications. One of the most well-known nanomaterials, colloidal gold, has been used in stained glass windows that date back to the fourth century. During the 1850s, Michael Faraday sparked the scientific community's interest in nanometer-sized particles after delivering a lecture entitled "Experimental Relations of Gold (and Other Metals) to Light," where he discussed his findings about the absorption or transmission properties of colloidal gold suspensions in relation to the size of suspended particles (Faraday 1857). Although Faraday did not call them "nanoparticles," he was one of the first experimentalists to explore the field we now refer to as "nanotechnology." However, interest in nanomaterials did not really take off until Richard Feynman's lecture in 1959 to the American Physical Society at the California Institute of Technology entitled "There's Plenty of Room at the Bottom" (Feynman 1960). Feynman discussed the advantages of manipulating material (even single atoms) on a smaller scale than what was currently achievable at the time and challenged scientists to advance the field of what is now known as nanotechnology. From the 1980s onward, nanotechnology has progressed at an extremely fast pace and nanomaterials have been developed for many industrial and clinical applications.

Nanomaterials are defined as materials in which one or more dimension is in the nanometer size range. Some classify a nanomaterial as having a size of less than 1000 nm, although it is common for nanomaterials to be categorized as 1–100 nm. The International Union of Pure and Applied Chemistry (IUPAC) defines a nanoparticle as a "particle of any shape with dimensions in the 1×10^{-9} and 1×10^{-7} m range" (Vert et al. 2012). Sometimes the word *nanoparticle* refers to a material that displays properties different from those of the bulk material, and for most materials, this phenomenon occurs at sizes less than 100 nm. Other times, *nano* is used as a prefix for structures with dimensions smaller than 500 nm. The latter definition will apply in this chapter.

Nanotoxicology is the study of the toxic effects caused by or associated with nanomaterials on human health and the environment. Because of the rapid growth of nanoscience, both commercially and industrially, over the past few decades, there has been a large focus on the safety issues accompanying nanosized materials. This is reflected in the drastic increase in the number of nanotoxicity-related articles published over the past 15 years (Figure 3.1). Besides the use of nanomaterials in scientific research, they are increasingly being used in many commercially available consumer products, such as toothpastes, shampoos, paints, sunscreens, cosmetics, food products, electronics, paper, and fabrics (Benn and Westerhoff 2008; Jin et al. 2008; Auffan et al. 2010; Benn et al. 2010; Shukla et al. 2011; Weir et al. 2012; Keller et al. 2014). Clearly, there is a large amount of human exposure to nanomaterials from consumer products alone, not to mention the rapidly developing nanomaterial-based biomedical applications, and

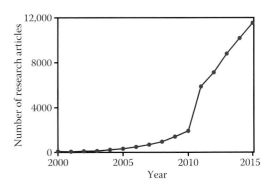

Figure 3.1. Number of research articles published on nanotoxicity from 2000 to 2015.

as a result, concerns have been raised about the short- and long-term implications of nanomaterial exposure on both human health and the environment. While these materials are appealing for many novel applications because of their unique physicochemical properties, which are drastically different from those of the bulk material (such as a high ratio of surface area to volume and surface reactivity), it is unclear how these properties impact human health and ecology.

3.2 NANOTECHNOLOGY-BASED MEDICINE

Over the last few decades, research and development of nanomaterial-based medicines (termed "nanomedicine") has led to the formation of hundreds of start-up companies aimed at commercializing novel therapies or drug delivery systems (Wagner et al. 2006; Etheridge et al. 2013). Although nanomaterials are being implemented in diagnostic and imaging devices among several other medical applications, drug delivery dominates the nanomedicine market. The incorporation of nanomaterials into drug delivery systems has the potential to improve medicine by allowing for controlled release of a drug in a sustained manner and the ability to target the drug to a specific organ or tissue. Many researchers are designing pH-sensitive materials in which the release profile of the drug from the nanocarrier is directly dependent on environmental pH (Jin et al. 2012; Li et al. 2014). These types of materials are commonly applied to cancer therapy because of the slightly acidic (pH 6.4–6.9) microenvironment of solid tumors compared with the neutral microenvironment of heathy cells (Ding and

Ma 2013; Nogueira et al. 2013; Liu et al. 2014). Nanomaterials also help overcome limitations in many pharmaceutical applications, one being the loading and delivery of poorly soluble drugs in an efficient manner (Parveen et al. 2012). Site-specific therapies can also be achieved by altering the physicochemical properties of nanomaterials in order to target drugs to certain cells (e.g., tumor cells) or tissues and to minimize the side effects of drugs to healthy tissues (Pan et al. 2012; Mackowiak et al. 2013; Xiao et al. 2014).

There are clear advantages to employing nanomaterials in biomedical applications; however, with the surge of nanotechnology in the past few decades, it is critical to evaluate the health effects of direct and indirect exposure to engineered nanomaterials. The study of nanotoxicity is vital to address these concerns and to aid in the development of safe, biocompatible nanomaterials.

3.3 MECHANISMS OF NANOMATERIAL TOXICITY

3.3.1 Oxidative Stress

There are many different mechanisms by which nanomaterials can cause toxicity in physiological systems. One of the major routes is through the generation of reactive oxygen species (ROS). ROS are chemically reactive compounds produced during biological processes such as mitochondrial respiration and inflammation (Manke et al. 2013). These compounds (such as hydrogen peroxide, superoxide, and hydroxyl radicals) are by-products of the metabolism of oxygen. ROS occur naturally in biological systems and are involved in cell signaling, as well as maintaining homeostasis. When molecular oxygen is reduced by the enzyme nicotinamide adenine dinucleotide phosphate (NADPH) oxidase, superoxide anions are formed. Although cells are equipped to be able to function with small amounts of ROS, too much can lead to a state of oxidative stress (Fu et al. 2014). Nanomaterials are known to cause oxidative stress as a mechanism of toxicity. This increase in ROS can be irreversibly harmful to cells, causing membrane damage, protein oxidation, DNA damage, mitochondria dysfunction, apoptosis, and genotoxicity.

There are many *in vitro* studies confirming that cells exposed to nanomaterials generate a higher level of ROS than healthy cells. Nonporous silica nanoparticles (50 nm in diameter) have been shown to induce significantly higher levels of ROS in RAW264.7 macrophages than untreated cells (Park and Park 2009; Lehman et al. 2015). Copper oxide nanoparticles increase the amount of both intracellular and mitochondrial ROS in A549 lung cells (Wongrakpanich et al. 2016). Multiwalled (MW) carbon nanotubes, both short (0.1–5 nm in length) and long (0.1–20 nm in length), titanium dioxide nanoparticles, and aluminum oxide nanoparticles caused higher levels of ROS in rat kidney cells (NRK-52E) than in control cells (Barillet et al. 2010). *In vivo* studies also support the theory that nanomaterial-induced ROS are damaging to physiological systems. The same silica nanoparticles used in the study by Lehman et al. were evaluated in a murine model of inhalation (Morris et al. 2016). A significant amount of ROS were observed in the lung cells of mice that were administered 0.5 mg of nanoparticles intratracheally. In this case, higher levels of ROS directly correlated to other markers of toxicity, such as reduced body weight in mice, membrane damage, and immune cell recruitment to the lungs, although the specific role of ROS on the inflammatory response is not clear. There is evidence suggesting that increased ROS can lead to membrane damage via lipid peroxidation and promote pro-inflammatory signaling (Nel et al. 2006; Nishanth et al. 2011; Mendoza et al. 2014).

3.3.2 DNA Damage and Genotoxicity

DNA damage is another important parameter that can be used to assess nanotoxicity. As mentioned earlier, DNA damage can occur indirectly through the action of ROS; however, nanomaterials can also physically damage DNA upon entering the nucleus. Nanomaterials can enter the nucleus via nuclear pores or by being entrapped in the nucleus during mitosis, both of which can lead to DNA damage (Singh et al. 2009). Determining the potential of a nanomaterial to promote DNA damage is a key aspect of nanotoxicity, as this damage can lead to genotoxicity (generation of mutations), which may ultimately result in cancer. Multiple studies have demonstrated that silver nanoparticles can cause damage to DNA both *in vitro* and *in vivo* (AshaRani et al. 2009; Ghosh et al. 2012). Copper oxide nanoparticles were also shown to have genotoxic effects in A549 lung cells as indicated by elevated concentrations of

DNA repair proteins, RAD51 and MSH2 (Ahamed et al. 2010). Other nanomaterials that have been linked to genotoxicity include cobalt nanoparticles, fullerenes, and zinc oxide nanoparticles, although for these and several other types of materials, there are conflicting results published in the literature about their genotoxic potential (Dufour et al. 2006; Colognato et al. 2008; Xu et al. 2009; Singh et al. 2010).

3.3.3 Mitochondrial Dysfunction

Loss of mitochondrial function is another cause of nanotoxicity since mitochondria are involved in the production of ATP and provide cells with the energy they need to survive and maintain biological functions. Excess ROS generated during oxidative stress is thought to damage mitochondria by altering the mitochondrial membrane potential and permeability (Li et al. 2003). Yu et al. (2013) reported that zinc oxide nanorods (15 nm in width and 82 nm in length) were cytotoxic to normal mouse skin epidermal cells (JB6 Cl 41-5a) and induced autophagy by means of excess ROS and mitochondrial damage. Other examples of materials that disrupt mitochondrial function *in vitro* and/or *in vivo* include silica nanoparticles (Sun et al. 2011), silver nanoparticles (Teodoro et al. 2011), and titanium dioxide nanoparticles (Huerta-Garcia et al. 2014).

3.3.4 Inflammation

Inflammation is a natural biological defense mechanism that is triggered by a range of factors, including toxins, the introduction of foreign matter or pathogens, or damaged tissue. Inflammation is a complex process that involves the recruitment of immune cells to the site of injury or infection by pro-inflammatory cytokines (Padmanabhan and Kyriakides 2015). Numerous studies have shown that nanomaterials can cause inflammation when introduced to a physiological system. For example, when nonporous silica nanoparticles were instilled into the lungs of mice, there was an increase in the number of immune cells, such as neutrophils and macrophages, present in the bronchoalveolar lavage (BAL) fluid compared with the control group (Morris et al. 2016). Another study demonstrated that pulmonary exposure of single-walled carbon nanotubes in mice caused an increase in the number of macrophages,

lymphocytes, and neutrophils in the BAL fluid, as well as elevated levels of pro-inflammatory cytokines (tumor necrosis factor [TNF]-α, interleukin [IL]-1β, and transforming growth factor [TGF]-β1) (Shvedova et al. 2005).

3.4 PHYSICOCHEMICAL-RELATED NANOTOXICITY

Since the early 2000s, researchers have recognized the need for extensive nanotoxicity studies of various commonly used nanomaterials. These studies are important not only for developing new biomedical applications that could be implemented in a clinical setting, but also for evaluating the safety of nanosized materials in the workplace, where large-scale synthesis of nanomaterials takes place and knowledge of safe working doses is necessary to keep workers free of harmful side effects (Wang et al. 2015). There are many factors that can contribute to the nanotoxicity of a particular material, including chemical composition, size, shape, dissolution, and surface chemistry (Figure 3.2).

3.4.1 Composition

The chemical composition of nanomaterials used in medicine varies widely depending on the application. For example, magnetic nanoparticles (i.e., iron oxide), gold nanoparticles, and quantum dots are commonly used in biomedical imaging applications as contrast agents (Lee et al. 2007, 2014), whereas silver nanoparticles are often used in medical devices because of their antimicrobial properties (Xiu et al. 2012; Tran et al. 2013). In terms of biochemistry, nanomaterials of various compositions can have very different effects on physiological systems and resultantly very different toxicities. Therefore, when designing systems where there is the potential for human exposure, it is important to consider the inherent toxicity of the material. Metal and metal oxide nanoparticles are present in a wide variety of consumer products, such as sunscreens and cosmetics, as well as actively being developed for biomedical applications; however, there is a large amount of scientific evidence that demonstrates their toxic effects on biological systems (Conde et al. 2012). An article published in 2008 compared the toxicities of several types of metal oxide (CuO, TiO_2, ZnO, $CuZnFe_2O_4$, Fe_3O_4, and Fe_2O_3)

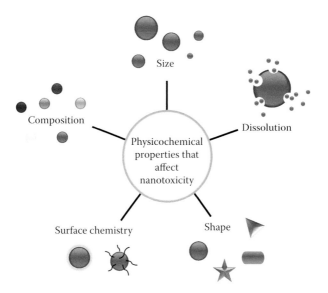

Figure 3.2. Physicochemical properties that play a role in the toxicity of nanomaterial.

nanoparticles, carbon nanoparticles, and carbon nanotubes (Karlsson et al. 2008) on human lung epithelial cells (A549). Of all the materials tested, copper oxide nanoparticles were the most cytotoxic, causing significant amounts of DNA damage and ROS production. Zinc oxide nanoparticles also decreased the cell viability of A549 cells due in part to DNA damage; however, there was no increase in intracellular ROS production. In contrast, both types of iron oxide nanoparticles displayed only small amounts of cytotoxicity and were considered the least harmful to A549 cells. Although many metal nanomaterials promote nanotoxicity, gold nanomaterials are generally considered safe, although their long-term toxic effects are yet to be sufficiently characterized (Soenen et al. 2012). Biodegradable nanomaterials such as poly(lactic-co-glycolic acid) (PLGA) and chitosan tend to be less toxic than inorganic nanomaterials (Anderson et al. 2008; Kohane and Langer 2010).

3.4.2 Size

Size is a major factor that influences the interactions of nanomaterials with cells. Many researchers have reported size-dependent nanotoxicity of materials of the same composition (Carlson et al. 2008; Passagne et al. 2012; Wongrakpanich et al. 2016). A number of factors can be responsible for such size-related differences in toxicity, including

the well-described example where cellular uptake rate is a primary influence (Gatoo et al. 2014). In general, nanoparticles with sizes greater than 500 nm enter cells via phagocytosis, primarily mediated by specialized phagocytes, whereas smaller particles are internalized by receptor-mediated or nonspecific endocytosis (Alkilany et al. 2010). Chithrani et al. (2006) investigated the cellular uptake of gold nanoparticles of various sizes (14–100 nm) in human epithelial (HeLa) cells and observed that maximum uptake was observed for 50 nm nanoparticles.

Gold nanoparticles are considered relatively nontoxic; however, there are numerous published results suggesting that the size of gold nanoparticles has an impact on the nanotoxicity. For example, Coradeghini et al. (2013) reported significant differences in the biological response of mouse fibroblast cells when treated with differently sized gold nanoparticles. Specifically, 5 nm gold nanoparticles inhibited colony formation in mouse fibroblast cells to a greater extent than 15 nm gold particles at concentrations above 50 µM. The same researchers also concluded that 5 nm gold nanoparticles had a greater effect on cell morphology than the larger 15 nm nanoparticles. One study published by Zhang et al. (2011) showed that the trend of smaller nanoparticles being more toxic in vitro does not always translate to in vivo models but reinforces the fact that size is an important factor when trying to minimize toxicity. This group evaluated several different sizes (5, 10, 30, and 60 nm) of polyethylene glycol (PEG)–coated gold nanoparticles on toxicity in mice. The results indicated that 10 and 60 nm PEG-coated gold nanoparticles caused a greater amount of liver damage in mice than 5 or 30 nm PEG-coated gold nanoparticles, and therefore were considered more toxic.

Several studies have reported that silica nanoparticles exhibit size-dependent nanotoxicity. Based on these size-dependent studies, the trend was that the smaller the silica nanoparticles, the more toxic they become. In one investigation, amorphous silica nanoparticles ranging from 16 to 335 nm in diameter were evaluated for their cytotoxicity in human endothelial cells (Napierska et al. 2009). The 16 nm particles were the most cytotoxic based on several different assays, and the cytotoxicity decreased as particle sized increased. In another publication, a similar trend was seen with silica nanoparticles of 20 and 100 nm in size, where the smaller nanoparticles were more cytotoxic to two separate kidney cell lines (LLC-PK$_1$ and HK-2) (Passagne et al. 2012).

3.4.3 Shape

As nanotechnology advances, the types of nanomaterials being developed are growing rapidly and diversifying. One physical parameter that is being manipulated is shape. Spherical particles are the most common geometry of nanomaterials used today; however, there are a variety of unconventional shapes emerging, such as rods, tubes, and stars. As an example, mesoporous silica can be synthesized into spheres or rod-like shapes, which can alter the nanotoxicity of the material. Rod-like shapes have higher aspect ratios than spheres. In one study, three types of mesoporous silica nanomaterials with varying aspect ratios were investigated for their effect on nanotoxicity (Huang et al. 2010). It was found that mesoporous nanostructures with larger aspect ratios (nanorods) had a greater impact on the proliferation and overall function of human melanoma cells than spherical nanoparticles.

3.4.4 Dissolution

Depending on the composition of the material, dissolution can play a large role in the nanotoxicity. Nanoparticles made of metals or metal oxides,

such as copper oxide, zinc oxide, and silver, release toxic ions from the surface of the material, and these ions can disrupt normal cellular functions (Wang et al. 2015). A study by Wongrakpanich et al. (2016) reported that both 4 and 24 nm copper oxide nanoparticles release Cu^{2+} ions into the cell culture medium, and after 48 hours, almost 60% of the copper was in the form of Cu^{2+} ions. Studer et al. (2010) investigated the effects of intracellular solubility on cytotoxicity in Chinese hamster oocytes and HeLa cells. To do this, two types of nanoparticles were used: degradable copper oxide nanoparticles and copper oxide nanoparticles that were stabilized with a carbon layer on the surface. In doing so, the direct effect of dissolved metal ions on cytotoxicity could be established. The results indicated that applying a carbon coating to the surface of copper oxide nanoparticles significantly reduced the amount of soluble Cu^{2+} released from the nanoparticles, and the cytotoxicity was reduced in both cell lines for the stabilized carbon-coated nanoparticles, compared with bare copper oxide nanoparticles. Cytotoxicity caused by Cu^{2+} has been linked to an increase in intracellular and mitochondrial ROS, as well as DNA damage (Midander et al. 2009; Wongrakpanich et al. 2016).

3.4.5 Surface Chemistry

Surface chemistry plays a large role in the interaction of nanomaterials with biological systems and, subsequently, the toxicity of nanomaterials. This has been demonstrated through numerous *in vitro* and *in vivo* experiments. For example, surface silanols on silica nanoparticles are considered to be very reactive, and thus contribute in large part to the toxicity of the material. Several studies have shown that functionalization of silica nanoparticles with amine-functional organic molecules, such as aminopropyltriethoxysilane (APTES), significantly increased the cell viability of RAW267.4 macrophage cells by reducing the amount of nanoparticle-mediated ROS (Lehman et al. 2015). The same APTES-functionalized nanoparticles caused less lung inflammation in mice compared to bare silica nanoparticles, as indicated by a lower number of immune cells present in BAL fluid and a lower amount of ROS in lung cells (Morris et al. 2016). In the same experiment, bare silica nanoparticles caused a significant weight drop in mice (a sign of toxicity), whereas the APTES-functionalized silica nanoparticles caused no statistically significant differences in mouse weight compared with untreated controls. As previously mentioned, coating metal oxide nanoparticles with a more inert material such as carbon can act to prevent toxic metal ions from dissociating from the surface and, as a result, reduce nanotoxicity (Studer et al. 2010). Other commonly used coatings to alter the surface properties of nanomaterials include polymers, such as chitosan, dextran, and PEG, and biologically relevant molecules, like proteins and peptides (Yu et al. 2012; Wang et al. 2014).

3.5 METHODS FOR MEASURING NANOTOXICITY

As the potential for human exposure to nanomaterials rises, there is a need to establish reliable methods, both *in vitro* and *in vivo*, for evaluating nanotoxicity. *In vitro* assays are generally used for toxicity screening of chemicals. They are less time-consuming and labor-intensive than *in vivo* models and can provide mechanistic details of nanotoxicity (Kroll et al. 2009). However, *in vitro* analyses do not always correlate with *in vivo* studies and are often too simplistic to accurately predict the toxicological outcomes in animal models. Moreover, *in vitro* methods are limited in the depth of information they can provide. For example, they cannot predict the biodistribution or fate of nanomaterials once they are administered to animals. Many researchers employ a combination of *in vitro* and *in vivo* methods in order to get a more comprehensive toxicity profile of the nanomaterial. Listed below are several commonly used methods for evaluating nanotoxicity *in vitro* and *in vivo*.

3.5.1 *In Vitro* Assays

3.5.1.1 *Cell Viability*

In vitro cell viability assays are among the most commonly used techniques for evaluating nanotoxicity. 3-(4,5-Dimethylthiazol-2-yl)-2,5-diphenyltetrazolium bromide (MTT) and (3-(4,5-dimethylthiazol-2-yl)-5-(3-carboxymethoxyphenyl)-2-(4-sulfophenyl)-2H-tetrazolium) (MTS) are two widely used reagents to assess the mitochondrial metabolic activity of cells. Both assays are based on the reduction of tetrazolium salts by the enzymes present in metabolically active cells. Upon reduction of the tetrazolium salts, a formazan product is produced, accompanied by a measurable color change. The relative amount of purple formazan product formed can be determined using spectrophotometry. During necrosis or apoptosis, the metabolic activity of cells decreases, resulting in a decrease in formazan production compared to viable cells. Once the linear relationship of formazan production versus number of cells is determined for a specific cell type, then this reaction can be used to evaluate nanotoxicity.

The trypan blue exclusion test can also be used to measure cell viability. This method is based on the fact that viable cells contain intact cell membranes, while dead or dying cells have damaged cell membranes (Strober 2001). During this procedure, a cell suspension is mixed with trypan blue and imaged under an optical microscope. Cells that are stained by trypan blue are considered dead, while viable cells are able to exclude the dye and have a clear cytoplasm. Other dyes that are used in a similar manner are propidium iodide and eosin (Hanks and Wallace 1958; Yeh et al. 1981).

The clonogenic assay is a technique used to determine the ability of cells to form colonies (groups of at least 50 cells) *in vitro* (Franken et al. 2006). For this method, cells are treated with nanomaterials, usually at a variety of concentrations, after which the cells are harvested and seeded into a new culture dish at a low density. The reproductive capability of the cells is evaluated by the number of colonies that form over a given amount of time (anywhere from 10 days to 3 weeks) (Hillegass et al. 2010).

3.5.1.2 *Membrane Damage*

One of the most common methods to detect membrane damage is the lactate dehydrogenase (LDH) assay. LDH is an enzyme present in the cytosol of cells that is released when the cell membrane is damaged (Korzeniewski and Callewaert 1983; Decker and Lohmann-Matthes 1988). The concentration of LDH leaked from the cell can be quantified indirectly using an enzymatic reaction. In this reaction, LDH catalyzes the conversion of pyruvate to lactate. During this process, nicotinamide adenine dinucleotide (NAD^+) is converted into its reduced form (NADH). NADH can then be reacted with a tetrazolium salt to form a colored formazan product that is detected by spectrophotometry. The concentration of formazan present in the cell culture medium is directly proportional to the amount of LDH leakage from the cells.

3.5.1.3 Detection of ROS

Measurement of ROS is frequently used as an indicator of nanotoxicity. Most ROS assays are based on the oxidation of cell-permeable probes such as 2′,7′-dichlorofluorescein diacetate (DCFH-DA) and dihydroethidium (DHE). For example, DCFH-DA is a nonionic compound that can diffuse into cells (LeBel et al. 1992). After entering cells, DCFH-DA can be enzymatically hydrolyzed by esterases into the DCFH form. Then, DCFH can be oxidized by intracellular ROS into a highly fluorescent compound (DCF). The amount of DCF present in cells can be quantified using analytical techniques such as flow cytometry and is directly proportional to the concentration of intracellular ROS. Although DCFH-DA is the most widely used probe for ROS, there can be several factors that can lead to inaccurate results (Zielonka and Kalyanaraman 2008; Kalyanaraman et al. 2012). For example, many researchers claim that DCFH-DA can be used to detect hydrogen peroxide (H_2O_2); however, there are numerous other species that can be involved in the conversion of DCFH-DA to its fluorescent form, including hydroxyl radicals ($^{\bullet}$OH) and redox-active metal ions such as Fe^{2+}. Because these dye-based assays are sensitive to numerous intracellular species, precautions should be taken to avoid interference and misinterpretation of data.

3.5.1.4 Genotoxicity

The comet assay (often referred to as single-cell gel electrophoresis) is used to measure DNA damage as an indication of genotoxicity. In this method, cells are embedded in agarose gel in order to immobilize the DNA and subsequently lysed with salt and detergent (Collins 2004; Kim et al. 2013). Then, electrophoresis is used to promote migration of DNA through the agarose gel. After staining with a fluorescent dye, the gel can be analyzed with fluorescence microscopy. The comet assay gets its name from the comet-like shape that the DNA forms in the gel. Undamaged DNA appears as a circle or the head of the comet, and the fragmented (or damaged) DNA appears

as the tail. The comet shape is formed due to the faster migration of DNA fragments through the gel compared with undamaged DNA (Ostling and Johanson 1984). Data analysis can be performed by measuring parameters such as the length of the tail and the relative fluorescence intensity of the head and tail. Tail length is commonly used to assess genotoxicity, but it does not accurately reflect the amount of DNA fragmentation that has taken place, as both low and high levels of DNA damage could produce similar tail lengths but have different fluorescent intensities. Therefore, relative fluorescence intensity is a useful parameter, as it allows for the semiquantitative comparison of intact DNA versus damaged DNA.

3.5.2 In Vivo Assays

A large amount of nanotoxicity data published in scientific journals is comprised of cell culture systems, but it is common for these simplistic assays to be inaccurate at predicting toxicity outcomes in animal models. There are a variety of techniques that can be used to determine nanomaterial toxicity *in vivo*.

3.5.2.1 Methods for Measuring Lung Inflammation

Because inhalation is a major route of human exposure, many studies have aimed to evaluate nanoparticle-induced lung inflammation. In an inhalation study where the animals are exposed to nanomaterials, the lungs can be washed (termed "bronchoalveolar lavage") and the fluid collected for analysis. The number of total cells and immune cells (such as macrophages, neutrophils, and eosinophils) in BAL fluid is counted and compared with control animals that were not exposed to nanomaterials. An increased concentration of immune cells in the lungs is a sign of inflammation and often related to nanotoxicity. Another assessment for lung inflammation is the quantification of inflammatory cytokines in BAL fluid (such as IL-6, TNF-α, and interferon [INF]-γ) (Worthington et al. 2013).

3.5.2.2 Liver Toxicity

When nanomaterials enter the bloodstream, they accumulate in the major organs of the animal, depending on their size, shape, and surface chemistry (Sa et al. 2012). After administration, nanomaterials can localize in the liver. Therefore, a technique for assessing nanotoxicity is to determine if there are any abnormalities in liver function. There are several markers for liver damage. Bilirubin, for example, is a liver enzyme that occurs naturally in the body and is involved in heme metabolism. Elevated levels of bilirubin in the blood plasma or serum are an indication that the liver is not functioning normally (Sedlak et al. 2009) and can be a result of nanotoxicity (Liu et al. 2013). Other markers for liver damage are alanine aminotransferase (ALT) and aspartate aminotransferase (AST) enzymes. All the liver enzymes mentioned here can be detected by well-established analytical methods, the most common being colorimetric analysis using spectrophotometric detection (Rutkowsk and Debaare 1966; Walters and Gerarde 1970; Hafkenscheid and Dijt 1979; Huang et al. 2006).

3.5.2.3 Histological Analysis

Histology can be used as a visual method to determine damage and/or inflammation in organs or tissues as a result of nanomaterial-dependent toxicity. After tissues or organs are extracted from the animal, they must be preserved with a chemical fixative (e.g., formaldehyde), embedded in a medium such as paraffin, and then cut into very thin sections. Afterward, the sections are stained and imaged with light microscopy. Hematoxylin and Eosin (H&E) stains are of the most widely used stains for histological analysis. In H&E staining, the nuclei of cells are stained blue by hematoxylin, while eosin stains the cytoplasm and extracellular matrix pink. This dye is often used to examine the infiltration of immune cells (e.g., macrophages) into a tissue after nanomaterial exposure (Naya et al. 2012). Immunohistochemistry is another histology method in which antibodies are used to tag cytokines (or other proteins) located within tissue sections (Cho et al. 2009).

3.6 CONCLUSIONS

Currently, there are more than 120,000 research articles published in the scientific literature (according to a search using Thompson Reuters Web of Science™) that aim to develop nanomaterials for drug delivery. Clearly with the advances in nanomedicine, the field of nanotoxicology is growing in parallel. However, there is a need for accurate and reproducible methods for testing nanotoxicity, as the findings for any one particular nanomaterial toxicity can return a plethora of confounding results that are dependent on the methods (in vitro or in vivo) used for risk assessment. As a general consensus, nondegradable nanomaterials induce more toxicity than synthetic, biodegradable nanomaterials, such as PLGA nanoparticles or nanoparticles made from natural polymers such as chitosan (Ai et al. 2011). Although there is a large amount of cumulative nanotoxicity data in the literature on these materials, it is critical to establish comprehensive guidelines for evaluating nanotoxicity that could be used to predict human health risks and aid in the development of safe systems for drug delivery or other biomedical applications.

GLOSSARY

genotoxicity: generation of mutations within a cell that can eventually lead to cancer.

inflammation: a natural biological defense mechanism that is triggered by a range of factors, including toxins, the introduction of foreign matter or pathogens, or damaged tissue.

nanomaterial: a material with one or more dimensions below 500 nm.

nanomedicine: the use of nanomaterials for medical applications.

nanotoxicology: the study of the toxicity of nanomaterials.

oxidative stress: a chemical imbalance between intracellular antioxidants and ROS where the damaging effects of ROS cannot be counteracted by naturally occurring antioxidants.

reactive oxygen species (ROS): chemically reactive compounds containing oxygen that are produced naturally during biological processes but can be elevated during toxicity. Examples include hydroxyl radicals, superoxide anion, and hydrogen peroxide.

REFERENCES

Ahamed, M., M. A. Siddiqui, M. J. Akhtar, I. Ahmad, A. B. Pant, and H. A. Alhadlaq. 2010. Genotoxic potential of copper oxide nanoparticles in human lung epithelial cells. *Biochem Biophys Res Commun* 396 (2):578–83.

Ai, J., E. Biazar, M. Jafarpour, M. Montazeri, A. Majdi, S. Aminifard, M. Zafari, H. R. Akbari, and H. G. Rad. 2011. Nanotoxicology and nanoparticle safety in biomedical designs. *Int J Nanomed* 6:1117–27.

Alkilany, A. M., P. K. Nagaria, M. D. Wyatt, and C. J. Murphy. 2010. Cation exchange on the surface of gold nanorods with a polymerizable surfactant: Polymerization, stability, and toxicity evaluation. *Langmuir* 26 (12):9328–33.

Anderson, J. M., A. Rodriguez, and D. T. Chang. 2008. Foreign body reaction to biomaterials. *Semin Immunol* 20 (2):86–100.

AshaRani, P. V., G. L. K. Mun, M. P. Hande, and S. Valiyaveettil. 2009. Cytotoxicity and genotoxicity of silver nanoparticles in human cells. *ACS Nano* 3 (2):279–90.

Auffan, M., M. Pedeutour, J. Rose, A. Masion, F. Ziarelli, D. Borschneck, C. Chaneac, C. Botta, P. Chaurand, J. Labille, and J. Y. Bottero. 2010. Structural degradation at the surface of a TiO2-based nanomaterial used in cosmetics. *Environ Sci Technol* 44 (7):2689–94.

Barillet, S., A. Simon-Deckers, N. Herlin-Boime, M. Mayne-L'Hermite, C. Reynaud, D. Cassio, B. Gouget, and M. Carriere. 2010. Toxicological consequences of TiO2, SiC nanoparticles and multi-walled carbon nanotubes exposure in several mammalian cell types: An in vitro study. *J Nanopart Res* 12 (1):61–73.

Benn, T., B. Cavanagh, K. Hristovski, J. D. Posner, and P. Westerhoff. 2010. The release of nanosilver from consumer products used in the home. *J Environ Qual* 39 (6):1875–82.

Benn, T. M., and P. Westerhoff. 2008. Nanoparticle silver released into water from commercially available sock fabrics. *Environ Sci Technol* 42 (11):4133–9.

Carlson, C., S. M. Hussain, A. M. Schrand, L. K. Braydich-Stolle, K. L. Hess, R. L. Jones, and J. J. Schlager. 2008. Unique cellular interaction of silver nanoparticles: Size-dependent generation of reactive oxygen species. *J Phys Chem B* 112 (43):13608–19.

Chithrani, B. D., A. A. Ghazani, and W. C. W. Chan. 2006. Determining the size and shape dependence of gold nanoparticle uptake into mammalian cells. *Nano Lett* 6 (4):662–8.

Cho, W. S., M. J. Cho, J. Jeong, M. Choi, H. Y. Cho, B. S. Han, S. H. Kim, H. O. Kim, Y. T. Lim, B. H. Chung, and J. Jeong. 2009. Acute toxicity and pharmacokinetics of 13 nm-sized PEG-coated gold nanoparticles. *Toxicol Appl Pharmacol* 236 (1):16–24.

Collins, A. R. 2004. The comet assay for DNA damage and repair—Principles, applications, and limitations. *Mol Biotechnol* 26 (3):249–61.

Colognato, R., A. Bonelli, J. Ponti, M. Farina, E. Bergamaschi, E. Sabbioni, and L. Migliore. 2008. Comparative genotoxicity of cobalt nanoparticles and ions on human peripheral leukocytes in vitro. *Mutagenesis* 23 (5):377–82.

Conde, J., G. Doria, and P. Baptista. 2012. Noble metal nanoparticles applications in cancer. *J Drug Deliv* 2012:751075.

Coradeghini, R., S. Gioria, C. P. Garcia, P. Nativo, F. Franchini, D. Gilliland, J. Ponti, and F. Rossi. 2013. Size-dependent toxicity and cell interaction mechanisms of gold nanoparticles on mouse fibroblasts. *Toxicol Lett* 217 (3):205–16.

Decker, T., and M. L. Lohmann-Matthes. 1988. A quick and simple method for the quantitation of lactate dehydrogenase release in measurements of cellular cytotoxicity and tumor necrosis factor (TNF) activity. *J Immunol Methods* 115 (1):61–9.

Ding, H. M., and Y. Q. Ma. 2013. Controlling cellular uptake of nanoparticles with pH-sensitive polymers. *Sci Rep* 3:2804.

Dufour, E. K., T. Kumaravel, G. J. Nohynek, D. Kirkland, and H. Toutain. 2006. Clastogenicity, photo-clastogenicity or pseudo-photo-clastogenicity: Genotoxic effects of zinc oxide in the dark, in pre-irradiated or simultaneously irradiated Chinese hamster ovary cells. *Toxicol Lett* 164:S290–1.

Etheridge, M. L., S. A. Campbell, A. G. Erdman, C. L. Haynes, S. M. Wolf, and J. McCullough. 2013. The big picture on nanomedicine: The state of investigational and approved nanomedicine products. *Nanomedicine* 9 (1):1–14.

Faraday, M. 1857. AuNP117—The Bakerian lecture: Experimental relations of gold (and other metals) to light. *Philos Trans R Soc Lond* 147:145–81.

Feynman, R. P. 1960. There's plenty of room at the bottom: An invitation to enter a new field of physics. *Eng Sci* 23:22–35.

Franken, N. A. P., H. M. Rodermond, J. Stap, J. Haveman, and C. van Bree. 2006. Clonogenic assay of cells in vitro. *Nat Protoc* 1 (5):2315–9.

Fu, P. P., Q. Xia, H. M. Hwang, P. C. Ray, and H. Yu. 2014. Mechanisms of nanotoxicity: Generation of reactive oxygen species. *J Food Drug Anal* 22 (1):64–75.

Gatoo, M. A., S. Naseem, M. Y. Arfat, A. M. Dar, K. Qasim, and S. Zubair. 2014. Physicochemical properties of nanomaterials: Implication in associated toxic manifestations. *Biomed Res Int* 2014:498420.

Ghosh, M., J. Manivannan, S. Sinha, A. Chakraborty, S. K. Mallick, M. Bandyopadhyay, and A. Mukherjee. 2012. In vitro and in vivo genotoxicity of silver nanoparticles. *Mutat Res Genet Toxicol Environ Mutagen* 749 (1–2):60–69.

Hafkenscheid, J. C. M., and C. C. M. Dijt. 1979. Determination of serum aminotransferases—Activation by pyridoxal-5′-phosphate in relation to substrate concentration. *Clin Chem* 25 (1):55–59.

Hanks, J. H., and J. H. Wallace. 1958. Determination of cell viability. *Proc Soc Exp Biol Med* 98 (1):188–92.

Hillegass, J. M., A. Shukla, S. A. Lathrop, M. B. MacPherson, N. K. Fukagawa, and B. T. Mossman. 2010. Assessing nanotoxicity in cells in vitro. *Wiley Interdiscip Rev Nanomed Nanobiotechnol* 2 (3):219–31.

Huang, X., X. Teng, D. Chen, F. Tang, and J. He. 2010. The effect of the shape of mesoporous silica nanoparticles on cellular uptake and cell function. *Biomaterials* 31 (3):438–48.

Huang, X. J., Y. K. Choi, H. S. Im, O. Yarimaga, E. Yoon, and H. S. Kim. 2006. Aspartate aminotransferase (AST/GOT) and alanine aminotransferase (ALT/GPT) detection techniques. *Sensors* 6 (7):756–82.

Huerta-Garcia, E., J. A. Perez-Arizti, S. G. Marquez-Ramirez, N. L. Delgado-Buenrostro, Y. I. Chirino, G. G. Iglesias, and R. Lopez-Marure. 2014. Titanium dioxide nanoparticles induce strong oxidative stress and mitochondrial damage in glial cells. *Free Radic Biol Med* 73:84–94.

Jin, C. Y., B. S. Zhu, X. F. Wang, and Q. H. Lu. 2008. Cytotoxicity of titanium dioxide nanoparticles in mouse fibroblast cells. *Chem Res Toxicol* 21 (9):1871–7.

Jin, Y. H., H. Y. Hu, M. X. Qiao, J. Zhu, J. W. Qi, C. J. Hu, Q. Zhang, and D. W. Chen. 2012. pH-sensitive chitosan-derived nanoparticles as doxorubicin carriers for effective anti-tumor activity: Preparation and in vitro evaluation. *Colloids Surf B Biointerfaces* 94:184–91.

Kalyanaraman, B., V. Darley-Usmar, K. J. Davies, P. A. Dennery, H. J. Forman, M. B. Grisham, G. E. Mann, K. Moore, L. J. Roberts 2nd, and H. Ischiropoulos. 2012. Measuring reactive oxygen and nitrogen species with fluorescent probes: Challenges and limitations. *Free Radic Biol Med* 52 (1):1–6.

Karlsson, H. L., P. Cronholm, J. Gustafsson, and L. Moller. 2008. Copper oxide nanoparticles are highly toxic: A comparison between metal oxide nanoparticles and carbon nanotubes. *Chem Res Toxicol* 21 (9):1726–32.

Keller, A. A., W. Vosti, H. T. Wang, and A. Lazareva. 2014. Release of engineered nanomaterials from personal care products throughout their life cycle. *J Nanopart Res* 16 (7):2489.

Kim, H. R., Y. J. Park, Y. Shin da, S. M. Oh, and K. H. Chung. 2013. Appropriate in vitro methods for genotoxicity testing of silver nanoparticles. *Environ Health Toxicol* 28:e2013003.

Kohane, D. S., and R. Langer. 2010. Biocompatibility and drug delivery systems. *Chem Sci* 1 (4):441–6.

Korzeniewski, C., and D. M. Callewaert. 1983. An enzyme-release assay for natural cytotoxicity. *J Immunol Methods* 64 (3):313–20.

Kroll, A., M. H. Pillukat, D. Hahn, and J. Schnekenburger. 2009. Current in vitro methods in nanoparticle risk assessment: Limitations and challenges. *Eur J Pharm Biopharm* 72 (2):370–7.

LeBel, C. P., H. Ischiropoulos, and S. C. Bondy. 1992. Evaluation of the probe 2′,7′-dichlorofluorescin as an indicator of reactive oxygen species formation and oxidative stress. *Chem Res Toxicol* 5 (2):227–31.

Lee, J. H., Y. M. Huh, Y. W. Jun, J. W. Seo, J. T. Jang, H. T. Song, S. Kim, E. J. Cho, H. G. Yoon, J. S. Suh, and J. Cheon. 2007. Artificially engineered magnetic nanoparticles for ultra-sensitive molecular imaging. *Nat Med* 13 (1):95–9.

Lee, S. H., B. H. Kim, H. B. Na, and T. Hyeon. 2014. Paramagnetic inorganic nanoparticles as T1 MRI contrast agents. *Wiley Interdiscip Rev Nanomed Nanobiotechnol* 6 (2):196–209.

Lehman, S. E., A. S. Morris, P. S. Mueller, A. K. Salem, V. H. Grassian, and S. C. Larsen. 2015. Silica nanoparticle-generated ROS as a predictor of cellular toxicity: Mechanistic insights and safety by design. *Environ Sci Nano* 3:56–66.

Li, M. Q., Z. H. Tang, S. X. Lv, W. T. Song, H. Hong, X. B. Jing, Y. Y. Zhang, and X. S. Chen. 2014. Cisplatin crosslinked pH-sensitive nanoparticles for efficient delivery of doxorubicin. *Biomaterials* 35 (12):3851–64.

Li, N., C. Sioutas, A. Cho, D. Schmitz, C. Misra, J. Sempf, M. Y. Wang, T. Oberley, J. Froines, and A. Nel. 2003. Ultrafine particulate pollutants induce oxidative stress and mitochondrial damage. *Environ Health Perspect* 111 (4):455–60.

Liu, J., F. Erogbogbo, K. T. Yong, L. Ye, J. Liu, R. Hu, H. Chen et al. 2013. Assessing clinical prospects of silicon quantum dots: Studies in mice and monkeys. *ACS Nano* 7 (8):7303–10.

Liu, J., Y. R. Huang, A. Kumar, A. Tan, S. B. Jin, A. Mozhi, and X. J. Liang. 2014. pH-Sensitive nanosystems for drug delivery in cancer therapy. *Biotechnol Adv* 32 (4):693–710.

Mackowiak, S. A., A. Schmidt, V. Weiss, C. Argyo, C. von Schirnding, T. Bein, and C. Brauchle. 2013. Targeted drug delivery in cancer cells with red-light photoactivated mesoporous silica nanoparticles. *Nano Lett* 13 (6):2576–83.

Manke, A., L. Wang, and Y. Rojanasakul. 2013. Mechanisms of nanoparticle-induced oxidative stress and toxicity. *Biomed Res Int* 2013:942916.

Mendoza, A., J. A. Torres-Hernandez, J. G. Ault, J. H. Pedersen-Lane, D. H. Gao, and D. A. Lawrence. 2014. Silica nanoparticles induce oxidative stress and inflammation of human peripheral blood mononuclear cells. *Cell Stress Chaperones* 19 (6):777–90.

Midander, K., P. Cronholm, H. L. Karlsson, K. Elihn, L. Moller, C. Leygraf, and I. O. Wallinder. 2009. Surface characteristics, copper release, and toxicity of nano- and micrometer-sized copper and copper(II) oxide particles: A cross-disciplinary study. *Small* 5 (3):389–99.

Morris, A. S., A. Adamcakova-Dodd, S. E. Lehman, A. Wongrakpanich, P. S. Thorne, S. C. Larsen, and A. K. Salem. 2016. Amine modification of nonporous silica nanoparticles reduces inflammatory response following intratracheal instillation in murine lungs. *Toxicol Lett* 241:207–15.

Napierska, D., L. C. Thomassen, V. Rabolli, D. Lison, L. Gonzalez, M. Kirsch-Volders, J. A. Martens, and P. H. Hoet. 2009. Size-dependent cytotoxicity of monodisperse silica nanoparticles in human endothelial cells. *Small* 5 (7):846–53.

Naya, M., N. Kobayashi, M. Ema, S. Kasamoto, M. Fukumuro, S. Takami, M. Nakajima, M. Hayashi, and J. Nakanishi. 2012. In vivo genotoxicity study of titanium dioxide nanoparticles using comet assay following intratracheal instillation in rats. *Regul Toxicol Pharmacol* 62 (1):1–6.

Nel, A., T. Xia, L. Madler, and N. Li. 2006. Toxic potential of materials at the nanolevel. *Science* 311 (5761):622–7.

Nishanth, R. P., R. G. Jyotsna, J. J. Schlager, S. M. Hussain, and P. Reddanna. 2011. Inflammatory responses of RAW 264.7 macrophages upon exposure to nanoparticles: Role of ROS-NF kappa B signaling pathway. *Nanotoxicology* 5 (4):502–16.

Nogueira, D. R., L. Tavano, M. Mitjans, L. Perez, M. R. Infante, and M. P. Vinardell. 2013. In vitro antitumor activity of methotrexate via pH-sensitive chitosan nanoparticles. *Biomaterials* 34 (11):2758–72.

Ostling, O., and K. J. Johanson. 1984. Microelectrophoretic study of radiation-induced DNA damages in individual mammalian-cells. *Biochem Biophys Res Commun* 123 (1):291–8.

Padmanabhan, J., and T. R. Kyriakides. 2015. Nanomaterials, inflammation, and tissue engineering. *Wiley Interdiscip Rev Nanomed Nanobiotechnol* 7 (3):355–70.

Pan, L. M., Q. J. He, J. N. Liu, Y. Chen, M. Ma, L. L. Zhang, and J. L. Shi. 2012. Nuclear-targeted drug delivery of tat peptide-conjugated monodisperse mesoporous silica nanoparticles. *J Am Chem Soc* 134 (13):5722–5.

Park, E. J., and K. Park. 2009. Oxidative stress and pro-inflammatory responses induced by silica nanoparticles in vivo and in vitro. *Toxicol Lett* 184 (1):18–25.

Parveen, S., R. Misra, and S. K. Sahoo. 2012. Nanoparticles: A boon to drug delivery, therapeutics, diagnostics and imaging. *Nanomedicine* 8 (2):147–66.

Passagne, I., M. Morille, M. Rousset, I. Pujalte, and B. L'Azou. 2012. Implication of oxidative stress in size-dependent toxicity of silica nanoparticles in kidney cells. *Toxicology* 299 (2–3):112–24.

Rutkowsk, R. B., and L. Debaare. 1966. An ultramicro colorimetric method for determination of total and direct serum bilirubin. *Clin Chem* 12 (7):432.

Sa, L. T. M., M. D. Albernaz, B. F. D. Patricio, M. V. Falcao, B. F. Coelho, A. Bordim, J. C. Almeida, and R. Santos-Oliveira. 2012. Biodistribution of nanoparticles: Initial considerations. *J Pharm Biomed Anal* 70:602–4.

Sedlak, T. W., M. Saleh, D. S. Higginson, B. D. Paul, K. R. Juluri, and S. H. Snyder. 2009. Bilirubin and glutathione have complementary antioxidant and cytoprotective roles. *Proc Natl Acad Sci USA* 106 (13):5171–6.

Shukla, R. K., V. Sharma, A. K. Pandey, S. Singh, S. Sultana, and A. Dhawan. 2011. ROS-mediated genotoxicity induced by titanium dioxide nanoparticles in human epidermal cells. *Toxicol In Vitro* 25 (1):231–41.

Shvedova, A. A., E. R. Kisin, R. Mercer, A. R. Murray, V. J. Johnson, A. I. Potapovich, Y. Y. Tyurina et al. 2005. Unusual inflammatory and fibrogenic pulmonary responses to single-walled carbon nanotubes in mice. *Am J Physiol Lung Cell Mol Physiol* 289 (5):L698–708.

Singh, N., G. J. Jenkins, R. Asadi, and S. H. Doak. 2010. Potential toxicity of superparamagnetic iron oxide nanoparticles (SPION). *Nano Rev* 1.

Singh, N., B. Manshian, G. J. S. Jenkins, S. M. Griffiths, P. M. Williams, T. G. G. Maffeis, C. J. Wright, and S. H. Doak. 2009. NanoGenotoxicology: The DNA damaging potential of engineered nanomaterials. *Biomaterials* 30 (23–24):3891–914.

Soenen, S. J., B. Manshian, J. M. Montenegro, F. Amin, B. Meermann, T. Thiron, M. Cornelissen et al. 2012. Cytotoxic effects of gold nanoparticles: A multiparametric study. *ACS Nano* 6 (7):5767–83.

Strober, W. 2001. Trypan blue exclusion test of cell viability. *Curr Protoc Immunol*, Appendix 3B.

Studer, A. M., L. K. Limbach, L. Van Duc, F. Krumeich, E. K. Athanassiou, L. C. Gerber, H. Moch, and W. J. Stark. 2010. Nanoparticle cytotoxicity depends on intracellular solubility: Comparison of stabilized copper metal and degradable copper oxide nanoparticles. *Toxicol Lett* 197 (3):169–74.

Sun, L., Y. Li, X. M. Liu, M. H. Jin, L. Zhang, Z. J. Du, C. X. Guo, P. L. Huang, and Z. W. Sun. 2011. Cytotoxicity and mitochondrial damage caused by silica nanoparticles. *Toxicol In Vitro* 25 (8):1619–29.

Teodoro, J. S., A. M. Simoes, F. V. Duarte, A. P. Rolo, R. C. Murdoch, S. M. Hussain, and C. M. Palmeira. 2011. Assessment of the toxicity of silver nanoparticles in vitro: A mitochondrial perspective. *Toxicol In Vitro* 25 (3):664–70.

Tran, Q. H., V. Q. Nguyen, and A. T. Le. 2013. Silver nanoparticles: Synthesis, properties, toxicology, applications and perspectives. *Adv Nat Sci Nanosci Nanotechnol* 4 (3).

Vert, M., Y. Doi, K. H. Hellwich, M. Hess, P. Hodge, P. Kubisa, M. Rinaudo, and F. Schue. 2012. Terminology for biorelated polymers and applications (IUPAC Recommendations 2012). *Pure Appl Chem* 84 (2):377–408.

Wagner, V., A. Dullaart, A. K. Bock, and A. Zweck. 2006. The emerging nanomedicine landscape. *Nat Biotechnol* 24 (10):1211–7.

Walters, M. I., and H. W. Gerarde. 1970. An ultramicromethod for determination of conjugated and total bilirubin in serum or plasma. *Microchem J* 15 (2):231.

Wang, X., Z. X. Ji, C. H. Chang, H. Y. Zhang, M. Y. Wang, Y. P. Liao, S. J. Lin et al. 2014. Use of coated silver nanoparticles to understand the relationship of particle dissolution and bioavailability to cell and lung toxicological potential. *Small* 10 (2):385–98.

Wang, Y., A. Santos, A. Evdokiou, and D. Losic. 2015. An overview of nanotoxicity and nanomedicine research: Principles, progress and implications for cancer therapy. *J Mater Chem B* 3:7153–72.

Weir, A., P. Westerhoff, L. Fabricius, K. Hristovski, and N. von Goetz. 2012. Titanium dioxide nanoparticles in food and personal care products. *Environ Sci Technol* 46 (4):2242–50.

Wongrakpanich, A., I. A. Mudunkotuwa, S. M. Geary, A. S. Morris, K. A. Mapuskar, D. R. Spitz, V. H. Grassian, and A. K. Salem. 2016. Size-dependent cytotoxicity of copper oxide nanoparticles in lung epithelial cells. *Environ Sci Nano* 3:365–74.

Worthington, K. L. S., A. Adamcakova-Dodd, A. Wongrakpanich, I. A. Mudunkotuwa, K. A. Mapuskar, V. B. Joshi, C. A. Guymon, D. R. Spitz, V. H. Grassian, P. S. Thorne, and A. K. Salem. 2013. Chitosan coating of copper nanoparticles reduces in vitro toxicity and increases inflammation in the lung. *Nanotechnology* 24 (39):395101.

Xiao, D., H. Z. Jia, J. Zhang, C. W. Liu, R. X. Zhuo, and X. Z. Zhang. 2014. A dual-responsive mesoporous silica nanoparticle for tumor-triggered targeting drug delivery. *Small* 10 (3):591–8.

Xiu, Z. M., Q. B. Zhang, H. L. Puppala, V. L. Colvin, and P. J. Alvarez. 2012. Negligible particle-specific antibacterial activity of silver nanoparticles. *Nano Lett* 12 (8):4271–5.

Xu, A., Y. F. Chai, T. Nohmi, and T. K. Hei. 2009. Genotoxic responses to titanium dioxide nanoparticles and fullerene in gpt delta transgenic MEF cells. *Particle Fibre Toxicol* 6 (3).

Yeh, C. J. G., B. L. Hsi, and W. P. Faulk. 1981. Propidium iodide as a nuclear marker in immunofluorescence. 2. Use with cellular-identification and viability studies. *J Immunol Methods* 43 (3):269–75.

Yu, K. N., T. J. Yoon, A. Minai-Tehrani, J. E. Kim, S. J. Park, M. S. Jeong, S. W. Ha, J. K. Lee, J. S. Kim, and M. H. Cho. 2013. Zinc oxide nanoparticle induced autophagic cell death and mitochondrial damage via reactive oxygen species generation. *Toxicol In Vitro* 27 (4):1187–95.

Yu, M., S. H. Huang, K. J. Yu, and A. M. Clyne. 2012. Dextran and polymer polyethylene glycol (PEG) coating reduce both 5 and 30 nm iron oxide nanoparticle cytotoxicity in 2D and 3D cell culture. *Int J Mol Sci* 13 (5):5554–70.

Zhang, X. D., D. Wu, X. Shen, P. X. Liu, N. Yang, B. Zhao, H. Zhang, Y. M. Sun, L. A. Zhang, and F. Y. Fan. 2011. Size-dependent in vivo toxicity of PEG-coated gold nanoparticles. *Int J Nanomed* 6:2071–81.

Zielonka, J., and B. Kalyanaraman. 2008. ROS-generating mitochondrial DNA mutations can regulate tumor cell metastasis—A critical commentary. *Free Radic Biol Med* 45 (9):1217–9.

CHAPTER FOUR

Translational Nanomedicine

*Lara Scheherazade Milane**

CONTENTS

4.1 TRANSLATING TRANSLATION

I have a colleague who has an obvious aversion to new terminology in pharmaceutical science. He is a seasoned German fellow who vocalizes his thoughts in multiple languages, especially when he is passionate about a topic. I discovered his distain when I mentioned the term *nanomedicine*. In two (perhaps more) languages, he proceeded to tell me how ridiculous the concept was—making up a word to claim something is new. He further confirmed (more calmly, in English)

that liposomes have been around for some time and the idea of nanomedicine as a new field was (two languages) sabotage of science. Although I do not share his outrage, I can see why he would be frustrated; scientific vernacular is under continual metamorphism and can suffer from contradictions and redundancy. On the other hand, when an area that has been around for some time suddenly starts to explode with progress, that progress often demands new terminology, categorizations, paradigms, and regulations. This was the case for nanomedicine. The coining of *nanomedicine* appears to have been in Drexler et al.'s book in 1991, *Unbounding the Future: The Nanotechnology*

* Email: lara.milane@gmail.com

Revolution (Drexler et al. 1991). Nanomedicine did not emerge as a new field overnight when the term was first coined in 1991, but this coining did hallmark the rampant progress being made and the need for further advancement. Four years later, the first Food and Drug Administration (FDA)–approved nanomedicine (Doxil®) hit the market in the United States.

4.1.1 Why Has Translation Emerged?

Just as *nanomedicine* was coined due to the rate of progress in the field, *translational medicine* was similarly coined due to the volume of work and resources dedicated to the effort of translation. Considered in simple terms, translational medicine seems to apply to all medicine—is not the goal of all developmental medicine to eventually be applied as a clinical therapeutic? Perhaps this was true at one time, but this is no longer the case. Distinguishing translational medicine as a specific area identifies that there must be other areas of medicine that are distinct from translational medicine. The spectrum includes clinical medicine, translational medicine, and experimental therapeutics. Clinical medicine can be considered medicine already marketed and in clinical use. Experimental therapeutics include pharmaceutics used in the laboratory with no intent for clinical progression. Translational medicine would fall in the middle of the spectrum between these two categories. Translational medicine can be defined as pharmaceutics designed and developed with the intent of clinical application, often with an expedited and clear pathway for transitioning from laboratory bench research to patient bedside use. Figure 4.1 demonstrates the path of each of the three lines of medicine. Experimental therapeutics never leave the laboratory and are not optimized or destined for human applications. Clinical medicine includes currently approved therapeutics, whereas translational medicine is developed in the laboratory, evaluated in preclinical cellular and animal models with the goal of evaluation in human clinical trials.

4.1.2 Why Has Translational Nanomedicine Emerged?

The rate of experimental nanomedicine research has increased so dramatically, yet this increase is not meeting the demand for new therapies. The unique benefits of nanomedicine have long been recognized, and although the field of nanomedicine is studded with FDA approvals, this approval rate does not parallel the immense experimental development and funding that is taking place. This disconnect has highlighted the emergence of translational nanomedicine, to align experimental nanomedicine on a clear pathway out of the laboratory door and toward the clinic. Understanding that translational nanomedicine emerged out of need for a clear pathway, it is easy to conceive that (beyond whimsical exceptions) translational nanomedicine needs to begin in the design phase of experimentation.

Although traditional drug development processes can take one of many approaches, including backward design (disease state or target selected first) and forward design (drug candidate identified and screening for therapeutic applications), nanomedicine is a deliberate process that very rarely succeeds with forward design. The drugs (and application) are preselected during the nanodesign phase of development, before nanosynthesis occurs. Figure 4.2 (Pritchard et al. 2003) demonstrates the key decision gates in traditional (non-nanotechnology, non-biotechnology based) drug development. The gears represent the decision points that determine progression, more development, or cessation of a project. The boxes represent the activities that should lead to assessment of the decision points. According to Pritchard et al. (Figure 4.2), the key decision points for traditional drug development are

1. Are the targets validated?
2. Are leads identified?
3. Are the leads developable?
4. Give to humans?
5. Does the drug work in humans? Clinical proof of concept.
6. Approved for marketing?
7. Does the drug sell?

These decision points are still valid for translational medicine; however, the process is tailored toward the goal of clinical application and use. The primary purpose of translational medicine is to quickly identify clinical drug candidates and fast-track those candidates. Translational

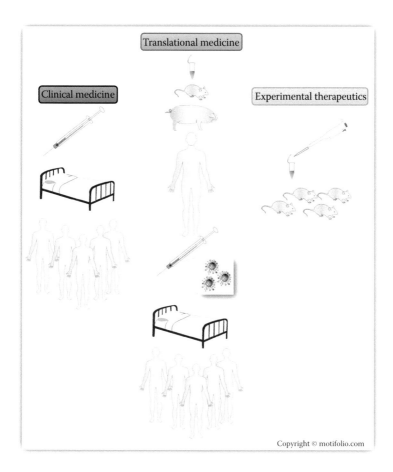

Figure 4.1. The three lines of medicine. Although historically most therapeutics developed in the laboratory were intended for eventual clinical application, this is not the case today. There are many therapeutics that are purely experimental and are used to study molecular pathways or knock out certain targets in the laboratory, with the aim of biological or molecular exploration. Clinical medicine, on the other side of the spectrum, is therapeutics that are in current use on the market. Translational medicine is medicine in current development with the intent of clinical application, such as systemically administered nanoparticles. Translational medicine must complete the pathway from animal studies, to FIH studies and clinical trials, before progressing to market approval.

medicine consists of three phases: early translational research, clinical research, and late-stage translational research. Early translational research includes identification of a therapeutic target, optimization of a drug target, *in vitro* and *in vivo* validation, and toxicology. The clinical research phase includes first-in-human (FIH) studies and clinical trials. Late-stage translational research includes Phase 4 clinical trials (postmarket analysis), policy reform, implementation, and characterization research.

The path for a drug candidate to enter translational research can be serendipitous or deliberate. Serendipitous entry is from experimental research that distinguishes the candidate as a prime agent for human application. Deliberate entry is from initial explicit design as a translational therapeutic. Deliberate design has a very different approach than experimental research that serendipitously ends up in clinical translation. Deliberate design of translational research can incorporate a modular approach so that if selection criteria eliminate a certain track, the design process can backtrack to the previous step instead of to the starting point. Designing specifically for translational nanomedicine is described shortly.

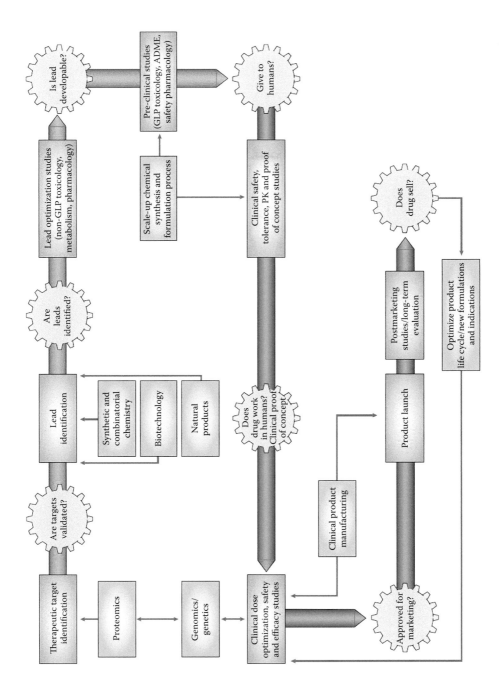

Figure 4.2. Drug development decision gates. The gears in the decision gate are the key questions that mark whether a drug should proceed through development. The boxes symbolize activities or disciplines. See text for a discussion of the process. Good laboratory practice (GLP); pharmacokinetics (PK). (From Pritchard, J. F. et al., Nat. Rev. Drug Discov., 2(7), 542–553, 2003. Reprinted with permission from Nature Publishing Group.)

NANOMEDICINE FOR INFLAMMATORY DISEASES

4.1.3 Biomaterials, Biotechnology, and Nanotechnology

4.1.3.1 Bionanotechnology

Bionanotechnology has recently emerged as a term in the literature in an attempt to cut out some vowels (Valic and Zheng 2016). Although my German colleague who becomes livid at the mention of *nanomedicine* would surely throw his arms in protest to the coining of *bionanotechnology*, I agree with the coining of the term. We live in an era where science and medicine are becoming progressively specialized, and we need to adapt our vocabulary to the velocity of progress. Biotechnology refers to a wide range of science, from the production of genetically modified foods to the production of synthetic, humanized insulin in *Escherichia coli*. Nanotechnology refers to a range of nanosciences, from nanomaterials to integrated chip nanotechnology. Although biotechnology focuses on cells or biological systems for technological applications, bionanotechnology more appropriately streamlines the field of nanotechnology to refer to nanotechnology with biological foundations. Nanomedicine is appropriately designated as therapeutics of nanoscale dimensions. In this regard, nanomedicine could be considered a sub-discipline of bionanotechnology.

4.1.3.2 Bionanomaterials

Advances in biomaterials (and dare I say bionanomaterials) have enabled the progression of translational nanomedicine. Bold advances address the desire to create *renaissance* nanomaterials that are capable of targeting, imaging, and treating multiple diseases. The desire is to create nanomaterials that are nanoplatforms for multiple applications. One such material is the porphysome nanoplatform (Valic and Zheng 2016). This platform consists of a porphyrin phospholipid (PoP) monomer that is light absorbing, capable of metal chelation, modifiable (allowing for active targeting), and amphipathic, allowing for drug encapsulation and self-assembly in aqueous solution (Valic and Zheng 2016). The porphyrin residue is derived from chlorophyll purified from cyanobacteria; it is the porphyrin residue that is light active and that enables metal ion chelation (Valic and Zheng 2016). The PoP monomer also self-assembles with other polymers and lipids in aqueous buffer to form dynamic particles (Valic and Zheng

2016). For nanomedicine applications, the PoP monomers are assembled with other constituents of traditional liposomes (such as polyethylene glycol lipid constructs) to maximize stability and "stealth" character (Valic and Zheng 2016). The porphysome platform clearly enables multifunctional nanomedicine, but even though it is a dynamic platform, there will never be a "one size fits all" or "universal solution" to drug delivery and nanomedicine. The applications and demands for treating diverse diseases are too varied to expect one platform to address every nanomedicine need.

Second to liposomes, polymers are really taking the lead as the most actively advancing area of biomaterials for nanotechnology. Perhaps it is just that lipid chemistry was the first arena to boom with development (as this was the first application of approved nanomedicine), followed by metals (until toxicity and safety concerns limited clinical potential), and now the polymer chemistry boom. Clinical interest in polymers began with the creation of hydrogels and soft contact lenses (Yang and Kopeček 2015). The current heightened appeal of polymeric nanoparticles is most likely due to their ease of modification to achieve active targeting and appreciable drug delivery of a wide range of agents, biocompatibility, stability, and potential for extended-drug-release formulations (Cheng et al. 2015).

4.1.3.3 Nanoappeal

Nanomaterials have distinct properties from their parent-derived or related material, and likewise, nanoparticles have distinct properties from the nanomaterial itself. For example, an amazing nano-feature is the inherent antibacterial properties of some nanoformulations (Campoccia et al. 2013). Unsurprisingly, many of these nanoformulations are metallic or have metal components, but some are nonmetals, such as silica nanoparticles grafted with acrylate and polyethylene glycol, silica nanoparticles coated with didodecyldimethylammonium bromide, cross-linked quaternary ammonium polyethyleneimine nanoparticles, single-walled carbon nanotubes incorporated with poly-lactic glycolic acid, single-walled carbon nanotubes assembled with poly(L-lysine) and poly(L-glutamic acid), zinc oxide nanostructured surfaces and coatings, magnesium oxide nanostructured coatings and particles, magnesium fluoride coatings, and hydroxyapatite nanoparticle coatings (Campoccia et al. 2013). The most common metals exploited for inherent antibacterial properties used in nanostructured materials are silver, gold, and titanium (Campoccia et al. 2013). The antibacterial properties of some nanoformulations can be exploited in therapeutic design or used as an added advantage. Assuredly, many nanoformulations being explored in translational medicine may not have been characterized yet for antibacterial traits. Broader characterization of nanoparticles in this way could lead to the identification and optimization of distinguishing characteristics that enable a formulation to have antibacterial properties.

The appeal of nanomedicine is multifold; nanoparticles enable the delivery of agents that would otherwise not be able to be used clinically, such as the decrease in toxicity and improvement in pharmacokinetics demonstrated by lonidamine

delivered in polymeric nanoparticles (Milane et al. 2011a,b). One of the most appealing aspects of using nanotechnology in medicine is that the functionality of a nanocarrier can be deliberately engineered. Figure 4.3 illustrates how the physical characteristics of a nanocarrier and the design parameters can be tailored to achieve a desired functionality. The physical characteristics include the size, shape, and ratio of surface area to volume.

Nanoparticles have a high surface area to volume ratio, which allows substantial drug encapsulation with a large area for surface modification to achieve active targeting and/or avoidance of immune clearance. The shape and size of the carrier (as well as the type of nanocarriers) determine the ratio of surface area to volume. Design parameters include the core platform material (i.e., the polymers, lipids, or metals), the biologically active agents (therapeutics), and the surface properties

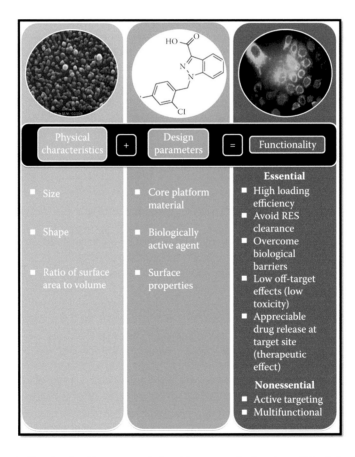

Figure 4.3. Nanocarrier functionality. The true appeal of applying nanotechnology in medicine is the ability to tailor functionality (including complex multifunctionality, active targeting, and combination drug delivery) by controlling the physical and design parameters of a drug delivery system. This process is described further in the text.

NANOMEDICINE FOR INFLAMMATORY DISEASES

(such as PEGylation for reticuloendothelial system [RES] protection and active targeting residues). Functionality includes essential and nonessential parameters. Essential functional parameters include high loading efficiency (for a nanocarrier to have a superior effect relative to drug solution, the nanocarrier must be able to carry substantial amounts of payload), avoiding RES clearance (if a nanocarrier is cleared by the immune system, it is not available to have a therapeutic effect), overcoming biological barriers (a nanocarrier must be able to transfer body-level, tissue-level, cellular-level, and molecular-level barriers to deliver a drug to the extracellular or intracellular site of action), low toxicity (a nanocarrier must offer comparable or superior safety relative to solution forms of a drug), and a high therapeutic effect (for a nanocarrier formulation to be clinically relevant, the formulation must have a higher or comparable therapeutic effect to solution forms of the therapeutics).

Nonessential functional parameters include active targeting (molecular-based therapies and personalized medicine) and multifunctionality (such as theranostics and theragnostics and the ability to achieve combination therapy with one formulation). The combination of nanoproperties with the properties of the biomaterial defines the characteristics of the drug delivery formulation. Manipulation of the physical and design characteristics of a nanomedicine formulation can control the functionality of the therapeutic.

4.2 HISTORY, PROGRESS, AND CURRENT STATUS OF TRANSLATION

4.2.1 Critical Path Initiative

A landmark in the history of nanomedicine occurred in 2004 when the FDA launched the Critical Path Initiative (CPI). The initial report addresses the apparent lack of progress in translational medicine and medical innovation, which was in stark contrast to expectations. The intention of the report was to analyze the challenges in the biomedical revolution that were causing the lack of expected progress. The report mentions how the sequencing of the human genome was followed by a globally decreased rate of new drug submissions and discusses the high cost of translating a medicine into the clinic as a barrier to innovation and medical advancement. Perhaps the most important issue discussed in the report is that our ability to

predict a drug candidate's success has not improved over the years, in reference to novel agents entering Phase 1 clinical trials in 1985 concerns and 2000. Not surprisingly, the report lists safety concerns and lack of efficacy as the main causes of failures. My German colleague would probably ask, as if there is nothing else to be concerned with, what else is there? But the report does go on to identify that there are three dimensions of the critical path from bench to market: assessing safety, demonstrating medical utility, and industrialization. The report discusses the key insight that FDA scientists have, being privy to the many failures that do not reach the market. These insightful FDA scientists suggest that a better "toolkit" is needed for better product development, tools for assessing safety, demonstrating medical utility, evaluating effectiveness, and manufacturing. The "better toolkit" conceptualizes the demand for more predictive safety assessments, more predictive preclinical screening models, identification and classification of more biomarkers, improved clinical trial design, and improved molecular imaging, and the development of technical standards for characterizing and manufacturing emerging pharmaceuticals, biologicals, and devices.

4.2.2 Critical Path Opportunities List

This initiative that began in 2004 later led to the generation of a Critical Path Opportunities List in 2006. The list was compiled by FDA scientists, and stakeholders in the public and private sector; the opportunities list identifies specific prospects for speeding and aiding the development and approval of pharmaceuticals, biologicals, and medical devices. The Critical Path Opportunities List is divided into the following topics: better evaluation tools, streamlining clinical trials, harnessing bioinformatics, moving manufacturing into the twenty-first century, developing products to address urgent public health needs, and specific at-risk populations (pediatrics). Nanotechnology is listed as a subsection of moving manufacturing into the twenty-first century. This section explains the need for better techniques and methods to characterize, standardize, and qualify nanotechnology to more effectively move nanomedicine through the pipeline from the bench to the clinic. The Critical Path Initiative (CPI) resulted in the creation of the Critical Path Institute. The CPI allocated funds for modestly supporting novel therapies (outlined in the opportunities list) through development. The directives of the CPI assisted in the development of nanomedicine-based therapies and establishing standards since 2006. The CPI also assisted indirectly through efforts to modernize clinical trials.

4.2.3 National Nanotechnology Initiative

The National Nanotechnology Initiative (NNI) Strategic Plan was recently released for public review by the National Science and Technology Council Committee on Technology, Subcommittee on Nanoscale Science, Engineering, and Technology. The NNI is closely related to the Materials Genome Initiative and the Precision Medicine Initiative. Important agencies with interests in the NNI include the Consumer Product Safety Commission, the National Institute of Standards and Technology (NIST), the U.S. Patent and Trademark Office, the FDA, the National Institute for Occupational Safety and Health, the National Institutes of Health (NIH), the National Science Foundation (NSF), and the Environmental Protection Agency (EPA). The plan identifies four goals: to advance nanotechnology research and development, promote the commercialization of nanotechnology products, maintain and expand resources to advance nanotechnology, and ensure the responsible development of nanotechnology.

4.2.4 The FDA's Nanotechnology Task Force

The FDA's Nanotechnology Task Force (established in 2006) is participating in the International Pharmaceutical Regulators Forum Nanomedicines Working Group, in the U.S.-EU Communities of Research (CORs), and in the International Organization for Standardization (ISO) and American Society for Testing and Materials (ASTM) development of standards, and is organizing the Global Summit on Regulatory Science: Nanotechnology Standards and Applications. Agencies that are members of the Nanotechnology Task Force include the Center for Drug Evaluation and Research (CDER) Nanotechnology Programs, the Center for Biologics Evaluation and Research Nanotechnology Programs (focuses on pathogen detection and blood safety), the Center for Devices and Radiological Health Nanotechnology Programs, the Center for Food Safety and Applied Nutrition Nanotechnology Programs, the Center for Veterinary Medicine Nanotechnology Programs, the National Center for Toxicological Research Nanotechnology Programs, the Office of Regulatory Affairs Nanotechnology Programs, and the Office of the Commissioner Nanotechnology Programs. The Nanotechnology Task Force coordinates the FDA's regulation and guidance for nanotechnology by supporting nanotechnology core facilities, establishing training, and establishing a nanotechnology Collaborative Opportunities for Research Excellence in Science (CORES) Program. The continued coordination of efforts to advance and characterize nanotechnology is exceedingly important for the translation of nanomedicine. Although the Nanotechnology Task Force is active, progress in establishing nanomedical standards has been slow and considerably unfruitful.

4.2.5 Center for Drug Evaluation and Research Nanotechnology Programs

The CDER Nanotechnology Programs began exploring the regulations, safety, and efficacy of nanomaterials with a 2010 study of zinc oxide and titanium dioxide nanoparticles in sunblock. The study concluded that the nanoparticles did not cross the dermis in a pig animal model. However, independent research has demonstrated that these particles have pulmonary toxicity (Vandebriel and De Jong 2012). There is no CDER study examining the aerosol deposition of the particles in the lungs, even though spray sunblocks are popular, especially for pediatric populations. It could be detrimental for translational nanomedicine for the FDA and CDER not to conduct necessary studies such as this. If future data demonstrate that nanoparticles from spray sunblock lead to compromised lung function and lung cancer, there will be public distrust in anything nano (including useful nanomedicines). Studies such as these are necessary to ensure the safety of the public and maintain trust in a field that is not well understood by the public realm.

4.3 PRECEDENT FOR SUCCESS: THE NCI'S NANOTECHNOLOGY CHARACTERIZATION LABORATORY

There is an undoubtable leader in translational nanomedicine: the National Cancer Institute (NCI).

The NCI is pioneering the pathway for the clinical translation of nanomedicine through the Nanotechnology Characterization Laboratory (NCL). The NCL was created in 2004 as a collaboration between the NCI, the FDA, and the NIST. Researchers can apply to have their nanomedicine characterized by the NCL, and within 1 year, the NCL conducts characterization studies, in vitro studies, and in vivo studies. The NCL will even assist with scale-up of nanomedicine. As listed on the NCL website (About the NCL 2016a), the six objectives of the NCL are

1. Establish and standardize an analytical cascade for nanomaterial characterization

2. Facilitate the clinical development and regulatory review of nanomaterials for cancer clinical trials

3. Identify and characterize critical parameters related to nanomaterials' absorption, distribution, metabolism, excretion (ADME) and toxicity profiles using animal models

4. Examine the biological and functional characteristics of multicomponent and combinatorial aspects of nanoscaled therapeutic, molecular, and clinical diagnostics, and detection platforms

5. Engage and facilitate academic and industrial-based knowledge sharing of nanomaterial performance data and behavior resulting from preclinical testing (i.e., physical characterization, in vitro testing, and in vivo pharmaco- and toxicokinetics)

6. Interface with other nanotechnology efforts

The NCL has done exactly what is needed for effective translation; they have developed a standardized assay cascade for evaluating nanomedicines from any origin (government, industry, academia), and they assist with progression into clinical trials and filing of an investigational new drug (IND) application. The physical characterization includes stability studies, surface characterization, and evaluation of batch-to-batch reproducibility. In vitro characterization includes binding, uptake, coagulation, safety, and efficacy studies. In vivo characterization studies are geared toward filing of an IND and include toxicity and pharmacokinetic studies (ADME) (Assay Cascade Protocols 2017).

The NCL truly is revolutionizing translational nanomedicine for cancer. Just 4 years after its creation, three of the NCL's methods for evaluating the biocompatibility of nanoparticles were included in the ASTM International standards (Resources 2016). The first standard (E2526) focuses on toxicity in kidney and liver cells, the second standard (E2524) evaluates hemolysis, and the third standard (E2525) evaluates the nanoparticle effect (stimulation or inhibition) on macrophages (Resources 2016). The NCL supports the NCI's Cancer Nanotechnology Plan (CaNanoPlan), and as of 2014, the NCL had characterized more than 300 nanomedicines (Resources 2016). The NCL is providing advisement to the EU, big pharma, and start-ups; they are a valued resource for translational nanomedicine in cancer. The NCL is a clear model for translational success. Now the question is, how do we expand this resource to nanomedicine in general? Is the NCL's success partially due to its disease focus? Would it be best to have disease-centered translational nanomedicine resources (such as an NCL for rheumatoid arthritis, an NCL for diabetes, etc.), or is it possible to "scale up" what the NCL has done for the NCI to the NIH? Is the NCL of the NIH the key to the future of translation?

4.4 DESIGNING FOR TRANSLATION

The design schema for translational nanomedicine is illustrated in Figure 4.4. Early translational research actually begins with inception of the idea and continues through clinical trials. The first questions to be answered are, what is the disease or condition that will be treated, and what is the goal of therapy? After the disease and aim of therapy are identified, the next considerations are what the therapeutic target will be and if this target is druggable. If the selected target is not druggable, a new target must be selected. Some targets may already have known ligands or even structure–activity relationship (SAR) studies of molecules with the binding sites of a receptor. These SAR studies could have generated a drug scaffold that would serve as a potential starting point for optimizing a lead compound. If there are no known ligands or SAR studies of the target, drug screening studies must be conducted to identify lead

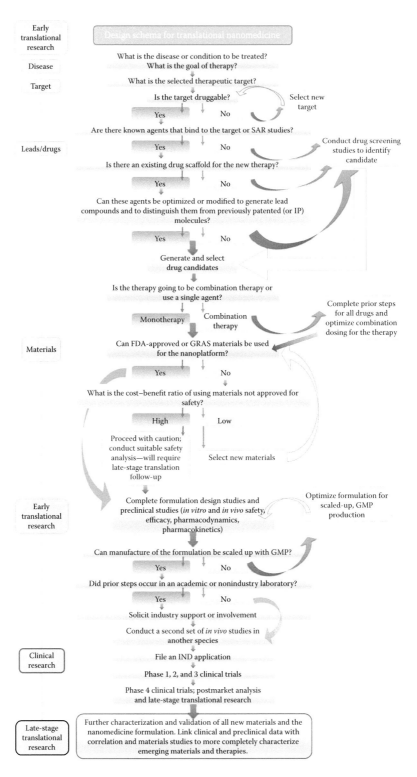

Figure 4.4. Design schema for translational nanomedicine. The key decision points following the pathway from disease identification to late-stage translational research are outlined in the schema; the major areas are disease identification, target selection, identification of leads and drug candidates, selection and characterization of materials, early translational research, clinical research, and late-stage translational research. GMP, good manufacturing practice; IP, intellectual property (protected).

candidates. For combination therapies, each active drug must be characterized and combination dosing must be optimized. If FDA-approved or generally recognized as safe (GRAS) materials cannot be used, the cost–benefit ratio of using material that does not have safety approval must be considered. If the cost–benefit ratio of using a non-approved material is high, then the development should proceed with caution with assurance for late-stage characterization of the material. If the cost–benefit ratio of using a non-approved material is low, then new materials with greater biocompatibility should be selected. The second major milestone (following the selection of drug candidates) is to optimize formulation design and conduct preclinical studies (major milestones in the schema are depicted by gray boxes). If the prior steps were not conducted in an industrial laboratory, then industry support or involvement should be solicited at this time. Before filing an IND application with the FDA, a second set of *in vivo* studies should be completed in another species. After filing the IND, Phase 1 clinical trials can progress. Phase 4 trials include postmarket analysis and begin late-stage translational research. This is perhaps the most important phase of translational nanomedicine, as data from preclinical and clinical studies should be correlated to more completely characterize emerging materials and therapies. This is a critical phase of translational nanomedicine, as it can accelerate the pace of future research and generate data to address public health or emerging concerns regarding nanomedicine therapies.

4.5 CHALLENGES

4.5.1 Cost

Although the cost of nanomedicine development is somewhat nebulous, it is absolutely daunting. Pushing a traditional drug into market can stagger a gasping $800 million, and the cost is undoubtedly higher for nanomedicine development and marketing (Webster 2012). The difference between traditional drug development and marketing costs versus those of nanomedicine is difficult to quantify at this early stage of translation. Insight into this discrepancy is provided by the 2009 average cost of Doxil per dose, waging in at more than $5000, compared with less than $200

for doxorubicin, and similarly, the average cost of Abraxane per dose was around $5000 compared with less than $500 for paclitaxel (Goldberg et al. 2013). The cost is a challenge, but the real challenge is to not let the cost debilitate the clinical translation of nanomedicine. In the early stages of the CPI, the FDA allocated minor funds to support the translation of critical path opportunities that they identified. At this critical stage of the field of nanomedicine, it would be an impetus for success if a more specific agency (such as the Nanotechnology Task Force) could provide core funds for clinical translation and bringing nanomedicines to market. Of course, this is a given: funds propel and enable progress. Without this external funding, translation is still possible. As depicted in Figure 4.3, what is needed are solid partnerships between academia and industry so that revolutionary nanomedicine therapies do not collect dust and expire on a laboratory shelf. An obvious obstacle to these partnerships is that academia and industry laboratories are already often competing in an area; concerns over intellectual property and patents can often obstruct a partnership. Specific FDA funding for translating nanomedicines would aid in this partnership, but these partnerships can (and are) established without FDA support.

The NCL discusses that improving the prediction of failures before clinical trial failure due to safety or efficacy is the key to cost reduction for drug development (About the NCL 2016b). For nanomedicine, establishing standards to improve the predictive value of preclinical data is imperative for cost reduction of nanomedicine development.

4.5.2 Extrapolation of Animal Data to First-in-Human Studies

FIH trials are globally recognized as the first transition from (laboratory) animals to humans (clinic). In July 2016, the European Medicines Agency of the European Union proposed changes to the regulatory guidelines for FIH trials involving healthy volunteers following the death of one volunteer and hospitalization of others in France following treatment with BIA 10-2474 in a FIH study (Feldwisch-Drentrup 2016). BIA 10-2474 is a fatty acid amide hydrolase inhibitor that targets the endocannabinoid system. A hallmark flaw of the study was excessive dosing and dose escalation without data analysis by external reviewers (Feldwisch-Drentrup 2016). Martin Enserink (2016) reported in a *Science* Insider article that there were three main mistakes made by Biotrial, the company responsible for the BIA 10-2474 study. The first mistake was that when "volunteer 2508" (the volunteer who later died) was hospitalized for headaches and blurry vision, Biotrial continued treatment of the other seven volunteers as planned without inquiry to the status of volunteer 2508 (Enserink 2016). Volunteer 2508 died that day, and four of the seven who were treated that morning were later hospitalized (Enserink 2016). The second significant mistake made by Biotrial in conducting the study was that they did not inform the returning volunteers of volunteer 2508's hospitalization (Enserink 2016). The third error that Biotrial made was not reporting the incident to the authorities (the French National Agency for Medicines and Health Products Safety) until 3 days after the study ceased (Enserink 2016). What were they thinking? Actions like these demoralize volunteers to just a study number and discredit the process. It was imperative that the European Union edit guidelines for FIH trials after this atrocity; they were obligated to act. The European Medicines Agency suggested modifications (which will be submitted as a revised guideline after public comments are received) that address multistage trial designs, such as the progressive dosing in the BIA 10-2474 study (Feldwisch-Drentrup 2016). The suggestions also address dose determination and stopping criteria (Feldwisch-Drentrup 2016). Unfortunately, mishaps and deaths tend to prompt change in policy. The guidelines were modified previously in 2007 after volunteers were hospitalized during the trial of TGN1412 (monoclonal antibody) in London (Feldwisch-Drentrup 2016). Hopefully as translational medicine progresses, policy will progress, with natural periodic review and modification to prevent tragedies such as the Biotrial mistakes.

4.5.3 Animal Models

Animal models are the necessary evil in drug development. There really is no way to win here; on the one hand, there is ethical objection from the nonscientific and even from within the scientific community, and on the other hand, there is no perfect animal model and every animal model has validity and predictive flaws. Yet, there is no other way to transition from cell culture to humans without the use of animal models. It would be less ethical to not execute due diligence in preclinical drug development. In 2014, Paul McGonigle and Bruce Ruggeri (2014) wrote a comprehensive review about the challenges of animal models in translation. In this review, McGonigle and Ruggeri (2014) identify a wish list of six parameters for the ideal animal model: (1) comparable anatomy and physiology, (2) similar genetics of the disease or condition, (3) comparable molecular mechanisms and pathological responses, (4) comparable phenotype outcomes to anticipated clinical studies, (5) demonstrated responsiveness to approved drugs with known efficacy, and (6) predictive of clinical efficacy in humans. The real challenge in translation is the sixth wish list item—the predictive value of an animal model. The model has to be good for both the disease and the new therapy; a laboratory may develop a great model for inflammatory bowel disease in mice but struggle with the technique of administering a nanomedicine by oral gavage in the mice. If the technique of administration is not consistent or perfected, then there is no merit to the study or model. If your model introduces variables that will undoubtedly affect the results and basal physiology, then perhaps an alternative model should be explored. For example, developing a model of estrogen-dependent cancer that requires the implantation of estrogen pellets in the animals may introduce a new level of variables that must be assessed. However, selecting a different model may eliminate the potential of studying that particular disease. There is no win–win situation with animal models. Every model, every study has its flaws.

Transgenic animal models are a significant advance for preclinical studies. Transgenic animals (or cells) are animals (or cells) that have a genomic alteration through the use of genetic engineering. There are many independent companies that will create customized knockout, knockin, humanized, and reporter models. Genetically engineered mouse models (GEMMs) of cancer have been invaluable in preclinical trials (Cook et al. 2012). Although transgenic models can offer better predictive value, there are limitations to these models as well, especially since the newer models lack characterization themselves and phenotyping is critical.

A true obstacle of translational medicine is transitioning from the mouse to the human. Although the transition from animals to humans is a challenge, the best way to optimize a study is to obtain clear and comprehensive pharmacokinetic and pharmacodynamics data while being realistic about the limitations of the selected model (McGonigle and Ruggeri 2014). Second to focusing on solid pharmacokinetic and pharmacodynamics data, having a biomarker to validate efficacy can increase the predictive value of a model (McGonigle and Ruggeri 2014). As identified in a special issue of the *European Journal of Pharmacology* on the translational value of animal models, regardless of the disease and therapy, the selected animal model should be (1) reliable, (2) reproducible, and (3) accurate (Groenink et al. 2015).

4.5.4 Characterization, Reproducibility, and Upscaling Manufacture

As previously mentioned, the characterization, reproducibility, and upscale manufacture of nanomedicine is a translational challenge. However, as demonstrated by the success of the NCI's NCL, there is promise in addressing this challenge. The NCL has established a standardized process for characterization. A majority of the characterization protocols established by the NCL actually apply to any nanomedicine, regardless of therapeutic application. The entire physicochemical characterization schema applies to any nanomedicine seeking translation, and these protocols are joint NIST-NCL protocols. They are openly accessible through the NCL's website (Working with the NCL 2016). The NCL outlines the following physicochemical characterization parameters (the starred parameters have accessible protocols on the NCL website) (Working with the NCL 2016):

Size and size distribution*

Topology

Molecular weight

Aggregation

Purity

Chemical composition*

Surface characteristics

Functionality

Zeta potential*

Stability

Solubility

Using existing NCL protocols as standards and further developing standards for these physicochemical parameters will aid in the translation of nanomedicine for inflammatory disease. Table 4.1 identifies the NCL's *in vitro* characterization parameters, while Table 4.2 identifies the NCL's *in vivo* characterization parameters. Although some of the parameters (such as efficacy) need to be tailored for particular diseases, they are a solid guide for standardization, and many of the parameters apply universally to nanomedicines seeking translation. The NCL assay cascade assumes intravenous administration, which, as of the current state of the field, is the route of administration for the majority of nanomedicines in experimental development, regardless of therapeutic application.

Although some of the cell-based assays for assessing *in vitro* immunology would need to be tailored to suit the therapeutic application, the parameters and, for the most part, the assays and subdivisions apply to all nanomedicines.

Although substantially less granular than the outlines for *in vitro* parameters, the *in vivo* parameters do serve as an efficient framework for disease-tailored characterization assays.

Reproducibility and scale-up manufacture of nanomedicine are two additional challenges for nanomedicine translation. As the NCL offers itself as a resource for evaluating batch-to-batch variability, an NIH resource such as this would be very helpful to the progression of translating nanomedicines for diseases other than cancer. Expansion of the standardized protocols developed by the NCL and NIST for physicochemical characterization would allow an individual laboratory to conduct a more thorough examination of batch-to-batch variability internally and optimize reproducibility by troubleshooting through the assay cascade. Similarly, as the NCL offers advisement for scale-up manufacture, a broad NIH resource that could mirror this advisement for non-cancer applications would assist with the translation of nanomedicine for inflammatory disease. The NCL facilitates partnerships with academia, government, and industry. An NIH resource that serves as a pathway to the clinic for nanomedicine development could also foster partnerships for manufacturing and scale-up of products.

TABLE 4.1
NCL's in vitro characterization parameters.

Parameter	Assay or subdivision
Sterility	Detection of endotoxin contamination
	Detection of microbial contamination
	Detection of bacterial contamination
	Detection of mycoplasma contamination
Targeting	Cell binding/internalization
Drug release	In vitro blood partitioning assay
In vitro immunology	Blood contact properties
	Analysis of hemolytic properties of nanoparticles
	Analysis of platelet aggregation by cell counting
	Analysis of platelet aggregation by light transmission aggregometry
	Analysis of nanoparticle interaction with plasma proteins by two-dimensional polyacrylamide gel electrophoresis
	Qualitative analysis of total complement activation by Western blot
	Quantitative analysis of complement activation
	Coagulation assay
	Cell-based assays
	Mouse granulocyte–macrophage colony-forming unit assay
	Leukocyte proliferation assay
	Macrophage/neutrophil function
	Analysis of cytokines, chemokines, and interferons
	Measurement of nanoparticle effects on cytotoxic activity of natural killer cells by label-free real-time cell electronic system (RT-CES) system
	Analysis of nanoparticle effects on maturation of monocyte-derived dendritic cells *in vitro*
	In vitro induction of leukocyte pro-coagulant activity by nanoparticles
	Human leukocyte proliferation assay
Toxicity	Oxidative stress
	Cytotoxicity (necrosis)
	Cytotoxicity (apoptosis)
	Autophagy

TABLE 4.2
NCL's in vivo characterization parameters.

Parameter	Assay or subdivision
Efficacy	Therapeutic
	Imaging
Disposition study	Tissue distribution
	Clearance
	Half-life
	Systemic exposure (plasma area under the curve)
Single- and repeat-dose toxicity	Immunotoxicity

4.6 FUTURE OF TRANSLATION

The clinical translation of nanomedicine has come a long way since the coining of *nanomedicine* in 1991. A 2014 analysis determined that there were 43 nanomedicines approved and on the market (Weissig et al. 2014). There is no better area of drug development to complement the rapid advancements in the molecular foundations of disease than nanomedicine. Nanomedicine is ideal for the design of complex molecularly targeted, combination therapy systems. Nanomedicine is the "multitasker" in medicine. Although advances in biomaterials and bionanotechnology have enabled heightened progress in experimental nanomedicine development and enactments such as the FDA's 2004 CPI have supported the development of nanomedicine, the challenges and obstacles stand tall. The challenges include the cost of development, extrapolation of animal data to FIH studies, limitations of animal models, and characterization, reproducibility, and upscaling manufacture. Deliberate design for translation can assist in overcoming the challenges. The future of nanomedicine translation has the potential to be amazing. A great model for success has been created: the NCI's NCL. The mission, schema, and assay cascades are applicable to nanomedicine in general. Expanding the NCL to be an NIH resource (or creating an independent NCL for the NIH) for all nanomedicine development (not just cancer) would transform the current arena of nanomedicine translation. The NCL has truly brightened the future of translational nanomedicine.

REFERENCES

Assay Cascade Protocols/Nanotechnology Characterization Lab (NCL). 2017. https://nanolab.cancer.gov/resources/assay-cascade-protocols (accessed January 26).

Campoccia, D., L. Montanaro, and C. R. Arciola. 2013. A review of the biomaterials technologies for infection-resistant surfaces. *Biomaterials* 34 (34): 8533–54.

Cheng, C. J., G. T. Tietjen, J. K. Saucier-Sawyer, and W. M. Saltzman. 2015. A holistic approach to targeting disease with polymeric nanoparticles. *Nature Reviews Drug Discovery* 14 (4): 239–47.

Cook, N., D. I. Jodrell, and D. A. Tuveson. 2012. Predictive in vivo animal models and translation to clinical trials. *Drug Discovery Today* 17 (5–6): 253–60.

Drexler, E., C. Peterson, and G. Pergamit. 1993. *Unbounding the Future: The Nanotechnology Revolution*. New York: Quill.

Enserink, M. 2016. French company bungled clinical trial that led to a death and illness, report says. *Science*, February 5.

Feldwisch-Drentrup, H. 2016. Europe overhauls rules for 'first-in-human' trials in wake of French disaster. *Science*, July 25.

Goldberg, M. S., S. S. Hook, A. Z. Wang, J. W. M. Bulte, A. K. Patri, F. M. Uckun, V. L. Cryns et al. 2013. Biotargeted nanomedicines for cancer: Six tenets before you begin. *Nanomedicine (London, England)* 8 (2): 299–308.

Groenink, L., G. Folkerts, and H.-J. Schuurman. 2015. *European Journal of Pharmacology*, Special issue on translational value of animal models: Introduction. *European Journal of Pharmacology* 759: 1–2.

McGonigle, P., and B. Ruggeri. 2014. Animal models of human disease: Challenges in enabling translation. *Biochemical Pharmacology* 87 (1): 162–71.

Milane, L., Z. Duan, and M. Amiji. 2011a. Therapeutic efficacy and safety of paclitaxel/lonidamine loaded EGFR-targeted nanoparticles for the treatment of multi-drug resistant cancer. *PLoS One* 6 (9): e24075.

Milane, L., Z.-F. Duan, and M. Amiji. 2011b. Pharmacokinetics and biodistribution of lonidamine/paclitaxel loaded, EGFR-targeted nanoparticles in an orthotopic animal model of multi-drug resistant breast cancer. *Nanomedicine: Nanotechnology, Biology, and Medicine* 7 (4): 435–44.

Pritchard, J. F., M. Jurima-Romet, M. L. J. Reimer, E. Mortimer, B. Rolfe, and M. N. Cayen. 2003. Making better drugs: Decision gates in non-clinical drug development. *Nature Reviews Drug Discovery* 2 (7): 542–53.

Process Overview/Nanotechnology Characterization Lab (NCL). 2017. https://nanolab.cancer.gov/working-ncl/process-overview (accessed January 26).

Resources—Nanotechnology related links—Nanotechnology Characterization Laboratory. 2017. https://nanolab.cancer.gov/resources (accessed January 26).

Valic, M. S., and G. Zheng. 2016. Rethinking translational nanomedicine: Insights from the "bottom-up" design of the porphysome for guiding the clinical development of imageable nanomaterials. *Current Opinion in Chemical Biology* 33: 126–34.

Vandebriel, R. J., and W. H. De Jong. 2012. A review of mammalian toxicity of ZnO nanoparticles. *Nanotechnology, Science and Applications* 5: 61–71.

Webster, T. J. 2012. *Nanomedicine: Technologies and Applications*. Amsterdam: Elsevier.

Weissig, V., T. K. Pettinger, and N. Murdock. 2014. Nanopharmaceuticals (Part 1): Products on the market. *International Journal of Nanomedicine* 9: 4357–73.

Yang, J., and J. Kopeček. 2015. Polymeric biomaterials and nanomedicines. *Journal of Drug Delivery Science and Technology* 30 (Pt B): 318–30.

PART TWO

Introduction

Inflammation is a protective response to endogenous or exogenous stimuli with the goals of eliminating the cause of injury or infection, clearing damaged cells, and initiating repair. Although inflammation can be acute or chronic and local or systemic, there are four components of the inflammatory pathway: inducers, sensors, mediators, and target tissues (Medzhitov 2010). Inducers can be exogenous (pathogen) or endogenous (cell stress signaling) (Medzhitov 2010). Sensors are receptors (on cells such as macrophages, dendritic cells, and mast cells) that detect the inducer and trigger a cascade of mediators (Medzhitov 2010). Mediators include an array of biomolecules, such as histamine, bradykinin, cytokines (tumor necrosis factor [TNF] and interleukin [IL]-6), chemokines (CCL2 and CXCL8), and prostaglandins, that exert an effect on target tissue (Medzhitov 2010). As discussed in Part 1, once the goals of the acute inflammatory response (mentioned above) are achieved, the inflammatory response enters a phase known as resolution of inflammation, which marks the termination of the inflammatory response (Medzhitov 2010). When resolution does not occur, acute inflammation transitions into chronic inflammation. Chronic inflammation can result in primary inflammatory disease, such as allergen-induced inflammation resulting in asthma, or occur as the foundation of an autoimmune disease, such as rheumatoid arthritis (RA). Chronic inflammation can also result from less defined triggers, such as for neurodegenerative disease, cancer, and diabetes (the focus of Part 3 of the book, secondary inflammatory disease). Chronic inflammation is central to primary inflammatory disease, whereas the role of chronic inflammation in secondary inflammatory disease is currently under investigation.

RA is a leading example of a primary chronic inflammatory disease. RA is an autoimmune disease where the synovium (membranes lining non-weight-bearing joints) become infiltrated with inflammatory cells such as T_H17 cells, dendritic cells, macrophages, and B cells (Isaacs 2010). The inflammation in the synovial joint transforms the synovial fibroblasts into malignant-like fibroblasts (similar to the transformation of cancer-associated fibroblasts in solid tumors; discussed in Part 3). These transformed fibroblasts contribute to joint damage through the release of factors such as matrix metalloproteinases, cytokines, chemokines, and cathepsins (Isaacs 2010). The inflammatory signaling in the synovial microenvironment alters osteoclast precursors, leading to bone damage (Isaacs 2010). In 2015, the American College of Rheumatology updated the guidelines for the treatment of RA to address the robust development of biologics used as disease-modifying anti-rheumatic drugs (DMARDs). The 2009 Food and Drug Administration (FDA) approval of certolizumab pegol (a PEGylated anti-TNFα biologic) for RA is promising for translational nanomedicine for RA. Although the challenges for biologics and nanomedicine are distinct, they are both unlike those for traditional small-molecule drugs, and the fact that the road to FDA approval of nontraditional (and PEGylated) formulations for RA has already begun is very positive. The FDA-approved nanomedicines for cancer could easily be used as a foundation for developing nanomedicines for RA. Clinicians have already envisioned the translation of nanomedicine for RA; in 2012, Rubinstein and Weinberg (2012) suggested the promise of DMARDs. Readers are directed to Rubinstein and Weinberg's (2012) perspective on the benefits nanomedicine has to offer RA treatment, such as reduced toxicity and active targeting and the potential for less drug desensitization. Readers are also directed to a more recent 2015 review covering nanomedicines in development for the treatment of RA (Prasad et al. 2015). This review includes a comprehensive discussion of nanomedicines for RA, including those in current clinical trials, such as PEGylated liposomes for prednisolone delivery (Prasad et al. 2015). The future does seem bright and close for the translation of nanomedicines for RA; as such, we selected three other primary inflammatory diseases for the focus of this section (inflammatory bowel disease, multiple sclerosis, and asthma).

To understand translational nanomedicine for inflammatory disease, it is imperative to gain insight into how the current standard of care in the clinic is related to the current experiments being conducted on the laboratory benchtops of nanomedicine investigators. To achieve this perspective, Part 2 has been designed with the following approach; each chapter in Part 2 is divided into three sections:

Section 1: The biology and clinical treatment of the disease

Section 2: Nanotherapeutics for the disease

Section 3: Bridging the gap between the bench and the clinic

By dividing each chapter into these three parts, we hope to provide a deeper insight into the translational challenges that must be overcome for treating inflammatory disease with nanomedicine. Part 2 focuses on three distinct inflammatory diseases (inflammatory bowel disease, multiple sclerosis, and asthma) to provide a dynamic perspective of translational nanomedicine for inflammatory disease.

REFERENCES

Isaacs, J. D. 2010. The changing face of rheumatoid arthritis: Sustained remission for all? *Nature Reviews Immunology* 10 (8): 605–11.

Medzhitov, R. 2010. Inflammation 2010: New adventures of an old flame. *Cell* 140 (6): 771–76.

Prasad, L. K., H. O'Mary, and Z. Cui. 2015. Nanomedicine delivers promising treatments for rheumatoid arthritis. *Nanomedicine (London, England)* 10 (13): 2063–74.

Rubinstein, I., and G. L. Weinberg. 2012. Nanomedicines for chronic non-infectious arthritis: The clinician's perspective. *Maturitas* 73 (1): 68–73.

Biology and Clinical Treatment of Inflammatory Bowel Disease

Christopher J. Moran and Bobby J. Cherayil

CONTENTS

5.1.1 INTRODUCTION

Inflammatory bowel disease (IBD) is an immune-mediated disorder that is associated with chronic, relapsing, and remitting inflammation of the gastrointestinal tract. It can occur in two forms—ulcerative colitis (UC) and Crohn's disease (CD)—that have overlapping mechanisms of pathogenesis but distinct patterns of histology, regional distribution, and clinical features. Like other diseases linked to dysfunction of the immune system, IBD has been increasing in incidence and prevalence over the last several decades and is estimated to affect close to 1.5 million individuals in the United States (Cosnes et al., 2011; Molodecky et al., 2012). It is currently one of the leading chronic gastrointestinal diseases and represents a significant economic and quality of life burden, with direct healthcare costs of $10,000–$20,000 per year per patient (Cohen et al., 2010; Floyd et al., 2015). The magnitude and chronicity of the problem, as well as the shortcomings of the treatment options that are currently available, are motivating factors in the ongoing search for improvements in IBD therapy.

5.1.2 PATHOGENESIS

A key aspect of IBD pathogenesis is the fact that the gastrointestinal tract is home to an extremely large and diverse community of commensal microorganisms—the gut microbiota—that reside in close proximity to host cells (Lozupone et al., 2012). In the healthy state, these organisms carry out a number of functions that are beneficial to the host, including processing of otherwise indigestible food components, generation of important metabolites, modulation of immune function, and provision of resistance to colonization by microbial pathogens (Erturk-Hasdemir and Kasper, 2013). In return, the host provides a protected environment rich in nutrients. The mutualistic and peaceful coexistence of host and microbiota is maintained by mechanisms that have coevolved to allow the microorganisms to thrive within defined niches, such as the intestinal lumen or epithelial surface, without inciting potentially damaging immune responses. IBD is generally thought to develop when this delicate state of equilibrium is disturbed, either by a derangement in the kinds or proportions of the organisms that make up the microbiota (a condition known as dysbiosis), or by abnormalities of host immune function that lead to impaired or excessive responses to the microbiota. Under such circumstances, cells of the intestinal immune system become inappropriately activated by components of the microbiota, including broadly distributed microbial molecules (microbe-associated molecular patterns [MAMPs]), such as lipopolysaccharide, that stimulate innate immunity, as well as specific antigens that trigger adaptive immune responses. An inflammatory response develops as a result, leading to local recruitment of neutrophils, monocytes, and lymphocytes, with associated release of a wide array of inflammatory mediators, including cytokines such as tumor necrosis factor α (TNFα), interleukin-1 (IL-1), IL-6, IL-17, and IL-23, and consequent damage to the intestinal epithelium and surrounding tissues. There are differences between CD and UC with respect to the character of the intestinal inflammation, with CD being generally more granulomatous in nature and dominated by Th1- and Th17-type helper T cells, while UC is more diffuse and dominated by Th2- and Th17-type helper T cells (de Souza and Fiocchi, 2016). Damage to the intestinal epithelial barrier that accompanies the inflammatory process in either CD or UC can exacerbate the pathology by allowing increased translocation of microbial molecules and further activation of cells of the immune system. Inflammation-associated dysbiosis, characterized by greater abundance of Enterobacteriaceae and increased production of lipopolysaccharide and virulence factors, also contributes to this pro-inflammatory positive feedback loop (Kostic et al., 2014). What exactly initiates the pathologic vicious cycle of inflammation, tissue damage, and microbial translocation in IBD is not clear, but host genetics, the microbiota, and the environment are all contributors (de Souza and Fiocchi, 2016). Each of these factors will be considered in the discussion that follows.

5.1.2.1 Genetic Susceptibility

Genetic susceptibility is an important factor in IBD pathogenesis, an idea suggested by long-standing observations that the disease runs in families, occurs with elevated frequency in certain ethnic groups, such as Ashkenazi Jews, and has higher concordance rates in monozygotic than in dizygotic twins (Yang et al., 1993; Halme et al., 2006). Over the last 15–20 years, considerable effort has been invested in characterizing the genetic architecture of IBD, most successfully by means of genome-wide association studies (GWASs). As of 2015, these studies have identified about 200 risk loci, most with a relatively small effect size and most involving noncoding variants (Jostins et al., 2012; Knights et al., 2013; Farh et al., 2015). About two-thirds of the loci influence risk for both CD and UC, pointing to shared mechanisms of pathogenesis, while a small minority affects only one type of disease. There is also considerable overlap between IBD risk loci and loci associated with susceptibility to other immune-mediated diseases (e.g., ankylosing spondylitis, psoriasis, and rheumatoid arthritis) and to mycobacterial infections (Jostins et al., 2012; Parkes et al., 2013).

The IBD susceptibility genes that have been identified by GWASs have helped to shed light on the mechanisms that appear to be malfunctioning in this disease. Many of the genes can be assembled into functional pathways that can be plausibly linked to preserving the host–microbiota equilibrium. The pathways that have been implicated include those involved in mucosal innate immunity (maintenance and repair of the intestinal epithelial barrier, antimicrobial defense, inflammatory response, autophagy, and cell death), as well as activation and regulation of adaptive immunity (IL-23 receptor signaling, antigen presentation, T cell activation, Th17 differentiation, and regulation of T and B cell responses) (Knights et al., 2013). It is easy to imagine (at least in relatively broad terms) how malfunctioning of one or more of these pathways could contribute to intestinal inflammation: abnormal handling of commensal organisms or opportunistic pathogens could lead to inappropriate immune system activation, or the immune response to even the normal commensal population or its products may not be properly controlled and may become excessive. Given the small effect sizes of most IBD risk loci, it is likely that multiple genetic variants have to act in concert with each other or with environmental factors in order for the disease phenotype to occur. Two of the best-studied IBD-associated genetic variants will be discussed below to illustrate these ideas.

5.1.2.1.1 NOD2

Polymorphisms in the nucleotide-binding oligomerization domain 2 (NOD2) gene represent the first identified and strongest link to the development of IBD, specifically CD, with three coding variants being among those most frequently associated with increased risk (Hugot et al., 2001; Ogura et al., 2001). Individuals with homozygous or compound heterozygous NOD2 variants have a 20- to 40-fold elevated risk of developing CD, while those who harbor only one variant allele have only a slight increase in susceptibility (Hugot et al., 2007). The NOD2 protein is expressed predominantly in the hematopoietic compartment, including myeloid and lymphoid cells, but can also be found in Paneth cells and intestinal epithelial cells, particularly in states of inflammation. Like its more widely distributed relative NOD1, NOD2 is localized to the cytosol and is involved in sensing and responding to peptidoglycan fragments released from the cell wall of commensal and pathogenic bacteria (Caruso et al., 2014). NOD1 responds specifically to γ-D-glutamyl-mesodiaminopimelic acid (iE-DAP) found in the peptidoglycan of Gram-negative and a few Gram-positive bacteria, while NOD2 responds to muramyl dipeptide (MDP), which is a component of both Gram-negative and Gram-positive peptidoglycan (Chamaillard et al., 2003; Girardin et al., 2003a,b; Inohara et al., 2003). These molecules can contaminate the cytosol when pathogens such as *Shigella* invade this compartment, or they can be specifically transported into the cytosol from extracellular or phagosomal sources. The binding of iE-DAP and MDP to NOD1 and NOD2, respectively, induces conformational changes in the proteins that lead to their oligomerization and the consequent activation of signals that increase expression of inflammatory cytokines, neutrophil and monocyte chemoattractants, and antimicrobial molecules, such as Paneth cell α-defensins (Caruso et al., 2014). The NOD proteins can also interact with the ATG16L1 protein, and thus help to eliminate microbial pathogens by inducing autophagy (Homer et al., 2010; Travassos et al., 2010). The importance of this interaction is indicated by the fact that CD-associated NOD2 variants are impaired in ATG16L1 recruitment (Homer et al., 2010; Travassos et al., 2010). However, it should be mentioned that mice with a hypomorphic mutation of *ATG16L1* have increased resistance to the bacterial enteropathogen *Citrobacter rodentium*, in contrast to NOD2-deficient animals, which have heightened susceptibility to this pathogen (Kim et al., 2011; Marchiando et al., 2013). Thus, the exact role of NOD2 and ATG16L1 in the response to infection remains to be clarified. In addition to having putative cell-intrinsic antimicrobial effects, ATG16L1 and autophagy have been implicated in the exocytosis of Paneth cell granules, an important source of antimicrobial molecules released into the gut lumen (Cadwell et al., 2008, 2010). Interestingly, mutations in *ATG16L1*, as well as other genes involved in autophagy, have been linked to CD susceptibility, emphasizing the pathogenic importance of defects in this pathway (Brain et al., 2012).

The responses elicited by NOD1 and NOD2 activation are all potential contributors to clearance of infection. Consistent with this idea, mice that are deficient in either NOD1 or NOD2 have increased susceptibility to a variety of microbial pathogens, both intestinal and extraintestinal (Kim et al., 2011; Caruso et al., 2014). NOD1 and NOD2 are also probably involved in the handling of commensal organisms. NOD2-deficient mice have a significant dysbiosis of the gut microbiota, and the altered microbiota predisposes to colitis (Couturier-Maillard et al., 2013). A recent study also showed that the intestinal microbiota of NOD2-deficient mice have a specific expansion of the common commensal *Bacteroides vulgatus* in association with inflammatory abnormalities in the epithelium (Ramanan et al., 2014). However, it should be mentioned that some studies have failed to demonstrate any changes in baseline microbiota composition in NOD1 or NOD2 knockout mice (Robertson et al., 2013). In humans with IBD, *NOD2* variants have been associated with significant changes in microbiota composition, including an increase in mucosa-adherent bacteria, decreased abundance of *Faecalibacterium*, and increased numbers of *Escherichia* (Swidinski et al., 2002; Frank et al., 2011; Rehman et al., 2011; Li et al., 2012).

What is the mechanistic link between *NOD2* mutations and the pathogenesis of CD? There is no definitive answer to this question currently, but several hypotheses have been suggested (Caruso et al., 2014). Common to most of the hypotheses is the idea that the CD-associated *NOD2* mutations compromise functions that are required, either directly or indirectly, for bacterial recognition and clearance in the gastrointestinal tract. The failure to control commensals or pathogens then leads to intestinal inflammation mediated by NOD2-independent pathways. Another idea (which does not exclude the first) is that

mutations in NOD2 impair the protein's ability to suppress pro-inflammatory responses. This was first suggested by the finding that Toll-like receptor (TLR)–induced IL-12 production was elevated in NOD2-deficient macrophages and dendritic cells (Watanabe et al., 2006). This observation was called into question by the results of subsequent studies (Caruso et al., 2014). However, more recent work has renewed support for the idea that NOD2 is a negative regulator of inflammatory responses. Experiments with human dendritic cells demonstrated that signals activated by NOD2 increased expression of the microRNA miR-29, which in turn acted to decrease expression of several mediators, particularly IL-23 (Brain et al., 2013). In addition, the NOD2-recruited protein ATG16L1 was shown to negatively regulate NOD1- and NOD2-induced inflammatory responses independent of its role in autophagy (Sorbara et al., 2013).

5.1.2.1.2 IL23R

Polymorphisms in the gene for the IL-23 receptor (IL23R) were first linked to CD and UC susceptibility in 2006 (Duerr et al., 2006). Several subsequent studies confirmed the association, extended it to other autoimmune diseases, such as psoriasis and ankylosing spondylitis, and demonstrated that genetic variants of signaling molecules downstream of the IL-23 receptor also influenced IBD risk (Cho and Feldman, 2015; Lubberts, 2015). Both coding and noncoding polymorphisms of IL23R have been associated with IBD, with an uncommon coding variant (R381Q) conferring a strong protective effect against CD, as well as against psoriasis and ankylosing spondylitis (Duerr et al., 2006; Capon et al., 2007; Cargill et al., 2007; Rueda et al., 2008). *In vitro* experiments using T cells from healthy donors showed that the protective R381Q variant attenuated IL-23-activated signals and reduced production of the pro-inflammatory cytokine IL-17 (Di Meglio et al., 2011; Pidasheva et al., 2011; Sarin et al., 2011). These results suggest that excessive activation of IL-23-induced responses contributes to the pathogenesis of IBD and other autoimmune diseases.

IL23R associates with IL12Rβ1 to form the heterodimeric receptor for IL-23 (Eken et al., 2014). It is expressed on multiple cell types in the gastrointestinal tract, including Th17 cells, natural killer (NK) T cells, γδ T cells, group 3 innate lymphoid cells (ILC3s), dendritic cells, macrophages, and neutrophils. IL-23 is also a heterodimer and consists of two subunits: p40, which is shared with IL-12, and p19. It is expressed mainly by monocytes, macrophages, and dendritic cells in response to the sensing of microbial molecules by various pattern recognition receptors, including NOD2. IL-23 is found constitutively in the terminal ileum of conventionally reared but not germ-free mice, indicating an important role for the gut microbiota in inducing intestinal expression of this cytokine (Becker et al., 2003). Several studies in mice have demonstrated that IL-23 plays a crucial role in the pathogenesis of inflammation in the gut and other tissues (McGovern and Powrie, 2007). One of the important mechanisms underlying this role involves the ability of IL-23 to promote the differentiation of so-called pathogenic Th17 cells. These cells are characterized by the production of high levels of IL-17, IL-22, and interferon γ (IFNγ), and are able to induce tissue inflammation following adoptive transfer (Burkett et al., 2015). Mucosal tissues like the gastrointestinal tract are important sites of Th17 differentiation, and a large number of the helper T cells in the small intestinal lamina propria at steady state are Th17 cells, presumably of the nonpathogenic variety (characterized by expression of IL-17 and IL-10). The gut microbiota is required for the development and maintenance of intestinal Th17 cells, with segmented filamentous bacteria (SFB) playing a key role in these processes in mice (Ivanov et al., 2008). Indeed, many murine intestinal Th17 cells are specific for SFB-derived antigens, although it is not clear if the antigens are required for differentiation of the cells. Circumstances that result in excess IL-23 stimulation, together with cytokines such as IL-1β and IL-6, favor the development of the pathogenic subset of Th17 cells (Burkett et al., 2015). IL-23 also influences the differentiation and function of ILC3s, an innate lymphoid population that resembles Th17 cells in producing IL-17, IL-22 and IFNγ (Eken et al., 2014). IL-17 produced by Th17 cells and ILC3s induces an inflammatory cascade, including the recruitment or activation of innate and adaptive immune cells and the release of various cytokines, that can help in the elimination of certain pathogens, but can also lead to tissue damage when uncontrolled.

Based on the observations summarized above, it seems plausible that genetic polymorphisms that increase signaling through the IL-23 receptor pathway could increase the risk of IBD by promoting a pathologic IL-17-mediated inflammatory response to otherwise innocuous microbial stimuli. This idea is consistent with observations that IL-17 is found at high levels in inflamed intestinal mucosa from IBD patients, and that a monoclonal antibody against IL-23 p40 provides therapeutic benefits in CD (Hundorfean et al., 2012; Sandborn et al., 2012a). Interestingly, antibodies to IL-17 or the IL-17 receptor did not have efficacy in CD and worsened disease in some patients (Hueber et al., 2012), suggesting that IL-17 may be protective in some situations and that IL-23 may have pathogenic effects that are not mediated by IL-17.

5.1.2.1.3 Monogenic IBD

In contrast to typical adult-onset IBD with its complex genetic architecture, there is a rare form of the disease that presents with aggressive intestinal inflammation in infancy and that is monogenic in nature. Patients with this type of infantile-onset IBD, many of whom are from populations with high rates of consanguinity, have homozygous or compound heterozygous mutations in genes encoding the anti-inflammatory cytokine IL-10 or the subunits of the IL-10 receptor (Kotlarz et al., 2012; Moran et al., 2013). The severe clinical problems associated with these mutations highlight the importance of IL-10 as a regulator of inflammatory responses in the gut (Kole and Maloy, 2014). Infantile-onset IBD is often refractory to conventional therapy but has been treated successfully with hematopoietic stem cell transplantation (HSCT). IBD-like manifestations can also occur early in life in patients with monogenic disorders caused by mutations in several other genes, including X-linked inhibitor of apoptosis (XIAP), tetratricopeptide repeat domain 7A (TTC7A), and genes encoding components of nicotinamide adenine dinucleotide phosphate (NADPH) oxidases (Aguilar et al., 2014; Avitzur et al., 2014; Dhillon et al., 2014; Okou and Kugathasan, 2014). Such mutations are likely to affect the same functional pathways (intestinal epithelial barrier properties, antimicrobial functions, etc.) that have been implicated in IBD pathogenesis in adults. In this context, it is interesting that mutations in XIAP have been associated with impairments of NOD1/2 signaling (Zeissig et al., 2015). Thus, the infantile-onset and adult-onset forms of IBD can be viewed as representing two ends of the disease spectrum, with the former being caused by single-gene mutations with large effects and the latter involving interactions between multiple genetic variants, each with a small effect, as well as interactions between genetic and environmental factors (Okou and Kugathasan, 2014). Very early-onset (VEO) IBD (age at diagnosis <10 years) and early-onset IBD (age at diagnosis 10–17 years) may represent intermediate forms in this spectrum with respect to genetic contribution and clinical presentation (Benchimol et al., 2014; Okou and Kugathasan, 2014).

5.1.2.1.4 Genetics and Disease Course

In addition to influencing IBD susceptibility per se, genetic variation can also affect the course of the disease. Recent studies found that a noncoding polymorphism in the *FOXO3A* gene had a significant impact on the severity of CD without affecting disease risk itself (Lee et al., 2013). The minor allele of *FOXO3A*, which was associated with a milder clinical course, was shown to attenuate monocyte inflammatory responses, in part by increasing production of IL-10. Another group of investigators found that *NOD2* variants also influenced disease severity, with effects that were stronger than those of *FOXO3A* in their population (Schnitzler et al., 2014). This is an emerging area of investigation and will no doubt yield additional insights into pathogenesis as more data is obtained.

5.1.2.2 Microbiota

IBD is characterized by significant changes in the composition of the intestinal microbiota (Kostic et al., 2014). The pathogenic importance of these alterations is illustrated by the ability of antibiotic treatment to reduce inflammation in some forms of IBD (Sartor, 2004). Based on the results of multiple studies that have used DNA sequence–based approaches to characterize the microbiota, including an analysis of a large number of treatment-naïve CD patients, several consistent patterns of abnormalities have been noted in IBD (Frank et al., 2011; Gevers et al., 2014; Kostic et al., 2014). There is an overall decrease in species richness (α-diversity), especially in inflamed tissues, with shifts in the relative abundances of major groups of bacteria. The proportions of Enterobacteriaceae and Fusobacteria, including organisms that are invasive and have pro-inflammatory potential, are increased. One such organism is adherent-invasive *Escherichia coli* (AIEC), which has been found at elevated levels in biopsies of ileal CD (Darfeuille-Michaud et al., 2004; Martin et al., 2004; Carriere et al., 2014). These bacteria are able to invade and survive within intestinal epithelial cells and macrophages, particularly in the context of impaired autophagy, and can induce inflammatory cytokine expression (Carvalho et al., 2009; Lapaquette et al., 2010). Other groups of bacteria are relatively depleted in patients with IBD. A well-studied example is *Faecalibacterium prausnitzii*, an organism with anti-inflammatory characteristics that is decreased in abundance in ileal biopsies of CD patients (Sokol et al., 2008, 2009; Willing et al., 2009). Higher levels of *F. prausnitzii* have been correlated with better prognosis in both CD and UC (Sokol et al., 2008; Varela et al., 2013). One of the mechanisms by which *F. prausnitzii* limits inflammation is by generating short-chain fatty acids (SCFAs) (acetate, butyrate, and propionate) from dietary fiber (Benus et al., 2010). SCFAs have a number of beneficial effects, including promoting the development of regulatory T cells (Tan et al., 2014). Interestingly, SCFA-producing bacteria other than *F. prausnitzii* are also decreased in the microbiota of IBD patients (Kostic et al., 2014). Illustrating the clinical relevance of such studies, a recent analysis identified ileal microbial communities and gene expression profiles in untreated pediatric CD patients that appeared to be independent of inflammation and that were predictive of disease course (Haberman et al., 2014).

Recent studies of the microbiota have made use of metagenomic sequencing, which provides both taxonomic information and insights into the metabolic capabilities of the microbial community. The results of these analyses have confirmed that the IBD-associated microbiota has a reduction in genes that function in SCFA production, and that it also has increases in genes involved in autotrophy, metabolism of sulfur-containing amino acids, resistance to oxidative stress, toxin secretion, and expression of virulence factors (Kostic et al., 2014). All these traits suggest that the microbiome in IBD has undergone a shift toward a more inflammatory character. It should be mentioned that most studies of the microbiota have focused so far on bacteria. However, large numbers of bacteriophages, other viruses, and fungi also contribute to the microbial communities of the gastrointestinal tract and other body surfaces, and the role of these organisms in IBD remains to be elucidated. A more complete understanding of the composition, population dynamics, and functions of the human intestinal microbiota will ultimately prove useful in devising new ways to manipulate these characteristics for the purpose of controlling the inflammatory process in IBD.

5.1.2.3 Environmental Factors

Genetic variation contributes to only a fraction of IBD risk, as indicated by observations that concordance rates in monozygotic twins are less than 50%, that most individuals who have homozygous risk variants of IBD-associated genes like NOD2 do not develop the disease, and that the currently identified IBD-associated loci account for less than 20% of CD and UC variance (Caruso et al., 2014; Ananthakrishnan, 2015; Lee et al., 2015a). Moreover, epidemiologic studies have shown that there has been a relatively recent and dramatic increase in IBD incidence in multiple ethnic populations that corresponds to changes in lifestyle, and that migration from one geographic region to another can modify disease risk (Molodecky et al., 2012). Taken together, these findings strongly suggest that nongenetic (environmental) factors contribute significantly to IBD pathogenesis. A number of environmental influences have been shown to affect IBD incidence, relapse, or disease course, but the mechanisms involved are unclear in most cases (Ananthakrishnan, 2015).

One of the best-studied nongenetic risk factors is cigarette smoking, which increases the risk of CD incidence by about twofold and also increases the severity of established disease (Cosnes et al., 1996). Interestingly, smoking is protective against UC incidence, although the reason for this distinction from CD is not known (Mahid et al., 2006). Individuals with variants in certain genes involved in the metabolism of nicotine and oxidative radicals may be particularly vulnerable to the effects of smoking, a finding that highlights the importance of gene–environment interactions (Ananthakrishnan et al., 2014). Diet has also been shown to have a significant impact on IBD incidence or recurrence, with certain food components, such as meat and omega-6 polyunsaturated fatty acid (n-6 PUFA), increasing risk, while others, such as fiber and n-3 PUFA, decrease risk

(Lee et al., 2015a). A particularly intriguing idea about diet and IBD, for which there is epidemiologic and experimental support, is that increasing consumption of emulsifiers like polysorbate 80 in processed foods is one contributor to the rising incidence of CD and UC (Roberts et al., 2013; Chassaing et al., 2015). Some of the other nongenetic factors implicated in IBD include breast feeding (protective), sunlight and vitamin D (protective, especially for CD), physical activity (protective for CD), early-life exposure to antibiotics (increased risk), appendectomy (protective for UC and increased risk for CD), and psychosocial stress (increased risk for CD) (Andersson et al., 2001, 2003; Barclay et al., 2009; Kronman et al., 2012; Ananthakrishnan, 2015).

The mechanisms that link environmental factors to IBD pathogenesis are not well understood. However, for many of the factors, a plausible connection can be made via alterations of the microbiota. Certainly, the type of food consumed can have a major impact on the composition of the intestinal microbial community (Albenberg and Wu, 2014). Diets that are associated with relative protection from IBD (rich in fruits, vegetables, and fiber) give rise to a gut microbiota composition that is quite distinct from those associated with increased IBD risk (rich in saturated fat and refined sugar) (De Filippo et al., 2010). Moreover, the former type of diet selects for a microbial community that produces anti-inflammatory SCFAs, while the latter promotes the outgrowth of pathobionts that have pro-inflammatory effects (Tan et al., 2014; Devkota and Chang, 2015). The deleterious effects of food emulsifiers have also been attributed to altered microbiota composition, including reduced diversity; an increase in mucosa-associated, inflammation-inducing Proteobacteria; and a bloom of Verrucomicrobia (Chassaing et al., 2015). Similarly, psychosocial stress and smoking have been associated with clear changes in microbiota composition (Biedermann et al., 2013; Galley et al., 2014). It is possible that other environmental factors that influence IBD risk may also act by inducing alterations in the resident microbial community of the intestine. Given the multiple ways in which the gut microbiota can influence the development and function of various cells of the immune system (Erturk-Hasdemir and Kasper, 2013), it should be no surprise that changes in the composition of this community can have effects on the development and course of immune-mediated diseases like IBD.

5.1.2.4 Summary of IBD Pathogenesis

To summarize current thinking, IBD pathogenesis involves complex interactions between multiple genetic loci, nongenetic influences, and the intestinal microbiota. Figure 5.1 illustrates a potential model of these interactions. In a genetically susceptible individual, environmental factors, possibly acting through the microbiota, trigger a series of immunological events that lead to a dysregulated intestinal inflammatory response to microbial stimuli. The tissue damage and microbiota alterations that result from the inflammation may further activate the immune system so that the process becomes self-perpetuating and chronic. Once initiated, the immunopathology rarely, if ever, resolves completely. However, both genetics and the environment can influence the course of the disease, including its severity, frequency of relapses, development of complications, and response to treatment. Continued elucidation of the molecular details of IBD pathogenesis will be required if we are to develop new and improved therapeutic strategies for this condition.

5.1.3 CLINICAL MANIFESTATIONS

Patients with IBD typically present with predominantly gastrointestinal symptoms. Abdominal pain (25%–43%) and diarrhea (35%–39%)—either bloody or nonbloody—are the most common symptoms (Gupta et al., 2008). Weight loss is seen in 8%–23% for a multitude of reasons—decreased nutrient absorption in the inflamed intestine, diminished caloric intake secondary to pain with eating, and increased caloric demands given the active inflammation (Gupta et al., 2008). Oral aphthous ulcers are seen in 21% of patients (Vavricka et al., 2015a). Beyond the gastrointestinal manifestations of IBD, inflammation can occur in various extraintestinal tissues. Ocular manifestations include episcleritis, scleritis, and anterior uveitis and occur in ~10% of patients (Taylor et al., 2006). Patients with IBD may develop dermatologic involvement, such as erythema nodosum (10%–15%) and pyoderma gangrenosum (0.4%–2%) (Vavricka et al., 2015b). Joint involvement includes peripheral arthritis that can be either

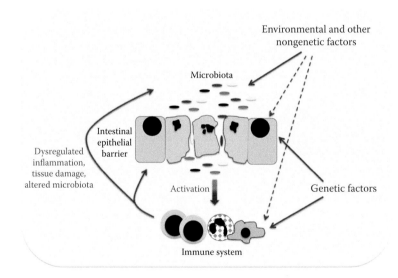

Figure 5.1. Schematic outline of IBD pathogenesis. In individuals who have genetic variants that affect functioning of the immune system or intestinal epithelial barrier, environmental and other nongenetic influences induce a microbiota-driven, dysregulated inflammatory response in the gut. The intestinal inflammation damages local tissues, including the epithelium, and also alters microbiota composition. These effects lead to further activation of the immune system as a result of increased exposure to microbial components. The nongenetic factors may exert their influence via changes in the microbiota or, possibly (as indicated by the dashed arrows), through alterations in epithelial barrier properties or immune function.

pauciarticular (involving <5 joints) or polyarticular (involving ≥5 joints), as well as arthritis involving the spine (Vavricka et al., 2015b). Most extraintestinal manifestations (with the exceptions of axial arthritis and polyarticular peripheral arthritis) are associated with active colonic disease, and the treatment of these complications is tightly related to the treatment of the underlying IBD (Vavricka et al., 2015b). Interestingly, as many as 25% of patients with IBD develop these extraintestinal manifestations prior to the onset of gastrointestinal symptoms (Vavricka et al., 2015a).

IBD can lead to a number of secondary problems, which in CD include the development of intestinal strictures, with associated obstructive symptoms, and the formation of fistulae (Gupta et al., 2008). One of the long-term complications of IBD, particularly UC, is the development of colitis-associated cancer, which occurs in about 10% of UC patients after 30 years (Foersch and Neurath, 2014). Pediatric-specific complications of IBD include growth failure (in 10%–56% of CD), pubertal delay, and bone density compromise (Abraham et al., 2012). Anemia is a commonly encountered problem in IBD and is multifactorial in origin. Vitamin B_{12} deficiency leads to a macrocytic anemia and occurs in 6%–33% of CD patients (especially when the ileum is involved) and 3%–16% of UC patients (Ward et al., 2015). Iron deficiency anemia can occur in IBD because of macroscopic or occult blood loss, as well as small bowel inflammation and compromised iron absorption (Stein et al., 2010). Anemia can also occur secondary to inflammation as a result of elevated levels of the hormone hepcidin, which downregulates the transporter involved in iron recycling and absorption (Murawska et al., 2016). Regardless of the exact cause, persisting anemia in IBD is associated with more aggressive disease (Koutroubakis et al., 2015).

5.1.4 DIAGNOSIS

The diagnosis of IBD depends on identification of classic findings by endoscopic evaluation (of both the upper and lower gastrointestinal tract), histologic examination of mucosal biopsies from intestinal segments, and radiologic assessment of the gastrointestinal tract. The timely identification of patients who require endoscopic evaluation is vital, as one must balance the need to avoid delays in diagnosis with the relative invasiveness of endoscopy. The screening process often includes a laboratory evaluation of the blood (hematocrit, platelet count, erythrocyte sedimentation rate, and C-reactive protein), although this testing is normal at the time of diagnosis in 9% of UC patients and 19% of CD patients (Mack et al., 2007). Testing of the stool for occult blood can improve the diagnostic accuracy (Moran et al., 2015).

The use of fecal markers of inflammation to screen for the presence of IBD has been recently developed. Calprotectin is a cytosolic protein predominantly produced by neutrophils that is an attractive noninvasive marker to screen for IBD given its stability at room temperature and resistance to degradation in the colon. Although this marker can be elevated during acute infectious conditions, measurement of calprotectin as a screening tool for IBD has recently been shown to be accurate in both adult and pediatric patients (Yang et al., 2014; Mosli et al., 2015). Other studies have shown the utility of measuring lactoferrin as an additional stool inflammatory marker (Mosli et al., 2015). These markers have also been shown to predict disease relapse in IBD and can act as more sensitive indicators of the return of mucosal inflammation than clinical symptoms (De Vos et al., 2013; Boschetti et al., 2015).

The macroscopic appearance of UC is continuous inflammation in the large intestine from the rectum to a point more proximal, whereas CD often includes skip lesions and can affect any segment of the gastrointestinal tract. Figure 5.2 shows the endoscopic appearance of UC and CD. The definition of IBD location by the Paris classification system relies on macroscopic findings during endoscopic and radiologic assessment. CD location is classified as ileal (L1), colonic (L2), or ileocolonic (L3) (Levine et al., 2011). Location is further classified based on the presence of perianal disease (p), inflammation in the upper gastrointestinal tract proximal to the ligament of Treitz (L4a), or inflammation in the small intestine proximal to the terminal ileum (L4b). The CD phenotype is divided into inflammatory (B1), stricturing (B2), penetrating with internal fistula (B3), or a combination of stricturing and penetrating disease (B2B3). The B1 phenotype is present in 62%–73% of CD patients at the time of diagnosis, although this declines to 30% at 10 years postdiagnosis, as patients progress to more complicated phenotypes (Solberg et al., 2007). UC is defined

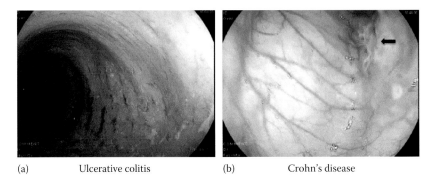

| (a) | Ulcerative colitis | (b) | Crohn's disease |

Figure 5.2. Endoscopic appearance of UC and CD. (a) The typical appearance of UC includes loss of normal colonic mucosa vasculature, loss of colonic folds with featureless colon, diffuse involvement, and contact bleeding. Note the small areas of ulcerated mucosa surrounded by inflamed mucosa. (b) The typical appearance of CD includes deep ulcerations (arrow) surrounded by normal-appearing mucosa with normal blood vessels visible.

by extent of involvement in the large intestine: proctitis (E1), left-sided colitis (distal to the splenic flexure; E2), extensive colitis (distal to the hepatic flexure; E3), and pancolitis (macroscopic inflammation proximal to the hepatic flexure; E4) (Levine et al., 2011).

Histologic findings of IBD vary depending on the segment and should include some features of chronicity to help differentiate the inflammation from that of acute, self-limited infections. The traditional hallmark of CD histopathology is the granuloma, but it is only identified in 37%–49% of cases at the time of diagnosis (Heresbach et al., 2005). Esophageal inflammation characterized as granulomatous, lymphocytic, neutrophilic, or eosinophilic has been described in CD (Ebach et al., 2011; Sutton et al., 2014). Gastric findings include chronic active gastritis and focally enhancing gastritis, which is a lesion initially thought to signify CD, although studies have shown that this is a common finding in UC as well (Ushiku et al., 2013). Ileal biopsies often show villous blunting and a neutrophilic infiltrate with chronic ileal changes that include pseudopyloric metaplasia. Features of chronic colitis include disarray of the colonic crypts, basal layer lymphocytosis, and Paneth cell metaplasia in the left colon. Despite a variable constellation of histologic findings, few studies have identified prognostic value in these features.

As IBD is commonly thought to involve aberrant immune responses to nonpathogenic stimuli, it is not surprising that patients with IBD may demonstrate immune reactivity to commensal flora. Anti-*Saccharomyces cerevisiae* antibodies (ASCAs) are traditionally thought to occur in CD, while antineutrophil cytoplasmic antibodies (ANCAs) occur commonly in UC. Although these antibodies have previously been used as both a screen for IBD and a potential discriminating factor between CD and UC, they are not consistently accurate, as 33%–36% of IBD patients lack both ASCAs and ANCAs (Benor et al., 2010). Further, ASCAs associate with ileal CD, while ANCAs associate with colonic IBD involvement (UC or CD) (Benor et al., 2010). The use of additional antibodies directed against microbes (e.g., CBir-1 and OmpC) helps to improve the sensitivity of diagnosis. Perhaps the most useful application of serology is to predict aggressive CD. Prospective data demonstrates that patients with CD who have particular NOD2 single-nucleotide polymorphisms (R702W, G908R, or 1007fs) or remarkably high antibody titers (ASCA, anti-OmpC, anti-CBir1, anti-I2) are more likely to develop complicated (B2, B3, or B2B3) disease (Lichtenstein et al., 2011).

5.1.5 TREATMENT

Long-standing goals of therapy for IBD focus on reduction of clinical symptoms and prevention of complications. Pediatric-specific therapy goals also include prevention of pubertal delay, maximizing bone density accrual, and maintaining age-appropriate weight and height growth. An evolving concept is the need to attain mucosal healing (either macroscopic or histologic). Mucosal healing has recently been shown to be a strong predictor of prolonged remission (Ferrante et al., 2013; Shah et al., 2016). Although clinical activity indices exist for both UC and CD, the correlation between these indices and the activity of mucosal inflammation is relatively poor (Efthymiou et al., 2008). Currently available therapies for IBD largely focus on reducing inflammation (both locally and systemically) with the use of broadly acting anti-inflammatory agents, antimetabolites, specific immunomodulators, nutritional support, and manipulation of the intestinal microbiome. Given the limited ability to predict aggressive disease in IBD (except for some evidence of NOD2 variants and excessive immune response to commensal flora) and the inability to predict response to individual therapies for a specific patient, treatment decisions are often based on the perception of the aggressiveness of disease.

5.1.5.1 Anti-Inflammatory Agents

5.1.5.1.1 5-Aminosalicylates

5-Aminosalicylates (5-ASAs) are commonly prescribed in UC because of their efficacy and favorable side effect profile. There are numerous potential mechanisms of action of 5-ASAs that include activation of the anti-inflammatory transcription factor PPAR-γ, inhibition of prostaglandin and leukotriene synthesis, and inhibition of cytokine-induced activation of the pro-inflammatory transcription factor NF-κB (Desreumaux et al., 2006). 5-ASAs are largely used for their local effects within the gastrointestinal tract, although 53% of the active drug is found in the urine. Mesalamine, a commonly used 5-ASA, is rapidly absorbed in the proximal gastrointestinal tract, with limited benefit if it is administered in uncoated form. As a result, numerous release mechanisms have been developed based on the desired site of action, including timed release, pH-dependent release, and release that requires bacterial lysis of an azo bond between mesalamine and a carrier. Mesalamine has been shown to be effective in 50%–72% of mild to moderate UC (Sandborn et al., 2009). The efficacy of mesalamine-based products has shown more variable results in CD, with some studies displaying advantage over placebo, while others show no benefit. However, results from a meta-analysis of 5-ASAs in CD showed a possible modest benefit (Ford et al., 2011). The mechanistic basis for the differential effect of 5-ASAs in CD and UC is not understood. Based on the available data, current treatment guidelines for pediatric CD recommend consideration of 5-ASAs only in mild disease, while adult guidelines avoid recommending their use in CD (Terdiman et al., 2013; Ruemmele et al., 2014).

5.1.5.1.2 Corticosteroids

Corticosteroids remain a commonly used medication for induction of remission in IBD, as well as for a number of other chronic inflammatory conditions. Potential mechanisms of action for corticosteroids include interactions with glucocorticoid response elements in gene promoters that reduce expression of pro-inflammatory cytokines (e.g., IL-2, IL-8, and TNFα) and cell surface adhesion molecules (e.g., intercellular adhesion molecule-1 [ICAM-1] and vascular cell adhesion molecule-1 [VCAM-1]), inhibition of protein synthesis by action at the RNA level, and direct interaction with NF-κB (Hayashi et al., 2004). Numerous studies have shown corticosteroids to be effective at inducing remission in IBD, with advantages over both placebo and 5-ASA (Benchimol et al., 2008).

Corticosteroids have not been shown to induce mucosal healing, nor are they an effective treatment for maintaining remission (Steinhart et al., 2003). Corticosteroid-dependent patients are those who demonstrate an initial response but develop symptoms with decreased corticosteroid dosing and require escalation of therapy. Corticosteroids carry significant adverse effects, including immediate side effects (mood and sleep disturbances), as well as concerns over bone density loss with chronic use (Azzopardi and Ellul, 2013). Further, the risk of infection for IBD patients on corticosteroids appears to be higher than with other immunosuppressive medications used in IBD (Lichtenstein et al., 2012). Although corticosteroids are effective at inducing clinical remission, their inability to induce mucosal healing, along with their side effects and potentially inferior long-term efficacy, limits their use in IBD.

5.1.5.2 Antimetabolites

5.1.5.2.1 Thiopurines

Thiopurines are broadly acting immunomodulators that are commonly used in IBD and include 6-mercaptopurine (6-MP) and azathioprine (a prodrug of 6-MP). Thiopurines act as purine analogs and inhibit purine biosynthesis, thereby inhibiting T cell activation. Metabolism of 6-MP and azathioprine proceeds down three pathways. Xanthine oxidase converts a large proportion of 6-MP to inert 6-thiouric acid. Thiopurine methyltransferase (TPMT) converts 6-MP to 6-methyl-mercaptopurine (6-MMP), which can be hepatotoxic at high levels and offers no immunosuppressive effects. The third pathway involves 6-MP conversion to 6-thioguanine (6-TG), which has the desired immunosuppressive effects. Genetic variants of TPMT that occur in 0.3% of individuals result in the absence of TPMT enzymatic activity and lead to severe myelosuppression (because of disproportionate shunting of 6-MP substrate toward 6-TG) (Weinshilboum and Sladek, 1980). Patients with IBD considering use of thiopurines should be screened for TPMT enzyme activity, and those with absent activity should not receive thiopurines due to the risk of severe myelosuppression (Ruemmele et al., 2014). Drug monitoring is possible with thiopurine use, as there is some correlation between thiopurine efficacy and 6-TG levels. Metabolite monitoring also allows for identification of individuals who disproportionately shunt metabolism toward 6-MMP (and away from 6-TG). Such individuals might benefit from the use of allopurinol and low-dose (25%–30% of standard dosing) thiopurine since allopurinol can inhibit both xanthine oxidase and TPMT (Curkovic et al., 2013). Early use of 6-MP was shown to be superior to placebo in maintaining remission in pediatric CD after initiation of remission with corticosteroids. As a result of this data, 80% of children with moderate to severe CD are placed on thiopurines within 1 year of their diagnosis to limit corticosteroid use (Punati et al., 2008). Despite evidence supporting early use of azathioprine in pediatric CD, studies in adults with newly diagnosed CD have not shown a benefit from early initiation of azathioprine in all forms of adult CD (Cosnes et al., 2013; Panes et al., 2013).

5.1.5.2.2 Methotrexate

Methotrexate is an antimetabolite that is used in chemotherapy for malignant disease that acts as a competitive inhibitor of dihydrofolate reductase. Methotrexate is also commonly used in inflammatory diseases such as rheumatoid arthritis and lupus. However, the mechanism of the anti-inflammatory action is not clear (Chan and Cronstein, 2013). In addition to the folate depletion mechanism, methotrexate has also been shown to increase extracellular adenosine and induce apoptosis of activated T cells. Methotrexate is effective in inducing and maintaining remission in CD (Feagan et al., 1995, 2000). Although most studies have examined subcutaneous methotrexate dosing, long-term subcutaneous administration may only offer a modest benefit over oral administration in CD, and pharmacokinetic studies have shown that the bioequivalence is only slightly less in the oral route (Stephens et al., 2005; Wilson et al., 2013; Turner et al., 2015). Despite efficacy in CD, studies of methotrexate in either inducing or maintaining remission in UC have shown it to be no better than placebo (Oren et al., 1996; Chande et al., 2014; Wang et al., 2015).

5.1.5.3 Specific Immunomodulators

5.1.5.3.1 Calcineurin Inhibitors

Severe exacerbation of IBD often requires hospitalization and use of intravenous corticosteroids, but some severe cases do not respond to corticosteroids. Rather than advancing to surgical treatment, calcineurin inhibitor use began in the 1980s for these steroid-refractory cases of IBD. Cyclosporine is a calcineurin inhibitor that acts by forming a complex with cyclophilin that directly binds calcineurin and prevents calcineurin-induced dephosphorylation and activation of the transcription factor nuclear factor of activated T cells (NFAT) (Ruhlmann and Nordheim, 1997). NFAT activation is a key step in T cell receptor–mediated IL-2 production (Ruhlmann and Nordheim, 1997). Cyclosporine is an effective salvage therapy for patients with UC and CD failing intravenous corticosteroids (Laharie et al., 2012; Lazarev et al., 2012). Cyclosporine in severe acute UC can lead to short-term colectomy sparing in up to 95% of patients (Laharie et al., 2012). Despite its efficacy during a severe exacerbation for UC or CD in adults and children, the majority of patients requiring cyclosporine will progress to colectomy because of loss of response within 1 year (Lazarev et al., 2012). Long-term use of high-dose cyclosporine for CD can lead to hypertension (due to nephrotoxicity), paresthesias, and high cholesterol levels, while low-dose cyclosporine use in CD has not been shown to be effective. As such, most clinicians consider cyclosporine a bridge to alternative medical therapy or to delay the need for surgery.

Tacrolimus (FK506) is a calcineurin inhibitor that binds calcineurin by forming a complex with FK-binding proteins (FKBPs) and subsequently prevents calcineurin-induced NFAT dephosphorylation (Ruhlmann and Nordheim, 1997). Although not used for mild to moderate UC, tacrolimus has been used for hospitalized steroid-refractory severe UC to avoid in-hospital colectomy and appears to offer better long-term colectomy-free outcomes than cyclosporine (Yamamoto et al., 2011; Ogata et al., 2012). Similar to cyclosporine, tacrolimus is often used as a bridge to other therapies rather than as a long-term maintenance treatment since sustained remission rates at 6 months with tacrolimus in UC are only 54% (Watson et al., 2011; Yamamoto et al., 2011). However, given the low number of clinical trials involving tacrolimus, its use is still limited.

5.1.5.3.2 Anti-TNFα Agents

Given the role that TNFα plays in IBD pathogenesis, therapies directed against this key proinflammatory cytokine have been developed. Anti-TNFα agents possess the ability to neutralize soluble TNFα, although that action does not appear to be the predominant therapeutic mechanism. Binding to membrane-bound TNFα on T cells by anti-TNFα agents induces apoptosis and is likely to play a key role in the biologic action (Shen et al., 2005). Infliximab is a chimeric monoclonal antibody directed against TNFα that is administered intravenously and has been shown to be effective in both inducing and maintaining remission for both CD and UC in children and adults (Hanauer et al., 2002; Rutgeerts et al., 2005; Hyams et al., 2007; McGinnis and Murray, 2008). Although infliximab is an effective therapy for IBD, about 12% of initial responders to infliximab discontinue use by 12 months due to allergic reactions or loss of efficacy, and attrition rates continue to rise over time (Vahabnezhad et al., 2014). This is partially explained by the immunogenicity of infliximab, which can lead to the development of anti-infliximab antibodies. Efficacy of infliximab at 12 months can be improved by combining it with a thiopurine (Armuzzi et al., 2014; Panaccione et al., 2014). Early use (within the first 3 months of diagnosis) of infliximab in pediatric CD was recently shown to be better than early use of drugs such as thiopurines (Walters and Hyams, 2014).

Beyond infliximab, additional anti-TNFα monoclonal antibodies have been used in IBD. Adalimumab is an injectable, human monoclonal antibody against TNFα that is effective in UC and CD (Hanauer et al., 2006; Reinisch et al., 2011; Hyams et al., 2012). Certolizumab, a pegylated, humanized monoclonal antibody (based on a Fab fragment), has also shown efficacy in adult CD (Sandborn et al., 2007). Most recently, golimumab, a fully human monoclonal antibody against TNFα, has been shown to be an effective therapy for UC (Sandborn et al., 2014a). Initial anti-TNFα responders who lose response can benefit from a change to another anti-TNFα medication, although patients who are initial anti-TNFα nonresponders are unlikely to benefit from further anti-TNFα agents (de Silva et al., 2012). Anti-TNFα agents are not universally effective in IBD. Etanercept, a fusion protein of the TNFα receptor and immunoglobulin G1 (IgG1) Fc region, is ineffective in CD, although it has shown benefit in rheumatoid arthritis (Sandborn et al., 2001).

5.1.5.3.3 Anti-IL-12 and -IL-23 Agents

The use of monoclonal antibody therapy directed against cytokines beyond TNFα is limited. IL-12 and IL-23 have been shown to play roles in the pathogenesis of CD (Duerr et al., 2006; Cargill et al., 2007). Ustekinumab is a monoclonal IgG1 antibody against the p40 subunit that is a component of both IL-12 and IL-23. Ustekinumab was shown to be effective in CD in the CERTIFI trial, a Phase 2b randomized, placebo-controlled study of adult subjects who failed to respond to anti-TNFα therapy (Sandborn et al., 2012a). However, its use has not yet been approved for CD.

5.1.5.3.4 Anti-Integrin Agents

The generation of the inflammatory response in the gastrointestinal tract in IBD requires trafficking of lymphocytes from the systemic circulation to the intestinal mucosa. In order to facilitate this trafficking, β_7-based integrins on lymphocytes bind to ligands on vascular endothelium, with $\alpha_4\beta_7$ binding to mucosal addressin cell adhesion molecule-1 (MAdCAM-1), while $\alpha_E\beta_7$ binds to E-cadherin (Gorfu et al., 2009). Blocking these interactions with anti-integrin therapy has been shown to ameliorate colitis in animal models.

Natalizumab is a humanized IgG4 monoclonal antibody that binds α4 integrin. This biologic agent is active in blocking both α4β1 (which interacts with VCAM-1) and α4β7 (which interacts with MAdCAM-1). Natalizumab was shown to be effective in adult CD (Ghosh et al., 2003; Sandborn et al., 2005; Targan et al., 2007). However, the use of natalizumab has been linked to the rare development of progressive multifocal leukoencephalopathy (PMLE) (Van Assche et al., 2005). The pathogenesis of PMLE is related to the reactivation of the John Cunningham (JC) virus in the brain secondary to the inhibition of lymphocyte recruitment by α4β1 blockade. This rare side effect has limited the broad use of natalizumab in CD, but interest in utilizing anti-integrin antibodies in IBD has persisted.

Given the benefits and risks of natalizumab, the use of a more gut-selective anti-integrin antibody is desirable. Vedolizumab was developed as a humanized monoclonal antibody against a gut-homing receptor, α4β7. Vedolizumab was shown to be effective for both UC (in the GEMINI I trial) and CD (in the GEMINI II trial) (Feagan et al., 2013; Sandborn et al., 2013). Patients who are treated with vedolizumab frequently have already failed at least one anti-TNFα agent, and data from GEMINI III demonstrate a benefit of vedolizumab over placebo in patients with CD who have failed anti-TNFα therapy (Sands et al., 2014). To date, there have not been any cases of PMLE in patients treated with vedolizumab.

Alternative anti-integrin monoclonal antibodies are also in development. Etrolizumab is a humanized monoclonal antibody directed against the β7 integrin subunit and has activity directed against both $\alpha_4\beta_7$ and $\alpha_E\beta_7$. Phase 1 and 2 trials involving patients with moderate to severe UC showed that etrolizumab induces clinical remission (Vermeire et al., 2014). Study subjects in the Phase 2 trial of etrolizumab who had high levels of expression of α_E integrin and granzyme A were more likely to have a clinical response to etrolizumab, identifying these molecules as potential biomarkers for future study (Tew et al., 2016). Additional therapies that target integrin–leukocyte interactions that are currently under study include vercirnon (an anti-CCR9 antibody), which has shown promise in Phase 2 studies in CD, and PF-00547659 (an anti-MAdCAM-1 antibody), which failed to show benefit in UC (Vermeire et al., 2011; Keshav et al., 2013).

5.1.5.3.5 SMAD7 Antagonists

Small-molecule therapy has been largely underutilized in IBD. However, recent studies have shown a benefit of small-molecule therapy that modulates the transforming growth factor β (TGFβ) axis. TGFβ is a multifunctional cytokine with immunoregulatory properties that signals by inducing phosphorylation of SMAD3, and this pathway is inhibited by SMAD7 (Wahl, 1994). Inflamed tissue from UC and CD patients demonstrates increased expression of SMAD7 compared with controls. Moreover, mononuclear cells extracted from inflamed colonic tissue demonstrated reduced levels of SMAD3 phosphorylation during in vitro stimulation with TGFβ. This effect was reversed with the use of an antisense SMAD7 oligonucleotide. Based on these observations and a small clinical study demonstrating safety, the use of an antisense SMAD7 oligonucleotide in humans became an attractive therapy. An oral antisense oligonucleotide against SMAD7, Mongersen, was recently shown in a Phase 2 trial to be effective at inducing remission in CD (Monteleone et al., 2015). Further studies are required to determine the broad utility of this approach in CD.

5.1.5.3.6 JAK Inhibitors

Given the clear involvement of the adaptive immune system in IBD pathogenesis, inhibition of cytokine signaling pathways involved in lymphocyte activation is another attractive therapeutic strategy. Tofacitinib is an oral inhibitor of Janus kinase (JAK) activation (Flanagan et al., 2010). Tofacitinib preferentially inhibits JAK1 and JAK3, which are involved in signals activated by IL-2, IL-4, IL-7, IL-9, IL-15, and IL-21, and is effective in the treatment of rheumatoid arthritis, psoriasis, and renal transplant rejection. In an 8-week Phase 2 study of moderate to severe UC, tofacitinib induced a clinical response in 78% of patients using a daily dose of 15 mg, compared with 42% in placebo ($p < 0.001$) (Sandborn et al., 2012b). A 4-week Phase 2 study of tofacitinib in moderate to severe CD failed to show induction of clinical response or remission using similar doses, although markers of inflammation (fecal calprotectin and C-reactive protein) did improve compared with placebo. The failure to reach the primary endpoint may have been due to the short nature of the study and the high placebo response

(Sandborn et al., 2014b). Further studies in IBD are necessary prior to the routine use of tofacitinib in either UC or CD.

5.1.5.4 Nutrition

Given the malnourished state that frequently accompanies IBD, nutrition is a required adjunct to all therapeutic regimens. However, specific nutritional plans can also serve as a primary means to treat IBD.

5.1.5.4.1 Exclusive Enteral Nutrition

Exclusive enteral nutrition (EEN) therapy focuses on ingestion of a single caloric source and was first shown to be effective for CD in the 1990s, although it was likely used prior to that time. Since then, numerous studies have demonstrated the benefit of EEN in CD, with the attractive advantage over other therapies of lacking immunosuppression. EEN requires daily caloric intake in the form of a formula (either orally or by nasogastric tube) and must comprise ≥90% of the total daily caloric intake. It does not appear that there is benefit in choosing an amino acid–based formula compared with a polymeric formula. EEN is as effective as anti-TNF therapy at reducing inflammation in CD (Lee et al., 2015b). Children with CD who are initially treated with EEN have improved linear growth at 2 years compared with those treated with corticosteroids (Grover and Lewindon, 2015). EEN is a preferred first-line therapy for active inflammatory (B1) CD in children (Ruemmele et al., 2014). Although the mechanism of action of EEN is not well understood, the use of EEN leads to significant changes in both the fecal microbiome and the concentrations of intestinal SCFAs (Kaakoush et al., 2015; Quince et al., 2015).

5.1.5.4.2 Specific Carbohydrate Diet

Beyond EEN therapy, a number of other nutritional therapies have been proposed for use in IBD. The specific carbohydrate diet (SCD) limits complex carbohydrates and refined sugar. It has been used in both UC and CD, as well as in nongastrointestinal disorders (e.g., autism and schizophrenia). The SCD has been shown recently to improve clinical symptoms and to induce mucosal healing in children with CD

(Cohen et al., 2014). Partial enteral nutrition (PEN), in which the formulas utilized in EEN comprise ~50% of daily caloric intake, have been shown to be ineffective, although the combination of a diet excluding specific foods and PEN can induce clinical remission (Sigall-Boneh et al., 2014).

5.1.5.5 Manipulation of the Microbiota

Based on the important role played by the microbiota in IBD pathogenesis, the idea of altering the composition and/or functionality of the microbiota for therapeutic purposes is gaining increasing attention. The microbiota can be altered by using antibiotics, probiotics, or fecal microbial transplantation (FMT).

5.1.5.5.1 Antibiotics

The expansive literature demonstrating a dysbiosis of the gut microbiota in patients with IBD would suggest that antimicrobial therapy could be an effective strategy (Kostic et al., 2014). A number of antimicrobial agents have been studied in IBD, including antituberculosis agents, clarithromycin, ciprofloxacin, metronidazole, and rifaximin, and the results of these studies have been mixed. Although a recent meta-analysis demonstrated a modest benefit of antibiotics in active UC, they are generally not recommended unless there is evidence of an enteric infection (Khan et al., 2011; Turner et al., 2012). However, there is recent data to suggest that combination antibiotic therapy can be effective in pediatric fulminant UC (Turner et al., 2014). Metronidazole and ciprofloxacin (either in isolation or in combination) have been shown to be effective for fistulizing perianal CD and can be combined with other therapies (Khan et al., 2011; Dewint et al., 2014). There is limited data from placebo-controlled studies showing benefit from using ciprofloxacin, rifaximin, and a combination of antituberculosis agents in inflammatory (B1) CD, although other studies have failed to replicate the benefit. The results of these individual trials, as well as the meta-analysis data pointing to an overall modest benefit from antibiotics in inflammatory CD, suggest that there may be a subset of CD patients for whom antibiotic therapy would be helpful (Khan et al., 2011; Ruemmele et al., 2014).

5.1.5.5.2 Probiotics

Similar to the potential of using antimicrobials to combat the dysbiosis of IBD, probiotics offer an alternative route to modulate the microbiome. VSL#3 (which is a probiotic mixture containing *Streptococcus*, *Lactobacillus*, and *Bifidobacterium* species) has shown the most promising effect in inducing remission in patients with active UC (Shen et al., 2014). Experimental evidence shows that this mixture reduces pro-inflammatory cytokine production and improves colonic mucin production. VSL#3 has also been shown to be effective at preventing inflammation in the surgically created ileoanal pouch after colectomy in UC (Shen et al., 2014). Another probiotic bacterial strain, *E. coli* Nissle 1917, has shown efficacy in maintaining remission similar to that of mesalamine in patients with UC (Rembacken et al., 1999; Kruis et al., 2004). Despite the benefit of certain probiotics in UC, the data supporting the use of probiotics in CD has not been encouraging. In general, the results of randomized, placebo-controlled trials have not shown benefits from the use of probiotics in CD (Shen et al., 2014). However, future studies of probiotics in IBD would likely benefit from investigation into the fecal microbiome before and after initiation of probiotics to predict response.

5.1.5.5.3 Fecal Microbial Transplantation

Modulation of the microbiome in IBD has expanded beyond simple probiotic administration after the success of FMT for recurrent *Clostridium difficile* colitis. *C. difficile* colitis is an infectious condition whose initial therapy involves antibiotics. Despite appropriate antibiotic therapy, 15%–26% of patients experience a recurrence of the infection, and repeat antibiotics are not as effective as with the initial course of therapy. FMT is highly effective for recurrent *C. difficile* colitis and involves the administration of donor feces orally (via capsule or nasoduodenal tube) or topically (via either retention enema or colonoscopy) (Russell et al., 2010; van Nood et al., 2013). Studies utilizing FMT in UC (in the absence of *C. difficile*) have yielded conflicting results (Moayyedi et al., 2015; Rossen et al., 2015; Suskind et al., 2015). However, there do seem to be specific differences between individual FMT-responsive and FMT-refractory UC patients (e.g., transplant donor species richness)

that may help to identify ways to improve this therapy (Moayyedi et al., 2015; Vermeire et al., 2016). Methods to identify patients and donors who are the optimal candidates for FMT in IBD require additional investigation.

5.1.5.6 Stem Cell Transplantation

HSCT is a therapeutic modality usually reserved for oncologic processes, although it has been utilized in select cases of IBD given the concept of IBD as a defect in immunoregulation. Initial evidence supporting the use of HSCT in IBD derives from case reports of patients with IBD receiving HSCT for lymphoma who had long-term resolution of their IBD, and conversely, other patients have developed *de novo* IBD following HSCT. However, allogeneic HSCT is a standard therapy for patients with immunodeficiency syndromes presenting with VEO-IBD, including chronic granulomatous disease, IL-10 pathway defects, and XIAP defects (Glocker et al., 2009; Moran et al., 2013).

Beyond the specific subsets of VEO-IBD and IBD patients with malignancy requiring allogeneic HSCT, autologous HSCT has been studied in IBD refractory to standard medical treatment. Autologous HSCT is based on the idea that the significant myeloablation that is involved depletes the body of abnormally primed lymphocytes that contribute to the disease process and allows reconstitution with a less autoinflammatory population of B and T cells. Preliminary studies have shown that autologous HSCT is fairly well tolerated in CD and can lead to disease remission. Long-term follow-up of these patients showed that the majority remained in remission from CD, and many did not require chronic use of immunosuppressants (Burt et al., 2010). However, a more recent randomized controlled trial in patients with refractory CD failed to show a benefit of autologous HSCT over standard therapy (Hawkey et al., 2015). HSCT carries the risk of side effects due to the strong myeloablation that is required.

The use of mesenchymal stem cells (MSCs) is an attractive alternative given the relatively low expression of human leukocyte antigen (HLA) Class I and II antigens on these cells. Local injection of MSCs into perianal fistulae in CD has shown clinical efficacy (Ciccocioppo et al., 2011). Weekly systemic administration of allogeneic MSCs in patients with anti-TNFα-refractory CD showed both clinical and endoscopic healing in a recent Phase 2 trial (Forbes et al., 2014). Larger studies are required to determine whether MSCs offer a therapeutic option for those with refractory CD.

5.1.5.7 Future Therapeutic Directions

The loss of mucosal immune homeostasis in IBD is likely caused by dysfunctional interactions between the host immune system and the intestinal microbiota. Current therapies are mainly designed to suppress the immune system (either locally or systemically), and therapeutic decisions are often based on perceived disease severity. Current approaches to IBD management leave much to be desired with respect to tailoring treatment to individual patients. Future studies must better delineate disease subsets to help stratify patients based on their likelihood of response to a given therapy. Potential biomarkers that may be informative include genetic polymorphisms in key genes, tissue expression of integrins, serologic measures (such as antibodies directed at microbes), and stool indicators (such as inflammatory molecules or specific bacteria). Finally, given the consequences of systemic immunosuppression (i.e., oncologic and infectious risks), the ability to deliver appropriate immunosuppression via gut-directed formulations will be a key component of IBD therapies that are safer and more effective than those currently available.

GLOSSARY

antigen: molecules that are able to elicit a B or T lymphocyte response as a result of interactions with specific receptors on the surface of these cells.

cytokine: secreted proteins that play important roles in mediating or regulating various cellular processes, including those involved in immune and inflammatory responses.

dysbiosis: a state of abnormal microbiota composition and/or function.

granuloma: a characteristic histopathological entity seen in certain chronic inflammatory conditions, including some cases of CD, typically consisting of a collection of activated macrophages ("epithelioid cells") surrounded by lymphocytes.

GWAS: genome-wide association study. A method of analysis in which multiple genetic variants are examined in large numbers of individuals in order to determine if any of the variants are associated with the disease or trait of interest.

integrin: cell surface protein complex consisting of α and β chains that mediates interactions with other cells or with components of the extracellular matrix.

interleukin: cytokine secreted by cells of the immune system.

MAMP: microbe-associated molecular pattern. Molecules such as lipopolysaccharide, peptidoglycan, and flagellin that are expressed broadly by microorganisms.

metagenomics: analysis of the DNA sequence and other properties of the microbial genomes represented in the microbiota.

microbiota: the community of microorganisms that reside in the gastrointestinal tract and at other mucosal and cutaneous surfaces.

placebo: a simulated but inactive medical treatment that is used for comparison purposes in clinical trials.

polymorphism: naturally occurring variations in genomic DNA sequence (synonymous with genetic variant).

probiotic: a live microorganism that is administered for its beneficial effects.

REFERENCES

Abraham, B.P., Mehta, S., El-Serag, H.B. 2012. Natural history of pediatric-onset inflammatory bowel disease: A systematic review. *J. Clin. Gastroenterol.* 46: 581–589.

Aguilar, C., Lenoir, C., Lambert, N. et al. 2014. Characterization of Crohn disease in X-linked inhibitor of apoptosis-deficient male patients and female symptomatic carriers. *J. Allergy Clin. Immunol.* 134: 1131–1141.

Albenberg, L.G., Wu, G.D. 2014. Diet and the intestinal microbiome: Associations, functions, and implications for health and disease. *Gastroenterology* 146: 1564–1572.

Ananthakrishnan, A.N. 2015. Epidemiology and risk factors for IBD. *Nat. Rev. Gastroenterol. Hepatol.* 12: 205–217.

Ananthakrishnan, A.N., Nguyen, D.D., Sauk, J., Yajnik, V., Xavier, R.J. 2014. Genetic polymorphisms in metabolizing enzymes modifying the association between smoking and inflammatory bowel diseases. *Inflamm. Bowel Dis.* 20: 783–789.

Andersson, R.E., Olaison, G., Tysk, C., Ekbom, A. 2001. Appendectomy and protection against ulcerative colitis. *N. Engl. J. Med.* 344: 808–814.

Andersson, R.E., Olaison, G., Tysk, C., Ekbom, A. 2003. Appendectomy is followed by increased risk of Crohn's disease. *Gastroenterology* 124: 40–46.

Armuzzi, A., Pugliese, D., Danese, S. et al. 2014. Long-term combination therapy with infliximab plus azathioprine predicts sustained steroid-free clinical benefit in steroid-dependent ulcerative colitis. *Inflamm. Bowel Dis.* 20: 1368–1374.

Avitzur, Y., Guo, C., Mastropaolo, L.A. et al. 2014. Mutations in tetratricopeptide repeat domain 7A result in a severe form of very early onset inflammatory bowel disease. *Gastroenterology* 146: 1028–1039.

Azzopardi, N., Ellul, P. 2013. Risk factors for osteoporosis in Crohn's disease: Infliximab, corticosteroids, body mass index and age of onset. *Inflamm. Bowel Dis.* 19: 1173–1178.

Barclay, A.R., Russell, R.K., Wilson, M.L. et al. 2009. Systematic review: The role of breastfeeding in the development of pediatric inflammatory bowel disease. *J. Pediatr.* 155: 421–426.

Becker, C., Wirtz, S., Blessing, M. et al. 2003. Constitutive p40 promoter activation and IL-23 production in the terminal ileum mediated by dendritic cells. *J. Clin. Invest.* 112: 693–706.

Benchimol, E.I., Mack, D.R., Nguyen, G.C. et al. 2014. Incidence, outcomes, and health services burden of very early onset inflammatory bowel disease. *Gastroenterology* 147: 803–813.

Benchimol, E.I., Seow, C.H., Steinhart, A.H., Griffiths, A.M. 2008. Traditional corticosteroids for induction of remission in Crohn's disease. *Cochrane Database Syst. Rev.* 2: CD006792.

Benor, S., Russell, G.H., Silver, M. et al. 2010. Shortcomings of the inflammatory bowel disease Serology 7 panel. *Pediatrics* 125: 1230–1236.

Benus, R.F., van der Werf, T.S., Welling, G.W. et al. 2010. Association between *Faecalibacterium prausnitzii* and dietary fiber in colonic fermentation in healthy subjects. *Br. J. Nutr.* 104: 693–700.

Biedermann, L., Zeitz, J., Mwinyi, J. et al. 2013. Smoking cessation induces profound changes in the composition of the intestinal microbiota in humans. *PLoS One* 8: e59260.

Boschetti, G., Garnero, P., Moussata, D. et al. 2015. Accuracies of serum and fecal S100 proteins (calprotectin and calgranulin C) to predict the response to TNF antagonists in patients with Crohn's disease. *Inflamm. Bowel Dis.* 21: 331–336.

Brain, O., Cooney, R., Simmons, A., Jewell, D. 2012. Functional consequences of mutations in the autophagy genes in the pathogenesis of Crohn's disease. *Inflamm. Bowel Dis.* 18: 778–781.

Brain, O., Owens, B.M., Pichulik, T. et al. 2013. The intracellular sensor NOD2 induces microRNA-29 expression in human dendritic cells to limit IL-23 release. *Immunity* 39: 521–536.

Burkett, P.R., zu Horste, G.M., Kuchroo, V.K. 2015. Pouring fuel on the fire: Th17 cells, the environment and autoimmunity. *J. Clin. Invest.* 125: 2211–2219.

Burt, R.K., Craig, R.M., Milanetti, F. et al. 2010. Autologous nonmyeloablative hematopoietic stem cell transplantation in patients with severe anti-TNF refractory Crohn disease: Long-term follow-up. *Blood* 116: 6123–6132.

Cadwell, K., Liu, J.Y., Brown, S.L. et al. 2008. A key role for autophagy and the autophagy gene *ATG16L1* in mouse and human intestinal Paneth cells. *Nature* 456: 259–263.

Cadwell, K., Patel, K.K., Maloney, N.S. et al. 2010. Virus-plus-susceptibility gene interaction determines Crohn's disease gene *ATG16L1* phenotype in intestine. *Cell* 141: 1135–1145.

Capon, F., Di Meglio, P., Szuab, J. et al. 2007. Sequence variants in the genes for the interleukin-23 receptor (*IL23R*) and its ligand (*IL12B*) confer protection against psoriasis. *Hum. Genet.* 122: 201–206.

Cargill, M., Schrodi, S.J., Chang, M. et al. 2007. A large-scale genetic association study confirms *IL12B* and leads to the identification of *IL23R* as psoriasis-risk genes. *Am. J. Hum. Genet.* 80: 273–290.

Carriere, J., Darfeuille-Michaud, A., Nguyen, H.T.T. 2014. Infectious etiopathogenesis of Crohn's disease. *World J. Gastroenterol.* 20: 12102–12117.

Caruso, R., Warner, N., Inohara, N., Nunez, G. 2014. NOD1 and NOD2: Signaling, host defense and inflammatory disease. *Immunity* 41: 898–908.

Carvalho, F.A., Barnich, N., Sivignon, A. et al. 2009. Crohn's disease adherent-invasive *Escherichia coli* colonize and induce strong gut inflammation in transgenic mice expressing human CEACAM. *J. Exp. Med.* 206: 2179–2189.

Chamaillard, M., Hashimoto, M., Horie, Y. et al. 2003. An essential role for NOD1 in host recognition of bacterial peptidoglycan containing diaminopimelic acid. *Nat. Immunol.* 4: 702–707.

Chande, N., Wang, Y., MacDonald, J.K., McDonald, J.W. 2014. Methotrexate for induction of remission in ulcerative colitis. *Cochrane Database Syst. Rev.* 8: CD006618.

Chan, E.S., Cronstein, B.N. 2013. Mechanisms of action of methotrexate. *Bull. Hosp. Joint Dis.* 71: S5–S8.

Chassaing, B., Koren, O., Goodrich, J.K. et al. 2015. Dietary emulsifiers impact the mouse gut microbiota promoting colitis and metabolic syndrome. *Nature* 519: 92–96.

Cho, J.H., Feldman, M. 2015. Heterogeneity of autoimmune diseases: Pathophysiologic insights from genetics and implications for new therapies. *Nat. Med.* 21: 730–738.

Ciccocioppo, R., Bernardo, M.E., Sgarella, A. et al. 2011. Autologous bone marrow-derived mesenchymal stromal cells in the treatment of fistulising Crohn's disease. *Gut* 60: 788–798.

Cohen, R.D., Yu, A.P., Wu, E.Q. et al. 2010. Systematic review: The costs of ulcerative colitis in Western countries. *Aliment. Pharmacol. Ther.* 31: 693–707.

Cohen, S.A., Gold, B.D., Oliva, S. et al. 2014. Clinical and mucosal improvement with specific carbohydrate diet in pediatric Crohn disease. *J. Pediatr. Gastroenterol. Nutr.* 59: 516–521.

Cosnes, J., Bourrier, A., Laharie, D. et al. 2013. Early administration of azathioprine vs conventional management of Crohn's disease: A randomized controlled trial. *Gastroenterology* 145: 758–765.

Cosnes, J., Carbonnel, F., Beaugerie, L., Le Quintrec, Y., Gendre, J.P. 1996. Effects of cigarette smoking on the long-term course of Crohn's disease. *Gastroenterology* 110: 424–431.

Cosnes, J., Gower-Rousseau, C., Seksik, P., Cortot, A. 2011. Epidemiology and natural history of inflammatory bowel diseases. *Gastroenterology* 140: 1785–1794.

Couturier-Maillard, A., Secher, T., Rehman, A. et al. 2013. NOD2-mediated dysbiosis predisposes mice to transmissible colitis and colorectal cancer. *J. Clin. Invest.* 123: 700–711.

Curkovic, I., Rentsch, K.M., Frei, P. et al. 2013. Low allopurinol doses are sufficient to optimize azathioprine therapy in inflammatory bowel disease patients with inadequate thiopurine metabolite concentrations. *Eur. J. Clin. Pharmacol.* 69: 1521–1531.

Darfeuille-Michaud, A., Boudeau, J., Bulois, P. et al. 2004. High prevalence of adherent-invasive *Escherichia coli* associated with ileal mucosa in Crohn's disease. *Gastroenterology* 127: 412–421.

De Filippo, C., Cavalieri, D., Di Paola, M. et al. 2010. Impact of diet in shaping gut microbiota revealed by a comparative study in children from Europe and rural Africa. *Proc. Natl. Acad. Sci. U.S.A.* 107: 14691–14696.

de Silva, P.S., Nguyen, D.D., Sauk, J. et al. 2012. Long-term outcome of a third anti-TNF monoclonal antibody after the failure of two prior anti-TNFs in inflammatory bowel disease. *Aliment. Pharmacol. Ther.* 36: 459–466.

de Souza, H.S., Fiocchi, C. 2016. Immunopathogenesis of IBD: Current state of the art. *Nat. Rev. Gastroenterol. Hepatol.* 13: 13–27.

Desreumaux, P., Ghosh, S. 2006. Review article: Mode of action and delivery of 5-aminosalicylic acid—New evidence. *Aliment. Pharmacol. Ther.* 24 (Suppl. 1): 2–9.

Devkota, S., Chang, E.B. 2015. Interactions between diet, bile acid metabolism, gut microbiota and inflammatory bowel diseases. *Dig. Dis.* 33: 351–356.

De Vos, M., Louis, E.J., Jahnsen, J. et al. 2013. Consecutive fecal calprotectin measurements to predict relapse in patients with ulcerative colitis receiving infliximab maintenance therapy. *Inflamm. Bowel Dis.* 19: 2111–2117.

Dewint, P., Hansen, B.E., Verhey, E. et al. 2014. Adalimumab combined with ciprofloxacin is superior to adalimumab monotherapy in perianal fistula closure in Crohn's disease: A randomised, double-blind, placebo controlled trial (ADAFI). *Gut* 63: 292–299.

Dhillon, S.S., Fattouch, R., Elkadri, A. et al. 2014. Variants in nicotinamide adenine dinucleotide phosphate oxidase complex components determine susceptibility to very early onset inflammatory bowel disease. *Gastroenterology* 147: 680–689.

Di Meglio, P., Di Cesare, A., Laggner, U. et al. 2011. The IL23R R381Q gene variant protects against immune-mediated diseases by impairing IL-23-induced Th17 effector response in humans. *PLoS One* 6: e17160.

Duerr, R.H., Taylor, K.D., Brant, S.R. et al. 2006. A genome-wide association study identifies IL23R as an inflammatory bowel disease gene. *Science* 314: 1461–1463.

Ebach, D.R., Vanderheyden, A.D., Ellison, J.M. et al. 2011. Lymphocytic esophagitis: Possible manifestation of pediatric upper gastrointestinal Crohn's disease. *Inflamm. Bowel Dis.* 17: 45–49.

Efthymiou, A., Viazis, N., Mantzaris, G. et al. 2008. Does clinical response correlate with mucosal healing in patients with Crohn's disease of the small bowel? A prospective, case series study using wireless capsule endoscopy. *Inflamm. Bowel Dis.* 14: 1542–1547.

Eken, A., Singh, A.K., Oukka, M. 2014. Interleukin-23 in Crohn's disease. *Inflamm. Bowel Dis.* 20: 587–595.

Erturk-Hasdemir, D., Kasper, D.L. 2013. Resident commensals shaping immunity. *Curr. Opin. Immunol.* 25: 450–455.

Farh, K.K., Marson, A., Zhu, J. et al. 2015. Genetic and epigenetic fine mapping of causal autoimmune disease variants. *Nature* 518: 337–343.

Feagan, B.G., Fedorak, R.N., Irvine, E.J. et al. 2000. A comparison of methotrexate with placebo for the maintenance of remission for Crohn's disease. North American Crohn's Study Group Investigators. *N. Engl. J. Med.* 342: 1627–1632.

Feagan, B.G., Rochon, J., Fedorak, R.N. et al. 1995. Methotrexate for the treatment of Crohn's disease. North American Crohn's Study Group Investigators. *N. Engl. J. Med.* 332: 292–297.

Feagan, B.G., Rutgeerts, P., Sands, B.E. et al. 2013. Vedolizumab as induction and maintenance therapy for ulcerative colitis. *N. Engl. Med. J.* 369: 699–710.

Ferrante, M., Colombel, J.F., Sandborn, W.J. et al. 2013. Validation of endoscopic activity scores in patients with Crohn's disease based on a post hoc analysis of data from SONIC. *Gastroenterology* 145: 978–986.

Flanagan, M.E., Blumenkopf, T.A., Brissette, W.H. et al. 2010. Discovery of CP-690,550: A potent and selective Janus kinase (JAK) inhibitor for the treatment of autoimmune diseases and organ transplant rejection. *J. Med. Chem.* 53: 8468–8484.

Floyd, D.N., Langham, S., Severac, H.C., Levesque, B.G. 2015. The economic and quality-of-life burden of Crohn's disease in Europe and the United States, 2000 to 2013: A systematic review. *Dig. Dis. Sci.* 60: 299–312.

Foersch, S., Neurath, M.F. 2014. Colitis-associated neoplasia: Molecular basis and clinical translation. *Cell. Mol. Life Sci.* 71: 3523–3535.

Forbes, G.M., Sturm, M.J., Leong, R.W. et al. 2014. A phase 2 study of allogeneic mesenchymal stromal cells for luminal Crohn's disease refractory to biologic therapy. *Clin. Gastroenterol. Hepatol.* 12: 64–71.

Ford, A.C., Kane, S.V., Khan, K.J. et al. 2011. Efficacy of 5-aminosalicylates in Crohn's disease: Systematic review and meta-analysis. *Am. J. Gastroenterol.* 106: 617–629.

Frank, D.N., Robertson, C.E., Hamm, C.M. et al. 2011. Disease phenotype and genotype are associated with shifts in intestinal-associated microbiota in inflammatory bowel diseases. *Inflamm. Bowel Dis.* 17: 179–184.

Galley, J.D., Nelson, M.C., Yu, Z. et al. 2014. Exposure to a social stressor disrupts the community structure of the colonic mucosa-associated microbiota. *BMC Microbiol.* 14: 189.

Gevers, D., Kugathasan, S., Denson, L.A. et al. 2014. The treatment-naïve microbiome in new-onset Crohn's disease. *Cell Host Microbe* 15: 382–392.

Ghosh, S., Goldin, E., Gordon, F.H. et al. 2003. Natalizumab for active Crohn's disease. *N. Engl. J. Med.* 348: 24–32.

Girardin, S.E., Boneca, I.G., Carneiro, L.A. et al. 2003a. NOD1 detects a unique muropeptide from Gram-negative bacterial peptidoglycan. *Science* 300: 1584–1587.

Girardin, S.E., Boneca, I.G., Viala, J. et al. 2003b. NOD2 is a general sensor of peptidoglycan through muramyl dipeptide (MDP) detection. *J. Biol. Chem.* 278: 8869–8872.

Glocker, E.O., Kotlarz, D., Boztug, K. et al. 2009. Inflammatory bowel disease and mutations affecting the IL-10 receptor. *N. Engl. J. Med.* 361: 2033–2045.

Gorfu, G., Rivera-Nieves, J., Ley, K. 2009. Role of beta7 integrins in intestinal lymphocyte homing and retention. *Curr. Mol. Med.* 9: 836–850.

Grover, Z., Lewindon, P. 2015. Two-year outcomes after exclusive enteral nutrition induction are superior to corticosteroids in pediatric Crohn's disease treated early with thiopurines. *Dig. Dis. Sci.* 60: 3069–3074.

Gupta, N., Bostrom, A.G., Kirschner, B.S. et al. 2008. Presentation and disease course in early- compared to later-onset pediatric Crohn's disease. *Am. J. Gastroenterol.* 103: 2092–2098.

Haberman, Y., Tickle, T.L., Dexheimer, P.J. et al. 2014. Pediatric Crohn disease patients exhibit specific ileal transcriptome and microbiome signature. *J. Clin. Invest.* 124: 3617–3633.

Halme, L., Paavola-Saki, P., Turunen, U., Lappalainen, M., Farkkila, M., Kontula, K. 2006. Family and twin studies in inflammatory bowel disease. *World J. Gastroenterol.* 12: 3668–3672.

Hanauer, S.B., Feagan, B.G., Lichtenstein, G.R. et al. 2002. Maintenance infliximab for Crohn's disease: The ACCENT I randomised trial. *Lancet* 359: 1541–1549.

Hanauer, S.B., Sandborn, W.J., Rutgeerts, P. et al. 2006. Human anti-tumor necrosis factor monoclonal antibody (adalimumab) in Crohn's disease: The CLASSIC-I trial. *Gastroenterology* 130: 323–333.

Hawkey, C.J., Allez, M., Clark, M.M. et al. 2015. Autologous hematopoetic stem cell transplantation for refractory Crohn disease: A randomized clinical trial. *JAMA* 314: 2524–2534.

Hayashi, R., Wada, H., Ito, K., Adcock, I.M. 2004. Effects of glucocorticoids on gene transcription. *Eur. J. Pharmacol.* 500: 51–62.

Heresbach, D., Alexandre, J.L., Branger, B. et al. 2005. Frequency and significance of granulomas in a cohort of incident cases of Crohn's disease. *Gut* 54: 215–222.

Homer, C.R., Richmond, A.L., Rebert, N.A., Achkar, J.P., McDonald, C. 2010. ATG16L1 and NOD2 interact in an autophagy-dependent antibacterial pathway implicated in Crohn's disease pathogenesis. *Gastroenterology* 139: 1630–1641.

Hueber, W., Sands, B.E., Lewitzky, S. et al. 2012. Secukinumab, a human anti-IL-17A monoclonal antibody, for moderate to severe Crohn's disease: Unexpected results of a randomized, double-blind placebo-controlled trial. *Gut* 61: 1693–700.

Hugot, J.P., Chamaillard, M., Zouali, H. et al. 2001. Association of NOD2 leucine-rich repeat variants with susceptibility to Crohn's disease. *Nature* 411: 599–603.

Hugot, J.P., Zaccaria, I., Cavanaugh, J. et al. 2007. Prevalence of CARD15/NOD2 mutations in Caucasian healthy people. *Am J. Gastroenterol.* 102: 1259–1267.

Hundorfean, G., Neurath, M.F., Mudter, J. 2012. Functional relevance of Th17 cells and the IL-17 cytokine family in inflammatory bowel disease. *Inflamm. Bowel Dis.* 18: 180–186.

Hyams, J., Crandall, W., Kugathasan, S. et al. 2007. Induction and maintenance infliximab therapy for the treatment of moderate-to-severe Crohn's disease in children. *Gastroenterology* 132: 863–873.

Hyams, J.S., Griffiths, A., Markowitz, J. et al. 2012. Safety and efficacy of adalimumab for moderate to severe Crohn's disease in children. *Gastroenterology* 143: 365–374.

Inohara, N., Ogura, Y., Fontalba, A. et al. 2003. Host recognition of bacterial muramyl dipeptide mediated through NOD2. Implications for Crohn's disease. *J. Biol. Chem.* 278: 5509–5512.

Ivanov, I.I., Frutos Rde, L., Manel, N. et al. 2008. Specific microbiota direct the differentiation of IL-17-producing T-helper cells in the mucosa of the small intestine. *Cell Host Microbe* 4: 337–349.

Jostins, L., Ripke, S., Weersma, R.K. et al. 2012. Host-microbe interactions have shaped the genetic architecture of inflammatory bowel disease. *Nature* 491: 119–124.

Kaakoush, N.O., Day, A.S., Leach, S.T. et al. 2015. Effect of exclusive enteral nutrition on the microbiota of children with newly diagnosed Crohn's disease. *Clin. Transl. Gastroenterol.* 6: e71.

Keshav, S., Vanasek, T., Niv, Y. et al. 2013. A randomized controlled trial of the efficacy and safety of CCX282-B, an orally-administered blocker of chemokine receptor CCR9, for patients with Crohn's disease. *PLoS One* 8: e60094.

Khan, K.J., Ullman, T.A., Ford, A.C. et al. 2011. Antibiotic therapy in inflammatory bowel disease: A systematic review and meta-analysis. *Am. J. Gastroenterol.* 106: 661–673.

Kim, Y.G., Kamada, N., Shaw, M.H. et al. 2011. The NOD2 sensor promotes intestinal pathogen eradication via the chemokine CCL2-dependent recruitment of inflammatory monocytes. *Immunity* 34: 769–780.

Knights, D., Lassen, K.G., Xavier, R.J. 2013. Advances in inflammatory bowel disease pathogenesis: Linking host genetics and the microbiome. *Gut* 62: 1505–1510.

Kole, A., Maloy, K.J. 2014. Control of intestinal inflammation by IL-10. *Curr. Top. Microbiol. Immunol.* 380: 19–38.

Kostic, A.D., Xavier, R.J., Gevers, D. 2014. The microbiome in inflammatory bowel disease: Current status and the future ahead. *Gastroenterology* 146: 1489–1499.

Kotlarz, D., Beier, R., Murugan, D. et al. 2012. Loss of IL-10 signaling and infantile inflammatory bowel disease: Implications for diagnosis and therapy. *Gastroenterology* 143: 347–355.

Koutroubakis, I.E., Ramos-Rivers, C., Regueiro, M. et al. 2015. Persistent or recurrent anemia is associated with severe and disabling inflammatory bowel disease. *Clin. Gastroenterol. Hepatol.* 13: 1760–1766.

Kronman, M.P., Zaoutis, T.E., Haynes, K., Feng, R., Coffin, S.E. 2012. Antibiotic exposure and IBD development among children: A population-based cohort study. *Pediatrics* 130: e794–e803.

Kruis, W., Fric, P., Pokrotnieks, J. et al. 2004. Maintaining remission of ulcerative colitis with the probiotic *Escherichia coli* Nissle 1917 is as effective as with standard mesalazine. *Gut* 53: 1617–1623.

Laharie, D., Bourreille, A., Branche, J. et al. 2012. Cyclosporine versus infliximab in patients with severe ulcerative colitis refractory to intravenous steroids: A parallel, open-label randomized controlled trial. *Lancet* 380: 1909–1915.

Lapaquette, P., Glaser, A.L., Huett, A., Xavier, R.J., Darfeuille-Michaud, A. 2010. Crohn's disease-associated adherent-invasive *E. coli* are selectively favored by impaired autophagy to replicate intracellularly. *Cell. Microbiol.* 12: 99–113.

Lazarev, M., Present, D.H., Lichtiger, S. et al. 2012. The effect of intravenous cyclosporine on rates of colonic surgery in hospitalized patients with severe Crohn's colitis. *J. Clin. Gastroenterol.* 46: 764–767.

Lee, D., Albenberg, L., Compher, C. et al. 2015a. Diet in the pathogenesis and treatment of inflammatory bowel diseases. *Gastroenterology* 148: 1087–1106.

Lee, D., Baldassano, R.N., Otley, A.R. et al. 2015b. Comparative effectiveness of nutritional and biological therapy in North American children with active Crohn's disease. *Inflamm. Bowel Dis.* 21: 1786–1793.

Lee, J.C., Espeli, M., Anderson, C.A. et al. 2013. Human SNP links differential outcomes in inflammatory and infectious disease to a FOXO3-regulated pathway. *Cell* 155: 57–69.

Levine, A., Griffiths, A., Markowitz, J. et al. 2011. Pediatric modification of the Montreal classification for inflammatory bowel disease: The Paris classification. *Inflamm. Bowel Dis.* 17: 1314–1321.

Li, E., Hamm, C.M., Gulati, A.S. et al. 2012. Inflammatory bowel diseases phenotype, *C. difficile* and NOD2 genotype are associated with shifts in human ileum-associated microbial composition. *PLoS One* 7: e26284.

Lichtenstein, G.R., Feagan, B.G., Cohen, R.D. et al. 2012. Serious infection and mortality in patients with Crohn's disease: More than 5 years of follow-up in the TREAT registry. *Am. J. Gastroenterol.* 107: 1409–1422.

Lichtenstein, G.R., Targan, S.R., Dubinsky, M.C. et al. 2011. Combination of genetic and quantitative serological immune markers are associated with complicated Crohn's disease behavior. *Inflamm. Bowel Dis.* 17: 2488–2496.

Lozupone, C.A., Stombaugh, J.I., Gordon, J.I., Jansson, J.K., Knight, R. 2012. Diversity, stability and resilience of the human gut microbiota. *Nature* 489: 220–230.

Lubberts, E. 2015. The IL-23-IL-17 axis in inflammatory arthritis. *Nat. Rev. Rheumatol.* 11: 415–429.

Mack, D.R., Langton, C., Markowitz, J. et al. 2007. Laboratory values for children with newly diagnosed inflammatory bowel disease. *Pediatrics* 119: 1113–1119.

Mahid, S.S., Minor, K.S., Soto, R.E., Hornung, C.A., Galandiuk, S. 2006. Smoking and inflammatory bowel disease: A meta-analysis. *Mayo Clin. Proc.* 81: 1462–1471.

Marchiando, A.M., Ramanan, D., Ding, Y. et al. 2013. A deficiency in the autophagy gene Atg16L1 enhances resistance to enteric bacterial infection. *Cell Host Microbe* 14: 216–224.

Martin, H.M., Campbell, B.J., Hart, C.A. et al. 2004. Enhanced *Escherichia coli* adherence and invasion in Crohn's disease and colon cancer. *Gastroenterology* 127: 80–93.

McGinnis, J.K., Murray, K.F. 2008. Infliximab for ulcerative colitis in children and adolescents. *J. Clin. Gastroenterol.* 42: 875–879.

McGovern, D., Powrie, F. 2007. The IL-23 axis plays a key role in the pathogenesis of IBD. *Gut* 56: 1333–1336.

Moayyedi, P., Surette, M.G., Kim, P.T. et al. 2015. Fecal microbiota transplantation induces remission in patients with active ulcerative colitis in a randomized controlled trial. *Gastroenterology* 149: 102–109.

Molodecky, N.A., Soon, I.S., Rabi, D.M. et al. 2012. Increasing incidence and prevalence of the inflammatory bowel diseases with time, based on systematic review. *Gastroenterology* 142: 46–54.

Monteleone, G., Neurath, M.F., Ardizzone, S. et al. 2015. Mongersen, an oral SMAD7 antisense oligonucleotide, and Crohn's disease. *N. Engl. J. Med.* 372: 1104–1113.

Moran, C.J., Kaplan, J.L., Winter, H.S., Masiakos, P.T. 2015. Occult blood and perianal examination: Value added in pediatric inflammatory bowel disease screening. *J. Pediatr. Gastroenterol. Nutr.* 61: 52–55.

Moran, C.J., Walters, T.D., Guo, C.H. et al. 2013. IL-10R polymorphisms are associated with very-early-onset ulcerative colitis. *Inflamm. Bowel Dis.* 19: 115–123.

Mosli, M.H., Zou, G., Garg, S.K. et al. 2015. C-reactive protein, fecal calprotectin and stool lactoferrin for detection of endoscopic activity in symptomatic inflammatory bowel disease patients: A systematic review and meta-analysis. *Am. J. Gastroenterol.* 110: 802–819.

Murawska, N., Fabisiak, A., Fichna, J. 2016. Anemia of chronic disease and iron deficiency anemia in inflammatory bowel diseases: Pathophysiology, diagnosis, and treatment. *Inflamm. Bowel Dis.* 22: 1198–1208.

Ogata, H., Kato, J., Hirai, F. et al. 2012. Double-blind, placebo-controlled trial of oral tacrolimus (FK506) in the management of hospitalized patients with steroid-refractory ulcerative colitis. *Inflamm. Bowel Dis.* 18: 803–808.

Ogura, Y., Bonen, D.K., Inohara, N. et al. 2001. A frameshift mutation in NOD2 associated with susceptibility to Crohn's disease. *Nature* 411: 603–606.

Okou, D.T., Kugathasan, S. 2014. The role of genetics in pediatric inflammatory bowel disease. *Inflamm. Bowel Dis.* 20: 1878–1884.

Oren, R., Arber, N., Odes, S. et al. 1996. Methotrexate in chronic active ulcerative colitis: A double-blind, randomized, Israeli multicenter trial. *Gastroenterology* 110: 1416–1421.

Panaccione, R., Ghosh, S., Middleton, S. et al. 2014. Combination therapy with infliximab and azathioprine is superior to monotherapy with either agent in ulcerative colitis. *Gastroenterology* 146: 392–400.

Panes, J., Lopez-Sanroman, A., Bermejo, F. et al. 2013. Early azathioprine therapy is no more effective than placebo for newly diagnosed Crohn's disease. *Gastroenterology* 145: 766–774.

Parkes, M., Cortes, A., van Heel, D.A., Brown, M.A. 2013. Genetic insights into common pathways and complex relationships among immune-mediated diseases. *Nat. Rev. Genet.* 14: 661–673.

Pidasheva, S., Trifari, S., Phillips, A. et al. 2011. Functional studies on the IBD susceptibility gene IL23R implicate reduced receptor function in the protective genetic variant R381Q. *PLoS One* 6: e25038.

Punati, J., Markowitz, J., Lerer, T. et al. 2008. Effect of early immunomodulator use in moderate to severe pediatric Crohn's disease. *Inflamm. Bowel Dis.* 14: 949–954.

Quince, C., Ijaz, U.Z., Loman, N. et al. 2015. Extensive modulation of the fecal metagenome in children with Crohn's disease during exclusive enteral nutrition. *Am. J. Gastroenterol.* 110: 1718–1729.

Ramanan, D., Tang, M.S., Bowcutt, R., Loke, P., Cadwell, K. 2014. Bacterial sensor NOD2 prevents inflammation of the small intestine by restricting the expansion of the commensal *Bacteroides vulgatus*. *Immunity* 41: 311–324.

Rehman, A., Sina, C., Gavrilova, O. et al. 2011. NOD2 is essential for temporal development of intestinal microbial communities. *Gut* 60: 1354–1362.

Reinisch, W., Sandborn, W.J., Hommes, D.W. et al. 2011. Adalimumab for induction of clinical remission in moderately to severely active ulcerative colitis: Results of a randomised controlled trial. *Gut* 60: 780–787.

Rembacken, B.J., Snelling, A.M., Hawkey, P.M. et al. 1999. Non-pathogenic *Escherichia coli* versus mesalazine for the treatment of ulcerative colitis: A randomised trial. *Lancet* 354: 635–639.

Roberts, C.L., Rushworth, S.L., Richman, E., Rhodes, J.M. 2013. Hypothesis: Increased consumption of emulsifiers as an explanation for the rising incidence of Crohn's disease. *J. Crohns Colitis* 7: 338–341.

Robertson, S.J., Zhou, J.Y., Geddes, K. et al. 2013. NOD1 and NOD2 signaling does not alter the composition of intestinal bacterial communities at homeostasis. *Gut Microbes* 4: 222–231.

Rossen, N.G., Fuentes, S., van der Spek, M.J. et al. 2015. Findings from a randomized controlled trial of fecal transplantation for patients with ulcerative colitis. *Gastroenterology* 149: 110–118.

Rueda, B., Orozco, G., Raya, E. et al. 2008. The IL23R Arg381Gln non-synonymous polymorphism confers susceptibility to ankylosing spondylitis. *Ann. Rheum. Dis.* 67: 1451–1454.

Ruemmele, F.M., Veres, G., Kolho, K.L. et al. 2014. Consensus guidelines of ECCO/ESPGHAN on the medical management of pediatric Crohn's disease. *J. Crohns Colitis* 8: 1179–1207.

Ruhlmann, A., Nordheim, A. 1997. Effects of the immunosuppressive drugs CsA and FK506 on intracellular signalling and gene regulation. *Immunobiology* 198: 192–206.

Russell, G., Kaplan, J., Ferraro, M., Michelow, I.C. 2010. Fecal bacteriotherapy for relapsing *Clostridium difficile* infection in a child: A proposed treatment protocol. *Pediatrics* 126: e239–e242.

Rutgeerts, P., Sandborn, W.J., Feagan, B.G. et al. 2005. Infliximab for induction and maintenance therapy for ulcerative colitis. *N. Engl. J. Med.* 353: 2462–2476.

Sandborn, W.J., Colombel, J.F., Enns, R. et al. 2005. Natalizumab induction and maintenance therapy for Crohn's disease. *N. Engl. J. Med.* 353: 1912–1925.

Sandborn W.J., Feagan, B.G., Marano C. et al. 2014a. Subcutaneous golimumab induces clinical response and remission in patients with moderate-to-severe ulcerative colitis. *Gastroenterology* 146: 85–95.

Sandborn, W.J., Feagan, B.G., Rutgeerts, P. et al. 2013. Vedolizumab as induction and maintenance therapy for Crohn's disease. *N. Engl. J. Med.* 369: 711–721.

Sandborn, W.J., Feagan, B.G., Stoinov, S. et al. 2007. Certolizumab pegol for the treatment of Crohn's disease. *N. Engl. J. Med.* 357: 228–238.

Sandborn, W.J., Gasink, C., Gao, L.L. et al. 2012a. Ustekinumab induction and maintenance therapy in refractory Crohn's disease. *N. Engl. J. Med.* 367: 1519–1528.

Sandborn, W.J., Ghosh, S., Panes, J. et al. 2012b. Tofacitinib, an oral Janus kinase inhibitor, in active ulcerative colitis. *N. Engl. J. Med.* 367: 616–624.

Sandborn, W.J., Ghosh, S., Panes, J. et al. 2014b. A phase 2 study of tofacitinib, an oral Janus kinase inhibitor, in patients with Crohn's disease. *Clin. Gastroenterol. Hepatol.* 12: 1485–1493.

Sandborn, W.J., Hanauer, S.B., Katz, S. et al. 2001. Etanercept for active Crohn's disease: A randomized, double-blind, placebo-controlled trial. *Gastroenterology* 121: 1088–1094.

Sandborn, W.J., Regula, J., Feagan, B.G. et al. 2009. Delayed-release oral mesalamine 4.8 g/day (800 mg tablet) is effective for patients with moderately active ulcerative colitis. *Gastroenterology* 137: 1934–1943.

Sands, B.E., Feagan, B.G., Rutgeerts, P. et al. 2014. Effects of vedolizumab induction therapy for patients with Crohn's disease in whom tumor necrosis factor antagonist treatment failed. *Gastroenterology* 147: 618–627.

Sarin, R., Wu, X., Abraham, C. 2011. Inflammatory disease protective R381Q IL23 receptor polymorphism results in decreased primary CD4+ and CD8+ T-cell functional responses. *Proc. Natl. Acad. Sci. U.S.A* 108: 9560–9565.

Sartor, R.B. 2004. Therapeutic manipulation of the enteric microflora in inflammatory bowel diseases: Antibiotics, probiotics, and prebiotics. *Gastroenterology* 126: 1620–1633.

Schnitzler, F., Friedrich, M., Wolf, C. et al. 2014. The NOD2 p.Leu1007fsX1008 mutation (rs2066847) is a stronger predictor of the clinical course of Crohn's disease than the FOXO3A intron variant rs12212067. *PLoS One* 9: e108503.

Shah, S.C., Colombel, J.F., Sands, B.E., Narula, N. 2016. Systematic review with meta-analysis: Mucosal healing is associated with improved long-term outcomes in Crohn's disease. *Aliment. Pharmacol. Ther.* 43: 317–333.

Shen, C., Assche, G.V., Colpaert, S. et al. 2005. Adalimumab induces apoptosis of human monocytes: A comparative study with infliximab and etanercept. *Aliment. Pharmacol. Ther.* 21: 251–258.

Shen, J., Zuo, Z.X., Mao, A.P. 2014. Effect of probiotics on inducing remission and maintaining therapy in ulcerative colitis, Crohn's disease, and pouchitis: Meta-analysis of randomized controlled trials. *Inflamm. Bowel Dis.* 20: 21–35.

Sigall-Boneh, R., Pfeffer-Gik, T., Segal, I. et al. 2014. Partial enteral nutrition with a Crohn's disease exclusion diet is effective for induction of remission in children and young adults with Crohn's disease. *Inflamm. Bowel Dis.* 20: 1353–1360.

Sokol, H., Pigneur, B., Watterlot, L. et al. 2008. *Faecalibacterium prausnitzii* is an anti-inflammatory commensal bacterium identified by gut microbiota analysis of Crohn disease patients. *Proc. Natl. Acad. Sci. U.S.A.* 105: 16731–16736.

Sokol, H., Seksik, P., Furet, J.P. et al. 2009. Low counts of *Faecalibacterium prausnitzii* in colitis microbiota. *Inflamm. Bowel Dis.* 15: 1183–1189.

Solberg, I.C., Vatn, M.H., Hoie, O. et al. 2007. Clinical course in Crohn's disease: Results of a Norwegian population-based ten-year follow-up study. *Clin. Gastroenterol. Hepatol.* 5: 1430–1438.

Sorbara, M.T., Ellison, L.K., Ramjeet, M. et al. 2013. The protein ATG16L1 suppresses inflammatory cytokines induced by the intracellular sensors NOD1 and NOD2 in an autophagy-independent manner. *Immunity* 39: 858–873.

Stein, J., Hartmann, F., Dignass, A.U. 2010. Diagnosis and management of iron deficiency anemia in patients with IBD. *Nat. Rev. Gastroenterol. Hepatol.* 7: 599–610.

Steinhart, A.H., Ewe, K., Griffiths, A.M., Modigliani, R., Thomsen, O.O. 2003. Corticosteroids for maintenance of remission in Crohn's disease. *Cochrane Database Syst. Rev.* 4: CD000301.

Stephens, M.C., Baldassano, R.N., York, A. et al. 2005. The bioavailability of oral methotrexate in children with inflammatory bowel disease. *J. Pediatr. Gastroenterol. Nutr.* 40: 445–449.

Suskind, D.L., Singh, N., Nielson, H., Wahbeh, G. 2015. Fecal microbial transplant via nasogastric tube for active pediatric ulcerative colitis. *J. Pediatr. Gastroenterol. Nutr.* 60: 27–29.

Sutton, L.M., Heintz, D.D., Patel, A.S., Weinberg, A.G. 2014. Lymphocytic esophagitis in children. *Inflamm. Bowel Dis.* 20: 1324–1328.

Swidinski, A., Ladhoff, A., Pernthaler, A. et al. 2002. Mucosal flora in inflammatory bowel disease. *Gastroenterology* 122: 44–54.

Tan, J., McKenzie, C., Potamitis, M. et al. 2014. The role of short-chain fatty acids in health and disease. *Adv. Immunol.* 121: 91–119.

Targan, S.R., Feagan, B.G., Fedorak, R.N. et al. 2007. Natalizumab for the treatment of active Crohn's disease: Results of the ENCORE Trial. *Gastroenterology* 132: 1672–1683.

Taylor, S.R., McCluskey, P., Lightman, S. 2006. The ocular manifestations of inflammatory bowel disease. *Curr. Opin. Ophthalmol.* 17: 538–544.

Terdiman, J.P., Gruss, C.B., Heidelbaugh, J.J. et al. 2013. American Gastroenterological Association Institute guideline on the use of thiopurines, methotrexate and anti-TNFα biologic drugs for the induction and maintenance of remission in inflammatory Crohn's disease. *Gastroenterology* 145: 1459–1463.

Tew, G.W., Hackney, J.A., Gibbons, D. et al. 2016. Association between response to etrolizumab and expression of integrin αE and granzyme A in colon biopsies of patients with ulcerative colitis. *Gastroenterology* 150: 477–487.

Travassos, L.H., Carneiro, L.A., Ramjeet, M. et al. 2010. NOD1 and NOD2 direct autophagy by recruiting ATG16L1 to the plasma membrane at the site of bacterial entry. *Nat. Immunol.* 11: 55–62.

Turner, D., Levine, A., Escher, J.C. et al. 2012. Management of pediatric ulcerative colitis: Joint ECCO and ESPGHAN evidence-based consensus guidelines. *J. Pediatr. Gastroenterol. Nutr.* 55: 340–361.

Turner, D., Levine, A., Escher, J.C. et al. 2015. Efficacy of oral methotrexate in pediatric Crohn's disease: A multicenter propensity score study. *Gut* 64: 1898–1904.

Turner, D., Levine, A., Kolho, K.L. et al. 2014. Combination of oral antibiotics may be effective in severe pediatric ulcerative colitis: A preliminary report. *J. Crohns Colitis* 8: 1464–1470.

Ushiku, T., Moran, C.J., Lauwers, G.Y. 2013. Focally enhanced gastritis in newly diagnosed pediatric inflammatory bowel disease. *Am. J. Surg. Pathol.* 37: 1882–1888.

Vahabnezhad, E., Rabizadeh, S., Dubinsky, M.C. 2014. A 10-year, single tertiary care center experience on the durability of infliximab in pediatric inflammatory bowel disease. *Inflamm. Bowel Dis.* 20: 606–613.

Van Assche, G., Van Ranst, M., Sciot, R. et al. 2005. Progressive multifocal leukoencephalopathy after natalizumab therapy for Crohn's disease. *N. Engl. J. Med.* 353: 362–368.

van Nood, E., Vrieze, A., Nieuwdorp, M. et al. 2013. Duodenal infusion of donor feces for recurrent *Clostridium difficile*. *N. Engl. J. Med.* 368: 407–415.

Varela, E., Manichanh, C., Gallart, M. et al. 2013. Colonization by *Faecalibacterium prausnitzii* and maintenance of clinical remission in patients with ulcerative colitis. *Aliment. Pharmacol. Ther.* 38: 151–161.

Vavricka, S.R., Rogler, G., Gantenbein, C. et al. 2015a. Chronological order of appearance of extraintestinal manifestations relative to the time of IBD diagnosis in the Swiss inflammatory bowel disease cohort. *Inflamm. Bowel Dis.* 21: 1794–1800.

Vavricka, S.R., Schoepfer, A., Scharl, M. et al. 2015b. Extraintestinal manifestations of inflammatory bowel disease. *Inflamm. Bowel Dis.* 21: 1982–1992.

Vermeire, S., Ghosh, S., Panes, J. et al. 2011. The mucosal addressin cell adhesion molecule antibody PF-00547,659 in ulcerative colitis: A randomised study. *Gut* 60: 1068–1075.

Vermeire, S., Joossens, M., Verbeke, K. et al. 2016. Donor species richness determines faecal microbiota transplantation success in inflammatory bowel disease. *J. Crohns Colitis* 10: 387–394.

Vermeire, S., O'Byrne, S., Keir, M. et al. 2014. Etrolizumab as induction therapy for ulcerative colitis: A randomised, controlled, phase 2 trial. *Lancet* 384: 309–318.

Wahl, S.M. 1994. Transforming growth factor β: The good, the bad, and the ugly. *J. Exp. Med.* 180: 1587–1590.

Walters, T.D., Hyams, J.S. 2014. Can early anti-TNFα treatment be an effective therapeutic strategy in children with Crohn's disease? *Immunotherapy* 6: 799–802.

Wang, Y., MacDonald, J.K., Vandermeer, B. et al. 2015. Methotrexate for maintenance of remission in ulcerative colitis. *Cochrane Database Syst. Rev.* 8: CD007560.

Ward, M.G., Kariyawasam, V.C., Morgan, S.B. et al. 2015. Prevalence and risk factors for functional vitamin B12 deficiency in patients with Crohn's disease. *Inflamm. Bowel Dis.* 21: 2839–2847.

Watanabe, T., Kitani, A., Murray, P.J. et al. 2006. Nucleotide binding oligomerization domain 2 deficiency leads to dysregulated TLR2 signaling and induction of antigen-specific colitis. *Immunity* 25: 473–485.

Watson, S., Pensabene, L., Mitchell, P., Bousvaros, A. 2011. Outcomes and adverse events in children and young adults undergoing tacrolimus therapy for steroid-refractory colitis. *Inflamm. Bowel Dis.* 17: 22–29.

Weinshilboum, R.M., Sladek, S.L. 1980. Mercaptopurine pharmacogenetics: Monogenic inheritance of erythrocyte thiopurine methyltransferase activity. *Am. J. Hum. Genet.* 32: 651–662.

Willing, B., Halvarson, J., Dicksved, J. et al. 2009. Twin studies reveal specific imbalances in the mucosa-associated microbiota of patients with ileal Crohn's disease. *Inflamm. Bowel Dis.* 15: 653–660.

Wilson, A., Patel, V., Chande, N. et al. 2013. Pharmacokinetic profiles for oral and subcutaneous methotrexate in patients with Crohn's disease. *Aliment. Pharmacol. Ther.* 37: 340–345.

Yamamoto, S., Nakase, H., Matsuura, M. et al. 2011. Tacrolimus therapy as an alternative to thiopurines for maintaining remission in patients with refractory ulcerative colitis. *J. Clin. Gastroenterol.* 45: 526–530.

Yang, H., McElree, C., Roth, M.P. et al. 1993. Familial empirical risks for inflammatory bowel disease: Differences between Jews and non-Jews. *Gut* 34: 517–524.

Yang, Z., Clark, N., Park, K.T. 2014. Effectiveness and cost-effectiveness of measuring fecal calprotectin in diagnosis of inflammatory bowel disease in adults and children. *Clin. Gastroenterol. Hepatol.* 12: 253–262.

Zeissig, Y., Petersen, B.S., Milutinovic, S. et al. 2015. XIAP variants in male Crohn's disease. *Gut* 64: 66–76.

Nanotherapeutics for Inflammatory Bowel Disease

Bo Xiao and Didier Merlin

CONTENTS

5.2.1 INTRODUCTION

Inflammatory bowel disease (IBD) is a chronic relapsing disorder that is associated with uncontrolled inflammation in the small and/or large intestine, and can develop into colorectal cancer if the inflammation is not adequately suppressed (Nguyen et al. 2011). There are two major subtypes of IBD: Crohn's disease (CD) and ulcerative colitis (UC). About 1.4 million patients in the United States suffer from IBD, around half of whom have UC. Approximately 2.2 million Europeans suffer from IBD (Ingersoll et al. 2012), and the prevalence rate is on the rise in various low-incidence areas, including southern Europe, Asia, and most developing countries. It has been estimated that the total medical expense for IBD therapy will reach nearly $6.2 billion in 2017 (Talaei et al. 2013). Tremendous advances have already been made in identifying risk factors that predispose patients to IBD and assessing the biochemical profiles of the inflammatory process. The factors generally believed to contribute to the development of IBD include genetic background, environmental features (e.g., diet, cigarette smoking, sanitation, and infectious microbes), luminal antigens, and dysregulation of the immune response (Pithadia and Jain 2011). However, the primary etiology of IBD remains elusive (Ingersoll et al. 2012).

The main goal of IBD therapy is to achieve mucosal healing, reduce intestinal inflammation, maintain remission, and reduce the need for surgeries and hospitalizations (Iacucci et al. 2010; Pineton de Chambrun et al. 2010). Since no permanent cure has yet been developed, patients require lifelong drug therapy; moreover, ~70% of IBD patients will require at least one surgical intervention in their lifetime (Pithadia and Jain 2011). The utilized treatment strategy is based on the severity of IBD, the disease subtype, the existence of preexisting illness, and the patient's tolerance for drugs (Lautenschlager et al. 2014). The most common classes of drugs used against IBD are anti-inflammatory and immunosuppressive agents. Two of the mainstay treatments for IBD are 5-aminosalicylate (5-ASA) and corticosteroids; 5-ASA (as well as mesalazine and olsalazine) is mainly used to treat mild attacks and maintain remission in UC, whereas corticosteroids are more effective against moderate to severe IBD (Taylor and Irving 2011). Immunosuppressive agents (e.g., azathioprine and anti–tumor necrosis factor [TNF]-α antibodies, such as infliximab) are also important for treating more severe disease (Isaacs et al. 2005).

Conventionally, IBD is treated by daily administration of high doses of anti-inflammatory or immunosuppressive drugs, but these treatments are often complicated by serious adverse effects. For example, corticosteroids show short- and long-term side effects, including hypertension, hyperglycemia, and glaucoma (Buchman 2001), while immunosuppressive agents are associated with increased susceptibilities to infections and malignoma (Cunliffe and Scott 2002; Mason and Siegel 2013). Consequently, the treatment of IBD requires that high therapeutic efficacy be balanced against the risk of short- and long-term adverse drug reactions, which may deteriorate the patient's health-related quality of life, and thus counteract the therapeutic success (Blondel-Kucharski et al. 2001).

Nanoparticle (NP)-based medicine is a precise therapy that involves the use of nanotechnology. Previous reports have shown that NP-based medicines may outperform conventional medications via the enhanced targeting, excellent availability at disease tissues, and decreased adverse effects of drugs (Xiao et al. 2015a). Moreover, NP-based therapeutics can confer similar or better therapeutic impacts at lower drug concentrations than their conventional counterparts.

Some challenges remain, however, such as the need to ensure that NPs remain stable in systemic circulation or the gastrointestinal tract (GIT), the need to transport an adequate amount of active drug to the required sites, and the need to minimize systemic absorption of the drugs and lower the risk of adverse side effects (Wachsmann and Lamprecht 2012). The conventional formulations designed for targeted drug delivery in IBD use physiological features that are particular to the colon to trigger drug release (Meissner and Lamprecht 2008). However, physiological conditions can differ among patients and at various stages of IBD, making it very difficult to attain sufficient therapeutic efficiency using such methods. Parallel breakthroughs in our understanding of the molecular pathophysiology of IBD and the development of intelligent NPs offer tremendous promise for IBD therapy.

5.2.2 OBSTACLES FOR DRUG DELIVERY

Generally, systemic therapy benefits IBD patients with severe inflammation, whereas oral or rectal drug administration is efficacious in patients with mild to moderate inflammation (Lautenschlager et al. 2014). The obstacles for drug delivery depend on their administration approaches.

5.2.2.1 Intravenous Administration

5.2.2.1.1 Reticuloendothelial System

The reticuloendothelial system (RES), also known as the macrophage system or the mononuclear phagocyte system, is a part of the immune system. It consists of the phagocytic cells located in reticular connective tissues, primarily the monocytes and macrophages that help filter out dead and toxic particles, and work to identify foreign substances in both blood and tissues. These cell types can be found in most parts of the body, but are often particularly dense in the spleen, an organ tasked with blood balance and purification. After intravenous injection, NPs with hydrophobic surfaces are preferentially taken up by the liver, followed by the spleen and lung (Brigger et al. 2002). In these organs, NPs are internalized by macrophages and removed from systemic circulation. However, this process can be delayed by the surface modification of NPs with hydrophilic groups, which repel plasma proteins and allow NPs to evade RES-mediated clearance (Gaur et al. 2000). Researchers seeking to increase the circulation time of NPs, and hence their ability to target the site of interest, have found that such parameters are maximized by a diameter of 100 nm or less and the presence of a hydrophilic surface that reduces clearance by macrophages (Storm et al. 1995).

5.2.2.1.2 Enhanced Permeability and Retention Effect

The enhanced permeability and retention (EPR) effect, which is commonly observed in malignant tumor tissues, is a mechanism of passive tumor targeting by exploiting the local interstitial fluid pressure (Fang et al. 2011). In the context of intestinal inflammation, an effect similar to the EPR

effect occurs by defects in the vascular architecture of inflamed tissues (due to the rapid vascularization needed for the infiltration of immune cells) coupled with poor lymphatic drainage (Figure 5.3). NPs that can evade the RES show increased systemic circulation in the bloodstream, undergo more passages along the fenestrated vascular endothelium of the inflamed tissues, and thus enable selective accumulation of the drug at the site of action (Azzopardi et al. 2013). Surprisingly, little interest has focused on the passive targeting of NPs to inflammatory sites in the GIT. The first such investigation involved scintigraphic imaging of experimental colitis using intravenously injected [99]mTc-labeled liposomes (Oyen et al. 1997; Dams et al. 2000). These studies showed that liposomes coated with poly(ethylene glycol) (PEG) preferentially accumulated in the inflamed colon tissues of rats compared with normal regions of the control group (~13% vs. 0.1%, respectively). However, a majority of the NPs still accumulated in RES organs (Awasthi et al. 2002).

In terms of active targeting, delivery approaches based on the targeting of specific surface glycoproteins of endothelial cells (e.g., adhesion molecules such as vascular cell adhesion molecule 1 and intercellular adhesion molecule 1) provide the possibility of selectively targeting inflamed endothelium. Targeting molecules are attached to the NP surface, increasing the adhesion of NPs to inflamed endothelium relative to noninflamed

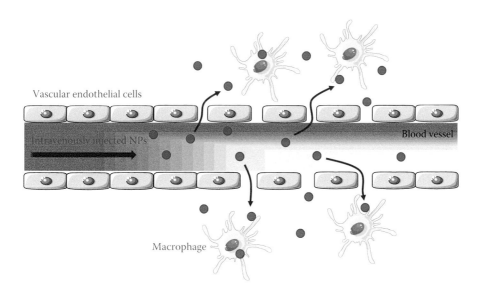

Figure 5.3. Schematic illustration of the passive targeting of intravenously administered NPs based on the EPR effect.

endothelium (Sakhalkar et al. 2003, 2005). However, it remains to be demonstrated whether this bioadhesion approach can increase the drug concentration at the site of inflammation to a degree sufficient to achieve a therapeutic effect at a low administration dose.

5.2.2.1.3 Cellular Uptake of NPs

Most drugs (e.g., plasmids and siRNAs) must enter cells in order to confer their therapeutic effects. If the drugs are not taken up, they are degraded or expelled from the body. In IBD, the responding immune cells include T cells, B cells, and antigen-presenting cells (e.g., macrophages and dendritic cells). Macrophages might be an important target for IBD therapy, as they contribute to disease manifestation via their production of pro-inflammatory cytokines (e.g., TNF-α and interleukin [IL]-6) (O'Neill et al. 2011). There are two main cellular uptake strategies in colitis tissues: phagocytosis and endocytosis. The former is defined as the internalization of cell fragments and large particles (from 25 nm to several micrometers in diameter [Vercauteren et al. 2012]), and is often restricted to specialized professional cells, such as macrophages, monocytes, neutrophils, and dendritic cells. In the case of pinocytosis, a form of endocytosis, an initial invagination brings small particles within cells, forming small pinocytotic vesicles that subsequently fuse with lysosomes. The contained particles are then hydrolyzed or otherwise broken down. Pinocytosis and endocytosis categories can be subclassified into macropinocytosis, clathrin-dependent endocytosis, and clathrin-independent endocytosis, which are characterized as fluid-phase, absorptive, and receptor-mediated forms of endocytosis, respectively (Vercauteren et al. 2012). Endocytosis of a NP may be triggered by its binding to the targeted receptor, provided that the structural analogy between the surface moiety and the natural substrate is adequate. The transport of NPs may be enhanced by the specific targeting of such receptors. The grafting or coating of particles with ligands that bind specific receptors can enhance their internalization and transportation (Khalil et al. 2006). Endocytosis of targeted NPs occurs principally via receptor-mediated endocytosis. Both pinocytosis and endocytosis may be followed by degradation of the absorbed substances.

5.2.2.1.4 Endosomal Escape

The endocytic pathway is the major route for NP uptake (Meier et al. 2002). The NPs become entrapped in early-stage endosomes. ATPase proton pump enzymes in the early endosome membrane transport protons from the cytosol into the endosome. This causes the pH to continuously decrease as the endosomes mature from early-stage organelles (pH 6.2–6.3) to late-stage endosomes (pH 5.0–5.5), which fuse with lysosomes (pH 4.8–5.4) containing various degradative enzymes (Garcia et al. 2005). Endosomes usually travel in a specific direction and converge around the nuclear membrane (Ganta et al. 2008). Most drugs trapped in NPs are degraded in the enzyme-rich acidic lysosome. To avoid such degradation, researchers have investigated several strategies through which the disruption of endosomal or lysosomal membranes can allow the drug to escape into the cytoplasm. Various mechanisms of endosomal escape have been employed, including the formation of pores in the endosomal membrane and the process explained by the "proton-sponge hypothesis" (Varkouhi et al. 2011).

To induce endosomal escape, a common strategy utilizes the proton-sponge hypothesis, which was first proposed by Jean-Paul Behr's group in 1995 (Boussif et al. 1995). Certain polymers, such as poly(ethylenimine) or those containing imidazole groups, absorb large amounts of proton ions (thus acting as a proton sponge). Protonation of these polymers in the acidic environment of an endosome or lysosome induces a charge gradient, which triggers a compensatory inflow of chloride ions. This further increases the osmolarity of the endosome or lysosome, triggers an influx of water to relieve the resulting gradient, and eventually leads to osmotic swelling, rupture of the endosomal membrane, and subsequent release of entrapped drugs into the cytoplasm (Markovsky et al. 2012).

5.2.2.1.5 Nuclear Localization

After the delivered drug has undergone cellular internalization and escape from the endosome, it often must enter a specific organelle in order to be effective. For example, nuclear entry is necessary for some drugs, including DNA-intercalating chemicals, drugs that alter chromatin structure, and inhibitors of transcription or the cell cycle

(Dean 2003). A plasmid, for example, must be transported into the nucleus to induce efficient gene therapy. In the absence of such transport, transcription cannot occur.

A drug can enter the nucleus via indirect delivery, endosome-mediated delivery, and/or active nuclear transport (Sui et al. 2011). In indirect delivery, drug-loaded NPs escape from endosomes or lysosomes into the cytoplasm, and then the loaded drug is released from the formulation. Small molecules can spontaneously pass through the nuclear pore complex (NPC) on the basis of a concentration gradient. However, drugs with diameters of >9 nm and/or molecular weights of >45 kDa cannot enter the nucleus through the NPC (Wagstaff and Jans 2009; Sui et al. 2011). Large molecules can access the nucleus during mitosis when the nuclear membrane breaks down, or through other means of delivery. Endosome-mediated nuclear transport enables the drug to directly enter the nucleus via the direct fusion of particle-loaded endosomes with the nuclear membrane (Akita et al. 2009). However, the most common method for delivering a drug into the nucleus is active nuclear transport. H. Y. Wang et al. (2011) synthesized a series of N-terminal stearylated nuclear localization signals (NLSs) and showed that such vectors with a diameter up to 500 nm could effectively deliver plasmids to nuclei. The maximum transfection efficiency of these carriers was up to 80% of that of jetPEI™, which is the optimized form of linear PEI available from Polyplus Transfection Co. Such nuclear delivery of plasmids was achieved via their conjugation with NLS, such as SV40 antigen. Continued progress in our understanding and exploitation of nuclear targeting should greatly increase the efficiency of plasmid delivery in the future (Luo and Saltzman 2000).

5.2.2.2 Oral Administration

Oral administration is considered to be the most convenient approach for colitis therapy–related drug delivery, as it avoids the pain and discomfort associated with injections, minimizes the potential for contamination, and is applicable for a self-medication that can be fully controlled by patients (Pinto 2010). However, drug-loaded NPs will encounter the harsh, acidic, and enzymatic environment of GIT after oral administration (Mrsny 2012). Thus, they have to be well protected until they reach the target sites.

5.2.2.2.1 Gastrointestinal Tract

Drug-loaded NPs pass through the esophagus and enter the stomach, where they encounter peristalsis, the highly acidic (pH 1.5–1.9) environment of the stomach (Dressman et al. 1990) and digestive enzymes, all of which can destabilize them and reduce the effectiveness of their loaded drugs (Loretz et al. 2006; Maroni et al. 2012). After passing through the stomach, drug-loaded NPs encounter (and may be destabilized by) pancreatic enzymes, bicarbonate, and bile salts, all of which are released from the common bile duct in the small intestine. The NPs continue through the GIT, the pH of which ranges from strongly acidic in the stomach, to almost neutral in the small intestine, and then to weakly acidic (pH 5–7) in the colon (Cook et al. 2012). Therefore, NPs must be stable over a wide pH range. The large intestine is characterized by the most abundant (10^{10} to 10^{12} CFU/ml [Goldin et al. 1996]) and diverse microbial population. Fermentation of proteins and carbohydrates by anaerobic bacteria renders the luminal pH weakly acidic. The large intestine contains numerous enzymes, including soluble proteases of pancreatic origin and enzymes derived from enterocytes and resident microorganisms (Maroni et al. 2012). Given the mechanical pressures applied in the colon, drug formulations are likely to mix with these enzymes, which contribute significantly to breaking down drugs in the colon. In addition, the semisolid contents of the lumen act as a physical barrier that may prevent drugs from moving toward inflamed areas.

5.2.2.2.2 Mucus

The mucus on the colon surface is highly viscoelastic and adhesive, and forms a thick layer (830 ± 110 μm) (Ensign et al. 2012). Mucus, which is primarily composed of mucins, lipids, and mucopolysaccharides, acts to trap and remove bacteria, viruses, and foreign matter via electrostatic and/or hydrophobic interactions (Y. Y. Wang et al. 2011). The viscosity of the mucus network is typically 1000- to 10,000-fold higher than that of water at low shear rates (Lai et al. 2009). Constant turnover of the adherent layer serves to remove potentially damaging compounds and organisms. In those with UC, the mucus layer is abnormally thin and the mucin content of adherent mucus is significantly decreased, whereas the mucus of CD patients is thicker than that commonly observed in healthy individuals (Pullan et al. 1994; Rankin et al. 1995). Biochemical abnormalities have also been identified in the mucins produced by IBD patients. The identified changes, which include variations in protein chain length and the extent of glycosylation, may decrease the viscosity and binding properties of the mucus (Corfield et al. 2001), potentially facilitating the penetration of drug formulations. The TFF3 peptide, which is secreted by goblet cells, is known to protect the epithelium (Dignass et al. 1994; Kindon et al. 1995). However, biochemical modifications of mucin may impair its ability to interact with TFF3, further decreasing the protective ability of the mucosa (Siccardi et al. 2005).

Mucus is a significant barrier with respect to the localized delivery of drugs to colitis tissues. PEG is an uncharged hydrophilic polymer that can be used to PEGylate both polymers and NPs. PEGylated drug delivery systems, which have a hydrophilic shield around the particle core, exhibit greatly reduced associations with mucus and increased penetration of the mucus matrix. Wang et al. (2008) noted that virus particles could efficiently traverse the mucus layer, and that a high-density coating of shorter PEG molecules could allow the NPs to "slip" through mucus with a greater diffusion ratio than that of unmodified NPs. Such modification could therefore greatly increase the efficacy of drug delivery to mucosal surfaces.

5.2.2.2.3 Epithelial EPR

Inflamed colon is associated with disruption of the intestinal epithelial layer, leading to loss of barrier function and increased epithelial permeability (Ingersoll et al. 2012). NPs may potentially accumulate in the gaps between cells, increasing the local drug concentration and thereby exerting therapeutic effects against IBD. This phenomenon is called the epithelial EPR (eEPR) effect (Figure 5.4) (Lamprecht 2010; Collnot et al. 2012). This effect is size dependent, showing a maximum efficacy in the nanorange of less than 500 nm. Furthermore, compared with drugs administered as a solution,

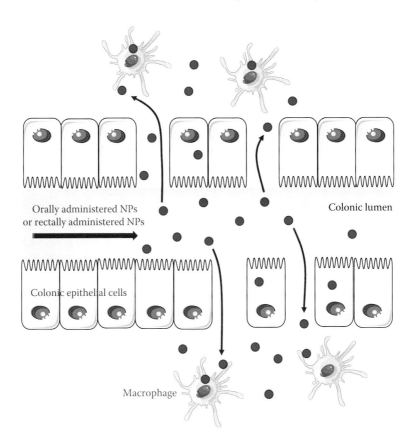

Figure 5.4. Schematic illustration of the passive targeting of orally or rectally administered NPs based on the eEPR effect.

NANOMEDICINE FOR INFLAMMATORY DISEASES

drug-loaded NPs were shown to accumulate to a greater extent in inflamed tissues, trigger increased mitigation, and prolong therapeutic effects (Lamprecht 2010). Small-molecule anti-inflammatory drugs loaded into NPs also showed enhanced penetration into colitis tissues, due to protection of the drug from efflux systems and mucosal metabolism.

As to oral administration, the obstacles for the following cellular uptake, lysosome escape, and nuclear localization are the same as those discussed in Section 5.2.1. The administration of NPs by the rectal route has also been investigated. However, many of the above-described obstacles (mucus, eEPR, cellular uptake, lysosomal escape, and nuclear localization) will need to be overcome in the future for this drug administration approach to be beneficial.

5.2.3 NANOTHERAPEUTICS FOR IBD

5.2.3.1 Conventional NPs

5.2.3.1.1 Polyester NPs

Polyesters, such as the Food and Drug Administration (FDA)–approved biodegradable copolymers, poly(lactic acid/glycolic acid) (PLGA) and poly(lactic acid) (PLA), can be used to efficiently encapsulate hydrophilic or hydrophobic drugs to form NPs, and thus have been widely used in drug delivery (Xiao et al. 2015a,b,d). Lamprecht et al. (2001) encapsulated an anti-inflammatory drug (rolipram) into PLGA NPs and used them to gavage rats with colitis once a day for five consecutive days. The NPs proved to be equally as efficient as the solution-administered drug in mitigating the experimental colitis. Moreover, the rolipram solution group had a high adverse effect index, whereas the rolipram-loaded NP-treated group had a significantly lower index, indicating that the drug is retained in GIT, perhaps via eEPR-mediated retention in the inflamed regions. A similar result was reported by Mahajan et al. (2011), who incorporated mesalamine into PLGA NPs and orally or rectally administered them once a day to rats suffering from acetic acid–induced colitis. These NPs were generally much more efficient in mitigating experimental colitis than the free drug in suspension, as assessed in terms of clinical activity, myeloperoxidase activity, and ulcer index. The authors also found that the mesalamine-loaded NPs exhibited selective adherence and enhanced drug penetration of inflamed tissues.

5.2.3.1.2 Liposomes

Liposomes are artificially fabricated drug vectors that comprise an aqueous core encased within one or more phospholipid layers. The encapsulation of hydrophilic drugs is based on the embedding of an aqueous phase of active agents within the liposome core, whereas hydrophobic compounds can be dissolved or inserted into the lipid membrane. Thus, liposomes can be used to simultaneously deliver hydrophilic and hydrophobic active agents using the same carrier (Lautenschlager et al. 2014). The application of orally administered liposomes for drug delivery to the GIT has not been extensively evaluated due to issues with instability and aggregation in the low-pH and enzyme-rich environment of the gut. Intracolonic application via enemas and intravenous injection can circumvent many of these issues, providing an opportunity for liposomes to be used in IBD therapy.

Jubeh et al. (2004) demonstrated in ex vivo experiments that positively charged liposomes adhered better to healthy mucosa, whereas anionically charged liposomes showed increased adhesion to inflamed mucosa. The authors (Jubeh et al. 2006) further encapsulated antioxidants into negatively charged liposomes and intracolonically administered drug-loaded liposomes in rat colon. The results showed that these liposomes were more effective than free drugs in treating experimental colitis. They speculated that the attachment of the negatively charged liposomes to the inflamed mucosa increased the residence time and uptake of the antioxidants. In addition to intracolonic administration, Arranz et al. (2013) intravenously injected amphoteric liposomes loaded with CD40-targeting antisense oligonucleotides (ASOs) into a 2,4,6-trinitrobenzenesulfonic acid (TNBS)–induced colitis model in mice. The authors found that administration of these liposomes inhibited the development of colitis, as assessed by weight loss, histology, and cytokine profiles. In contrast, unformulated CD40 ASOs or unrelated ASO-loaded liposomes were ineffective.

5.2.3.1.3 Silicon NPs

Silicon NPs have attracted wide interest for use in delivery systems, as they exhibit several attractive features, including the opportunity for researchers to control the particle size, relatively straightforward functionalization using well-established routes, good biocompatibility, acceptable degradability under physiological conditions, and a large surface area and pore volume that allow generous amounts of cargo to be incorporated (Rosenholm et al. 2010). Moulari et al. (2008) fabricated 5-ASA-loaded silica NPs with a particle diameter of around 140 nm. Following oral administration, these silicon NPs showed sixfold better adherence to inflamed tissues than tissues from healthy control mice. In mice subjected to TNBS-induced colitis, the NPs collected in inflamed regions and had positive impacts on the clinical activity score and myeloperoxidase activity at lower drug doses than those used for conventional delivery.

5.2.3.2 Complex NPs

The use of single-unit formulations for colonic drug delivery has been hampered by a number of challenges, such as a lack of targeting capacity and degradation of the formulation in the upper GIT, which leads to early drug release, serious side effects, and reduced therapeutic efficacy (Talaei et al. 2013). Numerous approaches have been proposed based on the fairly simple approaches of adding a surface coating or entrapping the drug within a secondary vehicle, including microparticles and hydrogels.

5.2.3.2.1 pH-Sensitive NPs

pH-sensitive NPs take advantage of the pH differences observed in various regions of the GIT (Xiao et al. 2015c). The pH in the terminal ileum and colon is generally higher than that in any other region of the GIT, such that dosage formulations that disintegrate preferentially at high pH levels have potential for colon-specific drug delivery. One of the simplest ways to modify dosage forms for pH-dependent drug delivery is to coat them with pH-sensitive biocompatible polymers. The most commonly used example of this is a methacrylic acid copolymer (Eudragit) whose side-group composition may be manipulated to be sensitive to certain pH values at which the copolymer is soluble. Eudragit L100 and Eudragit S100, which dissolve at pH 6 and 7, respectively, are commonly used in combination at various ratios to manipulate the release of the drug over this pH range. The more recently developed Eudragit FS 30D, which dissolves at pH > 6.5, is an ionic copolymer of methyl acrylate, methyl methacrylate, and methacrylic acid. It has been commonly used for oral colon-targeted drug delivery (Asghar and Chandran 2006). In addition to their pH-dependent releases, Eudragit coatings have also been suggested to have mucoadhesive properties. Ali et al. (2014) showed that budesonide-loaded PLGA NPs that were coated with Eudragit S100 and had a diameter around 240 nm significantly alleviated inflammation and demonstrated signs of regeneration in the dextran sulfate sodium (DSS)-, TNBS-, and oxazolone (OXA)-induced mouse models of colitis. Interestingly, the general endoscopic appearances of mice treated with the coated budesonide-loaded PLGA NPs were similar to those of healthy mice, with no signs of the severe inflammation seen in the inflamed control group, the budesonide solution-treated group, or the plain PLGA NP-treated group. The coated budesonide-loaded PLGA NPs also efficiently decreased the colonic expression levels of various pro-inflammatory cytokines (e.g., TNF-α, IL-6, IL-1β, and interferon [IFN]-γ).

A pH-dependent formulation may also be obtained by mixing a pH-sensitive polymer (e.g., Eudragit polymers) and other non-pH-sensitive polymers (e.g., PLGA) to obtain a more sustained drug release formulation. Beloqui et al. (2014) demonstrated that curcumin-loaded NPs prepared with PLGA and Eudragit S100 exhibited obvious pH sensitivity and were able to significantly reduce the secretion of TNF-α by lipopolysaccharide (LPS)-activated macrophages. Critically, oral administration of such NPs markedly decreased neutrophil infiltration and TNF-α secretion in a murine DSS-induced colitis model, and the mice maintained a colonic structure similar to that observed in the control group.

5.2.3.2.2 NP-in-Microparticle Oral Delivery System

A unique colon-targeted drug formulation called the NP-in-microparticle oral delivery system (NiMOS) was developed by Amiji's group (Bhavsar et al. 2006). Type B gelatin NPs containing plasmids or siRNAs were encapsulated into a poly(ϵ-caprolactone) (PCL) matrix to form MPs by a double emulsion-like technique (Kriegel and Amiji 2011). Type B gelatin, which is a biocompatible denatured protein that may be obtained from collagen via alkaline hydrolysis, has been widely used in the pharmaceutical, cosmetics, and food product industries (Elzoghby et al. 2012). Using a controlled precipitation technique, Type B gelatin can be formed into NPs containing plasmids or siRNAs. PCL is a type of synthetic hydrophobic polyester that resists acid degradation, and thus protects NPs during transit through the stomach. In addition, the coated MPs inhibit protein or enzyme adsorption, and thus avoid the harsh environment of the GIT. When the NPs reach the colon, lipases degrade the PCL coating and release the encapsulated NPs, which become available for uptake by colonic cells.

NiMOS is applicable for both reporter plasmids (e.g., those encoding enhanced green fluorescent protein [EGFP] or β-galactosidase) and therapeutic nucleic acids (e.g., plasmids encoding IL-10 or siRNAs targeting TNF-α or cyclin D1) (Bhavsar and Amiji 2007, 2008a,b; Kriegel and Amiji 2011). Bhavsar and Amiji (2008a) encapsulated an EGFP-encoding plasmid using Type B gelatin to form NPs (diameter ~150 nm), which were then embedded in a PCL matrix to form MPs (diameter 2–5 μm). Evaluation of radiolabeled NiMOSs for their biodistribution revealed that free gelatin NPs traversed the GIT more rapidly than did the NiMOSs. In vivo, naked plasmids and gelatin NPs failed to withstand the harsh environment of the GIT, and no EGFP expression was evident. In contrast, EGFP plasmid–loaded NiMOS triggered considerable GFP production, mainly in the small and large intestines.

Kriegel and Amiji (2011) recently loaded TNF-α siRNA (siTNF) into gelatin NPs and further entrapped them in a PCL matrix to yield NiMOS. In mice subjected to DSS-induced acute colitis, these NiMOSs suppressed the expression levels of TNF-α and other pro-inflammatory cytokines (e.g., IL-1β, IFN-γ, and monocyte chemotactic protein 1), and increased the body weight and colon length compared with DSS-treated control mice. These results indicated that siTNF-loaded NiMOS could be a valuable therapeutic option for IBD patients. Another study by the same group (Wan et al. 2009) showed that delivery of two anti-inflammatory siRNAs (targeting TNF-α and cyclin D1) in NiMOS yielded better outcomes than those obtained with the siRNAs alone. Thus, NiMOS is a promising option for colon-specific drug delivery.

5.2.3.2.3 NPs in Hydrogel

Various biocompatible and biodegradable polysaccharides, including chitosan and cellulose, have been approved by the FDA. They have been widely used in drug delivery, antibacteria, and tissue engineering (Xiao et al. 2011, 2012; Jiang et al. 2012). Given that the current IBD therapies are limited by a lack of efficient colon-targeted drug delivery, a hydrogel fabricated by chitosan and alginate was developed by Laroui et al. (2010). This hydrogel was sensitive to the colonic pH, and it would collapse when it arrived at the colon. The researchers used ions (Ca^{2+} and $SO4^{2-}$) to cross-link chitosan and alginate. The hydrogel was embedded with NPs containing an anti-inflammatory tripeptide (Lys-Pro-Val; KPV), an active protein (intestinal epithelial prohibitin 1; PHB), or siTNF (Laroui et al. 2010, 2011; Theiss et al. 2011).

Laroui et al. (2010) embedded KPV-containing PLA NPs in an alginate–chitosan hydrogel that was specifically degraded by colonic enzymes at pH 6.2. Thus protected, the NPs were able to pass through the stomach and upper small intestine and were degraded in the inflamed colon. Using this improved oral NP-based drug delivery system, a 1000-fold lower dose was sufficient to ameliorate mucosal inflammation in mice subjected to acute DSS-induced colitis.

Laroui et al. (2011) used hydrogel-embedded siRNA-loaded NPs for IBD therapy. TNF-α is known to play a central role in the pathogenesis and progression of IBD, and this disease has been successfully treated with antibodies against TNF-α in clinical trials. NPs containing siTNF efficiently downregulated the production of TNF-α by macrophages in vitro. Importantly, oral administration of the NP-embedded chitosan–alginate hydrogel significantly reduced TNF-α expression or secretion in colonic tissues from LPS-treated mice.

Among the many proteins known to exert anti-inflammatory effects, PHB is known to be downregulated in mucosal biopsies of animals experiencing colitis. Its expression can be suppressed by pro-inflammatory cytokines (including TNF-α) and oxidative stress induced by exogenous H_2O_2 in vitro and in vivo (Theiss et al. 2007). Based on the hypothesis that restoration of colonic epithelial PHB should be beneficial in the treatment of experimental colitis, Theiss et al.

(2011) tested the effects of orally delivered PHB-containing NPs embedded in hydrogel, using the DSS-induced mouse model of colitis. They found reductions in the TNF-α-induced activation of nuclear factor-κB, the extent of inflammatory reactions, and the severity of colitis.

5.2.3.2.4 Reactive Oxygen Species–Sensitive NPs

Abnormally high levels of reactive oxygen species (ROS) are produced at sites of intestinal inflammation (Kountouras et al. 2001), and excessive ROS can degrade the extracellular matrix and injure tissues (Eberlein et al. 2008). Biopsies taken from UC patients were found to have 10- to 100-fold increases in mucosal ROS concentrations, specifically at disease sites and in correlation with the progression of the disease. Activated neutrophils are known to be potential sources of free radicals in UC patients, whereas monocytes and macrophages produce free radicals in CD patients (Naito et al. 2011). ROS-induced injury causes both recruited and resident cells to lose their capacity to channel electrons through the mitochondrial electron transport chain in a controlled manner. This causes them to "leak" electrons, leading to the so-called "uncoupling" of the electron transport chain. Molecular oxygen readily accepts these electrons, leading to ROS generation in oxidatively stressed, focally hypoxic, and necrosing tissues (Winyard et al. 2000). Based on the fact that ROS accumulates in the colitis sites, ROS-sensitive NPs could target the release of drugs to inflammatory tissues.

Wilson et al. (2010) synthesized poly(1,4-phenyleneacetone dimethylenethioketal) (PPADT), a novel polymer that resists acidic, basic, and enzymatic degradation but is broken down by high levels of ROS (e.g., at inflamed colon sites). PPADT NPs were loaded with siTNF complexed with 1,2-dioleoyl-3-trimethylammonium-propane (DOTAP) to generate thioketal NPs (TKNs), and gavaged to a DSS-induced colitis mice model. The control PLGA formulation showed no significant therapeutic effect due to limited stability in the GIT, while the β-glucan microshells withstood gastric degradation but did not target the diseased tissue. Instead, the β-glucan system was found to be better suited for targeting M cells in the upper intestine, and for systemic delivery of siRNAs. After 5 days of oral gavage with 0.23 mg of siRNA/kg of body weight,

the mRNA expression levels of TNF-α and several other pro-inflammatory cytokines (IL-1, IL-6, and IFN γ) were decreased and DSS-induced acute colitis was efficiently ameliorated. This was comparable to the therapeutic effect of systemic treatment with siRNA-loaded NPs (2.5 mg of siRNA/kg) for 4 days (Peer et al. 2008).

In several experimental models, antioxidant compounds and free radical scavengers improved colitis, but showed significant bioavailability and retention issues (Millar et al. 1996; Araki et al. 2006). Therefore, Vong et al. (2012) designed a novel antioxidative nitroxide radical-containing NP (RNP^O) intended to deliver antioxidant compounds specifically to colitis tissues for the treatment of IBD. Each RNP^O contains a new redox polymer, methoxy-poly(ethylene glycol)-b-poly(4-(2,2,6,6-tetramethyl-piperidine-1-oxyl) oxymethylstyrene) (MeO-PEG-b-PMOT); this amphiphilic block copolymer has a stable nitroxide in a hydrophobic segment attached as a side chain via an ether linkage, and forms core shell-type micelles (40 nm) by self-assembly in aqueous environments regardless of pH. RNP^O showed significant accumulation in the colonic mucosa, especially in inflamed mucosal tissues, compared with control 4-hydroxyl-2,2,6,6-tetramethylpiperidine-1-oxyl (TEMPOL) or polystyrene latex particles. Moreover, although RNP^O showed ~50-fold higher retention in the colon than TEMPOL and exhibited long-term colonic retention, it did not undergo systemic absorption. Mice with DSS-induced colitis had significantly lower disease activity indexes and less inflammation following 7 days of oral dosing with RNP^O compared with TEMPOL or mesalamine. The accumulation of RNP^O in the colon was dependent on both the size and PEGylated character of the NPs; the latter parameter is thought to protect the nitroxide radicals in the hydrophobic core from the harsh conditions in the GIT following oral administration, resulting in significant colonic accumulation. Furthermore, PEG chains may achieve mucoadhesion due to their ability to interdiffuse within the mucus network and form polymer entanglements with mucin, which is composed of glycoproteins.

5.2.3.2.5 Ligand-Mediated Targeted Drug Delivery

To further reduce adverse reactions and improve the accumulation of drugs at inflamed sites within the colon, researchers have sought to induce active targeting (Lamprecht 2010), such as by exploiting disease-related changes in the expression levels of cell surface receptors, adhesion molecules, and other proteins. The existing studies of active targeting–based nanodelivery systems, which have most commonly used the parenteral route of administration, have targeted a multitude of conditions, such as cancers, infections, and inflammatory reactions (Forssen and Willis 1998; Sofou and Sgouros 2008). The positive results obtained in such studies prompted researchers to test this approach using orally administered NPs. Conceptually, interactions between targeting ligands and specific receptors expressed predominantly at inflamed sites are expected to improve the bioadhesion of the drug formulation to specific cells and increase the extent of endocytosis.

5.2.3.2.5.1 ENDOTHELIAL CELL–SPECIFIC DRUG DELIVERY

The endothelial cell–specific targeting of NPs is based on the "backside targeting" of inflamed intestinal areas (Lautenschlager et al. 2014). After intravenous administration, the NPs are distributed in the systemic circulation and adhere specifically to endothelial targets specifically overexpressed at sites of inflammation. The systemic route of administration is typically favorable for biologicals, as it avoids the stability issues created by passage through the GIT. Indeed, systemically applied NPs have been shown to efficiently target the inflamed endothelium. For example, NPs coated with an antibody against vascular cell adhesion molecule 1 showed a 12-fold increased adhesion to the inflamed endothelium compared with noninflamed areas when tested in an experimental model of colitis. The selectivity and efficiency of targeting were found to depend on the number of administered particles (Russell-Jones et al. 1999).

Other endothelial target molecules, such as selectins as cell adhesion molecules on the surfaces of activated endothelial cells, play an important role in the initial recruitment of leukocytes to the inflamed site. Lectin is a naturally occurring sugar binding protein that has a high capacity to bind to specific carbohydrate residues (Andrews et al. 2009). NPs surface coated with antiselectin were found to be captured out of the bloodstream by adhesion to selectin receptors, and to thereafter infiltrate into inflamed areas. The most highly investigated lectins are wheat germ agglutinin (WGA) (N-acetylglucosamine binding lectin) and tomato lectin (a poly-N-acetyllactosamine binding lectin) (Lehr et al. 1992; Lehr 2000). The diffusion of lectin-bound NPs through the mucus layer is a delicate situation, because the efficiency of passage is limited by adsorption to mucus and fecal elimination (similar to the situation with mucoadhesive delivery systems). Yin et al. (2006) fabricated lectin-conjugated PLGA NPs and evaluated their ability to facilitate oral drug delivery. In vitro experiments demonstrated that WGA-modified NPs exhibited 1.8- to 4.2-fold more interactions with pig mucin than unmodified NPs. Fluorescence photomicrographs showed that WGA-modified NPs adhered to the intestinal villous epithelium and Peyer's patches. In vivo, WGA-modified NPs showed 1.4- to 3.1-fold higher accumulation across the intestine than unmodified NPs. Together, these results suggested that lectin-modified NPs could yield improve intestinal bioadhesion and drug absorption for oral drug delivery (Yin et al. 2007).

5.2.3.2.5.2 TRANSFERRIN-MEDIATED TARGETED DRUG DELIVERY

The transferrin receptor (TfR) is ubiquitously expressed at low levels in most normal tissues, but it is overexpressed in inflamed colon tissue. Elevated expression was observed in both the basolateral and apical membranes of enterocytes from both colon biopsies of IBD patients and excised colon tissues of rats with experimentally induced colitis models (Tirosh et al. 2009). TfR levels are also elevated in activated immune cells, including lymphocytes and macrophages (Pallone et al. 1987; Tacchini et al. 2008), and the receptor is believed to play an important role in cellular signaling and proliferation (Cano et al. 1990; Levy et al. 1999). Harel et al. (2011) reported that TfR expression was elevated in both the basolateral and apical membranes of enterocytes. The authors conjugated anti-TfR antibodies to nanoliposomes and showed that the targeted formulation exhibited mucopenetration and a fourfold increase in uptake by inflamed colon tissues subjected to the TNBS-induced colitis model, compared with noninflamed colon tissues.

5.2.3.2.5.3 CD98-MEDIATED TARGETED DRUG DELIVERY

CD98 is a type II transmembrane protein in which a heavy chain (CD98hc or SLC3A2) and one of several versions of the L-type amino acid transporter 1 form a heterodimeric neutral amino acid transport system (Verrey et al. 1999). The cytoplasmic domains of CD98 can interact with β_1 integrin to regulate integrin signaling–mediated functions, such as cell homeostasis, epithelial adhesion or polarity, and immune responses (Hang and Merlin 2012). Previous studies have revealed that CD98 expression is highly upregulated in colonic tissues from mice with active colitis, in colonic biopsy specimens from patients with CD, and at the surface of intestinal B cells, CD4[+] T cells, and CD8[+] T cells isolated from patients with IBD (Schreiber et al. 1991; Kucharzik et al. 2005; Nguyen et al. 2010). Moreover, CD98 is highly expressed in intestinal macrophages, plays an important role in macrophage activation (MacKinnon et al. 2008), and is significantly upregulated in colonic epithelial cells from mice with active colitis (Kucharzik et al. 2005; Cantor and Ginsberg 2012).

Xiao and Merlin (2012) recently developed an orally delivered hydrogel that releases CD98 siRNA-loaded NPs surface conjugated with single-chain anti-CD98 antibodies (200 nm). In mice with DSS-induced colitis and colitis induced by transfer of CD4[+]CD45Rb[high] T cells to Rag[−/−] mice, oral administration of the targeted NPs significantly reduced the overexpression of CD98 by colonic epithelial cells and macrophages. Approximately 24% of colonic macrophages were found to have taken up the targeted NPs within 12 hours of administration, and the severity of colitis was significantly reduced compared with the control groups, as assessed by analyses of body weight, myeloperoxidase activity, inflammatory cytokine production, and histology. Overall, these studies demonstrate that active targeting is a very promising approach to enhancing the accumulation and uptake of drugs by inflamed tissues in IBD.

5.2.3.2.5.4 MANNOSE-MEDIATED TARGETED DRUG DELIVERY

The mannose receptor (MR), a 175 kDa transmembrane protein of the C-type lectin family, is exclusively expressed on the surfaces of macrophages (Wileman et al. 1986). MRs can undergo high-affinity binding to infectious agents containing terminal mannose residues; this triggers their transport via endocytic pathways, leading to major histocompatibility complex (MHC) presentation and subsequent T cell activation (East and Isacke 2002). This receptor system can bind to NPs coated with mannose residues, leading to their rapid internalization within membrane-bound vesicles (Sato and Beutler 1993). The introduction of mannose to various drug delivery carriers has been shown to provide selective macrophage targeting, improve cellular uptake, and yield high transfection efficiency (Jiang et al. 2008; Jain et al. 2010).

Xiao et al. (2013) synthesized a mannosylated bioreducible cationic polymer (PPM) that was formed into NPs with sodium triphosphate (TPP) and siTNF by electrostatic interaction. The generated TPP-PPM/siTNF NPs showed an enhanced ability to condense siRNA, a desirable size distribution of around 240 nm, and a significant macrophage targeting ability. Moreover, they significantly inhibited TNF-α synthesis and secretion in tissue samples from the DSS-induced colitis model *ex vivo*. Importantly, TPP-PPM/siTNF NPs were efficiently taken up by macrophages, with flow cytometry demonstrating that 29.5% of the NPs were internalized by colon macrophages, whereas there was no significant uptake by epithelial cells.

5.2.3.2.5.5 GALACTOSE-MEDIATED TARGETED DRUG DELIVERY

The galactose receptor is overexpressed on the surfaces of activated macrophages under inflammatory conditions (Coombs et al. 2006; van Vliet et al. 2008). The targeted drug delivery capacity of galactose-functionalized NPs into activated macrophages was demonstrated by Zuo et al. (2010). Recently, Zhang et al. (2013) prepared galactosylated trimethyl chitosan–cysteine (GTC) NPs for oral delivery of siRNAs against mitogen-activated protein kinase kinase kinase kinase 4 (Map4k4), which is a key upstream mediator of TNF-α production. siRNA-loaded GTC NPs were prepared based on ionic gelation of GTC with anionic cross-linkers, such as TPP. Cellular uptake in activated macrophages was significantly higher for GTC/TPP NPs than for trimethyl chitosan–cysteine (TC)/TPP NPs, owing to galactose receptor–mediated endocytosis. *In vitro* and *in vivo* studies showed that TNF-α production was effectively inhibited and the siRNAs were selectively biodistributed in ulcerative tissues. Critically, daily oral administration of siMap4k4-loaded GTC/TPP NPs significantly improved DSS-induced colitis, as measured by body weight, histology, and myelo-peroxidase activity.

F4/80 is an important macrophage-specific marker, and antibodies against F4/80 have been widely used in the detection of macrophages (Wellen and Hotamisligil 2003; van den Berg and Kraal 2005). A recent study by Laroui et al. (2014) demonstrated that siTNF can be efficiently loaded into NPs made of PLA-PEG block copolymer, and that grafting of a macrophage-specific ligand (F4/80 Ab Fab′) onto the NP surface via malei-mide or thiol group–mediated covalent bonding increased the specificity of targeting to intestinal macrophages. Oral administration of siTNF-loaded NPs significantly improved DSS-induced colitis in vivo, and higher therapeutic efficacy was obtained using Fab′-bearing NPs compared with nonconjugated NPs. Grafting of Fab′-bearing ligands also improved the endocytosis of these NPs, as well as their macrophage targeting ability, as indicated by flow cytometry. It should be noted that the NPs were loaded into a colon-specific bio-degradable hydrogel (chitosan–alginate), which also enhanced their specific delivery to the colon and protected the grafted ligand during progression through the GIT (Laroui et al. 2014).

5.2.4 DISCUSSION

IBD is a chronic disease characterized by periods of remission during which the disease is not active (Xiao and Merlin 2012). Thus, most patients need maintenance medication to ease symptoms and reduce the number and severity of flare-ups. Although there is no known cure, medicines can reduce inflammation and increase the number and length of periods of remission, lowering the risk of more serious health problems, such as colorectal cancer (Xiao et al. 2013). The intravenous, oral, and rectal administration of conventional medications used in IBD can induce severe side effects, trigger discomfort, and/or prove inefficient. Thus, we need to develop more efficient drug delivery systems capable of circumventing these issues for the treatment of IBD.

Such efforts have largely concentrated on the identification of anti-inflammatory cytokines, the development of agents that inhibit these inflammatory cytokines (e.g., infliximab), and the synthesis of selective adhesion molecule inhibitors that are capable of suppressing the trafficking of T lymphocytes into the gut epithelium (Xiao and Merlin 2012; Laroui et al. 2014; Xiao et al. 2014). Traditionally, the drugs capable of mediating these desired effects are administered at high doses and/or systemically, leading to significant adverse events. The ultimate goal of IBD therapy is to eliminate symptoms, maintain long-term remission, and restore the highest achievable quality of life (Pineton de Chambrun et al. 2010). To date, only enteric-modified colonic tablets have been used in the clinical treatment of IBD. Given that the pH values in the colon of IBD patients can vary over a wide range, enteric-modified colonic tablets may not always provide accurate drug release at the inflamed colon. A major drawback in the development of new therapeutic strategies for diseases such as IBD is our present inability to target sufficient quantities of drugs to sites of inflammation, such that the local drug concentration is maximized while systemic side effects are minimized. Another problem is that organs of the GIT, particularly the colon, differ in their drug absorption properties, and it is difficult to deliver the drug to the colon with minimal enzymatic degradation and/or systemic absorption.

Nanomedicine is a new field of investigation that may hold significant potential in the treatment of various diseases (Xiao et al. 2017). However, the behavior of drug-loaded NPs in humans is very complex and only partially understood at this time, which has limited the development of innovative nanosized medicine. Despite this, significant progress has been made toward developing a truly selective nanoscale drug delivery system that is capable of targeting the sites of inflammation in IBD (Si et al. 2016). NPs have great potential to improve IBD therapy by specifically accumulating in the inflamed intestinal regions. Conditions in the circulatory system or GIT are harsh and complicated, as are those in intracellular environments. Thus, various environmentally responsive NPs have been developed for drug delivery to the inflamed colon. Such targeting has been shown to improve the therapeutic profiles of such drugs by allowing them to accumulate and be released at the desired sites, with only minimal systemic exposure. Cell-specific targeting may be achieved by modifying drug-containing NPs with targeting molecules (Xiao et al. 2015a). After internalization by cells, NPs need to release the drug to the cytoplasm or enable its transport into the nucleus or another targeted organelle. In the future, advanced methodologies will be needed to construct NPs that are sensitive to many different environments, generate various particles of defined diameters and shapes that have the required surface properties and are stable during passage through the stomach and small intestine, and surface modify the particles with particular ligands to improve targeting. In summary, the successful treatment of IBD will require parallel developments in materials science (particularly in the field of environment-responsive polymers), particle preparation technology (especially that of NPs), and our pathophysiological understanding of IBD.

5.2.5 CONCLUSION

We must clearly understand the barriers impeding the specific delivery of drugs to inflamed colonic tissues. Advances in our knowledge of the pathophysiological features of IBD have clarified the approaches that must be used in our efforts to design the desired drug formulations. Drug-loaded NPs may be much more effective than conventional drug formulations when the aim is to specifically target the inflamed colon.

ACKNOWLEDGMENTS

This work was supported by grants from the Department of Veterans Affairs (merit award to D.M.), the National Institute of Diabetes and Digestive and Kidney Diseases (grant RO1-DK-071594 to D.M.), the American Heart Association (postdoctoral fellowship grant 13POST16400004 to B.X.), the National Natural Science Foundation of China (grants 51503172 and 81571807), Fundamental Research Funds for the Central Universities (SWU114086 and XDJK2015C067), and the Scientific Research Foundation for the Returned Overseas Chinese Scholars (State Education Ministry). D.M. is a recipient of a career scientist award from the Department of Veterans Affairs.

GLOSSARY

enhanced permeability and retention (EPR) effect: EPR is a phenomenon of the tending accumulation of NPs in the desired tissues, such as malignant tumor tissues and inflamed tissues. It is allowed by the defection of the vascular architecture and poor lymphatic drainage of disease tissues.

epithelial enhanced permeability and retention (eEPR) effect: eEPR is generally confined in inflamed colon, which is associated with disruption of the intestinal epithelial layer and penetration of immune cells, leading to loss of barrier function and increased epithelial permeability. NPs can potentially accumulate in the gaps between cells and be internalized by immune cells there, increasing the local drug concentration in inflamed colon.

hydrogel: hydrogel constitutes a group of polymeric materials, the hydrophilic structure of which renders them capable of holding large amounts of water in their three-dimensional networks.

inflammatory bowel disease (IBD): IBD involves chronic, relapsing inflammation in all or part of the digestive tract, which arises as a result of the interaction of environmental and genetic factors. It mainly includes UC and CD, and they usually involve severe diarrhea, pain, fatigue and weight loss.

liposome: liposome is made from phospholipids, and it has an aqueous solution core surrounded by a hydrophobic membrane.

nanomedicine: nanomedicine is the application of nanotechnology to the diagnosis, prevention, monitoring, and treatment of disease.

polyester: polyester is a category of polymers that contain the ester functional group in their main chain, such as poly(lactide-co-glycolide), poly(lactide), and PCL. It has been widely used in medicine because of its high strength, high modulus, and biocompatibility.

proton-sponge phenomenon: the proton-sponge phenomenon has been observed in polymers that usually contain protonatable secondary and/or tertiary amine groups with pKa close to endosomal or lysosomal pH. During the maturation of endosomes, the membrane-bound ATPase proton pumps actively translocate protons from the cytosol into the endosomes, leading to the acidification of endosomal compartments and activation of hydrolytic enzymes. In the meantime, polymers will become protonated and resist the acidification of endosomes. As a result, more protons will be continuously pumped into the endosomes in an attempt to lower the pH. The proton pumping action is followed by passive entry of chloride ions, increasing ionic concentration and leading to water influx. Eventually, the osmotic pressure causes swelling and rupture of endosomes, releasing their contents to the cytosol.

reticuloendothelial system (RES): RES, also termed the macrophage system or the mononuclear phagocyte system, is a part of the immune system. It consists of the phagocytic cells located in reticular connective tissues, primarily the monocytes and macrophages that help filter out dead and toxic particles, and work to identify foreign substances in both blood and tissues.

targeted drug delivery: targeted drug delivery refers to predominant drug accumulation within target sites, thereby avoiding interaction with healthy tissue.

REFERENCES

Akita, H., A. Kudo, A. Minoura, M. Yamaguti, I. A. Khalil, R. Moriguchi, T. Masuda, R. Danev, K. Nagayama, K. Kogure, and H. Harashima. 2009. Multi-layered nanoparticles for penetrating the endosome and nuclear membrane via a step-wise membrane fusion process. *Biomaterials* 30 (15):2940–9.

Ali, H., B. Weigmann, M. F. Neurath, E. M. Collnot, M. Windbergs, and C. M. Lehr. 2014. Budesonide loaded nanoparticles with pH-sensitive coating for improved mucosal targeting in mouse models of inflammatory bowel diseases. *Journal of Controlled Release* 183:167–77.

Andrews, G. P., T. P. Laverty, and D. S. Jones. 2009. Mucoadhesive polymeric platforms for controlled drug delivery. *European Journal of Pharmaceutics and Biopharmaceutics* 71 (3):505–18.

Araki, Y., H. Sugihara, and T. Hattori. 2006. The free radical scavengers edaravone and tempol suppress experimental dextran sulfate sodium-induced colitis in mice. *International Journal of Molecular Medicine* 17 (2):331–4.

Arranz, A., C. Reinsch, K. A. Papadakis, A. Dieckmann, U. Rauchhaus, A. Androulidaki, V. Zacharioudaki, A. N. Margioris, C. Tsatsanis, and S. Panzner. 2013. Treatment of experimental murine colitis with CD40 antisense oligonucleotides delivered in amphoteric liposomes. *Journal of Controlled Release* 165 (3):163–72.

Asghar, L. F. A., and S. Chandran. 2006. Multiparticulate formulation approach to colon specific drug delivery: Current perspectives. *Journal of Pharmacy and Pharmaceutical Sciences* 9 (3):327–38.

Awasthi, V. D., B. Goins, R. Klipper, and W. T. Phillips. 2002. Accumulation of PEG-liposomes in the inflamed colon of rats: Potential for therapeutic and diagnostic targeting of inflammatory bowel diseases. *Journal of Drug Targeting* 10 (5):419–27.

Azzopardi, E. A., E. L. Ferguson, and D. W. Thomas. 2013. The enhanced permeability retention effect: A new paradigm for drug targeting in infection. *Journal of Antimicrobial Chemotherapy* 68 (2):257–74.

Beloqui, A., R. Coco, P. B. Memvanga, B. Ucakar, A. des Rieux, and V. Preat. 2014. pH-sensitive nanoparticles for colonic delivery of curcumin in inflammatory bowel disease. *International Journal of Pharmaceutics* 473 (1–2):203–12.

Bhavsar, M. D., and M. M. Amiji. 2007. Gastrointestinal distribution and in vivo gene transfection studies with nanoparticles-in-microsphere oral system (NiMOS). *Journal of Controlled Release* 119 (3):339–48.

Bhavsar, M. D., and M. M. Amiji. 2008a. Development of novel biodegradable polymeric nanoparticles-in-microsphere formulation for local plasmid DNA delivery in the gastrointestinal tract. *AAPS PharmSciTech* 9 (1):288–94.

Bhavsar, M. D., and M. M. Amiji. 2008b. Oral IL-10 gene delivery in a microsphere-based formulation for local transfection and therapeutic efficacy in inflammatory bowel disease. *Gene Therapy* 15 (17):1200–9.

Bhavsar, M. D., S. B. Tiwari, and M. M. Amiji. 2006. Formulation optimization for the nanoparticles-in-microsphere hybrid oral delivery system using factorial design. *Journal of Controlled Release* 110 (2):422–30.

Blondel-Kucharski, F., C. Chircop, P. Marquis, A. Cortot, F. Baron, J. P. Gendre, J. F. Colombel, and Digestives Groupe d'Etudes Therapeutique des Affections Inflammatoires. 2001. Health-related quality of life in Crohn's disease: A prospective longitudinal study in 231 patients. *American Journal of Gastroenterology* 96 (10):2915–20.

Boussif, O., F. Lezoualc'h, M. A. Zanta, M. D. Mergny, D. Scherman, B. Demeneix, and J. P. Behr. 1995. A versatile vector for gene and oligonucleotide transfer into cells in culture and in vivo: Polyethylenimine.

Proceedings of the National Academy of Sciences of the United States of America 92 (16):7297–301.

Brigger, I., C. Dubernet, and P. Couvreur. 2002. Nanoparticles in cancer therapy and diagnosis. *Advanced Drug Delivery Reviews* 54 (5):631–51.

Buchman, A. L. 2001. Side effects of corticosteroid therapy. *Journal of Clinical Gastroenterology* 33 (4):289–94.

Cano, E., A. Pizarro, J. M. Redondo, F. Sanchezmadrid, C. Bernabeu, and M. Fresno. 1990. Induction of T-cell activation by monoclonal-antibodies specific for the transferrin receptor. *European Journal of Immunology* 20 (4):765–70.

Cantor, J. M., and M. H. Ginsberg. 2012. CD98 at the crossroads of adaptive immunity and cancer. *Journal of Cell Science* 125 (6):1373–82.

Collnot, E. M., H. Ali, and C. M. Lehr. 2012. Nano- and microparticulate drug carriers for targeting of the inflamed intestinal mucosa. *Journal of Controlled Release* 161 (2):235–46.

Cook, M. T., G. Tzortzis, D. Charalampopoulos, and V. V. Khutoryanskiy. 2012. Microencapsulation of probiotics for gastrointestinal delivery. *Journal of Controlled Release* 162 (1):56–67.

Coombs, P. J., M. E. Taylor, and K. Drickamer. 2006. Two categories of mammalian galactose-binding receptors distinguished by glycan array profiling. *Glycobiology* 16 (8):1–7C.

Corfield, A. P., D. Carroll, N. Myerscough, and C. S. Probert. 2001. Mucins in the gastrointestinal tract in health and disease. *Frontiers in Bioscience* 6: D1321–57.

Cunliffe, R. N., and B. B. Scott. 2002. Review article: Monitoring for drug side-effects in inflammatory bowel disease. *Alimentary Pharmacology & Therapeutics* 16 (4):647–62.

Dams, E. T., W. J. Oyen, O. C. Boerman, G. Storm, P. Laverman, P. J. Kok, W. C. Buijs, H. Bakker, J. W. van der Meer, and F. H. Corstens. 2000. 99mTc-PEG liposomes for the scintigraphic detection of infection and inflammation: Clinical evaluation. *Journal of Nuclear Medicine* 41 (4):622–30.

Dean, D. A. 2003. Nuclear transport: An emerging opportunity for drug targeting. *Advanced Drug Delivery Reviews* 55 (6):699–702.

Dignass, A., K. Lynch-Devaney, H. Kindon, L. Thim, and D. K. Podolsky. 1994. Trefoil peptides promote epithelial migration through a transforming growth factor beta-independent pathway. *Journal of Clinical Investigation* 94 (1):376–83.

Dressman, J. B., R. R. Berardi, L. C. Dermentzoglou, T. L. Russell, S. P. Schmaltz, J. L. Barnett, and K. M. Jarvenpaa. 1990. Upper gastrointestinal (GI) pH in young, healthy men and women. *Pharmaceutical Research* 7 (7):756–61.

East, L., and C. M. Isacke. 2002. The mannose receptor family. *Biochimica et Biophysica Acta—General Subjects* 1572 (2–3):364–86.

Eberlein, M., K. A. Scheibner, K. E. Black, S. L. Collins, Y. Chan-Li, J. D. Powell, and M. R. Horton. 2008. Anti-oxidant inhibition of hyaluronan fragment-induced inflammatory gene expression. *Journal of Inflammation (London)* 5:20.

Elzoghby, A. O., W. M. Samy, and N. A. Elgindy. 2012. Protein-based nanocarriers as promising drug and gene delivery systems. *Journal of Controlled Release* 161 (1):38–49.

Ensign, L. M., R. Cone, and J. Hanes. 2012. Oral drug delivery with polymeric nanoparticles: The gastrointestinal mucus barriers. *Advanced Drug Delivery Reviews* 64 (6):557–70.

Fang, J., H. Nakamura, and H. Maeda. 2011. The EPR effect: Unique features of tumor blood vessels for drug delivery, factors involved, and limitations and augmentation of the effect. *Advanced Drug Delivery Reviews* 63 (3):136–51.

Forssen, E., and M. Willis. 1998. Ligand-targeted liposomes. *Advanced Drug Delivery Reviews* 29 (3):249–71.

Ganta, S., H. Devalapally, A. Shahiwala, and M. Amiji. 2008. A review of stimuli-responsive nanocarriers for drug and gene delivery. *Journal of Controlled Release* 126 (3):187–204.

Garcia, E., M. Pion, A. Pelchen-Matthews, L. Collinson, J. F. Arrighi, G. Blot, F. Leuba, J. M. Escola, N. Demaurex, M. Marsh, and V. Piguet. 2005. HIV-1 trafficking to the dendritic cell-T-cell infectious synapse uses a pathway of tetraspanin sorting to the immunological synapse. *Traffic* 6 (6):488–501.

Gaur, U., S. K. Sahoo, T. K. De, P. C. Ghosh, A. Maitra, and P. K. Ghosh. 2000. Biodistribution of fluoresceinated dextran using novel nanoparticles evading reticuloendothelial system. *International Journal of Pharmaceutics* 202 (1–2):1–10.

Goldin, B. R., L. J. Gualtieri, and R. P. Moore. 1996. The effect of *Lactobacillus* GG on the initiation and promotion of DMH-induced intestinal tumors in the rat. *Nutrition and Cancer* 25 (2):197–204.

Hang, T. T. N., and D. Merlin. 2012. Homeostatic and innate immune responses: Role of the transmembrane glycoprotein CD98. *Cellular and Molecular Life Sciences* 69 (18):3015–26.

Harel, E., A. Rubinstein, A. Nissan, E. Khazanov, M. Nadler Milbauer, Y. Barenholz, and B. Tirosh. 2011. Enhanced transferrin receptor expression by proinflammatory cytokines in enterocytes as a means for local delivery of drugs to inflamed gut mucosa. *PLoS One* 6 (9):e24202.

Iacucci, M., S. de Silva, and S. Ghosh. 2010. Mesalazine in inflammatory bowel disease: A trendy topic once again? *Canadian Journal of Gastroenterology* 24 (2):127–33.

Ingersoll, S. A., S. Ayyadurai, M. A. Charania, H. Laroui, Y. Yan, and D. Merlin. 2012. The role and pathophysiological relevance of membrane transporter PepT1 in intestinal inflammation and inflammatory bowel disease. *American Journal of Physiology—Gastrointestinal and Liver Physiology* 302 (5):G484–92.

Isaacs, K. L., J. D. Lewis, W. J. Sandborn, B. E. Sands, and S. R. Targan. 2005. State of the art: IBD therapy and clinical trials in IBD. *Inflammation Bowel Disease* 11 (Suppl 1):S3–12.

Jain, A., A. Agarwal, S. Majumder, N. Lariya, A. Khaya, H. Agrawal, S. Majumdar, and G. P. Agrawal. 2010. Mannosylated solid lipid nanoparticles as vectors for site-specific delivery of an anti-cancer drug. *Journal of Controlled Release* 148 (3):359–67.

Jiang, H. L., M. L. Kang, J. S. Quan, S. G. Kang, T. Akaike, H. S. Yoo, and C. S. Cho. 2008. The potential of mannosylated chitosan microspheres to target macrophage mannose receptors in an adjuvant-delivery system for intranasal immunization. *Biomaterials* 29 (12):1931–9.

Jiang, X. B., X. L. Lu, B. Xiao, Y. Wan, and Y. X. Zhao. 2012. In vitro growth and activity of chondrocytes on three dimensional polycaprolactone/chitosan scaffolds. *Polymers for Advanced Technologies* 23 (1):99–107.

Jubeh, T. T., Y. Barenholz, and A. Rubinstein. 2004. Differential adhesion of normal and inflamed rat colonic mucosa by charged liposomes. *Pharmaceutical Research* 21 (3):447–53.

Jubeh, T. T., M. Nadler-Milbauer, Y. Barenholz, and A. Rubinstein. 2006. Local treatment of experimental colitis in the rat by negatively charged liposomes of catalase, TMN and SOD. *Journal of Drug Targeting* 14 (3):155–63.

Khalil, I. A., K. Kogure, H. Akita, and H. Harashima. 2006. Uptake pathways and subsequent intracellular trafficking in nonviral gene delivery. *Pharmacological Reviews* 58 (1):32–45.

Kindon, H., C. Pothoulakis, L. Thim, K. Lynch-Devaney, and D. K. Podolsky. 1995. Trefoil peptide protection of intestinal epithelial barrier function: Cooperative interaction with mucin glycoprotein. *Gastroenterology* 109 (2):516–23.

Kountouras, J., D. Chatzopoulos, and C. Zavos. 2001. Reactive oxygen metabolites and upper gastrointestinal diseases. *Hepatogastroenterology* 48 (39):743–51.

Kriegel, C., and M. Amiji. 2011. Oral TNF-alpha gene silencing using a polymeric microsphere-based delivery system for the treatment of inflammatory bowel disease. *Journal of Controlled Release* 150 (1):77–86.

Kucharzik, T., A. Lugering, Y. T. Yan, A. Driss, L. Charrier, S. Sitaraman, and D. Merlin. 2005. Activation of epithelial CD98 glycoprotein perpetuates colonic inflammation. *Laboratory Investigation* 85 (7):932–41.

Lai, S. K., Y. Y. Wang, and J. Hanes. 2009. Mucus-penetrating nanoparticles for drug and gene delivery to mucosal tissues. *Advanced Drug Delivery Reviews* 61 (2):158–71.

Lamprecht, A. 2010. IBD selective nanoparticle adhesion can enhance colitis therapy. *Nature Reviews: Gastroenterology & Hepatology* 7 (6):311–12.

Lamprecht, A., N. Ubrich, H. Yamamoto, U. Schafer, H. Takeuchi, P. Maincent, Y. Kawashima, and C. M. Lehr. 2001. Biodegradable nanoparticles for targeted drug delivery in treatment of inflammatory bowel disease. *Journal of Pharmacology and Experimental Therapeutics* 299 (2):775–81.

Laroui, H., G. Dalmasso, H. T. T. Nguyen, Y. T. Yan, S. V. Sitaraman, and D. Merlin. 2010. Drug-loaded nanoparticles targeted to the colon with polysaccharide hydrogel reduce colitis in a mouse model. *Gastroenterology* 138 (3):843–53.

Laroui, H., A. L. Theiss, Y. T. Yan, G. Dalmasso, H. T. T. Nguyen, S. V. Sitaraman, and D. Merlin. 2011. Functional TNF alpha gene silencing mediated by polyethyleneimine/TNF alpha siRNA nanocomplexes in inflamed colon. *Biomaterials* 32 (4):1218–28.

Laroui, H., E. Viennois, B. Xiao, B. S. B. Canup, D. Geem, T. L. Denning, and D. Merlin. 2014. Fab′-bearing siRNA TNF alpha-loaded nanoparticles targeted to colonic macrophages offer an effective therapy for experimental colitis. *Journal of Controlled Release* 186:41–53.

Lautenschlager, C., C. Schmidt, D. Fischer, and A. Stallmach. 2014. Drug delivery strategies in the therapy of inflammatory bowel disease. *Advanced Drug Delivery Reviews* 71:58–76.

Lehr, C. M. 2000. Lectin-mediated drug delivery: The second generation of bioadhesives. *Journal of Controlled Release* 65 (1–2):19–29.

Lehr, C. M., J. A. Bouwstra, W. Kok, A. B. J. Noach, A. G. Deboer, and H. E. Junginger. 1992. Bioadhesion by means of specific binding of tomato lectin. *Pharmaceutical Research* 9 (4):547–53.

Levy, J. E., O. Jin, Y. Fujiwara, F. Kuo, and N. C. Andrews. 1999. Transferrin receptor is necessary for development of erythrocytes and the nervous system. *Nature Genetics* 21 (4):396–9.

Loretz, B., F. Foger, M. Werle, and A. Bernkop-Schnurch. 2006. Oral gene delivery: Strategies to improve stability of pDNA towards intestinal digestion. *Journal of Drug Targeting* 14 (5):311–9.

Luo, D., and W. M. Saltzman. 2000. Synthetic DNA delivery systems. *Nature Biotechnology* 18 (1):33–7.

MacKinnon, A. C., S. L. Farnworth, P. S. Hodkinson, N. C. Henderson, K. M. Atkinson, H. Leffler, U. J. Nilsson, C. Haslett, S. J. Forbes, and T. Sethi. 2008. Regulation of alternative macrophage activation by galectin-3. *Journal of Immunology* 180 (4):2650–8.

Mahajan, N., D. Sakarkar, A. Manmode, V. Pathak, R. Ingole, and D. Dewade. 2011. Biodegradable nanoparticles for targeted delivery in treatment of ulcerative colitis. *Advanced Science Letters* 4 (2):349–56.

Markovsky, E., H. Baabur-Cohen, A. Eldar-Boock, L. Omer, G. Tiram, S. Ferber, P. Ofek, D. Polyak, A. Scomparin, and R. Satchi-Fainaro. 2012. Administration, distribution, metabolism and elimination of polymer therapeutics. *Journal of Controlled Release* 161 (2):446–60.

Maroni, A., L. Zema, M. D. Del Curto, A. Foppoli, and A. Gazzaniga. 2012. Oral colon delivery of insulin with the aid of functional adjuvants. *Advanced Drug Delivery Reviews* 64 (6):540–56.

Mason, M., and C. A. Siegel. 2013. Do inflammatory bowel disease therapies cause cancer? *Inflammatory Bowel Diseases* 19 (6):1306–21.

Meier, O., K. Boucke, S. V. Hammer, S. Keller, R. P. Stidwill, S. Hemmi, and U. F. Greber. 2002. Adenovirus triggers macropinocytosis and endosomal leakage together with its clathrin-mediated uptake. *Journal of Cell Biology* 158 (6):1119–31.

Meissner, Y., and A. Lamprecht. 2008. Alternative drug delivery approaches for the therapy of inflammatory bowel disease. *Journal of Pharmaceutical Sciences* 97 (8):2878–91.

Millar, A. D., D. S. Rampton, C. L. Chander, A. W. D. Claxson, S. Blades, A. Coumbe, J. Panetta, C. J. Morris, and D. R. Blake. 1996. Evaluating the antioxidant potential of new treatments for inflammatory bowel disease using a rat model of colitis. *Gut* 39 (3):407–15.

Moulari, B., D. Pertuit, Y. Pellequer, and A. Lamprecht. 2008. The targeting of surface modified silica nanoparticles to inflamed tissue in experimental colitis. *Biomaterials* 29 (34):4554–60.

Mrsny, R. J. 2012. Oral drug delivery research in Europe. *Journal of Controlled Release* 161 (2):247–53.

Naito, Y., M. Suematsu, and T. Yoshikawa. 2011. *Free Radical Biology in Digestive Diseases: Frontiers of Gastrointestinal Research*. Basel: Karger.

Nguyen, H. T., G. Dalmasso, L. Torkvist, J. Halfvarson, Y. Yan, H. Laroui, D. Shmerling, T. Tallone, M. D'Amato, S. V. Sitaraman, and D. Merlin. 2011. CD98 expression modulates intestinal homeostasis, inflammation, and colitis-associated cancer in mice. *Journal of Clinical Investigation* 121 (5):1733–47.

Nguyen, H. T. T., G. Dalmasso, Y. T. Yan, H. Laroui, S. Dahan, L. Mayer, S. V. Sitaraman, and D. Merlin. 2010. MicroRNA-7 modulates CD98 expression during intestinal epithelial cell differentiation. *Journal of Biological Chemistry* 285 (2):1479–89.

O'Neill, M. J., L. Bourre, S. Melgar, and C. M. O'Driscoll. 2011. Intestinal delivery of non-viral gene therapeutics: Physiological barriers and preclinical models. *Drug Discov Today* 16 (5–6):203–18.

Oyen, W. J., O. C. Boerman, E. T. Dams, G. Storm, L. van Bloois, E. B. Koenders, U. J. van Haelst, J. W. van der Meer, and F. H. Corstens. 1997. Scintigraphic evaluation of experimental colitis in rabbits. *Journal of Nuclear Medicine* 38 (10):1596–600.

Pallone, F., S. Fais, O. Squarcia, L. Biancone, P. Pozzilli, and M. Boirivant. 1987. Activation of peripheral blood and intestinal lamina propria lymphocytes in Crohn's disease. In vivo state of activation and in vitro response to stimulation as defined by the expression of early activation antigens. *Gut* 28 (6):745–53.

Peer, D., E. J. Park, Y. Morishita, C. V. Carman, and M. Shimaoka. 2008. Systemic leukocyte-directed siRNA delivery revealing cyclin D1 as an anti-inflammatory target. *Science* 319 (5863):627–30.

Pineton de Chambrun, G., L. Peyrin-Biroulet, M. Lemann, and J. F. Colombel. 2010. Clinical implications of mucosal healing for the management of IBD. *Nature Reviews: Gastroenterology & Hepatology* 7 (1):15–29.

Pinto, J. F. 2010. Site-specific drug delivery systems within the gastro-intestinal tract: From the mouth to the colon. *International Journal of Pharmaceutics* 395 (1–2):44–52.

Pithadia, A. B., and S. Jain. 2011. Treatment of inflammatory bowel disease (IBD). *Pharmacological Reports* 63 (3):629–42.

Pullan, R. D., G. A. Thomas, M. Rhodes, R. G. Newcombe, G. T. Williams, A. Allen, and J. Rhodes. 1994. Thickness of adherent mucus gel on colonic mucosa in humans and its relevance to colitis. *Gut* 35 (3):353–9.

Rankin, B. J., E. D. Srivastava, C. O. Record, J. P. Pearson, and A. Allen. 1995. Patients with ulcerative colitis have reduced mucin polymer content in the adherent colonic mucus gel. *Biochemical Society Transactions* 23 (1):104S.

Rosenholm, J., C. Sahlgren, and M. Linden. 2010. Cancer-cell targeting and cell-specific delivery by mesoporous silica nanoparticles. *Journal of Materials Chemistry* 20 (14):2707–13.

Russell-Jones, G. J., H. Veitch, and L. Arthur. 1999. Lectin-mediated transport of nanoparticles across Caco-2 and OK cells. *International Journal of Pharmaceutics* 190 (2):165–74.

Sakhalkar, H. S., M. K. Dalal, A. K. Salem, R. Ansari, A. Fu, M. F. Kiani, D. T. Kurjiaka, J. Hanes, K. M. Shakesheff, and D. J. Goetz. 2003. Leukocyte-inspired biodegradable particles that selectively and avidly adhere to inflamed endothelium in vitro and in vivo. *Proceedings of the National Academy of Sciences of the United States of America* 100 (26):15895–900.

Sakhalkar, H. S., J. Hanes, J. Fu, U. Benavides, R. Malgor, C. L. Borruso, L. D. Kohn, D. T. Kujiaka, and D. J. Goetz. 2005. Enhanced adhesion of ligand-conjugated biodegradable particles to colitic venules. *FASEB Journal* 19 (3):792.

Sato, Y., and E. Beutler. 1993. Binding, internalization, and degradation of mannose-terminated glucocerebrosidase by macrophages. *Journal of Clinical Investigation* 91 (5):1909–17.

Schreiber, S., R. P. MacDermott, A. Raedler, R. Pinnau, M. J. Bertovich, and G. S. Nash. 1991. Increased activation of isolated intestinal lamina propria mononuclear cells in inflammatory bowel disease. *Gastroenterology* 101 (4):1020–30.

Si, X. Y., D. Merlin, and B. Xiao. 2016. Recent advances in orally administered cell-specific nanotherapeutics for inflammatory bowel disease. *World Journal of Gastroenterology* 22 (34):7718–26.

Siccardi, D., J. R. Turner, and R. J. Mrsny. 2005. Regulation of intestinal epithelial function: A link between opportunities for macromolecular drug delivery and inflammatory bowel disease. *Advanced Drug Delivery Reviews* 57 (2):219–35.

Sofou, S., and G. Sgouros. 2008. Antibody-targeted liposomes in cancer therapy and imaging. *Expert Opinion on Drug Delivery* 5 (2):189–204.

Storm, G., S. O. Belliot, T. Daemen, and D. D. Lasic. 1995. Surface modification of nanoparticles to oppose uptake by the mononuclear phagocyte system. *Advanced Drug Delivery Reviews* 17 (1):31–48.

Sui, M., W. Liu, and Y. Shen. 2011. Nuclear drug delivery for cancer chemotherapy. *Journal of Controlled Release* 155 (2):227–36.

Tacchini, L., E. Gammella, C. De Ponti, S. Recalcati, and G. Cairo. 2008. Role of HIF-1 and NF-kappaB transcription factors in the modulation of transferrin receptor by inflammatory and anti-inflammatory signals. *Journal of Biological Chemistry* 283 (30):20674–86.

Talaei, F., F. Atyabi, M. Azhdarzadeh, R. Dinarvand, and A. Saadatzadeh. 2013. Overcoming therapeutic obstacles in inflammatory bowel diseases: A comprehensive review on novel drug delivery strategies. *European Journal of Pharmaceutical Sciences* 49 (4):712–22.

Taylor, K. M., and P. M. Irving. 2011. Optimization of conventional therapy in patients with IBD. *Nature Reviews: Gastroenterology & Hepatology* 8 (11):646–56.

Theiss, A. L., R. D. Idell, S. Srinivasan, J. M. Klapproth, D. P. Jones, D. Merlin, and S. V. Sitaraman. 2007. Prohibitin protects against oxidative stress in intestinal epithelial cells. *FASEB Journal* 21 (1):197–206.

Theiss, A. L., H. Laroui, T. S. Obertone, I. Chowdhury, W. E. Thompson, D. Merlin, and S. V. Sitaraman. 2011. Nanoparticle-based therapeutic delivery of prohibitin to the colonic epithelial cells ameliorates acute murine colitis. *Inflammatory Bowel Diseases* 17 (5):1163–76.

Tirosh, B., N. Khatib, Y. Barenholz, A. Nissan, and A. Rubinstein. 2009. Transferrin as a luminal target for negatively charged liposomes in the inflamed colonic mucosa. *Molecular Pharmaceutics* 6 (4):1083–91.

van den Berg, T. K., and G. Kraal. 2005. A function for the macrophage F4/80 molecule in tolerance induction. *Trends in Immunology* 26 (10):506–9.

van Vliet, S. J., E. Saeland, and Y. van Kooyk. 2008. Sweet preferences of MGL: Carbohydrate specificity and function. *Trends in Immunology* 29 (2):83–90.

Varkouhi, A. K., M. Scholte, G. Storm, and H. J. Haisma. 2011. Endosomal escape pathways for delivery of biologicals. *Journal of Controlled Release* 151 (3):220–8.

Vercauteren, D., J. Rejman, T. F. Martens, J. Demeester, S. C. De Smedt, and K. Braeckmans. 2012. On the cellular processing of non-viral nanomedicines for nucleic acid delivery: Mechanisms and methods. *Journal of Controlled Release* 161 (2):566–81.

Verrey, F., D. L. Jack, I. T. Paulsen, M. H. Saier Jr., and R. Pfeiffer. 1999. New glycoprotein-associated amino acid transporters. *Journal of Membrane Biology* 172 (3):181–92.

Vong, L. B., T. Tomita, T. Yoshitomi, H. Matsui, and Y. Nagasaki. 2012. An orally administered redox nanoparticle that accumulates in the colonic mucosa and reduces colitis in mice. *Gastroenterology* 143 (4):1027–36.

Wachsmann, P., and A. Lamprecht. 2012. Polymeric nanoparticles for the selective therapy of inflammatory bowel disease. *Methods in Enzymology* 508:377–97.

Wagstaff, K. M., and D. A. Jans. 2009. Nuclear drug delivery to target tumour cells. *European Journal of Pharmacology* 625 (1–3):174–80.

Wan, Y., B. Xiao, S. Dalai, X. Cao, and Q. Wu. 2009. Development of polycaprolactone/chitosan blend porous scaffolds. *Journal of Materials Science: Materials in Medicine* 20 (3):719–24.

Wang, H. Y., J. X. Chen, Y. X. Sun, J. Z. Deng, C. Li, X. Z. Zhang, and R. X. Zhuo. 2011. Construction of cell penetrating peptide vectors with N-terminal stearylated nuclear localization signal for targeted delivery of DNA into the cell nuclei. *Journal of Controlled Release* 155 (1):26–33.

Wang, Y. Y., S. K. Lai, C. So, C. Schneider, R. Cone, and J. Hanes. 2011. Mucoadhesive nanoparticles may disrupt the protective human mucus barrier by altering its microstructure. *PLoS One* 6 (6):e21547.

Wang, Y. Y., S. K. Lai, J. S. Suk, A. Pace, R. Cone, and J. Hanes. 2008. Addressing the PEG mucoadhesivity paradox to engineer nanoparticles that "slip" through the human mucus barrier. *Angewandte Chemie International Edition in English* 47 (50):9726–9.

Wellen, K. E., and G. S. Hotamisligil. 2003. Obesity-induced inflammatory changes in adipose tissue. *Journal of Clinical Investigation* 112 (12):1785–8.

Wileman, T. E., M. R. Lennartz, and P. D. Stahl. 1986. Identification of the macrophage mannose receptor as a 175-Kda membrane-protein. *Proceedings of the National Academy of Sciences of the United States of America* 83 (8):2501–5.

Wilson, D. S., G. Dalmasso, L. Wang, S. V. Sitaraman, D. Merlin, and N. Murthy. 2010. Orally delivered thioketal nanoparticles loaded with TNF-alpha-siRNA target inflammation and inhibit gene expression in the intestines. *Nature Materials* 9 (11):923–8.

Winyard, P. G., D. R. Blake, and C. H. Evans. 2000. *Free Radicals and Inflammation: Progress in Inflammation Research.* Boston: Birkhäuser Verlag.

Xiao, B., M. K. Han, E. Viennois, L. Wang, M. Zhang, X. Si, and D. Merlin. 2015a. Hyaluronic acid-functionalized polymeric nanoparticles for colon cancer-targeted combination chemotherapy. *Nanoscale* 7 (42):17745–55.

Xiao, B., H. Laroui, S. Ayyadurai, E. Viennois, M. A. Charania, Y. Zhang, and D. Merlin. 2013. Mannosylated bioreducible nanoparticle-mediated macrophage-specific TNF-alpha RNA interference for IBD therapy. *Biomaterials* 34 (30):7471–82.

Xiao, B., H. Laroui, E. Viennois, S. Ayyadurai, M. A. Charania, Y. C. Zhang, Z. Zhang, M. T. Baker, B. Y. Zhang, A. T. Gewirtz, and D. Merlin. 2014. Nanoparticles with surface antibody against CD98 and carrying CD98 small interfering RNA reduce colitis in mice. *Gastroenterology* 146 (5):1289–300.

Xiao, B., L. Ma, and D. Merlin. 2017. Nanoparticle-mediated co-delivery of chemotherapeutic agent and siRNA for combination cancer therapy. *Expert Opinion on Drug Delivery* 14 (1):65–73.

Xiao, B., and D. Merlin. 2012. Oral colon-specific therapeutic approaches toward treatment of inflammatory bowel disease. *Expert Opinion on Drug Delivery* 9 (11):1393–407.

Xiao, B., X. Y. Si, M. K. Han, E. Viennois, M. Z. Zhang, and D. Merlin. 2015b. Co-delivery of camptothecin and curcumin by cationic polymeric nanoparticles for synergistic colon cancer combination chemotherapy. *Journal of Materials Chemistry B* 3 (39):7724–33.

Xiao, B., X. Si, M. Zhang, and D. Merlin. 2015c. Oral administration of pH-sensitive curcumin-loaded microparticles for ulcerative colitis therapy. *Colloids and Surfaces B—Biointerfaces* 135:379–85.

Xiao, B., Y. Wan, X. Y. Wang, Q. C. Zha, H. M. Liu, Z. Y. Qiu, and S. M. Zhang. 2012. Synthesis and characterization of N-(2-hydroxy)propyl-3-trimethyl ammonium chitosan chloride for potential application in gene delivery. *Colloids and Surfaces B—Biointerfaces* 91:168–74.

Xiao, B., Y. Wan, M. Q. Zhao, Y. Q. Liu, and S. M. Zhang. 2011. Preparation and characterization of antimicrobial chitosan-N-arginine with different degrees of substitution. *Carbohydrate Polymers* 83 (1):144–50.

Xiao, B., M. Z. Zhang, E. Viennois, Y. C. Zhang, N. Wei, M. T. Baker, Y. J. Jung, and D. Merlin. 2015d. Inhibition of MDR1 gene expression and enhancing cellular uptake for effective colon cancer treatment using dual-surface-functionalized nanoparticles. *Biomaterials* 48:147–60.

Yin, Y., D. Chen, M. Qiao, Z. Lu, and H. Hu. 2006. Preparation and evaluation of lectin-conjugated PLGA nanoparticles for oral delivery of thymopentin. *Journal of Controlled Release* 116 (3):337–45.

Yin, Y. S., D. W. Chen, M. X. Qiao, H. Y. Hu, and J. Qin. 2007. Preparation of lectin-conjugated PLGA nanoparticles and evaluation of their in vitro bioadhesive activity. *Yao Xue Xue Bao* 42 (5):550–6.

Zhang, J., C. Tang, and C. H. Yin. 2013. Galactosylated trimethyl chitosan-cysteine nanoparticles loaded with Map4k4 siRNA for targeting activated macrophages. *Biomaterials* 34 (14):3667–77.

Zuo, L. S., Z. Huang, L. Dong, L. Q. Xu, Y. Zhu, K. Zeng, C. Y. Zhang, J. N. Chen, and J. F. Zhang. 2010. Targeting delivery of anti-TNF alpha oligonucleotide into activated colonic macrophages protects against experimental colitis. *Gut* 59 (4):470–9.

Bridging the Gap between the Bench and the Clinic

INFLAMMATORY BOWEL DISEASE

Susan Hua

CONTENTS

5.3.1 INTRODUCTION

Inflammatory bowel disease (IBD) is a localized, chronic inflammatory condition that affects the gastrointestinal (GI) wall. Currently, there is no cure for IBD, with therapeutic strategies aimed toward attaining and maintaining remission from inflammatory episodes. While conventional medications can temporarily induce and maintain remission, 70% of IBD patients will require at least one surgical intervention in their lifetime (Byrne et al. 2007; Talley et al. 2011). Due to the chronic nature of IBD and the impact the disease has on patient compliance to medications, oral formulations are the preferred route of administration (Saini et al. 2009). Conventional oral formulations have limited use in IBD, as they are generally designed to achieve systemic delivery of therapeutics, which results in adverse effects and toxicity following distribution of drug around the body (Hua et al. 2015). These formulations can be adversely affected during active IBD or following intestinal resection, and have limited efficacy and specificity for diseased GI tissue versus healthy GI tissue (Malayandi et al. 2014). In addition, despite coverage of the GI surface (including diseased tissue), there is no guarantee that the drug is effectively taken up into the tissue and cells at the site of inflammation (Hua 2014). Oral formulations achieving a localized effect are preferred in rational drug delivery design for IBD. This ensures that drug will be delivered to the site of action within the GI tract, but will not be absorbed or will be poorly absorbed to avoid unwanted side effects (Talaei et al. 2013; Hua et al. 2015).

Current therapeutic approaches specifically indicated for IBD are based on conventional dosage forms containing free drug that rely on delayed or controlled mechanisms to trigger drug release, such as enzymatic activity of colonic bacteria (Sandborn 2002; Hanauer and Sparrow 2004), GI transit times (Steed et al. 1997), or pH gradient of the GI tract (Yang et al. 2002) (see Chapter 5.1). Their design is generally based on exploiting physiological conditions in specific segments of the GI tract, in particular the colon (Yang et al. 2002). These approaches are associated with inconsistent efficacy and interpatient variability, with diverse pathophysiological changes and variability in the GI tract that present with chronic and active inflammation in IBD patients (e.g., GI transit time, pH, and colonic microbiome) (Hebden et al. 2000; Talley et al. 2011). Attempts to overcome these issues have focused on improving our understanding of the pathophysiology of the GI tract during acute disease, during chronic IBD, and following GI tract resection, as well as the rational design of oral formulations. Improved oral drug delivery design has drastically improved the biodistribution of therapeutics to specific segments of the GI tract. However, in order for a drug to have therapeutic efficacy, it must also have specific accumulation and cellular uptake within diseased tissue (Laroui et al. 2010; Hua 2014).

Recent pharmaceutical advances have applied nanotechnology to oral dosage form design in an effort to overcome the limitations of conventional formulations (Coco et al. 2013). Pharmaceutical strategies utilizing nanodelivery systems as carriers for active compounds have shown promising results in addressing the pathophysiological changes in IBD, and exploiting these differences to enhance specific delivery of drugs to inflammatory tissue (Collnot et al. 2012; Xiao and Merlin 2012). Therefore, the use of nanotechnology in formulation design may further improve the efficacy of therapeutics by allowing inflammation-specific targeting and uptake within diseased tissue in the GI tract. Nanodelivery systems have been designed to passively or actively target the site of inflammation (Hua et al. 2015). These systems have been shown to be more beneficial than conventional formulations, because their size leads to more effective targeting, better bioavailability at diseased tissues, and reduced systemic adverse effects (Lautenschlager et al. 2014; Viscido et al. 2014). Hence, nanodelivery systems have been found to have similar or improved therapeutic efficacy at lower drug concentrations in comparison with conventional formulations (Collnot et al. 2012; Xiao and Merlin 2012). Although size is an important factor in targeting inflammatory regions in the GI tract, additional strategies have been explored to further enhance drug delivery to and achieve maximal retention time in diseased tissues (see Chapter 5.2). This section describes the advances and challenges in the clinical translation of current nanomedicine formulations for IBD, and the future strategies that are required to move promising drug delivery platforms from the bench to the bedside.

5.3.2 PATHOPHYSIOLOGICAL CONSIDERATIONS FOR IBD DRUG DELIVERY

Disease-specific changes in the physiology of the GI tract are important factors to take into account when designing effective oral formulation strategies for IBD. Considerations should be made during formulation design for factors such as the residence time of the formulation in the GI tract, how the GI environment affects the delivery of the formulation and dissolution of the drug at the site of action, the intestinal fluid volume, and the propensity of the formulation or drug to be metabolized in the GI tract through enzymatic or microbial degradation (Lautenschlager et al. 2014; Hua et al. 2015). These pathophysiological changes can vary significantly, depending on the state of the disease, which includes active disease, chronic disease, and intestinal resection. Other considerations for drug delivery include the goal of treatment (induction or maintenance of remission), location, extent and severity of the disease, and presence of complications. These physiological factors are dynamic and interrelated, and remain an important challenge in dosage form design for IBD (Podolsky 2002; Talley et al. 2011). For example, mucosal inflammation not only disrupts the intestinal barrier with the formation of mucosal surface alterations, crypt distortions, and ulcers, but also increases the production of mucus and infiltration of immune cells (e.g., neutrophils, macrophages, lymphocytes, and dendritic cells) (Li and Thompson 2003; Antoni et al. 2014). Patients suffering from severe mucosal inflammation may exhibit altered GI motility and diarrhea, which in turn affects intestinal volume, pH, mucosal integrity, and the resident microbiome (Podolsky 2002; Hua et al. 2015). These factors can lead to inadequate response to treatment in patients, and therefore are important considerations in the rational design of formulations for the management of IBD.

5.3.2.1 Disease Location

As discussed in Chapter 5.1, there are two major types of IBD: ulcerative colitis (UC) and Crohn's disease (CD). Although both conditions cause chronic inflammation of the GI tract, the distinct differences in disease location and nature of the inflammation affect the specificity and efficacy of oral treatment strategies (Podolsky 2002; Sartor 2010). The inflammation that occurs in UC is confined to the colon and can extend proximally from the rectum, with some cases involving the entire colon (pancolitis). In addition, the inflammation is usually continuous and only affects the mucosal lining. Conversely, CD can affect any segment of the GI tract, from the mouth to the anus, with the terminal ileum and the colon commonly affected. The inflammation associated with CD is generally discontinuous and can affect the full thickness of the GI tract wall (transmural) (Podolsky 2002). In general, CD is more variable than UC, depending on which part of the GI tract is involved. These differences make rational drug delivery design challenging for IBD, especially with formulations predominantly based on exploiting pathophysiological changes in specific segments of the GI tract to trigger drug release (Talaei et al. 2013; Lautenschlager et al. 2014). Treatment strategies focused on colon-targeted drug delivery are commonly used in both UC and CD (Hua et al. 2015). This strategy is less likely to be effective for inflammation occurring in other segments of the GI tract, which occurs in CD. Furthermore, the discontinuous nature of the inflammation in CD would particularly benefit from inflammation-specific targeting to avoid drug exposure in healthy GI tissue.

5.3.2.2 GI Transit Time

Formulation transit time through the GI tract is critical to ensuring delivery of therapeutic compounds to the site of action and sufficient time for drug uptake (Asghar and Chandran 2006). In healthy individuals, transit time through the small intestine can vary from 2 to 6 hours (Hu et al. 2000). This range is significantly more variable with colonic transit times, which have been reported to range from 6 to 70 hours (Coupe et al. 1991; Rao et al. 2004). Additional confounders influencing GI transit time include gender, with females having significantly longer colonic transit times (Buhmann et al. 2007), and the time of dosing with respect to an individual's bowel movements (Sathyan et al. 2000). This variability in GI transit time is even more complex in GI disease, with changes occurring based on disease state (i.e., active or remission) and its association with other physiological GI factors. For instance, orocecal transit time (OCTT), which is the time taken for a meal to reach the cecum, has been shown to be delayed in both CD and UC patients compared with healthy controls (Rana et al. 2013). However, significantly faster OCTTs have been observed in IBD patients with the dysbiotic condition known as small intestinal bacterial overgrowth (SIBO) (Rana et al. 2013). These observations have been confirmed experimentally in humanized mice, following dietary manipulation of the gut microflora (Kashyap et al. 2013). In contrast to OCTT, colonic transit is considerably faster in IBD patients, which is likely due to the diarrhea that is associated with disease (Hebden et al. 2000; Podolsky 2002). UC patients may exhibit transit times twice as rapid as those of healthy individuals, leading to difficulties in targeting specific regions of the colon with conventional formulations. For example, studies using conventional delayed release formulations have shown asymmetric biodistribution in the colon, with higher retention of drug in the proximal colon and significantly lower drug concentrations in the distal colon (Hebden et al. 2000). Thus, transit time in itself may not be a reliable approach for targeted drug delivery in IBD.

5.3.2.3 Intestinal Microbiome

The GI tract plays host to more than 500 distinct bacterial species, with many estimating the number of species to be close to 2000 (Sartor 2008a). While there appears to be considerable variation in the composition of the microbiome between healthy individuals, which is influenced by both genetic and environmental factors (Sartor 2010; Albenberg and Wu 2014), the dominant Firmicutes, Bacteroidetes, Proteobacteria, and Actinobacteria species appear to be consistent and represent the majority of the colonic flora (Frank et al. 2007). These bacteria play pivotal roles in both digestion and intestinal health (Macfarlane and Macfarlane 2011). The majority of the intestinal microbiome resides in the anaerobic colon, and fermentation of carbohydrates is the main source of nutrition for this population (Macfarlane and Macfarlane 2011). This relatively exclusive fermentation of nonstarch polysaccharides by the colonic microbiome is exploited in formulations that use nonstarch polysaccharide coatings (Sinha and Kumria 2001).

Changes in the composition of the microbiome (dysbiosis) can occur with alterations in physiology or the inflammatory state, or by medications (e.g., antibiotics) (Sartor 2008b; Hua et al. 2015). While it is generally accepted that the bacterial load is relatively static in IBD, the diversity of the microbiome is reduced with increases in major species such as *Bacteroides*, *Eubacteria*, and *Peptostreptococcus* (Linskens et al. 2001). It is not known what precipitates the initial dysbiosis, and whether dysbiosis precedes or is a symptom of disease. However, there is some evidence to suggest that physiological factors, such as dysmotility and increased luminal fluid (diarrhea), may play a role. For example, studies in animal models have shown that prolonged water secretion into the bowel causes decreased colonic transit times, which can alter the colonic microbiome (Yang et al. 1998; Keely et al. 2012; Musch et al. 2013). Therefore, in IBD, both the mucosal inflammatory response and diarrhea disrupt the resident microbiome, which can alter microbial metabolism and transit times in the GI tract.

5.3.2.4 Changes in Colonic pH

Differences in pH along the GI tract have been exploited for the purposes of delayed release therapies (Lautenschlager et al. 2014). In general, the highly acidic stomach environment rises rapidly to pH 6 in the duodenum and increases along the small intestine to pH 7.4 at the terminal ileum (Fallingborg et al. 1993; Bratten and Jones 2006). The cecal pH drops below pH 6 and again rises in the colon, reaching pH 6.7 at the rectum (Sasaki et al. 1997; Nugent et al. 2001). These pH ranges can exhibit variability between individuals, and are influenced by factors such as water and food intake, microbial metabolism, and GI disease (Ibekwe et al. 2008). In patients with IBD, there is little evidence to suggest major alterations to the pH in the small intestine (Lucas et al. 1978; Barkas et al. 2013). In the large intestine, however, the pH is significantly lower in both UC and CD patients. In the colon, the intestinal pH is influenced by microbial fermentation processes, bicarbonate and lactate secretions, bile acid metabolism of fatty acids, and intestinal volume and transit times (Nugent et al. 2001). As all these factors may be disrupted during IBD, especially during active disease, this can alter the luminal pH in the colon. Normal colonic pH ranges from 6.8 in the proximal colon to 7.2 in the distal colon. In active UC patients, the pH in the colon can vary from 5.5 to as low as 2.3 (Fallingborg et al. 1993; McConnell et al. 2008). Similarly, the reported colonic pH values for CD patients are approximately 5.3, irrespective of disease activity (Sasaki et al. 1997). These pH changes are likely to affect the composition of the colonic microbiome, and thus colonic transit times, which can have an effect on the release of drug from formulations requiring bacterial fermentation or enzymatic activity. Likewise, pH changes can affect the release of compounds from pH-dependent release coatings (Hua et al. 2015).

5.3.2.5 Intestinal Volume

The composition of the intestinal volume (fluid–matter ratio) can alter various characteristics in the GI tract (Hua et al. 2015). For example, food intake can significantly alter free fluid volumes, bile salt and digestive enzyme levels, and luminal pH (Nyhof et al. 1985; Fatouros and Mullertz 2008)—this can affect the way drug formulations are processed following oral administration. Intestinal fluid secretion also affects the viscosity of the mucous–gel layer, which may influence the ability of drugs to be taken up by cells at the site of action (Keely et al. 2011). In addition, the composition of the intestinal biomass is altered in IBD and is directly related to changes in microbial metabolism, intestinal transit time, and luminal pH. In particular, increased fluid secretion and decreased reabsorption can dilute the digestive enzymes that control intestinal transit to allow nutrient absorption (Van Citters and Lin 2006). This in turn may influence the intestinal microbiome, which can alter carbohydrate and polysaccharide digestion (Yang 2008), as well as contribute to changes in intestinal transit times (Van Citters and Lin 2006). These changes in intestinal fluid volumes may alter the efficacy of conventional oral formulations for IBD.

5.3.2.6 Mucosal Integrity

The intestinal mucosa normally has a complex and multilayered defense system that forms an efficient barrier to provide protection against detrimental influences, such as invading microorganisms, from the GI lumen (Antoni et al. 2014). In both UC and CD, the intestinal barrier is compromised at different levels to allow an abnormally high number of microorganisms to come into contact with the host's immune system (Sartor 2008a). This is thought to initiate a process that leads to excessive immune reactions and eventually to chronic intestinal inflammation (Rath et al. 1996; Sellon et al. 1998). These barrier disturbances include alterations in pattern recognition receptors (PRRs), disturbed antimicrobial peptide (AMP) production, unresolved endoplasmic reticulum (ER) stress, alterations of the mucus layer, defects in the process of autophagy, and an increased epithelial barrier permeability (Antoni et al. 2014). These mucosal changes may be a source of new targets for inflammation-specific drug delivery, with the upregulation of disease-specific surface receptors and molecules.

Alteration of the intestinal mucus layer is an important consideration for the design of more effective oral formulations for IBD. This variation has been reported to be more pronounced in UC, with a reduction in the thickness and continuity of the mucus layer (Pullan et al. 1994). In addition, IBD affects the composition and structure of the mucus layer, including shorter glycans, less complex structure (Larsson et al. 2011), and decreased sulfation (Corfield et al. 1996). Sulfates confer negative charge to the mucins and enhance the resistance of the glycans against enzymatic degradation. From a drug delivery perspective, reduced sulfation may cause less effective retention of mucoadhesive drug delivery platforms that rely on positive–negative electrostatic charge interactions (Hua et al. 2015). Disturbances in intestinal permeability are caused by different molecular mechanisms, including tight junction abnormalities and epithelial apoptosis (Laukoetter et al. 2008; Wang et al. 2010; Cunningham and Turner 2012). To date, it is still unclear whether these changes play a primary role in disease pathogenesis or whether they are a secondary effect triggered by inflammation. Understanding of the complex mechanisms underlying the pathogenesis of IBD will enable the development of more effective therapeutic approaches for this condition.

5.3.2.7 Intestinal Resection in IBD Patients

Surgical resection of the intestinal tissue is common among IBD patients for either a complication of the disease or inadequate control of symptoms. It has been estimated that more than 70% of IBD patients will undergo at least one surgery in their lifetime (Byrne et al. 2007). Even with early introduction to antibody therapy, 1 in 5 patients with UC (Sandborn et al. 2009) and 7 in 10 patients with CD (Lazarev et al. 2010) will eventually require colectomy or small bowel resections. The goal of a resection is to keep as much of the healthy tissue as possible and to only remove portions of the bowel that are beyond healing. Removal of bowel tissue results in a shortening of the intestine and reduced transit distance through the GI tract, which potentially affects the way some conventional oral formulations are processed (Hua et al. 2015). Furthermore, resection can significantly change the physiology of the intestinal tract by altering pH, digestion, transit, and nutrient absorption (Spiller et al. 1988; Schmidt et al. 1996; Fallingborg et al. 1998). The latter occurs particularly when too much of the small intestine is removed. Resection of the terminal ileum alters water absorption and leads to dilution of residual bile acids in the colon, which reduces net colonic fatty acid concentrations (Thompson et al. 1998;

Gracie et al. 2012). This may profoundly alter microbial metabolism of fatty acids by hydroxylation to produce ricinoleic acid analogues that can drive diarrhea (Ammon and Phillips 1974; Gracie et al. 2012). Diarrhea significantly affects the therapeutic efficacy of conventional oral formulations (Watts et al. 1992). The reduction in fatty acids also reduces the "ileal brake," which is a nutrient feedback mechanism that slows transit times to allow nutrient absorption (Van Citters and Lin 1999, 2006). Fatty acids are the most potent stimulant of the ileal brake; therefore, a loss of both fatty acid receptors (from resected tissue) and fatty acids from digestion leads to a loss of the ileal brake (Lin et al. 2005), and a subsequent decrease in intestinal transit time. As resection surgery still plays an important part in the overall management of IBD patients, these physiological changes should be taken into account when devising drug delivery strategies.

5.3.3 CURRENT STATE OF NANOMEDICINES FOR IBD

In the last few decades, there has been considerable research undertaken to elucidate the mechanisms underlying the pathogenesis of IBD, in order to identify novel treatment strategies that are effective, specific, and safe. Conventional formulations have been demonstrated to have limited efficacy, as they mainly rely on nonstable parameters in the GI tract to trigger drug release, such as pH changes and the action of local enzymes (Malayandi et al. 2014). These parameters are highly variable in healthy and IBD patients, as well as in active disease and chronic disease in remission (Hua et al. 2015). Therefore, these pharmaceutical strategies tend to be nonspecific by also releasing active agents onto healthy mucosa, causing adverse effects, and are unable to ensure distribution over the entire area of inflamed GI tissue (Viscido et al. 2014; Hua et al. 2015). Nanosized therapeutics have shown increased effectiveness in drug delivery compared with conventional formulations for IBD, due to having prolonged retention, specific accumulation, and increased uptake in target inflammatory tissues in the GI tract (Collnot et al. 2012; Xiao and Merlin 2012). This longer-lasting retention of the drug within the mucosa may also reduce the dose to be taken and frequency of administration, thereby improving adherence to treatment. Nanoparticles have

the unique ability to discriminate between diseased and nondiseased sites by exploiting specific features of inflamed GI tissue, such as enhanced mucosal inflammatory cell infiltrate, disruption of the mucosal barrier, increased permeability, and increased production of mucus (Antoni et al. 2014; Viscido et al. 2014). Therefore, reformulation of active agents into nanomedicines has proven to be an effective strategy to increasing the local bioavailability, solubility, physical or chemical stability, safety, and efficacy of former drugs in conventional formulations (Teli et al. 2010; Diab et al. 2012). In addition, nanoencapsulation has allowed the delivery and protection of novel compounds that are sensitive to degradation in the GI tract (e.g., siRNA and peptides), which has opened up the possibilities for new therapeutic strategies for the treatment of IBD. This section discusses the current state of nanomedicines for IBD in the research and development (R&D) pipeline.

5.3.3.1 Nanomedicines for Clinical Use in IBD

Despite the advantages of nanoencapsulation, there are currently no nanomedicines that have been approved or marketed for clinical use in IBD. This is based on the definition of nanomedicines being the application of nanotechnology for medical purposes, in particular as diagnostic platforms (delivery systems incorporating imaging agents) or therapeutic platforms (delivery systems incorporating active compounds). These nanosized delivery systems generally consist of multiple components, including the carrier materials, encapsulated drug or imaging compounds, and surface modifications with coatings and/or targeting ligands (Hua et al. 2015). Although technically not a nanomedicine by this definition, the closest clinically approved platform that should be noted is certolizumab pegol (Cimzia®), which is a biological therapy administered by subcutaneous injection for the treatment of CD (Schreiber 2011; Schreiber et al. 2011; Sainz et al. 2015). Certolizumab pegol is a PEGylated Fab′ of a human monoclonal antibody with high affinity for tumor necrosis factor (TNF) α. The Fab′ of human anti-TNF is conjugated with a 40 kDa poly(ethylene glycol) (PEG) molecule, which enhances its distribution in inflamed tissue (Schreiber 2011). The use of biological agents has been a game changer in the treatment of a number of chronic inflammatory diseases, including IBD; however, they have been associated with serious systemic adverse effects, such as infections and malignancies (Elinav and Peer 2013; Nielsen and Ainsworth 2013). Other drawbacks have been the loss of response caused by antibody formation, patient compliance issues with the route of administration, and the significant costs associated with long-term therapy (Nielsen and Ainsworth 2013; Nielsen 2014). Around 33% of patients fail to respond to TNF inhibitors, and another third of all patients lose response over time (Nielsen 2014; Steenholdt et al. 2014). A combination of biological agents with nanodelivery system technology may be a strategy to achieve maximum targeted efficacy with lower doses and frequency of doses, as well as reduced side effects.

The number of clinical studies evaluating the effect of novel nanomedicines in the human GI tract for IBD have been limited and warrant further investigation. From a patient compliance viewpoint, it would make sense to transition orally administered nanomedicines over parenterally administered platforms for the chronic therapeutic management of IBD (Saini et al. 2009). The biological obstacles facing orally administered therapeutics in humans are challenging and highly variable (Diab et al. 2012). In addition, our understanding of the interaction of nanomedicines in the human GI tract is still unclear. This may also explain the reason for the lack of nanomedicine-based intervention studies in the clinical trial phase of the R&D process. The therapeutic use of nanomedicines is still an emerging area for IBD, with the majority of research still in the preclinical investigation phase. There are a number of key companies globally that are focusing on the development of new therapeutics for IBD (Table 5.1), the majority of which are focused on small-molecule drugs, peptide or protein-based therapeutics (e.g., biologics), and gene therapy. The development of nanomedicine products for IBD is mostly conducted by scientists in academia, research institutes, or industry who specialize in manufacturing nanotechnology platforms. This is due to unique challenges faced in the R&D process and manufacturing of nanomedicines, in addition to many of the traditional hurdles of product development (Sainz et al. 2015) (discussed in Section 5.3.4). Nanomedicines have huge potential for the management of IBD. In the near future, we should expect to see platforms transitioning to clinical trials, with results from head-to-head comparisons with existing therapies being of key interest.

5.3.3.2 Nanomedicines for Experimental Use in IBD

It has been well established that reducing the size of drug delivery systems improves their retention time and accumulation in inflamed intestinal tissue in IBD (Laroui et al. 2013; Talaei et al. 2013; Viscido et al. 2014; Hua et al. 2015). As discussed in Chapter 5.2, this is due to the epithelial enhanced permeability and retention (eEPR) effect (Collnot et al. 2012; Xiao and Merlin 2012), which also allows preferential uptake of nanosized therapeutics by the increased number of immune cells residing in inflamed regions (Lamprecht et al. 2005a). This increased cellular uptake avoids rapid carrier elimination by diarrhea, which is a particular advantage over conventional formulations for IBD (Beloqui et al. 2013). There are very few studies evaluating the deposition and uptake of nanoparticles in humans; therefore, the study by Schmidt et al. (2013) investigating the potential of nano- and microparticle uptake into the rectal mucosa of human IBD patients is of particular importance. The study found a significant accumulation of microparticles in active IBD in comparison with nanoparticles, which were only detectable in traces in the mucosa, after rectal administration in patients with CD and UC. The microparticles exhibited accumulation and bioadhesion to the inflamed mucosal wall, with no detectable absorption of these particles across the epithelial barrier. Conversely, nanoparticles were found to translocate to the serosal compartment, which would be beneficial for patients with CD; however, may also potentially lead to systemic absorption. Although particle accumulation in ulcerated areas was statistically significant, the total fraction of particles penetrating into the mucosa was relatively low in the study.

Interestingly, a similar study was conducted by Lamprecht et al. (2001a), which investigated the significance of particle size–dependent accumulation in the trinitrobenzenesulfonic acid (TNBS)–induced rat model of colitis. Studying fluorescent polystyrene particles ranging in size from 0.1 to 10 μm, administered orally for 3 days *in vivo*, they found that the highest binding to inflamed tissue was for 0.1 μm particles (control healthy group, 2.2% ± 1.6%; colitis, 14.5 ± 6.3%). The study showed that the ratio

TABLE 5.1
Companies focused on therapeutic development for IBD.

4SC AG	Eisai Co. Ltd.	Pepscan Therapeutics
AB Science	Eli Lilly and Company	Pfenex, Inc.
AbbVie, Inc.	Emergent BioSolutions, Inc.	Pfizer, Inc.
Advinus Therapeutics Ltd.	Enlivex Therapeutics Ltd.	Pieris AG
Aerpio Therapeutics, Inc.	Enzo Biochem, Inc.	Plexxikon, Inc.
Alba Therapeutics Corporation	Epirus Biopharmaceuticals, Inc.	Pluristem Therapeutics, Inc.
Albireo Pharma	Farmacija d.o.o. Tuzla	ProtAb Ltd.
Alfa Wassermann SpA	Ferring International Center SA	Protagonist Therapeutics, Inc.
Allozyne, Inc.	FibroGen, Inc.	Protalix BioTherapeutics, Inc.
Altheus Therapeutics, Inc.	Galapagos NV	Provid Pharmaceuticals, Inc.
Alvine Pharmaceuticals, Inc.	Genentech, Inc.	Qu Biologics, Inc.
Am-Pharma BV	Genfit SA	Receptos, Inc.
Amakem NV	Genor BioPharma Co. Ltd.	RedHill Biopharma Ltd.
Amgen, Inc.	Gilead Sciences, Inc.	ReveraGen BioPharma, Inc.
Amorepacific Corporation	GlaxoSmithKline Plc	Salix Pharmaceuticals Ltd.
Ampio Pharmaceuticals, Inc.	GlycoMar Ltd.	Sandoz International GmbH
Apogee Biotechnology Corporation	iCo Therapeutics, Inc.	Saniona AB
Ardelyx, Inc.	Idera Pharmaceuticals, Inc.	Sareum Holdings Plc
Arena Pharmaceuticals, Inc.	Immune Response BioPharma, Inc.	Selecta Biosciences, Inc.
Argos Therapeutics, Inc.	ImmusanT, Inc.	Selexys Pharmaceuticals Corporation
ASKA Pharmaceutical Co. Ltd.	Inbiopro Solutions Pvt. Ltd.	Sigmoid Pharma Ltd.
AstraZeneca Plc	Innovent Biologics, Inc.	Sitari Pharmaceuticals, Inc.
Avaxia Biologics, Inc.	Intrexon Corporation	Soligenix, Inc.
Axxam SpA	Jenrin Discovery, Inc.	Spherium Biomed SL
Basilea Pharmaceutica AG	Johnson & Johnson	Stelic Institute & Co.
BIOCAD	Kineta, Inc.	Sucampo Pharmaceuticals, Inc.
BioLineRx Ltd.	Kymab Ltd.	Swecure AB
Biotec Pharmacon ASA	Kyorin Pharmaceutical Co. Ltd.	Sylentis SA
BioTherapeutics, Inc.	Kytogenics Pharmaceuticals, Inc.	Synovo GmbH
Blueberry Therapeutics Ltd.	Lead Pharma Holding BV	Takeda Pharmaceutical Company Ltd.
Boehringer Ingelheim GmbH	Lpath, Inc.	Teva Pharmaceutical Industries Ltd.
Bristol-Myers Squibb Company	Lycera Corp.	Tiziana Life Sciences Plc
Calypso Biotech SA	Mabion SA	Tolerys SA
Catabasis Pharmaceuticals, Inc.	Medgenics, Inc.	Toray Industries, Inc.
Celgene Corporation	Mesoblast Ltd.	Torrent Pharmaceuticals Ltd.
Celltrion, Inc.	Mitsubishi Tanabe Pharma	Trino Therapeutics Ltd.
Celsus Therapeutics Plc	Corporation	Upsher-Smith Laboratories, Inc.
ChemoCentryx, Inc.	Momenta Pharmaceuticals, Inc.	Ventria Bioscience
ChironWells GmbH	Neovacs SA	VG Life Sciences, Inc.
CLL Pharma	Nivalis Therapeutics, Inc.	Virobay, Inc.
Cosmo Pharmaceuticals SpA	Novozymes A/S	Winston Pharmaceuticals, Inc.
Creabilis SA	Omni Bio Pharmaceutical, Inc.	Zedira GmbH
Daiichi Sankyo Company Ltd.	Oncobiologics, Inc.	
DBV Technologies SA	Oncodesign SA	
Delenex Therapeutics AG	Ono Pharmaceutical Co. Ltd.	
Dr. Falk Pharma GmbH	Onyx Pharmaceuticals, Inc.	
Effimune SAS	Opsona Therapeutics Ltd.	

of colitis to control deposition increased with smaller particle sizes. While passive targeting with size enables longer retention time and enhanced permeability, there have been contradictory findings with regard to specificity to disease versus nondiseased tissue. For example, nanoparticles prepared with cetyltrimethylammonium bromide (CTAB) with a size of 200 nm and a relatively neutral charge significantly adhered to both non-inflamed colonic tissue in healthy controls and inflamed colonic tissue in the TNBS colitis model following rectal administration (Wachsmann et al. 2013).

While cumulative results to date confirm that reduction in particle size is essential for effective IBD treatment, more studies are required to ascertain the effective particle size across various regions of the lower GI tract for clinical translation to human IBD. It may be that microparticles are more useful in the treatment of active disease by allowing mucosal penetration of larger particles, whereas nanoparticles are more effective under conditions of remission or minor inflammation where the mucosal barrier is less permeable (Viscido et al. 2014). The reason for the discrepancy of particle size between animal and human studies is unclear; however, it may be of major importance in the future treatment of human IBD. To improve the effectiveness of nanomedicines administered orally, it is likely that other strategies are also required to maximize accumulation and better differentiate delivery to diseased tissue.

Surface modification of nanosized therapeutics has been investigated in an attempt to improve adhesion and uptake to inflamed intestinal tissue in IBD. This includes (1) modification of surface charge, (2) incorporation of PEG, (3) use of pH-dependent coatings, (4) inclusion of biodegradable components to trigger release, (5) redox-dependent strategies, and (6) active targeting approaches (Hua et al. 2015). Once again, the number of preclinical and clinical studies assessing oral administration is limited, with the majority of the studies based on *ex vivo* tissue binding studies or *in vivo* studies following rectal administration. Of these strategies, the least complex in design are modification of surface charge, incorporation of PEG, and use of pH-dependent coatings.

Modifying the surface charge of nanosized therapeutics has been used to increase electrostatic interactions with components of diseased tissue in the GI tract (see Chapter 5.2). Cationic nanomedicines (mucoadhesive) may be useful for drugs that act on extracellular domains or only act after uptake into immune cells in active inflammation (Thirawong et al. 2008; Niebel et al. 2012; Coco et al. 2013; Lautenschlager et al. 2013). This is due to the nanoparticles being immobilized in the mucus, rather than penetrating the mucus layer and adhering to the inflamed mucosa for uptake into epithelial cells or immune cells. Immobilization may lead to premature drug release by an ion exchange mechanism. Conversely, anionic nanomedicines (bioadhesives) are able to interdiffuse among the mucus network due to less electrostatic interaction with the mucus (Jubeh et al. 2004). The results to date, especially in preclinical studies, are promising for the specificity of anionic nanodelivery systems to diseased tissue in IBD following oral administration (Lamprecht et al. 2001a; Meissner et al. 2006; Beloqui et al. 2013, 2014b). Studies in humans are warranted to determine if additional approaches are required to improve local bioavailability into diseased GI regions. It should be noted, however, that there is a potential for electrostatic interactions and subsequent binding of these nanoparticles with other charge-modifying substances during GI transit (e.g., bile acids and soluble mucins) (Hua et al. 2015). Therefore, it is likely that additional pharmaceutical strategies are needed, in addition to surface charge, in order to enhance localized drug delivery specifically to inflamed GI tissue.

The conjugation of PEG on the surface of nanomedicines creates a hydrophilic surface chemistry that minimizes strong interaction with the mucus constituents in the GI tract, and increases particle translocation through the mucus, as well as mucosa (Tobio et al. 2000; Cu and Saltzman 2008; Lai et al. 2009; Tang et al. 2009; Vong et al. 2012; Lautenschlager et al. 2013). This provides an accelerated translocation into the leaky inflamed intestinal epithelium, which is ideal for IBD (Lautenschlager et al. 2013). Despite a limited number of studies, those to date support PEGylation as a promising pharmaceutical strategy for accumulation in inflamed colonic mucosa in IBD (Vong et al. 2012; Lautenschlager et al. 2013). For example, Vong et al. (2012) designed a novel nitroxide radical-containing nanoparticle (RNP[O]), which possesses antioxidative nitroxide radicals in the core for treatment of mice with dextran sulfate sodium (DSS)–induced colitis (Vong et al. 2012). Accumulation of RNP^O in the colon was observed to be almost 50 times higher than that of the control 4-hydroxyl-2,2,6,6-tetramethyl-piperidine-1-oxyl (TEMPOL). Mice with DSS-induced colitis had a significantly lower disease activity index and less inflammation after 7 days of oral administration of RNP^O compared with mice given TEMPOL or mesalamine. The accumulation of RNP^O in the colon was dependent on both the size and PEGylated character of the nanoparticles.

Coating nanomedicines with pH-dependent coatings (e.g., Eudragit®) is a pharmaceutical strategy that takes advantage of the difference in pH in various regions of the GI tract to trigger drug release (Ashford et al. 1993). The enteric coating also protects the encapsulated active agents during GI transit. Several studies have thoroughly evaluated the clinical outcomes following treatment with pH-dependent nanomedicines in colitis (Lamprecht et al. 2005b; Ali et al. 2014; Beloqui et al. 2014a). For example, Ali et al. (2014) showed that budesonide-loaded poly(lactide-co-glycolide) (PLGA) nanoparticles coated with Eudragit S100 ($\sim$240 ± 14.7 nm) were able to significantly alleviate inflammation and demonstrated signs of regeneration in the DSS, TNBS, and oxazolone (OXA) *in vivo* models. Although preclinical studies have demonstrated translational applicability, a major concern has been the inherent interindividual and intraindividual variability of pH and emptying times from the GI tract, as well as the change in luminal pH due to disease state. A colonic delivery system that is based only on GI transit time or pH of the GI tract would simply not be reliable for IBD (Hua et al. 2015). Studies in human volunteers have shown that since the pH drops from 7.0 at the terminal ileum to 6.0 in the ascending colon, such systems sometimes fail to release the drug (Asghar and Chandran 2006). The potential for degradation of the Eudragit coating by bile acids in the duodenum also requires further investigation (Barea et al. 2010).

Having an understanding of the physiological variability in IBD (e.g., pH and GI transit time), biodegradable (Bhavsar and Amiji 2007; Moulari et al. 2008; Laroui et al. 2010, 2014a; Kriegel and Amiji 2011a,b; Xiao et al. 2014), redox (Wilson et al. 2010), and active targeting–based (Harel et al. 2011; Mane and Muro 2012; Coco et al. 2013; Xiao et al. 2013, 2014; Zhang et al. 2013; Laroui et al. 2014b) nanodelivery systems were

devised to take advantage of other factors that are known to be more consistent in IBD patients to allow efficient drug delivery. For example, Laroui et al. (2010) developed a biodegradable hydrogel that is specifically degraded by enzymes in the colon at pH 6.2, using ions (Ca^{2+} and SO4^{2-}) that cross-link chitosan and alginate (Laroui et al. 2010). The hydrogel has been used to embed nanoparticles loaded with active agents (Laroui et al. 2010) and siRNA (Laroui et al. 2014a; Xiao et al. 2014). Under the protection of the hydrogel, nanoparticles were able to pass through the stomach and upper small intestine unaffected, before being degraded in the inflamed colon. Another unique biodegradable nanodelivery system is the nanoparticle-in-microparticle oral delivery system (NiMOS) (Bhavsar and Amiji 2007; Kriegel and Amiji 2011a,b). NiMOSs are designed for oral administration of plasmid and siRNA by encapsulating them in type B gelatin nanoparticles, which are further entrapped in poly(ε-caprolactone) (PCL) microspheres. PCL is a synthetic hydrophobic polyester that is resistant to degradation by acid, therefore protecting nanoparticles during transit through the stomach. Release of the payload-carrying nanoparticles occurs over time at inflamed sites in the intestine, via controlled degradation of the outer PCL layer by action of lipases abundantly present at this location, after which they can be endocytosed by enterocytes or other cells at these sites (Bhavsar and Amiji 2007). Biodegradable systems generally take advantage of the increased enzymatic activity occurring in inflamed tissue and have shown promising therapeutic efficacy in IBD.

Similarly, redox nanodelivery systems take advantage of the abnormally high levels of reactive oxygen species (ROS) produced at the site of intestinal inflammation. For example, biopsies taken from patients suffering from UC have a 10- to 100-fold increase in mucosal ROS concentrations, which are confined to sites of disease and correlate with disease progression (Simmonds et al. 1992; Lih-Brody et al. 1996). The unusually high concentrations of ROS localized to sites of intestinal inflammation are generated by activated phagocytes (Mahida et al. 1989). The only key study in this area is by Wilson et al. (2010), which synthesized thioketal nanoparticles (TKNs) as a delivery vehicle for siRNA. TKNs are formulated from a polymer, poly(1,4-phenyleneacetone dimethylene thioketal) (PPADT), that degrades selectively in response to ROS. PPADT was used to encapsulate TNF-α siRNA complexed with the cationic lipid 1,2-dioleoyl-3-trimethylammonium-propane (DOTAP) to form nanoparticles. TKNs showed selective localization of orally delivered siRNA, against the pro-inflammatory cytokine TNF-α (0.23 mg of siRNA/kg/day), to sites of intestinal inflammation in the DSS-induced colitis model. These nanoparticles significantly suppressed mRNA levels of TNF-α and several other pro-inflammatory cytokines (interleukin [IL]-1, IL-6, and interferon [IFN]-γ) in colon tissues, while also reducing colonic inflammation as measured by histology. These results support the ability of TKNs to target inflamed tissues as an important factor for their in vivo efficacy.

Active targeting approaches using ligands coupled to the surface of nanodelivery systems may increase therapeutic efficiency and reduce adverse reactions, by further improving selective drug accumulation at inflamed sites within the GI tract. This pharmaceutical strategy is based on the concept that interactions between targeting ligands and specific receptors expressed predominantly at inflamed sites would improve bioadhesion of the drug formulation to specific cells and increase the extent of endocytosis (Hua 2013; Hua et al. 2015). Monoclonal antibodies and peptides are commonly used as targeting moieties, as they have been shown to have high specificity in targeting and potential mucopenetrative properties (Saltzman et al. 1994). An increasing number of targeting ligands that are overexpressed in inflamed GI tissue have been studied for oral colon-specific drug delivery strategies for IBD, including intercellular adhesion molecule [ICAM]-1 (Mane and Muro 2012), mannose receptors (Coco et al. 2013; Xiao et al. 2013), macrophage galactose-type lectin (MGL) (Zhang et al. 2013), macrophage-specific ligand (Laroui et al. 2014b), transferrin receptor (TfR) (Harel et al. 2011), and epithelial CD98 (Xiao et al. 2014). Only a few studies have assessed this strategy in vivo (Zhang et al. 2013; Laroui et al. 2014b). For example, Zhang et al. (2013) prepared galactosylated trimethyl chitosan–cysteine (GTC) nanoparticles loaded with mitogen-activated protein kinase kinase kinase kinase 4 (Map4k4) siRNA (siMap4k4), which is a key upstream mediator of TNF-α production, for targeting of MGL on activated macrophages in inflamed tissue (Zhang et al. 2013). Daily oral administration of siMap4k4-loaded GTC/TPP

nanoparticles significantly improved DSS-induced colitis, as measured by body weight, histology, and myeloperoxidase activity.

Similar results were demonstrated by Laroui et al. (2014b) with TNF-α-siRNA-loaded nanoparticles made of a poly(lactic acid)–poly(ethylene glycol) (PLA–PEG) block copolymer, conjugated with a macrophage-specific ligand (Fab′-portion of the F4/80 Ab—Fab′-bearing) onto the nanoparticle surface. TNF-α-siRNA-loaded nanoparticles significantly improved DSS-induced colitis in vivo, following oral administration, more efficiently when the nanoparticles were covered with Fab′-bearing ligands than nonconjugated nanoparticles. Grafting of Fab′-bearing ligands also improved nanoparticle endocytosis, as well as macrophage targeting ability, as indicated by flow cytometry. It should be noted that this study did load the nanoparticles into a colon-specific biodegradable hydrogel (chitosan–alginate), which also enhanced its specific delivery to the colon and protected the grafted ligand during GI transit. Overall, active targeting is a promising approach to enhance drug accumulation and uptake into inflamed tissue in IBD; however, further in vivo studies are required to assess the different targeting ligands and formulations for efficacy and stability in animal models of colitis (Hua et al. 2015). Oral administration of antibody and peptide-based formulations encounters many obstacles in the GI tract, in particular degradation by the stomach acid, as well as by enzymes. Therefore, it is likely that active targeting approaches will also require other formulation strategies to enhance regional deposition at the site of disease and to protect the conjugated ligands from the harsh environment during GI transit.

5.3.4 CHALLENGES WITH TRANSLATION OF CURRENT NANOMEDICINES FOR IBD

In the last several decades, the application of nanotechnology for medical purposes has received significant attention from researchers, academia, funding agencies, government, and regulatory bodies. Although nanoparticle-based formulations have demonstrated significant therapeutic advantages for a multitude of biomedical applications, including IBD, their clinical translation has not progressed as rapidly as the plethora of positive preclinical results would have suggested (Luxenhofer et al. 2014). Nanosized systems are far more complex in comparison with conventional formulations, which also makes it challenging to evaluate their pharmacodynamic and pharmacokinetic interaction with human cells and tissues following administration (Teli et al. 2010; Tinkle et al. 2014; Sainz et al. 2015). This leads to several key issues with the clinical development of nanomedicines, in particular large-scale manufacturing, biocompatibility and safety, government regulations and intellectual property (IP), and overall cost-effectiveness in comparison with current therapies (Allen and Cullis 2004, 2013; Zhang et al. 2008; Sawant and Torchilin 2012; Narang et al. 2013). These factors can impose high hurdles for nanotechnology-based formulations, regardless of therapeutic efficacy or advantage.

5.3.4.1 Complexity in Manufacturing and Evaluating Nanomedicines

One of the main reasons for the limited clinical translation of nanomedicines for IBD resides in the structural and physicochemical complexity of the formulation itself. Platforms that require complex and/or laborious synthesis procedures generally have limited clinical translation, as they are challenging to pharmaceutically manufacture on a large scale (Teli et al. 2010; Tinkle et al. 2014; Barz et al. 2015; Sainz et al. 2015). Limitations in pharmaceutical development are centered on quality assurance and cost. Quality assurance involves the manufacturing process and stability of the formulation, with nanomedicines being affected by issues surrounding (1) lack of quality control, (2) scalability complexities, (3) separation from undesired nanostructures (e.g., by-products and starting materials), (4) high material and/or manufacturing costs, (5) production rate to increase yield, (6) batch-to-batch reproducibility and consistency of the final product (e.g., size distribution, porosity, charge, and mass), (7) lack of infrastructure and/or in-house expertise, (8) chemical instability or denaturation of the encapsulated compound in the manufacturing process, (9) relative scarcity of venture funds and pharmaceutical industry investment, and (10) in vivo and storage stability issues (Teli et al. 2010; Narang et al. 2013; Hafner et al. 2014; Tinkle et al. 2014).

Suitable methods for the industrial-scale production of several basic nanomedicine platforms, such as liposomes, have been successfully developed without the need for numerous manufacturing steps or the use of organic solvents (Jaafar-Maalej et al. 2012; Kraft et al. 2014). The challenges arise when the functionality of the nanoparticle system becomes more complex, such as the addition of surface modification with coatings and/or ligands, multiple targeting components, or dual encapsulation of therapeutic agents. Integration of multiple components to a single nanosized carrier requires multiple chemical synthesis steps and formulation processes, which inevitably poses problems for large-scale good manufacturing (current good manufacturing practice [cGMP]) production, increases the cost of production, and makes the evaluation of such products more difficult (Teli et al. 2010; Svenson 2012; Tinkle et al. 2014).

Characteristics of the manufactured nanomedicine need to be well defined and reproducible to enable clinical translation. The characterization and validation of more complex nanomedicines is challenging due to the sheer number of variable components in the formulation (e.g., size, morphology, charge, purity, encapsulation efficiency of agents, coating efficiency, and conjugation of ligands) (Teli et al. 2010). Even small differences in molecular structure of small molecules can significantly change their physicochemical properties (e.g., polarity, size, and molecular mass), pharmacokinetic parameters (i.e., absorption, distribution, metabolism, and excretion), and pharmacodynamic interactions (e.g., cellular interaction and functionality) (Teli et al. 2010; Tinkle et al. 2014; Barz et al. 2015). Nanomedicines will also need to be stable following (1) the manufacturing process, (2) long-term storage, and (3) clinical administration en route to the site of action. Therefore, the key goal for clinical translation is to have a synthetic method that is scalable to the quantities of nanomedicine needed to treat thousands or even millions of patients (e.g., kilograms or tons), and that is manufactured at the same level of quality and is highly reproducible (Grainger 2013; Lammers 2013; Barz et al. 2015).

Finally, clinical trials of nanomedicines are generally more complex than conventional formulations, as a number of control groups are required to account for different aspects of the drug delivery system (Tinkle et al. 2014). In order to get to this stage, the formulation has to first pass pharmaceutical and commercial qualities, as discussed above. Then, we can determine whether therapeutic efficacy in preclinical animal studies translates to success in humans (Allen and Cullis 2004, 2013; Narang et al. 2013). Even at this stage, the cost–benefit analysis may be a limitation to the clinical translation of some nanomedicines when compared with an approved counterpart or existing therapies, in terms of efficacy or drug-related side effects.

5.3.4.2 Complexity in Toxicology and Safety Assessment

Detailed toxicology is essential for the clinical translation of nanomedicines to determine the therapeutic index of the formulation and overall safety in humans (Nystrom and Fadeel 2012). Studies focused on the nanotoxicology of these delivery systems in the GI tract of patients with IBD have been limited, and are likely to vary according to the composition, geometry, and size of the nanoparticles (Nystrom and Fadeel 2012; Pichai and Ferguson 2012). Increasing the complexity of nanomedicines introduces more physicochemical variables that complicate the pharmacokinetics, pharmacodynamics, and toxicology assessment of the formulation following administration into the human body (Teli et al. 2010; Tinkle et al. 2014). For example, the use of different synthetic coatings and ligands can have a significant effect on the biocompatibility, biodistribution, and toxicology profile of nanomedicines (Allen and Cullis 2004, 2013; Zhang et al. 2008; Sawant and Torchilin 2012; Narang et al. 2013; Tinkle et al. 2014). To appropriately assess the nanotoxicology of these formulations, there is a need for validated and standardized protocols incorporating *in vitro*, *ex vivo*, and *in vivo* studies (Dobrovolskaia and McNeil 2013). In particular, *in vitro* or *ex vivo* assays for nanosafety testing are essential to screen for potential hazards early in the development of nanomedicines (Gaspar 2007). In order to do this, standardized reference materials would need to be established and the testing would also need to be relevant for the intended route of administration (Tinkle et al. 2014). Rational characterization strategies should be developed to understand the interaction of nanoparticles with biological tissues and cells in organs they are distributed to following administration, in particular the mechanisms for intracellular trafficking, functionality, potential for cellular toxicity, and processing for degradation or clearance (Gaspar 2007; Nystrom and Fadeel 2012; Dobrovolskaia and McNeil 2013).

There is also a need to perform specialized toxicology studies in animal models to assess both short-term and long-term toxicity, as circulation half-lives and drug retention times are generally significantly increased with nanoencapsulation. A thorough understanding of the absorption, distribution, metabolism, and excretion of emerging nanomaterials *in vivo* is important to predict the biological and toxicological responses of nanomedicines (Dobrovolskaia and McNeil 2013; Tinkle et al. 2014). Importantly, biocompatibility, immunotoxicological, and inflammatory potential should be assessed, with functional outcomes correlated with mechanisms of tissue uptake and clearance (Gaspar 2007). These parameters need to be well investigated based on dose, dosage form, and route of administration to establish safe limits prior to clinical trials (Gaspar 2007; Nystrom and Fadeel 2012). This is of particular importance for nanomedicines composed of materials that have never been used before in clinical applications. Although materials such as phospholipids and biodegradable polymers have been studied previously, any significant changes to their physicochemical properties, such as charge, structure, surface modifications, or composition, would warrant the need for complete toxicology testing. Even in the clinical trial phase, regulatory protocols should be in place to detect any toxicity caused by novel mechanisms unique to nanotechnology (Gaspar 2007; Nystrom and Fadeel 2012). Addressing these issues is necessary to achieve safe application of emerging nanomedicines in the clinical setting.

5.3.4.3 Complexity of Nanomedicine Patents and Intellectual Property

Given the complexities of incorporating nanotechnology into biomedical applications, there need to be more precise definitions of what constitutes novel IP of a nanomedicine (Satalkar et al. 2015). Nanomedicines are complex, as they have a number of variable components, and bridge between the field of medicine and medical device (Paradise et al. 2009). Generally, the control of a nanomedicine product requires an IP position on (1) the encapsulated cargo, (2) the carrier technology, and (3) the characteristics of the drug and carrier together. Although this definition is straightforward, it does open up a number of problems with the issuing of patents to date

(Bawa 2007; Bawa et al. 2008), for example, nanomedicines that incorporate existing drugs with novel carrier technology, or those that incorporate existing drugs with existing carrier technology for a new disease indication. The IP situation becomes even more confusing with more complex drug delivery systems, such as those that incorporate existing targeting ligands (e.g., antibodies) or coatings (e.g., Eudragit) that are owned by other companies. IP strategies may likely involve multiple patents associated with any given technology and the need for cross-licensing arrangements (Murday et al. 2009). Therefore, it is important to simplify the pathway from invention to commercialization through new IP practices and protocols, so as to reduce the time and expense required for negotiating collaboration and licensing agreements (Murday et al. 2009).

With the significant increase in the number of nanotechnology patent applications over the last few decades, other key issues that need to be addressed include patent review delays, "patent thickets," and issuance of invalid patents (Bawa 2005, 2007; Bawa et al. 2005). There needs to be a universal nanonomenclature on identical or similar nanostructures or nanomaterials, and more refined search tools and commercial databases to avoid the issuing of multiple nanopatents on the same invention (Bawa et al. 2005; Bawa 2007). Databases used by the Patent and Trademark Office (PTO) need to be able to search through nanotech-related prior art that reside in scientific publications worldwide, including earlier publications that preceded the emergence of online publication databases (Tinkle et al. 2014). Patent examiners also require expertise and training with respect to the emerging fields of nanotechnology and nanomedicine. The complexities with nanotechnology have led to the so-called patent thickets, which can lead to costly litigation and halt commercialization efforts (Tinkle et al. 2014). Therefore, improved clarity on IP and patenting surrounding nanotechnology in health and medicine is required, and will need to involve implementation of universal regulations and policies that are tailored toward this niche commercialization field.

5.3.4.4 Complexity in the Regulation of Nanomedicines

Nanomedicines have significant potential to increase the growth of the pharmaceutical market and improve health benefits; however, the current scientific and regulatory gap for nanomedicines is large and challenging. Commercialization of nanomedicines is highly dependent on a number of regulatory factors based on government policies in the area of manufacturing practice, quality control, safety, and patent protection (Gaspar 2007; Tinkle et al. 2014; Sainz et al. 2015). The lack of clear regulatory and safety guidelines has affected the development of nanotherapeutic products toward timely and effective clinical translation (Gaspar 2007; Tinkle et al. 2014; Sainz et al. 2015). For example, polymers have been widely investigated as an effective platform for nanodelivery strategies; however, their safety and efficacy are highly dependent on the polymer molecular weight, polydispersity, molecular structure, and conjugation chemistry (Gaspar and Duncan 2009; Diab et al. 2012). Due to the increased number of novel polymeric materials and complex polymeric-based nanomedicine formulations, there is an urgent need for an appropriate regulatory framework to assist in evaluation (Gaspar and Duncan 2009). As each polymer-based nanomedicine is different, it is important to consider each individually based on doses, administration routes, dosing frequency, and proposed clinical use. This would be the same for most other nanomedicine platforms.

Nanomedicines are currently regulated within the conventional framework governed by the key regulatory authority of each country (e.g., Food and Drug Administration [FDA], Therapeutic Goods Administration [TGA], and European Medicines Agency [EMA]). Although nanomedicines have been on the market for nearly two decades, the first generation of nanomedicine products passed regulatory approval by only having to meet general standards, applicable to medicinal compounds. These regulations are no longer appropriate to confirm the quality, safety, and efficacy of nanotherapeutics for clinical use (Gaspar 2007; Tinkle et al. 2014; Sainz et al. 2015). Reasons for this are based on the complex structure of nanomedicines, their unclear interaction with cells and tissues within the human body, the increased complexity of clinical use, and the multifunctional nature of some formulations (e.g., integration of therapeutics with imaging diagnostics) (Gaspar 2007; Tinkle et al. 2014; Sainz et al. 2015). Regulatory standards and protocols validated specifically for nanoparticles are needed that bridge both medicine and medical device regulations. This should take into account a nanomedicine's complexity, route of administration, pharmacokinetics, pharmacodynamics, and safety profile, as well as provide information on the most appropriate clinical trial design and patient selection (Tinkle et al. 2014). There needs to be a fine balance to ensure the safety and quality of nanotherapeutics without overregulation, which can negatively affect the progress of innovative products to the market, by escalating costs for achieving regulatory approval and/or consuming a significant portion of the life of a patent.

Development of global nanopharmaceutical regulatory standards should be established alongside key countries with invested interest. Although major steps have been taken in the last 5 years, a closer collaboration between regulatory agencies, academia, research, and industry is needed (Gaspar 2007; Murday et al. 2009; Hafner et al. 2014). This is of particular importance due to the limited availability of contract manufacturing organizations worldwide that specialize in producing nanopharmaceutical products, in accordance with the requirements for GMP (Hafner et al. 2014). It should be noted that this limited number of manufacturing organizations may be further divided based on their infrastructure capabilities of producing specific nanomedicine platforms (e.g., liposomes, polymeric nanoparticles, dendrimers, and drug–polymer conjugates). Therefore, nanomedicines produced in these manufacturing organizations will likely be marketed in multiple countries, and thus should be governed under the same regulatory standards (Hafner et al. 2014). There will need to be complete evaluation and documentation of production processes for nanomedicines, incorporating appropriate industrial standards, for both quality control and prevention of environmental issues (Gaspar 2007). Manufactured nanomedicines will still need to meet general pharmaceutical standards, such as purity, sterility, stability, manufacturing operations, and related industrial control standards (Gaspar 2007). In addition, new analytical tools and standardized methods will need to be implemented to evaluate key physical characteristics of nanomedicines that can affect *in vivo* performance, such as particle size and size distribution, surface chemistry, morphology, surface area, surface coating,

hydrophilicity, porosity, and surface charge density (Gaspar 2007; Tinkle et al. 2014; Sainz et al. 2015). These methods will vary for different nanomaterials and nanostructures. Thus, regulatory authorities should work together to develop the testing methods and appropriate standardized protocols for toxicity studies and regulatory requirements, which will be needed to ensure the efficacy and safety of current and emerging nanomedicines.

5.3.5 PERSPECTIVES ON THE TRANSLATIONAL DEVELOPMENT OF NANOMEDICINES FOR IBD

Current treatment options for IBD remain disappointing and nonspecific, and are associated with multiple systemic and local adverse effects; therefore, there is an urgent medical need to develop novel treatment options (Elinav and Peer 2013; Talaei et al. 2013). Nanomedicine-based therapeutics hold great promise for advancing targeted therapeutic approaches for IBD, by offering the potential for unprecedented interactions with biomolecules on cell surfaces or within the cells. This allows the development of more effective therapies that enhance the delivery of active agents to diseased and inflamed tissue in the GI tract (Talaei et al. 2013; Hua et al. 2015). Although the clinical translation of nanomedicines for IBD is still in its infancy, with the majority of research still in the preclinical R&D phase, there are a number of novel nanomedicine platforms that have shown encouraging results in preclinical studies in animal models of experimental IBD (see Section 5.3.3). In order to accelerate the translation of nanomedicines into clinical application, there is a need for thorough evaluation of the efficacy and toxicities of these novel delivery systems in relevant animal models and in humans. In addition, issues related to long-term storage stability, scale-up, and manufacturing of the systems need to be addressed (Gaspar 2007; Tinkle et al. 2014; Sainz et al. 2015; Tran and Amiji 2015). With a lack of nanomedicines specifically for IBD in the clinical phase, it is important to use the information and lessons learned from the clinical translation of other nanomedicines to assist in the development of targeted therapeutics for IBD. Table 5.2 identifies some key considerations in developing nanomedicine for IBD.

TABLE 5.2
Considerations for the translational development of nanomedicines for IBD.

Nanopharmaceutical design

Key considerations
- Oral dosage form
- Reduced complexity in formulation design
- Final dosage form for human use
- Biocompatibility
- Stability during GI transit

Current obstacles
- Large-scale production according to GMP standards
  - E.g., reproducibility, infrastructure, techniques, expertise, and cost
- Quality control assays for characterization
  - E.g., size and polydispersity, morphology, charge, encapsulation, surface modifications, purity, and stability

Preclinical evaluation

Key considerations
- Need for validated and standardized assays for early detection of toxicity
- Evaluation in appropriate animal models of disease
- Improved understanding of cellular and molecular interactions
  - Pharmacokinetics (absorption, distribution, metabolism, and excretion)
  - Pharmacodynamics (intracellular trafficking, functionality, toxicity, and clearance)

Current obstacles
- Specialized toxicology studies *in vitro*, *ex vivo*, and in animal models
- Lack of understanding of the interaction of nanomedicines with biological tissues and cells
- Structural stability of nanomedicine during GI transit
- Improved specificity and accumulation of nanomedicines in diseased GI tissue

Clinical evaluation for commercialization

Key considerations
- Simplification of pathway from invention to commercialization to minimize time and expense
- Evaluation of safety and toxicity in humans (acute and chronic)
- Evaluation of therapeutic efficacy in IBD patients (including in active disease, chronic disease, and intestinal resection)

Current obstacles
- Lack of clear regulatory and safety guidelines specific for nanomedicines
- Complexity with nanomedicine patents and IP
- Optimal clinical trial design
- Lack of understanding of the interaction of nanomedicines in the GI tract of IBD patients

5.3.5.1 Pharmaceutical Design and Development

From a therapeutic perspective, increasing drug accumulation at target GI tissues and minimizing systemic side effects are still the biggest design challenges to developing new drug delivery systems for IBD. This is even more complicated with the preference for oral formulations for the treatment of IBD, in order to improve the quality of life of patients and to ensure their adherence to long-term treatment (Saini et al. 2009). It should be noted that rectal formulations are not effective for the treatment of more widespread inflammation of the colon or GI tract, which can occur in both UC and CD, with their distribution limited to the distal colon and rectum (Hua 2014). Therefore, nanoencapsulation has been shown to be an effective strategy for oral drug delivery in order to (1) enhance local drug accumulation to inflamed GI tissue, (2) increase solubility and permeability of active compounds (which is an essential factor for drug effectiveness), and (3) protect compounds from the harsh conditions during GI transit (Hua 2014; Lautenschlager et al. 2014; Viscido et al. 2014; Hua et al. 2015; Tran and Amiji 2015). Even though promising nanomedicines may demonstrate significant efficacy in *in vitro* or *ex vivo* studies, it is important to evaluate the platforms *in vivo* using appropriate animal models of the disease. It is here where many of the current GI-targeting nanomedicine platforms have shown limited specificity, accumulation, and/or stability, therefore providing unsatisfactory results to warrant progression in the R&D process (Hua et al. 2015). Efficacy in an animal model also does not necessarily equate to efficacy in humans, as drug delivery within the human body is complex and can be highly variable, especially when associated with disease (Hebden et al. 2000; Podolsky 2002; Talley et al. 2011). Therefore, this concept of designing nanomedicines that act like a "magic bullet," which refers to the exclusive delivery of active compounds to specific organs, tissues, or cells, is just not realistic when taking into account the pharmacokinetic and pharmacodynamic processes that occur following administration into the body. This term should refer to the development of realistic therapeutic platforms, in which doses are minimized, complexity in dosage form design is reduced, and therapeutic effects are maximized (Barz et al. 2015).

Complexity in dosage form design is a key factor in the ability of a formulation to be translated to the clinic, irrelevant of its therapeutic efficacy. Simplification in formulation design is required to allow efficient and reproducible large-scale manufacturing (Grainger 2013; Lammers 2013; Barz et al. 2015). In addition, when translating findings from animal models to humans, we need to determine how to modify these formulations so that they are appropriate for human administration (Hua et al. 2015). *In vivo* studies have been conducted in animal models of experimental IBD, especially in mice, which places limitations on the size and consistency of the dosage form that can be administered orally. The practicability of designing dosage forms that are both acceptable to humans and efficacious needs to be further explored for clinical studies. Thus, there needs to be a balance between complexity, therapeutic efficacy, and clinical translation.

Taking into consideration the basic foundation of nanomedicines, reducing the size of therapeutics to the nanoscale alone improves retention time in inflamed intestinal regions in the GI tract and provides benefits for IBD therapy (Laroui et al. 2013; Talaei et al. 2013; Viscido et al. 2014; Hua et al. 2015). To transition these nanomedicines to

the clinic, attention should be given to nanosized carriers that are stable during GI transit following oral administration, easily able to be scaled up for manufacturing with high control over their physicochemical properties (e.g., size and polydispersity, morphology, percent encapsulation efficiency, and charge), and composed of materials that are biocompatible, biodegradable, and nontoxic. Oral applications remain limited for basic lipid nanoformulations (e.g., liposomes and solid lipid nanoparticles) due to destabilization and degradation of lipids in the GI fluids, especially in the presence of bile salts (e.g., pancreatic lipase) (Dial et al. 2008; Barea et al. 2010). Although solid lipid nanoparticles have increased stability in comparison with liposomes, they have still been shown to undergo degradation during contact with GI fluids (Diab et al. 2012). Modifying the composition of the lipid matrix and stabilizing surfactant can alter the degradation rate, with fatty acids of longer hydrocarbon backbones and use of surfactants such as poloxamer 407 showing slower degradation rates (Olbrich and Muller 1999). Conversely, polymer-based carriers have shown promising results for IBD therapy. For example, polyester-based nanoparticles such as PLA, poly(glycolic acid) (PGA), and PLGA are commercially available and commonly used for nanoparticle manufacturing (Diab et al. 2012). These polymers are easily manufactured and are completely degraded into small molecules through hydrolytic and/or enzyme degradation, which are subsequently eliminated primarily by renal clearance (Xiong and Tam 2004). As nanoparticles are able to enter cells and interfere with molecular pathways, synthetic polymers should be carefully evaluated for potential short-term and long-term toxicity for clinical application (Gaspar and Duncan 2009). For example, recent studies have identified potentially toxic *in vitro* and *in vivo* effects with the use of cationic lipids and polymers, including cell shrinking, reduced number of mitoses, vacuolization of the cytoplasm, and detrimental effects on key cellular proteins (e.g., protein kinase C) (Lv et al. 2006).

A combination of pharmaceutical strategies will likely be required for the clinical translation of a nanomedicine for IBD. This is important to protect the encapsulated cargo from premature release, reduce the quantity of nanomedicine administered in each dose, maximize the proportion of nanomedicine deposited at the site of disease, and enhance mucosal retention and penetration in diseased tissue. The final oral dosage form for clinical use will also have to be considered. For predominantly lower GI targeting for IBD, the nanomedicine is unlikely to be clinically administered as a liquid suspension, but rather in tablet or capsule form for patient compliance, stability issues during storage, and to maximize therapeutic effects (Witticke et al. 2012; Hua et al. 2015), for example, encapsulating freeze-dried liposomes or polymer-based nanoparticles in capsules with a triggered-release coating for regional deposition at the disease site (Gupta et al. 2013; Hua 2014). This would mean that the ideal external coating would at least need to bypass the harsh GI conditions of the stomach and duodenum before releasing the encapsulated nanomedicine, which can then travel along the lower GI tract to enhance drug delivery to diseased inflamed tissue. Conventional formulations have used pH-specific soluble coatings based on the pH gradient of the GI tract, time-dependent release systems based on GI transit times, and nonstarch polysaccharide coatings based on enzymatic degradation by colonic bacteria. These physiological characteristics are highly variable in IBD, which can limit their efficacy (Hebden et al. 2000; Talley et al. 2011). Hence, other triggered-release strategies will need to be considered.

The addition of other pharmaceutical strategies to the basic nanomedicine platform has been shown to further enhance mucosal retention and penetration in diseased GI tissue. For clinical applicability, these strategies should be evaluated for significance in enhancing therapeutic outcomes in comparison with the basic nanomedicine platform, as well as complexity in manufacturing and characterization. Any addition to the basic nanomedicine platform would need to show improved benefits that are reliable and reproducible for IBD patients, due to the added costs in the manufacturing process. As discussed in Section 5.3.2, the strategies that involve the least complexity in pharmaceutical design are modification of the surface charge, incorporation of PEG, and use of pH-dependent coatings. Of these, anionic nanoparticles (Lamprecht et al. 2001; Meissner et al. 2006; Beloqui et al. 2013, 2014b) and PEGylated nanoparticles (Tobio et al. 2000; Cu and Saltzman 2008; Lai et al. 2009; Tang et al.

2009; Vong et al. 2012; Lautenschlager et al. 2013) have been shown to be the most effective in improving penetration and uptake into diseased mucosal tissue. These nanoparticles are likely to be more beneficial if designed to be released from tablets or capsules during transit through the lower GI tract in IBD patients.

Active targeting, biodegradable, and redox-dependent nanomedicines are more complex and sophisticated in design, therefore making scale-up manufacturing and characterization more difficult to meet regulatory requirements. There is no doubt that the preclinical studies to date have demonstrated improved efficacy with these strategies (Bhavsar and Amiji 2007; Moulari et al. 2008; Laroui et al. 2010, 2014a,b; Wilson et al. 2010; Harel et al. 2011; Kriegel and Amiji 2011a,b; Mane and Muro 2012; Coco et al. 2013; Xiao et al. 2013, 2014; Zhang et al. 2013). However, for active targeting approaches, further studies are required to examine their benefits in human IBD, in particular the reliability and consistency of the expression of the target across disease severity and in different patients. Although target cells may be accessible, cellular uptake still remains a process of probability, and thus cannot be fully specific (Barz et al. 2015). Active targeting can facilitate uptake, enhance accumulation, and direct intracellular translocalization (Barz et al. 2015). Biodegradable and redox pharmaceutical strategies, which rely on increased enzymatic and ROS activity in diseased GI tissue, should be considered for coating of the capsules or tablets to trigger the release of nanoparticles at diseased sites, for example, coating nanoparticle-loaded capsules with the polymer PPADT, which degrades selectively in response to ROS (Wilson et al. 2010). This may be more effective than strategies based on physiological characteristics that can be highly variable in IBD, such as pH and time-dependent release systems (Hebden et al. 2000; Talley et al. 2011). Overall, researchers need to consider minimizing the complexity of nanomedicines and take into account the final dosage form for human use, in order for a formulation to have the potential to be translated into a clinically applicable therapeutic. Reducing complexity to the minimum required for pathophysiological or medical need is paramount in nanoparticle design and synthesis to generate clinically translatable nanosized therapeutics.

5.3.5.2 Pathway to Translation and Commercialization

The experimental development of nanomedicines for disease is progressing at a fast pace; however, significant challenges still exist in promoting these platforms into clinically feasible therapies for IBD. The majority of nanomedicines in the clinic are for the treatment of cancer, predominantly by the parenteral route of administration. They are structurally based on simple nanomedicine platforms, in particular basic nanoparticles, surface charge–modified nanoparticles, and PEGylated nanoparticles (Hafner et al. 2014; Sainz et al. 2015). Although clinical applications of nanotechnology for IBD are on the pathway to becoming a reality based on promising experimental results, there are several barriers that have slowed progress in the preclinical and, especially, clinical stages of development. These include issues surrounding complexity in manufacturing and characterization, lack of understanding of *in vivo* pharmacokinetics and pharmacodynamics, acute and chronic toxicity, and cost-effectiveness (Gaspar 2007; Teli et al. 2010; Hafner et al. 2014; Tinkle et al. 2014; Sainz et al. 2015). These challenges are even greater with the increasing complexity of nanomedicine design. There has to be a clear positive benefit-to-risk ratio that will accompany the clinical implementation of products and procedures based on nanotechnology. Emerging nanotherapeutic products for IBD, which are more complex in structure and more expensive than conventional therapies, are designed to provide an overall reduction in health care costs (Hafner et al. 2014). This reduction in health care costs is likely to be obtained by increasing therapeutic efficacy, improving quality of life, and reducing the need for surgical interventions (Gandjour and Chernyak 2011). In addition, from a business perspective, nanopharmaceuticals can offer the ability to extend the economic life of proprietary drugs and create additional revenue streams (Tinkle et al. 2014). These factors should be taken into account when assessing the overall cost-effectiveness of IBD nanomedicines compared with existing therapies.

Nanomedicines generally face a number of regulatory approval hurdles. The control of materials in the nanosize range often presents greater scientific and technical challenges than conventional formulations (Gaspar 2007; Teli et al. 2010; Hafner et al. 2014; Tinkle et al. 2014; Sainz et al. 2015). Nanomedicines encompass a number of different types of nanomaterials and nanostructures, which make it even more challenging to establish appropriate regulatory protocols and tools to ensure standardized GMP manufacturing and characterization, safety and toxicology evaluation, and clinical trial design. These procedures are paramount to confirming therapeutic efficacy and safety prior to marketing approval for use in patients on a larger scale. Effective clinical translation of nanopharmaceutics for IBD will require an interdisciplinary approach to develop novel protocols, assays, and infrastructure for the manufacturing and characterization of nanomedicines (Gaspar 2007; Teli et al. 2010; Hafner et al. 2014; Tinkle et al. 2014; Sainz et al. 2015). This will need to involve experts from academia and industry with specialty in pharmaceutics, engineering, biology, medicine, and toxicology. Potential approaches to fast-track promising IBD nanotherapeutics to clinical trials include the establishment or coordination of laboratories and centers that have expertise in (1) characterizing nanomedicine platforms, (2) conducting preclinical studies on nanomedicines for submission to regulatory agencies, (3) scale-up laboratory preparation of nanomaterials according to regulatory and industry standards for early clinical trials, and (4) designing and conducting clinical trials of nanotherapeutic platforms (Hafner et al. 2014).

Overall, the use of nanotechnology in medicine has the potential to have a major impact on human health. It has been suggested to facilitate the development of personalized medicine, in which patient therapy is tailored by the patient's individual genetic and disease profile (Teli et al. 2010; Mura and Couvreur 2012; Laroui et al. 2013). For example, individualized IBD characteristics, such as cellular receptor expression and molecular pathway activation, could be analyzed and used to design personalized nanomedicines (Teli et al. 2010; Mura and Couvreur 2012; Laroui et al. 2013). The size and structure of the delivery system can also be modified according to the severity of the disease for optimal therapeutic benefits (Viscido et al. 2014). This concept would significantly advance the way in which we treat IBD patients. However, for this to occur, there are still a number of issues that need to be addressed—from our basic understanding of the interaction of nanomedicines in the GI tract of IBD patients, to commercialization hurdles related to manufacturing, financial, and regulatory standards.

5.3.6 CONCLUSION

Nanomedicine-based therapeutics have the potential to revolutionize the way we treat IBD by improving the selective targeting of active agents to diseased GI tissue. This drug delivery approach increases therapeutic efficacy, while reducing the therapeutically effective dose and risk of systemic side effects (Lautenschlager et al. 2014; Viscido et al. 2014). Nanoencapsulation also gives the opportunity to deliver novel compounds with poor oral physicochemical properties or fragile compounds that degrade easily in biological environments in the GI tract (Hua et al. 2015). In order to move a nanomedicine formulation from the bench to the bedside for IBD, several issues remain to be addressed. From a biological perspective, this includes studies focused on understanding the cellular and molecular interactions of nanoparticles in the human GI tract of healthy and IBD patients (Gaspar 2007; Nystrom and Fadeel 2012; Dobrovolskaia and McNeil 2013). For nanomedicines to have the potential to translate to the clinic, the complexity in the design and development of nanomedicines needs to be reduced in order to create systems that are able to be reproducibly synthesized and characterized (Lammers 2013; Barz et al. 2015). In the coming decade, it is likely that we will see a rich and progressive pipeline of novel nanosized therapeutic strategies for IBD entering the clinical R&D phase that bridge the gap between medicine and medical devices.

REFERENCES

Albenberg, L. G., and G. D. Wu. 2014. Diet and the intestinal microbiome: Associations, functions, and implications for health and disease. *Gastroenterology* 146 (6):1564–72.

Ali, H., B. Weigmann, M. F. Neurath, E. M. Collnot, M. Windbergs, and C. M. Lehr. 2014. Budesonide loaded nanoparticles with pH-sensitive coating for improved mucosal targeting in mouse models of inflammatory bowel diseases. *J Control Release* 183:167–77.

Allen, T. M., and P. R. Cullis. 2004. Drug delivery systems: Entering the mainstream. *Science* 303 (5665):1818–22.

Allen, T. M., and P. R. Cullis. 2013. Liposomal drug delivery systems: From concept to clinical applications. *Adv Drug Del Rev* 65 (1):36–48.

Ammon, H. V., and S. F. Phillips. 1974. Inhibition of ileal water absorption by intraluminal fatty acids. Influence of chain length, hydroxylation, and conjugation of fatty acids. *J Clin Invest* 53 (1):205–10.

Antoni, L., S. Nuding, J. Wehkamp, and E. F. Stange. 2014. Intestinal barrier in inflammatory bowel disease. *World J Gastroenterol* 20 (5):1165–79.

Asghar, L. F., and S. Chandran. 2006. Multiparticulate formulation approach to colon specific drug delivery: Current perspectives. *J Pharm Pharm Sci* 9 (3):327–38.

Ashford, M., J. T. Fell, D. Attwood, H. Sharma, and P. J. Woodhead. 1993. An in vivo investigation into the suitability of pH dependent polymers for colonic targeting. *Int J Pharm* 95 (1–3):193–9.

Barea, M. J., M. J. Jenkins, M. H. Gaber, and R. H. Bridson. 2010. Evaluation of liposomes coated with a pH responsive polymer. *Int J Pharm* 402 (1–2):89–94.

Barkas, F., E. Liberopoulos, A. Kei, and M. Elisaf. 2013. Electrolyte and acid-base disorders in inflammatory bowel disease. *Ann Gastroenterol* 26 (1):23–28.

Barz, M., R. Luxenhofer, and M. Schillmeier. 2015. Quo vadis nanomedicine? *Nanomedicine (Lond)* 10 (20):3089–91.

Bawa, R. 2005. Will the nanomedicine "patent land grab" thwart commercialization? *Nanomedicine* 1 (4):346–50.

Bawa, R. 2007. Patents and nanomedicine. *Nanomedicine (Lond)* 2 (3):351–74.

Bawa, R., S. R. Bawa, S. B. Maebius, T. Flynn, and C. Wei. 2005. Protecting new ideas and inventions in nanomedicine with patents. *Nanomedicine* 1 (2):150–8.

Bawa, R., S. Melethil, W. J. Simmons, and D. Harris. 2008. Nanopharmaceuticals—Patenting issues and FDA regulatory challenges. *American Bar Association SciTech Lawyer* 5:10–15.

Beloqui, A., R. Coco, M. Alhouayek, M. A. Solinis, A. Rodriguez-Gascon, G. G. Muccioli, and V. Preat. 2013. Budesonide-loaded nanostructured lipid carriers reduce inflammation in murine DSS-induced colitis. *Int J Pharm* 454 (2):775–83.

Beloqui, A., R. Coco, P. B. Memvanga, B. Ucakar, A. des Rieux, and V. Preat. 2014a. pH-sensitive nanoparticles for colonic delivery of curcumin in inflammatory bowel disease. *Int J Pharm* 473 (1–2):203–12.

Beloqui, A., M. A. Solinis, A. Delgado, C. Evora, A. Isla, and A. Rodriguez-Gascon. 2014b. Fate of nanostructured lipid carriers (NLCs) following the oral route: Design, pharmacokinetics and biodistribution. *J Microencapsul* 31 (1):1–8.

Bhavsar, M. D., and M. M. Amiji. 2007. Gastrointestinal distribution and in vivo gene transfection studies with nanoparticles-in-microsphere oral system (NiMOS). *J Control Release* 119 (3):339–48.

Bratten, J., and M. P. Jones. 2006. New directions in the assessment of gastric function: Clinical applications of physiologic measurements. *Dig Dis* 24 (3–4):252–9.

Buhmann, S., C. Kirchhoff, R. Ladurner, T. Mussack, M. F. Reiser, and A. Lienemann. 2007. Assessment of colonic transit time using MRI: A feasibility study. *Eur Radiol* 17 (3):669–74.

Byrne, C. M., M. J. Solomon, J. M. Young, W. Selby, and J. D. Harrison. 2007. Patient preferences between surgical and medical treatment in Crohn's disease. *Dis Colon Rectum* 50 (5):586–97.

Coco, R., L. Plapied, V. Pourcelle, C. Jerome, D. J. Brayden, Y. J. Schneider, and V. Preat. 2013. Drug delivery to inflamed colon by nanoparticles: Comparison of different strategies. *Int J Pharm* 440 (1):3–12.

Collnot, E. M., H. Ali, and C. M. Lehr. 2012. Nano- and microparticulate drug carriers for targeting of the inflamed intestinal mucosa. *J Control Release* 161 (2):235–46.

Corfield, A. P., N. Myerscough, N. Bradfield, A. Corfield Cdo, M. Gough, J. R. Clamp, P. Durdey, B. F. Warren, D. C. Bartolo, K. R. King, and J. M. Williams. 1996. Colonic mucins in ulcerative colitis: Evidence for loss of sulfation. *Glycoconj J* 13 (5):809–22.

Coupe, A. J., S. S. Davis, and I. R. Wilding. 1991. Variation in gastrointestinal transit of pharmaceutical dosage forms in healthy subjects. *Pharm Res* 8 (3):360–4.

Cu, Y., and W. M. Saltzman. 2008. Controlled surface modification with poly(ethylene) glycol enhances diffusion of PLGA nanoparticles in human cervical mucus. *Mol Pharm* 6:173–81.

Cunningham, K. E., and J. R. Turner. 2012. Myosin light chain kinase: Pulling the strings of epithelial tight junction function. *Ann NY Acad Sci* 1258:34–42.

Diab, R., C. Jaafar-Maalej, H. Fessi, and P. Maincent. 2012. Engineered nanoparticulate drug delivery systems: The next frontier for oral administration? *AAPS J* 14 (4):688–702.

Dial, E. J., S. H. Rooijakkers, R. L. Darling, J. J. Romero, and L. M. Lichtenberger. 2008. Role of phosphatidylcholine saturation in preventing bile salt toxicity to gastrointestinal epithelia and membranes. *J Gastroenterol Hepatol* 23 (3):430–6.

Dobrovolskaia, M. A., and S. E. McNeil. 2013. Understanding the correlation between in vitro and in vivo immunotoxicity tests for nanomedicines. *J Control Release* 172 (2):456–66.

Elinav, E., and D. Peer. 2013. Harnessing nanomedicine for mucosal theranostics—A silver bullet at last? *ACS Nano* 7 (4):2883–90.

Fallingborg, J., L. A. Christensen, B. A. Jacobsen, and S. N. Rasmussen. 1993. Very low intraluminal colonic pH in patients with active ulcerative colitis. *Dig Dis Sci* 38 (11):1989–93.

Fallingborg, J., P. Pedersen, and B. A. Jacobsen. 1998. Small intestinal transit time and intraluminal pH in ileocecal resected patients with Crohn's disease. *Dig Dis Sci* 43 (4):702–5.

Fatouros, D. G., and A. Mullertz. 2008. In vitro lipid digestion models in design of drug delivery systems for enhancing oral bioavailability. *Expert Opin Drug Metab Toxicol* 4 (1):65–76.

Frank, D. N., A. L. St. Amand, R. A. Feldman, E. C. Boedeker, N. Harpaz, and N. R. Pace. 2007. Molecular-phylogenetic characterization of microbial community imbalances in human inflammatory bowel diseases. *Proc Natl Acad Sci USA* 104 (34):13780–5.

Gandjour, A., and N. Chernyak. 2011. A new prize system for drug innovation. *Health Policy* 102 (2–3):170–7.

Gaspar, R. 2007. Regulatory issues surrounding nanomedicines: Setting the scene for the next generation of nanopharmaceuticals. *Nanomedicine (Lond)* 2 (2):143–7.

Gaspar, R., and R. Duncan. 2009. Polymeric carriers: Preclinical safety and the regulatory implications for design and development of polymer therapeutics. *Adv Drug Deliv Rev* 61 (13):1220–31.

Gracie, D. J., J. S. Kane, S. Mumtaz, A. F. Scarsbrook, F. U. Chowdhury, and A. C. Ford. 2012. Prevalence of, and predictors of, bile acid malabsorption in outpatients with chronic diarrhea. *Neurogastroenterol Motil* 24 (11):983–e538.

Grainger, D. W. 2013. Connecting drug delivery reality to smart materials design. *Int J Pharm* 454 (1):521–4.

Gupta, A. S., S. J. Kshirsagar, M. R. Bhalekar, and T. Saldanha. 2013. Design and development of liposomes for colon targeted drug delivery. *J Drug Target* 21 (2):146–60.

Hafner, A., J. Lovric, G. P. Lakos, and I. Pepic. 2014. Nanotherapeutics in the EU: An overview on current state and future directions. *Int J Nanomed* 9:1005–23.

Hanauer, S. B., and M. Sparrow. 2004. COLAL-PRED Alizyme. *Curr Opin Investig Drugs* 5 (11):1192–7.

Harel, E., A. Rubinstein, A. Nissan, E. Khazanov, M. Nadler Milbauer, Y. Barenholz, and B. Tirosh. 2011. Enhanced transferrin receptor expression by pro-inflammatory cytokines in enterocytes as a means for local delivery of drugs to inflamed gut mucosa. *PLoS One* 6 (9):e24202.

Hebden, J. M., P. E. Blackshaw, A. C. Perkins, C. G. Wilson, and R. C. Spiller. 2000. Limited exposure of the healthy distal colon to orally-dosed formulation is further exaggerated in active left-sided ulcerative colitis. *Aliment Pharmacol Ther* 14 (2):155–61.

Hu, Z., S. Mawatari, N. Shibata, K. Takada, H. Yoshikawa, A. Arakawa, and Y. Yosida. 2000. Application of a biomagnetic measurement system (BMS) to the evaluation of gastrointestinal transit of intestinal pressure-controlled colon delivery capsules (PCDCs) in human subjects. *Pharm Res* 17 (2):160–7.

Hua, S. 2013. Targeting sites of inflammation: Intercellular adhesion molecule-1 as a target for novel inflammatory therapies. *Front Pharmacol* 4:127.

Hua, S. 2014. Orally administered liposomal formulations for colon targeted drug delivery. *Front Pharmacol* 5:138.

Hua, S., E. Marks, J. J. Schneider, and S. Keely. 2015. Advances in oral nano-delivery systems for colon targeted drug delivery in inflammatory bowel disease: Selective targeting to diseased versus healthy tissue. *Nanomedicine* 11 (5):1117–32.

Ibekwe, V. C., H. M. Fadda, E. L. McConnell, M. K. Khela, D. F. Evans, and A. W. Basit. 2008. Interplay between intestinal pH, transit time and feed status on the in vivo performance of pH responsive ileo-colonic release systems. *Pharm Res* 25 (8):1828–35.

Jaafar-Maalej, C., A. Elaissari, and H. Fessi. 2012. Lipid-based carriers: Manufacturing and applications for pulmonary route. *Expert Opin Drug Deliv* 9 (9):1111–27.

Jubeh, T. T., Y. Barenholz, and A. Rubinstein. 2004. Differential adhesion of normal and inflamed rat colonic mucosa by charged liposomes. *Pharm Res* 21 (3):447–53.

Kashyap, P. C., A. Marcobal, L. K. Ursell, M. Larauche, H. Duboc, K. A. Earle, E. D. Sonnenburg et al. 2013. Complex interactions among diet, gastrointestinal transit, and gut microbiota in humanized mice. *Gastroenterology* 144 (5):967–77.

Keely, S., L. Feighery, D. P. Campion, L. O'Brien, D. J. Brayden, and A. W. Baird. 2011. Chloride-led disruption of the intestinal mucous layer impedes *Salmonella* invasion: Evidence for an 'enteric tear' mechanism. *Cell Physiol Biochem* 28 (4):743–52.

Keely, S., C. J. Kelly, T. Weissmueller, A. Burgess, B. D. Wagner, C. E. Robertson, J. K. Harris, and S. P. Colgan. 2012. Activated fluid transport regulates bacterial-epithelial interactions and significantly shifts the murine colonic microbiome. *Gut Microbes* 3 (3):250–60.

Kraft, J. C., J. P. Freeling, Z. Wang, and R. J. Ho. 2014. Emerging research and clinical development trends of liposome and lipid nanoparticle drug delivery systems. *J Pharm Sci* 103 (1):29–52.

Kriegel, C., and M. Amiji. 2011a. Oral TNF-alpha gene silencing using a polymeric microsphere-based delivery system for the treatment of inflammatory bowel disease. *J Control Release* 150 (1):77–86.

Kriegel, C., and M. M. Amiji. 2011b. Dual TNF-alpha/cyclin D1 gene silencing with an oral polymeric microparticle system as a novel strategy for the treatment of inflammatory bowel disease. *Clin Transl Gastroenterol* 2:e2.

Lai, S. K., Y. Y. Wang, and J. Hanes. 2009. Mucus-penetrating nanoparticles for drug and gene delivery to mucosal tissues. *Adv Drug Deliv Rev* 61 (2):158–71.

Lammers, T. 2013. Smart drug delivery systems: Back to the future vs. clinical reality. *Int J Pharm* 454 (1):527–9.

Lamprecht, A., U. Schäfer, and C.-M. Lehr. 2001a. Size-dependent bioadhesion of micro- and nanoparticulate carriers to the inflamed colonic mucosa. *Pharm Res* 18 (6):788–93.

Lamprecht, A., N. Ubrich, H. Yamamoto, U. Schafer, H. Takeuchi, P. Maincent, Y. Kawashima, and C. M. Lehr. 2001b. Biodegradable nanoparticles for targeted drug delivery in treatment of inflammatory bowel disease. *J Pharmacol Exp Ther* 299 (2):775–81.

Lamprecht, A., H. Yamamoto, H. Takeuchi, and Y. Kawashima. 2005a. Nanoparticles enhance therapeutic efficiency by selectively increased local drug dose in experimental colitis in rats. *J Pharmacol Exp Ther* 315 (1):196–202.

Lamprecht, A., H. Yamamoto, H. Takeuchi, and Y. Kawashima. 2005b. A pH-sensitive microsphere system for the colon delivery of tacrolimus containing nanoparticles. *J Control Release* 104 (2):337–46.

Laroui, H., G. Dalmasso, H. T. Nguyen, Y. Yan, S. V. Sitaraman, and D. Merlin. 2010. Drug-loaded nanoparticles targeted to the colon with polysaccharide hydrogel reduce colitis in a mouse model. *Gastroenterology* 138 (3):843–53.e1–2.

Laroui, H., D. Geem, B. Xiao, E. Viennois, P. Rakhya, T. Denning, and D. Merlin. 2014a. Targeting intestinal inflammation with CD98 siRNA/PEI-loaded nanoparticles. *Mol Ther* 22 (1):69–80.

Laroui, H., P. Rakhya, B. Xiao, E. Viennois, and D. Merlin. 2013. Nanotechnology in diagnostics and therapeutics for gastrointestinal disorders. *Dig Liver Dis* 45 (12):995–1002.

Laroui, H., E. Viennois, B. Xiao, B. S. Canup, D. Geem, T. L. Denning, and D. Merlin. 2014b. Fab′-bearing siRNA TNFalpha-loaded nanoparticles targeted to colonic macrophages offer an effective therapy for experimental colitis. *J Control Release* 186:41–53.

Larsson, J. M., H. Karlsson, J. G. Crespo, M. E. Johansson, L. Eklund, H. Sjovall, and G. C. Hansson. 2011. Altered O-glycosylation profile of MUC2 mucin occurs in active ulcerative colitis and is associated with increased inflammation. *Inflamm Bowel Dis* 17 (11):2299–307.

Laukoetter, M. G., P. Nava, and A. Nusrat. 2008. Role of the intestinal barrier in inflammatory bowel disease. *World J Gastroenterol* 14 (3):401–7.

Lautenschlager, C., C. Schmidt, D. Fischer, and A. Stallmach. 2014. Drug delivery strategies in the therapy of inflammatory bowel disease. *Adv Drug Deliv Rev* 71:58–76.

Lautenschlager, C., C. Schmidt, C. M. Lehr, D. Fischer, and A. Stallmach. 2013. PEG-functionalized microparticles selectively target inflamed mucosa in inflammatory bowel disease. *Eur J Pharm Biopharm* 85 (3 Pt A):578–86.

Lazarev, M., T. Ullman, W. H. Schraut, K. E. Kip, M. Saul, and M. Regueiro. 2010. Small bowel resection rates in Crohn's disease and the indication for surgery over time: Experience from a large tertiary care center. *Inflamm Bowel Dis* 16 (5):830–5.

Li, A. C., and R. P. Thompson. 2003. Basement membrane components. *J Clin Pathol* 56 (12):885–7.

Lih-Brody, L., S. R. Powell, K. P. Collier, G. M. Reddy, R. Cerchia, E. Kahn, G. S. Weissman et al. 1996. Increased oxidative stress and decreased antioxidant defenses in mucosa of inflammatory bowel disease. *Dig Dis Sci* 41 (10):2078–86.

Lin, H. C., C. Prather, R. S. Fisher, J. H. Meyer, R. W. Summers, M. Pimentel, R. W. McCallum, L. M. Akkermans, V. Loening-Baucke, and AMS Task Force Committee on Gastrointestinal Transit. 2005. Measurement of gastrointestinal transit. *Dig Dis Sci* 50 (6):989–1004.

Linskens, R. K., X. W. Huijsdens, P. H. Savelkoul, C. M. Vandenbroucke-Grauls, and S. G. Meuwissen. 2001. The bacterial flora in inflammatory bowel disease: Current insights in pathogenesis and the influence of antibiotics and probiotics. *Scand J Gastroenterol Suppl* (234):29–40.

Lucas, M. L., B. T. Cooper, F. H. Lei, I. T. Johnson, G. K. Holmes, J. A. Blair, and W. T. Cooke. 1978. Acid microclimate in coeliac and Crohn's disease: A model for folate malabsorption. *Gut* 19 (8):735–42.

Luxenhofer, R., M. Barz, and M. Schillmeier. 2014. Quo vadis nanomedicine? *Nanomedicine (Lond)* 9 (14):2083–6.

Lv, H., S. Zhang, B. Wang, S. Cui, and J. Yan. 2006. Toxicity of cationic lipids and cationic polymers in gene delivery. *J Control Release* 114 (1):100–9.

Macfarlane, G. T., and S. Macfarlane. 2011. Fermentation in the human large intestine: Its physiologic consequences and the potential contribution of prebiotics. *J Clin Gastroenterol* 45 (Suppl):S120–7.

Mahida, Y. R., K. C. Wu, and D. P. Jewell. 1989. Respiratory burst activity of intestinal macrophages in normal and inflammatory bowel disease. *Gut* 30 (10):1362–70.

Malayandi, R., P. K. Kondamudi, P. K. Ruby, and D. Aggarwal. 2014. Biopharmaceutical considerations and characterizations in development of colon targeted dosage forms for inflammatory bowel disease. *Drug Deliv Transl Res* 4:187–202.

Mane, V., and S. Muro. 2012. Biodistribution and endocytosis of ICAM-1-targeting antibodies versus nanocarriers in the gastrointestinal tract in mice. *Int J Nanomed* 7:4223–37.

McConnell, E. L., H. M. Fadda, and A. W. Basit. 2008. Gut instincts: Explorations in intestinal physiology and drug delivery. *Int J Pharm* 364 (2):213–26.

Meissner, Y., Y. Pellequer, and A. Lamprecht. 2006. Nanoparticles in inflammatory bowel disease: Particle targeting versus pH-sensitive delivery. *Int J Pharm* 316 (1–2):138–43.

Moulari, B., D. Pertuit, Y. Pellequer, and A. Lamprecht. 2008. The targeting of surface modified silica nanoparticles to inflamed tissue in experimental colitis. *Biomaterials* 29 (34):4554–60.

Mura, S., and P. Couvreur. 2012. Nanotheranostics for personalized medicine. *Adv Drug Deliv Rev* 64 (13):1394–416.

Murday, J. S., R. W. Siegel, J. Stein, and J. F. Wright. 2009. Translational nanomedicine: Status assessment and opportunities. *Nanomedicine* 5 (3):251–73.

Musch, M. W., Y. Wang, E. C. Claud, and E. B. Chang. 2013. Lubiprostone decreases mouse colonic inner mucus layer thickness and alters intestinal microbiota. *Dig Dis Sci* 58 (3):668–77.

Narang, A. S., R. K. Chang, and M. A. Hussain. 2013. Pharmaceutical development and regulatory considerations for nanoparticles and nanoparticulate drug delivery systems. *J Pharm Sci* 102 (11):3867–82.

Niebel, W., K. Walkenbach, A. Beduneau, Y. Pellequer, and A. Lamprecht. 2012. Nanoparticle-based clodronate delivery mitigates murine experimental colitis. J Control Release 160 (3):659–65.

Nielsen, O. H. 2014. New strategies for treatment of inflammatory bowel disease. Front Med (Lausanne) 1:3.

Nielsen, O. H., and M. A. Ainsworth. 2013. Tumor necrosis factor inhibitors for inflammatory bowel disease. N Engl J Med 369 (8):754–62.

Nugent, S. G., D. Kumar, D. S. Rampton, and D. F. Evans. 2001. Intestinal luminal pH in inflammatory bowel disease: Possible determinants and implications for therapy with aminosalicylates and other drugs. Gut 48 (4):571–7.

Nyhof, R. A., D. Ingold-Wilcox, and C. C. Chou. 1985. Effect of atropine on digested food-induced intestinal hyperemia. Am J Physiol 249 (6 Pt 1):G685–90.

Nystrom, A. M., and B. Fadeel. 2012. Safety assessment of nanomaterials: Implications for nanomedicine. J Control Release 161 (2):403–8.

Olbrich, C., and R. H. Muller. 1999. Enzymatic degradation of SLN-effect of surfactant and surfactant mixtures. Int J Pharm 180 (1):31–9.

Paradise, J., S. M. Wolf, J. Kuzma, A. Kuzhabekova, A. W. Tisdale, E. Kokkoli, and G. Ramachandran. 2009. Developing U.S. oversight strategies for nanobiotechnology: Learning from past oversight experiences. J Law Med Ethics 37 (4):688–705.

Pichai, M. V., and L. R. Ferguson. 2012. Potential prospects of nanomedicine for targeted therapeutics in inflammatory bowel diseases. World J Gastroenterol 18 (23):2895–901.

Podolsky, D. K. 2002. Inflammatory bowel disease. N Engl J Med 347 (6):417–29.

Pullan, R. D., G. A. Thomas, M. Rhodes, R. G. Newcombe, G. T. Williams, A. Allen, and J. Rhodes. 1994. Thickness of adherent mucus gel on colonic mucosa in humans and its relevance to colitis. Gut 35 (3):353–9.

Rana, S. V., S. Sharma, A. Malik, J. Kaur, K. K. Prasad, S. K. Sinha, and K. Singh. 2013. Small intestinal bacterial overgrowth and orocecal transit time in patients of inflammatory bowel disease. Dig Dis Sci 58 (9):2594–8.

Rao, K. A., E. Yazaki, D. F. Evans, and R. Carbon. 2004. Objective evaluation of small bowel and colonic transit time using pH telemetry in athletes with gastrointestinal symptoms. Br J Sports Med 38 (4):482–7.

Rath, H. C., H. H. Herfarth, J. S. Ikeda, W. B. Grenther, T. E. Hamm Jr., E. Balish, J. D. Taurog, R. E. Hammer, K. H. Wilson, and R. B. Sartor. 1996.

Normal luminal bacteria, especially Bacteroides species, mediate chronic colitis, gastritis, and arthritis in HLA-B27/human beta2 microglobulin transgenic rats. J Clin Invest 98 (4):945–53.

Saini, S. D., P. Schoenfeld, K. Kaulback, and M. C. Dubinsky. 2009. Effect of medication dosing frequency on adherence in chronic diseases. Am J Manag Care 15 (6):e22–33.

Sainz, V., J. Conniot, A. I. Matos, C. Peres, E. Zupancic, L. Moura, L. C. Silva, H. F. Florindo, and R. S. Gaspar. 2015. Regulatory aspects on nanomedicines. Biochem Biophys Res Commun. 468 (3):504–10.

Saltzman, W. M., M. L. Radomsky, K. J. Whaley, and R. A. Cone. 1994. Antibody diffusion in human cervical mucus. Biophys J 66 (2 Pt 1):508–15.

Sandborn, W. J. 2002. Rational selection of oral 5-aminosalicylate formulations and prodrugs for the treatment of ulcerative colitis. Am J Gastroenterol 97 (12):2939–41.

Sandborn, W. J., P. Rutgeerts, B. G. Feagan, W. Reinisch, A. Olson, J. Johanns, J. Lu et al. 2009. Colectomy rate comparison after treatment of ulcerative colitis with placebo or infliximab. Gastroenterology 137 (4):1250–60; quiz 1520.

Sartor, R. B. 2008a. Microbial influences in inflammatory bowel diseases. Gastroenterology 134 (2):577–94.

Sartor, R. B. 2008b. Therapeutic correction of bacterial dysbiosis discovered by molecular techniques. Proc Natl Acad Sci USA 105 (43):16413–4.

Sartor, R. B. 2010. Genetics and environmental interactions shape the intestinal microbiome to promote inflammatory bowel disease versus mucosal homeostasis. Gastroenterology 139 (6):1816–9.

Sasaki, Y., R. Hada, H. Nakajima, S. Fukuda, and A. Munakata. 1997. Improved localizing method of radiopill in measurement of entire gastrointestinal pH profiles: Colonic luminal pH in normal subjects and patients with Crohn's disease. Am J Gastroenterol 92 (1):114–8.

Satalkar, P., B. S. Elger, and D. M. Shaw. 2015. Defining nano, nanotechnology and nanomedicine: Why should it matter? Sci Eng Ethics 22 (5):1255–76.

Sathyan, G., S. Hwang, and S. K. Gupta. 2000. Effect of dosing time on the total intestinal transit time of non-disintegrating systems. Int J Pharm 204 (1–2):47–51.

Sawant, R. R., and V. P. Torchilin. 2012. Challenges in development of targeted liposomal therapeutics. AAPS J 14 (2):303–15.

Schmidt, C., C. Lautenschlaeger, E. M. Collnot, M. Schumann, C. Bojarski, J. D. Schulzke, C. M. Lehr, and A. Stallmach. 2013. Nano- and microscaled particles for drug targeting to inflamed intestinal mucosa: A first in vivo study in human patients. *J Control Release* 165 (2):139–45.

Schmidt, T., A. Pfeiffer, N. Hackelsberger, R. Widmer, C. Meisel, and H. Kaess. 1996. Effect of intestinal resection on human small bowel motility. *Gut* 38 (6):859–63.

Schreiber, S. 2011. Certolizumab pegol for the treatment of Crohn's disease. *Therap Adv Gastroenterol* 4 (6):375–89.

Schreiber, S., I. C. Lawrance, O. O. Thomsen, S. B. Hanauer, R. Bloomfield, and W. J. Sandborn. 2011. Randomised clinical trial: Certolizumab pegol for fistulas in Crohn's disease—Subgroup results from a placebo-controlled study. *Aliment Pharmacol Ther* 33 (2):185–93.

Sellon, R. K., S. Tonkonogy, M. Schultz, L. A. Dieleman, W. Grenther, E. Balish, D. M. Rennick, and R. B. Sartor. 1998. Resident enteric bacteria are necessary for development of spontaneous colitis and immune system activation in interleukin-10-deficient mice. *Infect Immun* 66 (11):5224–31.

Simmonds, N. J., R. E. Allen, T. R. Stevens, R. N. Van Someren, D. R. Blake, and D. S. Rampton. 1992. Chemiluminescence assay of mucosal reactive oxygen metabolites in inflammatory bowel disease. *Gastroenterology* 103 (1):186–96.

Sinha, V. R., and R. Kumria. 2001. Polysaccharides in colon-specific drug delivery. *Int J Pharm* 224 (1–2):19–38.

Spiller, R. C., I. F. Trotman, T. E. Adrian, S. R. Bloom, J. J. Misiewicz, and D. B. Silk. 1988. Further characterisation of the 'ileal brake' reflex in man—Effect of ileal infusion of partial digests of fat, protein, and starch on jejunal motility and release of neurotensin, enteroglucagon, and peptide YY. *Gut* 29 (8):1042–51.

Steed, K. P., G. Hooper, N. Monti, M. S. Benedetti, G. Fornasini, and I. R. Wilding. 1997. The use of pharmacoscintigraphy to focus the development strategy for a novel 5-ASA colon targeting system ("TIME CLOCK (R)" system). *J Control Release* 49 (2–3):115–22.

Steenholdt, C., J. Brynskov, O. O. Thomsen, L. K. Munck, J. Fallingborg, L. A. Christensen, G. Pedersen et al. 2014. Individualised therapy is more cost-effective than dose intensification in patients with Crohn's disease who lose response to anti-TNF treatment: A randomised, controlled trial. *Gut* 63 (6):919–27.

Svenson, S. 2012. Clinical translation of nanomedicines. *Curr Opin Solid State Mater Sci* 16 (6):287–94.

Talaei, F., F. Atyabi, M. Azhdarzadeh, R. Dinarvand, and A. Saadatzadeh. 2013. Overcoming therapeutic obstacles in inflammatory bowel diseases: A comprehensive review on novel drug delivery strategies. *Eur J Pharm Sci* 49 (4):712–22.

Talley, N. J., M. T. Abreu, J. P. Achkar, C. N. Bernstein, M. C. Dubinsky, S. B. Hanauer, S. V. Kane, W. J. Sandborn, T. A. Ullman, P. Moayyedi, and IBD Task Force American College of Gastroenterology. 2011. An evidence-based systematic review on medical therapies for inflammatory bowel disease. *Am J Gastroenterol* 106 (Suppl 1):S2–25; quiz S26.

Tang, B. C., M. Dawson, S. K. Lai, Y. Y. Wang, J. S. Suk, M. Yang, P. Zeitlin, M. P. Boyle, J. Fu, and J. Hanes. 2009. Biodegradable polymer nanoparticles that rapidly penetrate the human mucus barrier. *Proc Natl Acad Sci USA* 106 (46):19268–73.

Teli, M. K., S. Mutalik, and G. K. Rajanikant. 2010. Nanotechnology and nanomedicine: Going small means aiming big. *Curr Pharm Des* 16 (16):1882–92.

Thirawong, N., J. Thongborisute, H. Takeuchi, and P. Sriamornsak. 2008. Improved intestinal absorption of calcitonin by mucoadhesive delivery of novel pectin–liposome nanocomplexes. *J Control Release* 125 (3):236–45.

Thompson, J. S., E. M. Quigley, T. E. Adrian, and F. R. Path. 1998. Role of the ileocecal junction in the motor response to intestinal resection. *J Gastrointest Surg* 2 (2):174–85.

Tinkle, S., S. E. McNeil, S. Muhlebach, R. Bawa, G. Borchard, Y. C. Barenholz, L. Tamarkin, and N. Desai. 2014. Nanomedicines: Addressing the scientific and regulatory gap. *Ann NY Acad Sci* 1313:35–56.

Tobio, M., A. Sanchez, A. Vila, I. I. Soriano, C. Evora, J. L. Vila-Jato, and M. J. Alonso. 2000. The role of PEG on the stability in digestive fluids and in vivo fate of PEG-PLA nanoparticles following oral administration. *Colloids Surf B Biointerfaces* 18 (3–4):315–323.

Tran, T. H., and M. M. Amiji. 2015. Targeted delivery systems for biological therapies of inflammatory diseases. *Expert Opin Drug Deliv* 12 (3):393–414.

Van Citters, G. W., and H. C. Lin. 1999. The ileal brake: A fifteen-year progress report. *Curr Gastroenterol Rep* 1 (5):404–9.

Van Citters, G. W., and H. C. Lin. 2006. Ileal brake: Neuropeptidergic control of intestinal transit. *Curr Gastroenterol Rep* 8 (5):367–73.

Viscido, A., A. Capannolo, G. Latella, R. Caprilli, and G. Frieri. 2014. Nanotechnology in the treatment of inflammatory bowel diseases. *J Crohns Colitis* 8 (9):903–18.

Vong, L. B., T. Tomita, T. Yoshitomi, H. Matsui, and Y. Nagasaki. 2012. An orally administered redox nanoparticle that accumulates in the colonic mucosa and reduces colitis in mice. *Gastroenterology* 143 (4):1027–36.e3.

Wachsmann, P., B. Moulari, A. Beduneau, Y. Pellequer, and A. Lamprecht. 2013. Surfactant-dependence of nanoparticle treatment in murine experimental colitis. *J Control Release* 172 (1):62–8.

Wang, X., S. Maher, and D. J. Brayden. 2010. Restoration of rat colonic epithelium after in situ intestinal instillation of the absorption promoter, sodium caprate. *Ther Deliv* 1 (1):75–82.

Watts, P. J., L. Barrow, K. P. Steed, C. G. Wilson, R. C. Spiller, C. D. Melia, and M. C. Davies. 1992. The transit rate of different-sized model dosage forms through the human colon and the effects of a lactulose-induced catharsis. *Int J Pharm* 87:215–21.

Wilson, D. S., G. Dalmasso, L. Wang, S. V. Sitaraman, D. Merlin, and N. Murthy. 2010. Orally delivered thioketal nanoparticles loaded with TNF-alpha-siRNA target inflammation and inhibit gene expression in the intestines. *Nat Mater* 9 (11):923–8.

Witticke, D., H. M. Seidling, H. D. Klimm, and W. E. Haefeli. 2012. Do we prescribe what patients prefer? Pilot study to assess patient preferences for medication regimen characteristics. *Patient Prefer Adherence* 6:679–84.

Xiao, B., H. Laroui, S. Ayyadurai, E. Viennois, M. A. Charania, Y. Zhang, and D. Merlin. 2013. Mannosylated bioreducible nanoparticle-mediated macrophage-specific TNF-alpha RNA interference for IBD therapy. *Biomaterials* 34 (30):7471–82.

Xiao, B., H. Laroui, E. Viennois, S. Ayyadurai, M. A. Charania, Y. Zhang, Z. Zhang, M. T. Baker, B. Zhang, A. T. Gewirtz, and D. Merlin. 2014. Nanoparticles with surface antibody against CD98 and carrying CD98 small interfering RNA reduce colitis in mice. *Gastroenterology* 146 (5):1289–300.e1–19.

Xiao, B., and D. Merlin. 2012. Oral colon-specific therapeutic approaches toward treatment of inflammatory bowel disease. *Expert Opin Drug Deliv* 9 (11):1393–407.

Xiong, X., and K. Tam. 2004. Hydrolytic degradation of pluronic F127/poly(lactic acid) block copolymer nanoparticles. *Macromolecules* 37:3425–30.

Yang, H., W. Jiang, E. E. Furth, X. Wen, J. P. Katz, R. K. Sellon, D. G. Silberg, T. M. Antalis, C. W. Schweinfest, and G. D. Wu. 1998. Intestinal inflammation reduces expression of DRA, a transporter responsible for congenital chloride diarrhea. *Am J Physiol* 275 (6 Pt 1):G1445–53.

Yang, L. 2008. Biorelevant dissolution testing of colon-specific delivery systems activated by colonic microflora. *J Control Release* 125 (2):77–86.

Yang, L., J. S. Chu, and J. A. Fix. 2002. Colon-specific drug delivery: New approaches and in vitro/in vivo evaluation. *Int J Pharm* 235 (1–2):1–15.

Zhang, J., C. Tang, and C. Yin. 2013. Galactosylated trimethyl chitosan-cysteine nanoparticles loaded with Map4k4 siRNA for targeting activated macrophages. *Biomaterials* 34 (14):3667–77.

Zhang, L., F. X. Gu, J. M. Chan, A. Z. Wang, R. S. Langer, and O. C. Farokhzad. 2008. Nanoparticles in medicine: Therapeutic applications and developments. *Clin Pharmacol Ther* 83 (5):761–9.

The Biology and Clinical Treatment of Multiple Sclerosis

Mahsa Khayat-Khoei, Leorah Freeman, and John Lincoln

CONTENTS

6.1.1 OVERVIEW, RISK FACTORS, AND DIAGNOSIS OF MS

Multiple sclerosis (MS) affects nearly 400,000 people in the United States alone and more than 2.5 million people worldwide (Noseworthy et al. 2000; Reingold 2002), is the most common nontraumatic neurologic disease of young people leading to clinical disability, and reduces life span by approximately 7 years (Leray et al. 2015; Marrie et al. 2015). While numbers are variable, the average annual direct and indirect cost for the individual MS patient to society is estimated at more than $40,000, when combining treatments that modify disease course and manage clinical symptoms and time lost due to acute and chronic disability (Kolasa 2013).

6.1.1.1 Epidemiology

The incidence of MS is estimated at 5.2 (range 0.5–20.6) per 100,000 patient-years, with a median prevalence of 112/100,000 (Melcon et al. 2014). MS incidence peaks between 20 and 40 years of age, although childhood and late-onset disease have been described (Confavreux and Vukusic 2006). Relapsing forms of MS are nearly threefold more common in women than in men, while phenotypes with progressive onset are equally common among men and women (Noonan et al. 2010).

6.1.1.1.1 Genetics

MS is characterized by "familial aggregation" in that the risk to develop MS is higher in patient's relatives than in the total population. Risk is negatively correlated with genetic distance to the proband (Oksenberg 2013). Concordance rates vary, with 25%–30% risk in monozygotic twins and 3%–5% in dizygotic twins and nontwin siblings (Lin et al. 2012). This type of inheritance is more frequently seen in polygenic diseases where each gene polymorphism contributes only minimal risk for disease.

There are now nearly 100 candidate MS risk loci. Initial gene candidates were identified using linkage analysis. Of these, the association of combinations of various HLA-DRB1 alleles (human leukocyte antigen [HLA] class II genes) confers an increased relative risk of between 3 and 30 and remains the candidate adding the greatest risk (Ramagopalan and Ebers 2009). Genome-wide association studies (GWASs) have now become the most common method to search for new candidate genes. GWASs compare allele frequencies from microarrays of single-nucleotide polymorphisms (SNPs) distributed throughout the genome from large samples of affected patients and controls. Recent studies using this technique have evaluated more than 10,000 samples with more than 1 million comparisons. Stringent significance levels are set to take into account the Bonferroni correction for the million-plus comparisons. Based on GWAS studies, MS-associated SNPs were most numerous on chromosomes 1 and 6 and absent on sex chromosomes (Bashinskaya et al. 2015).

Many associated SNPs are located within introns with functional polymorphisms. These causative polymorphisms can affect the functional activity, level, location, or timing of the gene product. For example, several SNPs have been associated with cytokine receptor genes, including interleukin 7 receptor agonist (IL7RA), IL2RA, and tumor necrosis factor (TNF) and can affect proportions of soluble and membrane-bound receptor isoforms (Gregory et al. 2007; Gregory et al. 2012).

6.1.1.1.2 Epigenetics and the Environment

MS prevalence varies greatly between continents, with greater prevalence found in North America and Europe. In addition, epidemiologic studies suggest that there might be a latitudinal and altitudinal gradient possibly related to a combination of genetic and various environmental factors, such as vitamin D exposure, cigarette smoking, or late-onset Epstein–Barr virus (EBV) infection (Lincoln et al. 2008; Lincoln and Cook 2009).

Epidemiologic studies have shown lower incidence of infectious mononucleosis (IM), typically resultant from EBV infection later in life, in areas with lower compared with higher MS prevalence (Giovannoni and Ebers 2007). A large prospective population-based study found a greater than fivefold increased risk of developing MS in persons with IM (Marrie et al. 2000), while another study found odds ratios of 2.7–3.7 in persons with heterophile-positive IM (Haahr et al. 1995). Serological studies have shown EBV-specific antibodies in both adults (99%) and children (83%–99%) with MS compared with their respective controls without disease (84%–95% of non-MS adults and 42%–72% of non-MS children) (Pohl et al. 2006; Lünemann and Münz 2007). Finally, oligoclonal bands from the cerebrospinal fluid of some patients with MS have been shown to react with EBV-specific proteins (Cepok et al. 2005).

Vitamin D is known to either directly or indirectly interact with more than 200 genes and specific vitamin D receptors and is a potent modulator of the immune system by suppressing antibody production, decreasing pro-inflammatory cytokine production, and enhancing Th2 function (Holick 2007). It has long been postulated that decreased sun exposure or enteral vitamin D intake may be associated with the incidence of MS. A recent study by Munger et al. (2016) evaluated MS risk related to vitamin D exposure in offspring of mothers in the Finnish maternity cohort, assessed between January 1, 1983, and December 31, 1991. Maternal vitamin D in the first trimester of less than 12 ng/ml was associated with a nearly twofold increased risk of MS in offspring, although no significant association between higher levels of vitamin D and MS was observed (Munger et al. 2016).

There have been several case control, cohort, prospective studies that highlight an increased risk of MS in smokers. Participants in these studies who smoked prior to disease onset had between a 1.2- and 1.9-fold increased risk of subsequently developing MS and a nearly 4-fold increased hazard for secondary progression (Hernán et al. 2005).

Overall, genetic factors alone are inadequate to account for the recent variations in MS risk. Environmental agents might interact with genetic elements, potentially modifying gene expression and/or function. Giovannoni and Ebers (2007) postulated that the interactions between genes and various environmental agents more completely account for the differing MS risk in populations and the recent changes in MS incidence among women.

6.1.1.2 Diagnosis of Multiple Sclerosis

6.1.1.2.1 Clinical Features

Initial presentation can greatly vary from patient to patient. Common presenting symptoms include optic neuritis, brainstem or spinal cord manifestations, or in less frequent instances, hemispheric symptomatology. In up to one-fourth of cases, symptoms at presentation may be multifocal (Confavreux et al. 2000). When a patient presents with symptoms suggestive of white matter (WM) tract damage, the exclusion of an alternate diagnosis is imperative before a diagnosis of MS can be made. Such diagnosis will then rely on the demonstration of "dissemination in space" (DIS) and "dissemination in time" (DIT) based on clinical grounds alone (clinically definite MS [CDMS]) or a combination of clinical and radiological findings.

A "relapse" is defined as "patient-reported symptoms or objectively observed signs typical of acute inflammatory demyelinating event in the CNS ... with duration of at least 24 hours, in the absence of fever or infection" (Polman et al. 2011). Based on the McDonald criteria of the International Panel on Diagnosis of MS, initially published in 2001 (McDonald et al. 2001), and subsequently revised in 2005 (Polman et al. 2005) and 2010 (Polman et al. 2011), a diagnosis of MS can be reached on clinical findings alone if the patient presents with a history of two or more relapses and objective clinical evidence of two or more lesions. It should be expected for MRI findings to be consistent with a diagnosis of MS, although not mandatory in this case. In all other presentations (two attacks with objective evidence of only one lesion, single relapse, progressive course), MRI will play a central role in the demonstration of DIS and DIT.

6.1.1.2.2 Magnetic Resonance Imaging

MRI is currently the most useful paraclinical tool for the diagnosis of MS. MS WM plaques, the pathological hallmark of the disease, can be detected with great sensitivity, particularly on T2-weighted or fluid-attenuated inversion recovery (FLAIR) sequences (Figure 6.1). Their objective presence on MRI is considered an essential requirement for the diagnosis of MS. These lesions are often periventricular with a characteristic ovoid shape, but can also be seen in juxtacortical or infratentorial areas (Figure 6.2a and b).

MRI lesions enhancing after injection of gadolinium (Figure 6.3) reflect active inflammation and breakdown of the blood–brain barrier (BBB) and are thus considered more recent (4–6 weeks on average).

Spinal cord lesions have been reported in up to 90% of MS patients (Bot et al. 2004), and asymptomatic lesions have been detected in up to one-third of patients presenting with a demyelinating event suggestive of MS. Spinal cord MRI at the

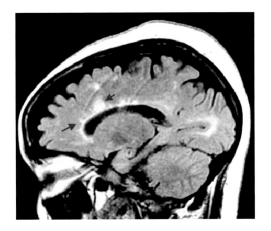

Figure 6.1 Sagittal FLAIR sequence showing classical "Dawson's fingers" (arrows).

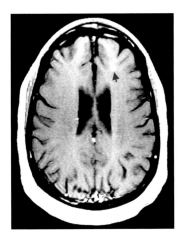

Figure 6.3 MS lesions with BBB damage related to active inflammation are often hyperintense (enhanced) on post-contrast T1 MRI.

time of diagnosis can thus be useful to demonstrate DIS. Spinal cord lesions, however, much less frequently present with contrast enhancement and are therefore rarely useful for demonstration of DIT. While not commonly used in routine monitoring of disease activity, spinal MRI might be important in identifying alternate causes in patients presenting with symptoms of myelopathy (Kearney et al. 2015).

Spinal MRI is particularly important when evaluating for neuromyelitis optica (NMO), a chronic demyelinating disease previously considered a variant of MS but now confirmed to have a dissimilar pathophysiology. Spinal cord lesions in MS are commonly short-segment lesions often located in the peripheral of the cord, as seen on axial views, while NMO lesions are central in location, involve spinal gray matter (GM), and are typically edematous and longitudinally expansive (more than three vertebral segments in length) on sagittal views.

While earlier diagnostic criteria using MRI were based on lesion number (Barkhof et al. 1997), revised and simplified criteria by Swanton and colleagues (2006) now focus on lesion location (periventricular, juxtacortical, infratentorial, and spinal cord) for demonstration of DIS. Still, the risk of overdiagnosing MS remains real, and as the Magnetic Resonance Imaging in MS (MAGNIMS) committee recently recommended, "MRI scans should be interpreted by experienced readers who are aware of the patient's clinical and laboratory information" (Rovira et al. 2015).

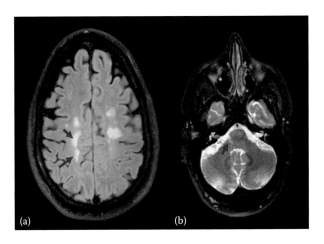

Figure 6.2 Juxtacortical (a) and infratentorial (b) MS lesions.

NANOMEDICINE FOR INFLAMMATORY DISEASES

6.1.1.3 Evolution and Prognosis

6.1.1.3.1 Clinical Phenotypes

Clarity and consistency in defining clinical phenotypes are essential for demographic studies, clinical trials, and management of therapy in clinical practice. A newly revised classification proposed by Lublin and colleagues (2014) recommends that patient phenotype be assessed on clinical grounds, with input from imaging studies when needed. According to the new consensus, three disease phenotypes can be defined: clinically isolated syndrome (CIS), relapsing–remitting (RR) disease, and progressive disease, including primary progressive (PP) and secondary progressive (SP).

CIS refers to the initial clinical presentation of the disease in patients with symptoms typical of demyelination of the central nervous system (CNS) WM tracts, but who fail to show evidence of DIT of the disease process. Patients with CIS are more likely to "convert" to definite MS if they meet criteria for DIS and DIT on radiological grounds.

A majority of patients diagnosed with definite MS will follow an RR disease course characterized by exacerbations (relapses) with intervening periods of clinical stability. Patients may recover fully or partially from relapses. Patients with an initial RR form of the disease may subsequently experience worsening disability progression unrelated to relapse activity. This clinical phenotype is termed SPMS. Between 10% and 15% experience a gradual worsening of clinical disability from onset with no initial exacerbations (PP course). It is important to note that progressive disease (SPMS or PPMS) does not progress in a uniform fashion, and patients may experience periods of relative clinical stability.

Current consensus recommendations also include disease "activity" as a modifier of the basic clinical phenotypes previously mentioned. Disease activity is defined by either clinical relapses or radiologic activity (presence of contrast-enhancing lesions, or new or unequivocally enlarged T2 lesions).

The widespread availability of MRI has resulted in an increase in incidental imaging findings not related to clinical presentation. Radiologically isolated syndrome (RIS) is defined as MRI findings suggestive of MS in persons without typical MS symptoms and with normal neurological signs. A scenario often encountered is a patient with headaches with a brain MRI showing incidental lesions suggestive of MS. The RIS Consortium presented results of a retrospective study of 451 RIS subjects from 22 databases in five countries (Okuda et al. 2014). This study showed that only 34% of RIS individuals develop an initial clinical event within 5 years of RIS diagnosis. Important predictors of symptom onset include age less than 37 years, male sex, and spinal cord involvement.

6.1.1.3.2 Prognosis and Prediction

Clinical phenotypes are a dynamic process. Patients with CIS may convert to RRMS, and patients with RRMS may subsequently follow an SP course. In addition, patients with SP or even PPMS might have ongoing radiologic or possibly even clinical activity.

Tintoré (2008) described a large cohort of patients presenting with CIS and followed for 20 years. Over the first 10-year follow-up period, nearly 80% of patients with more than one T2 lesion on MRI and nearly 90% of patients with more than three T2 lesions developed CDMS. In contrast, only 11% of patients without T2 lesions on baseline MRI "converted" to CDMS. By 14 years of follow-up, nearly 90% of patients with at least one T2 lesion on baseline MRI converted to CDMS. Several independent risks factors for conversion to MS have been identified: young age (Mowry et al. 2009), presence of cognitive impairment at onset (Feuillet et al. 2007), genetic factors such as HLA-DRB1 (Zhang et al. 2011), and vitamin D deficiency (Martinelli et al. 2014). As shown in Tintoré's (2008) work, the most significant predictor of conversion to MS from CIS is the presence of brain abnormalities on baseline MRI, with number, location, and activity of the lesions all providing prognostic information.

Scalfari et al. (2014) recently provided a review of the London Ontario MS database, which evaluated 806 patients annually or semiannually for 28 years (shortest follow-up = 16 years). None of the patients received Disease modifying therapies (DMTs). At the end of the study period, 66.3% of patients had developed an SP course. The authors demonstrated that the rate of conversion to SPMS increases proportionally to disease duration. However, they highlighted the fact that individual prognosis was highly variable. About 25% of patients will become progressive within 5 years of onset of the disease, while on the opposite end of the spectrum, 25% of patients will remain RR at 15 years. This natural history study confirmed previous findings suggesting that male sex (Vukusic and Confavreux 2003) and older age of onset (Stankoff et al. 2007) were significant risk factors for conversion to SPMS.

The role of early clinical activity in the probability and latency of secondary progression is still unclear. Annual relapse rates remain the primary endpoint of many controlled clinical trials and are believed to serve as a surrogate for disability progression (Sormani et al. 2010). However, total relapse numbers were found to have little or no significant effect on the risk of progression, the latency to onset of the SP phase, or attainment of high disability levels (Kremenchutzky et al. 2006; Scalfari et al. 2010).

Physical disability in the clinical setting or in research trials can be assessed using the Expanded Disease Severity Scale (EDSS), which quantifies disability in eight functional systems. EDSS is an ordinal scale with values ranging from 0 (normal neurological examination) to 10 (death due to MS). In a recent publication, Tintore et al. (2015) performed multivariate analyses incorporating not only demographic and clinical data, but also MRI and biological variables to determine the risk of attaining EDSS 3.0 in individual patients. Their comprehensive work on a prospective cohort of 1015 patients with CIS highlights the importance of radiological and biological metrics to more accurately assess early risk of disability.

Beyond the early stages of the disease, focal MS pathology appears less relevant to disease progression. Particularly, once a threshold of disability is reached, progression may not be influenced by relapses either before or after onset of the SP phase (Confavreux et al. 2003). Leray and colleagues (2010) proposed the concept of MS as a two-stage disease. The early phase is defined from clinical onset to irreversible EDSS 3.0 and is thought to be mainly dependent on focal damage in the WM. The second or late phase, from EDSS 3.0 to EDSS 6.0, is thought to be independent of focal inflammation and may instead be related to diffuse inflammatory and neurodegenerative changes. The authors were able to show that disability progression in the first phase of MS does not influence progression during the second phase, although it was able to delay time to second phase. The duration of the early phase was found to be highly variable, while the duration of the late phase was remarkably constant (Leray et al. 2010).

6.1.2 PATHOPHYSIOLOGY OF MS

The immune system is an essential mediator in MS disease pathology. Ultimately, over the course of the disease, inflammatory demyelination, loss of protective support of the myelin sheath, and loss of trophic support of oligodendrocytes to the axons lead to chronic demyelination, gliosis, axonal loss, and neurodegeneration, which manifests as progressive neurological dysfunction in patients (Franklin et al. 2012; von Büdingen et al. 2015). Both innate and adaptive immune responses play important roles in initiating injury and in disease progression. Indeed, there might be preferential roles for each immune arm in different disease stages.

6.1.2.1 Adaptive Immune Response

Adaptive immune responses are largely governed through the interplay between T and B lymphocytes. T lymphocytes are further divided into multiple helper (CD4$^+$) and cytotoxic T (CD8$^+$) cells. T and B cells express unique antigen-specific surface receptors (T cell [TCR] and B cell [BCR] receptors, respectively). Unique TCR and BCR are assembled by somatic rearrangement of genomic elements with random nucleotide insertions and can theoretically yield more than 10^{15} unique receptors, which after selection results in more than 25 million distinct clones (Arstila et al. 1999). B cell clones can adapt receptors during affinity maturation, resulting in potentially greater numbers of BCR clones (Eisen 2014). B cells can directly bind antigen, while T cells require antigenic peptides to be processed by antigen-presenting cells (APCs) and are presented bound with HLA. In addition to numerous innate immune cells, B cells can function as APCs. Most important to MS pathology, each TCR and BCR can recognize more than one antigen (antigenic polyspecificity), potentially leading to autoimmunity through molecular mimicry (Gran et al. 1999).

Autoreactive CD4$^+$ T cells are known to be a key player in experimental autoimmune encephalitis (EAE), an important mouse model of MS. In most MS models, effector CD4$^+$ cells that enhance inflammatory processes are either of the T helper

1 type (Th1) that secretes interferon γ (IFNγ) and IL2, or Th17 that secretes IL17, IL21, and IL22. By contrast, Th2-type CD4+ cells downregulate inflammation via secretion of IL4, IL5, IL10, and IL13. Subpopulations of regulatory T cells (Tregs), both induced in the periphery or originating in the thymus, are also CD4+ cells that play a prominent role in immune regulation and maintaining homeostasis (Pankratz et al. 2016).

Finally, in addition to helper T cells, cytotoxic T cells (CD8+ cells) are present in MS brain lesions, although their role in disease pathology has been controversial. Activated CD8+ cells are primed against antigen in the context of HLA class I and are directly cytotoxic. However, these cells may also serve a regulatory role. CD8+ T cell depletion prior to EAE induction results in worsened disease (Najafian et al. 2003).

6.1.2.2 Innate Immune Response

Innate immune responses are mediated through cells of myeloid origin, including dendritic cells (DCs), monocytes, macrophages, natural killer (NK) cells, granulocytes, and mast cells. Microglia and astrocytes are innate immune cells resident in the CNS without direct counterparts in the periphery, and might be involved in the pathology of progressive MS (Correale and Farez 2015).

Innate immune cells respond to diverse stimuli using an array of pattern recognition receptors (PRRs) that bind to diverse pathogen-associated molecular patterns (PAMPs). PRRs also recognize self-molecules such as heat-shock proteins, double-stranded DNA, and purine metabolites released after cell damage or death. Responses to endogenous host molecules may trigger inflammatory reactions, and therefore play an important role in autoimmunity.

6.1.2.2.1 Astrocytes

Astrocytes, the most abundant of brain cells, are distributed in both gray and white matter and serve various functions, including (1) formation and maintenance of the BBB and glial limitans, (2) regulation of local blood flow through prostaglandin E and water homeostasis through aquaporin 4, (3) trophic support for neurons and their processes, and (4) immune regulation through release of chemokines or cytokines (Lundgaard et al. 2014; Cheslow and Alvarez 2016).

Astrocytes can mediate innate immune responses through several mechanisms, as they express diverse PRRs. At the BBB, astrocytes have direct control of cell entry into the CNS. Astrocytes regulate expression of adhesion molecules, particularly intercellular adhesion molecule-1 (ICAM-1) and vascular cell adhesion molecule-1 (VCAM-1), which bind to lymphocyte receptors, such as lymphocyte function–associated antigen-1 (LFA-1) and antigen-4 (VLA4), respectively. In addition, astrocytes can regulate passage of immune cells through BBB by releasing factors such as IL6, IL1β, TNFα, and transforming growth factor β (TGFβ) that affect endothelial cells and tight junctions.

Moreover, astrocytes help to orchestrate immune-mediated demyelination and neurodegeneration by secreting different chemokines, such as CCL2 (MCP-1), CCL5 (RANTES), IP-10 (CXCL10), CXCL12 (SDF-1), and IL8 (CXCL8), which attract both peripheral immune cells (e.g., T cells, monocytes, and DCs) and as resident CNS cells (microglia) to lesion sites.

Astrocyte morphology and responses are determined by the state of injury. Inflammatory injury in MS can be either active or inactive. Activity can be subtle (prelesional), as seen in normal-appearing white matter (NAWM) or dirty-appearing white matter (DAWM), or clearly evident, as focal lesions. Similarly, inactive or chronic lesions can either be completely gliotic or have an inactive core and active rim.

In lesional tissue, astrocytes play both pro-inflammatory and regulatory roles. Increases in pro-inflammatory cytokines augment inflammatory injury and encourage glial scar formation, which inhibits remyelination and axon regeneration (Lassmann 2014a). Astrocytes may affect both the number and the phenotype of T cells present in the CNS. Astrocytes secrete certain cytokines that have the potential of committing T cells to a pro-inflammatory phenotype (Th1 and Th17) or to a regulatory phenotype (Treg). It has been shown that activated astrocytes secrete compounds with toxic effects on neurons, axons, and oligodendrocytes or myelin, including reactive oxygen and nitrogen species, ATP, and glutamate (Brosnan et al. 1994; Liu et al. 2001; Stojanovic et al. 2014). By contrast, regulatory cytokines secreted by astrocytes function to orchestrate macrophage and microglial-mediated clearance and provide support and protection for oligodendrocytes and neurons (Correale and Farez 2015).

Additionally, trophic factors such as ciliary neurotrophic factor, vascular endothelial growth factor (VEGF), insulin-like growth factor-1 (IGF-1) and neurotrophin-3 are important mediators for cellular support and remyelination.

6.1.2.2.2 Microglia

Microglia are the resident macrophages of the CNS and provide predominantly homeostatic function. Microglia share many macrophage functions, making it challenging to separate these cell types in CNS diseases. These "resting" microglia, at times referred to as an M0 phenotype, are important for debris clearance and secrete neurotrophic factors such as IGF-1 and brain-derived neurotrophic factor (BDNF). Resident microglia can also become "activated" with neurodegeneration, injury, or inflammation. Activated microglia, analogous to macrophage or monocytes in the periphery, can adopt either an M1 or M2 phenotype. Chhor et al. (2013) propose that M1 microglia secrete proinflammatory cytokines, including IL1, IL2, IFNγ, CXCL9, and CXCL10, which augment CD8$^+$ T cell and CD4$^+$ Th1 function. In contrast, M2 microglia can have various functions that are immune regulatory and anti-inflammatory. M2a cells function in repair and regeneration and express immune-regulatory molecules such as TGFβ. M2b/c microglia function as a "deactivating" phenotype and express various anti-inflammatory markers, such as IL4, IL10, and CXCL13.

Microglial activation occurs diffusely in normal-appearing WM and GM and is not necessarily restricted to MS lesions. Activated microglia also predominate at the edge of active lesions, likely worsening demyelination and tissue injury, contributing to an expanding lesion. As the disease advances, perilesional microglia and macrophages have been shown to accumulate iron liberated from oligodendroglial damage (Mehta et al. 2013). Iron overload in perilesional microglia promotes a proinflammatory M1 phenotype and might promote formation of redox radicals contributing to mitochondrial dysfunction and potentially disease progression (see Section 6.1.2.6.2) (Lassmann 2014a).

6.1.2.3 Focal Demyelination, Inflammation, and Neurodegeneration

The pathological hallmark of the disease is perivenular inflammation, associated with damage to the BBB, and demyelination resulting in the formation of WM plaques. WM plaques occurring in eloquent brain areas, regions important to clinical function, present as a clinical relapse. In the early stages of the disease, active WM tissue demyelination within plaques is associated with significant inflammation, BBB damage, and microglial activation. Inflammatory infiltrates composed of clonally activated T and B cells are characteristically detected around postcapillary venules or scattered throughout the brain parenchyma and correlate with the degree of demyelination in focal active lesions (Babbe et al. 2000). Remyelination of focal lesions, more extensive in animal models of the disease, is limited in a majority of MS patients. A study of 168 WM lesions showed that only 22% were completely remyelinated as "shadow plaques," 73% were partially remyelinated, and 5% were completely demyelinated (Patani et al. 2007).

In addition to focal demyelination, axonal transection has been shown to occur early in disease (Trapp et al. 1998; Kuhlmann et al. 2002). Axonal transection occurs not only as a direct result of acute inflammatory injury, but also due to indirect membrane dysfunction. Activated T cells initiate a pro-inflammatory cascade resulting in the production of IFNγ, macrophage activation, and production of peroxinitrate products, such as nitric oxide (NO). NO is a potent mitochondrial inhibitor. Excitotoxicity due to increased release of glutamate by microglial cells or macrophages during the inflammatory process may further hinder mitochondrial function. Glutamate release leads to overstimulation of glutamate receptors on the postsynaptic membrane of neurons, loss of calcium homeostasis, and increased intracellular calcium, leading to cytoskeleton disruption, all of which contribute to loss of axonal integrity (Su et al. 2009).

6.1.2.3.1 Evaluating WM Damage In Vivo

Focal damage to the WM is well appreciated using MRI. T2-weighted sequences can detect WM plaques with great sensitivity. As an adjunct to the clinical exam, MRI can help detect subclinical disease activity by the presence of contrast-enhancing lesions or the presence of new or enlarging lesions on serial scans. These markers of disease activity are particularly useful to the clinician to evaluate the response to therapies that currently target the inflammatory process of the disease (see discussion in Section 6.1.3.2).

However, conventional MRI cannot distinguish WM lesions that are fully or partly remyelinated from fully demyelinated ones. Remyelination may promote short-term neuronal function recovery and help prevent subsequent axonal degeneration, possibly via trophic effects of axon-myelin interactions (Franklin et al. 2012). A recent longitudinal PET study of MS patients using the radiotracer (Levin et al. 2005) PIB, a thioflavine derivative sensitive to changes in tissue myelin content, showed that patient-specific remyelination potential was strongly associated with clinical scores (Bodini et al. 2016).

6.1.2.4 Diffuse White Matter Damage

In the progressive phase of the disease, inflammation becomes much less pronounced within plaques. Overall, the percentage of an individual's lesions that are active declines as the disease evolves (Frischer et al. 2015). Lesions are either inactive or slowly expanding at the edges and frequently fail to enhance with contrast.

A characteristic feature of progressive MS is diffuse pathology of brain tissue, outside of focal lesions. Abnormalities have been described in the so-called NAWM, that is, WM tissue that appears normal on both gross examination and MRI (Mahad et al. 2015). Despite its normal appearance, as much as 75% of NAWM has been found to be histologically abnormal (Allen and McKeown 1979). Areas of DAWM have also been characterized on MRI as having an intensity higher than that of the NAWM, but lower than that of focal lesions. DAWM (Figure 6.4) can be found in direct proximity of focal lesions or in locations not related to WM lesions and may represent a separate pathologic entity (Seewann et al. 2009). Within these regions, axonal pathology is evident by the presence of axonal swellings, axonal end bulbs, and degenerating axons.

Scattered microglial activation is another significant component of NAWM pathology and is profound at the later stages of the disease. Microglial cells are the resident macrophages of the CNS and can be activated following tissue injury (Ciccarelli et al. 2014). Once activated, these cells can either be protective or drive the degenerative process of the disease.

Finally, both meningeal inflammation, present at all stages of the disease, and Wallerian

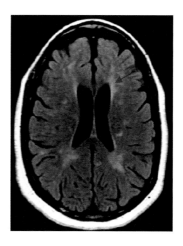

Figure 6.4 DAWM areas of intermediate signal intensity between those of focal lesions and NAWM.

degeneration may influence the degree of diffuse WM damage (Seewann et al. 2009).

6.1.2.5 Gray Matter Demyelination

Unlike WM lesions, demyelination of cortical neurons is not visible macroscopically in postmortem samples. In their seminal study, Brownell and Hughes (1962) showed that about 22% of all brain lesions were located at least partly in the cerebral cortex, and an additional 4% in the deep gray matter (DGM) structures. Immunocytochemical staining of myelin proteins has shown more extensive GM demyelination than initially suspected. Recent pathological studies reported that the extent of GM demyelination often exceeds that of the WM in progressive patients (Gilmore et al. 2009). GM demyelination is particularly extensive in the spinal cord, cerebellum, cingulate gyrus (Gilmore et al. 2009), thalamus (Vercellino et al. 2009), and hippocampus (Dutta et al. 2013) and likely contributes to the spectrum of both physical and cognitive MS symptoms.

Lesions found in the MS GM differ strikingly from their WM counterparts. Lymphocyte infiltration, complement deposition, and BBB disruption, all typical pathological hallmarks of WM lesions, are not usually found in cortical lesions (CLs).

Three different types of CLs have been described, leukocortical, intracortical, and subpial, based on their location and extent (Peterson et al. 2001; Bø et al. 2003). Leukocortical lesions consist of WM lesions that extend into the GM.

Intracortical lesions project along vessels within the cortical ribbon. Subpial lesions are band-like plaques that extend from the pial surface into cortical layer 3 or 4 and can involve several gyri.

At the earliest stages of the disease, leukocortical lesions are generally inflammatory in nature (Lucchinetti et al. 2011), with predominantly perivascular CD3+ and CD8+ T cell infiltrates and less commonly B cell infiltrates. These differ from CLs found at the latter stages of the disease, which are more frequently subpial and less inflammatory. It has been suggested that GM demyelination could be due to myelinotoxic factors diffusing from meninges. The presence of these meningeal B cell follicles has been associated with more extensive cortical damage and disease severity (Magliozzi et al. 2007).

6.1.2.6 Neurodegeneration

As previously described, degenerative changes in axons within acute WM lesions or NAWM are well documented. Similarly, postmortem studies have provided evidence of early and evolving GM injury. Neuronal loss was seen in chronic lesions without significant inflammation, suggesting that this phenomenon may not be directly linked to immune insult, but rather a consequence of chronic injury. Wegner et al. (2006) quantified neuronal damage in the MS neocortex. The authors found a 10% reduction in mean neuronal density in leukocortical lesions compared with normally myelinated cortex, with a decrease in neuronal size and significant changes in neuronal shape (Vercellino et al. 2005). Synaptic loss was significant in lesional cortex, suggesting that loss of dendritic arborization is an important feature in MS (Wegner et al. 2006). Pathologic changes in neuronal morphology, as well as reduced neuron size and axonal loss, were also detected in normal-appearing cortex compared with controls (Wegner et al. 2006; Popescu et al. 2015).

The neurodegenerative changes seen in the MS cortex are more subtle than those described in DGM structures, particularly the thalamus. Unlike neocortical structures, neuronal density in DGM was decreased in both demyelinated and nondemyelinated regions, although more pronounced in demyelinated areas (Vercellino et al. 2009). Neuronal atrophy and morphologic changes were also detected in the MS DGM regardless of myelination status and may precede or accompany neuronal loss. For example, in the hippocampus, neuronal counts were decreased by up to 30% depending on location (Papadopoulos et al. 2009). Dutta et al. (2011) reported substantial reduction in synaptic density in the hippocampus and found decreased expression of neuronal proteins involved in axonal transport, synaptic plasticity, and neuronal survival. These findings may explain, at least partly, some of the cognitive deficits observed in MS patients.

The mechanisms underlying neuronal pathology remain to be fully established. Of particular interest is the interplay between WM and GM pathology. It has been suggested that loss of myelin and reduction in axonal density observed diffusely in the NAWM plays a role in the neurodegenerative process by promoting retrograde or transsynaptic degeneration. Recent studies have provided evidence of neuronal dysfunction in connected GM neurons and correlated loss of integrity of WM tracts to histopathological measures of neurodegeneration in corresponding GM structures (Kolasinski et al. 2012). This is further supported by reports of tract-specific associations between cortical thinning patterns and MRI-derived metrics of NAWM integrity (Bergsland et al. 2015) and suggests a link between diffuse damage of the WM and neurodegenerative processes in connected GM.

Some argue that WM pathology cannot satisfactorily explain the full extent of diffuse GM damage observed in MS (Calabrese et al. 2015). Indeed, despite the relationship, neuronal damage may also occur independently of WM pathology. Neuronal changes in nondemyelinated areas have been reported both in the neocortex and in subcortical GM structures (Wegner et al. 2006; Klaver et al. 2015; Popescu et al. 2015), suggesting that focal GM demyelination and neurodegeneration are at least partly distinct phenomena in progressive MS.

6.1.2.6.1 Meningeal Follicles

A number of studies have drawn attention to the inflammatory process occurring in the meningeal compartment. In a proportion of patients with progressive MS, meningeal inflammation is precipitated by B cell follicles (Magliozzi et al. 2007; Howell et al. 2011). These lymphoid structures appear to spatially coincide with subpial demyelinating lesions and are associated with a quantitative increase in microglial activation within the GM (Howell et al. 2011). SPMS cases with B cell follicles presented a more severe disease course, with younger age at onset, younger age at irreversible disability, and earlier death, emphasizing the clinical significance of these findings (Magliozzi et al. 2007). The link between meningeal inflammation and GM damage is further corroborated by studies pointing to a specific role exerted by both meningeal T cells and activated microglia in diffuse axonal loss in the spinal cord (Androdias et al. 2010) and strengthen the hypothesis that meningeal inflammation is implicated in neurodegeneration in MS and contributes to clinical severity and progression.

6.1.2.6.2 Mitochondrial Dysfunction

Meningeal inflammatory cells and activated microglia in the GM induce the production of both oxygen and NO species, as well as peroxinitrates by enzymes, including nicotinamide adenine dinucleotide phosphate oxidase (Fischer et al. 2013). Reaction oxygen and nitrogen species amplify mitochondrial dysfunction and energy failure, which are increasingly being recognized as major pathways of neurodegeneration in MS (Lassmann 2014b; Witte et al. 2014). Neurons in MS GM exhibit decreased respiratory chain function, creating a mismatch between energy demand and ATP supply, thought to drive neuronal dysfunction or degeneration via excessive stimulation of calcium-dependent degradative pathways (Trapp and Stys 2009).

In addition to calcium-dependent processes, sodium channel redistribution along denuded axons can aggravate this imbalance by significantly increasing energy demand in a context of supply deficit, leading to a state of "virtual hypoxia" (Stys 2004).

Finally, iron stored within oligodendrocytes and myelin sheaths may be liberated following demyelination. In its extracellular form, iron generates reactive oxygen species and contributes actively to oxidative damage. The physiologic accumulation of iron is well described and plateaus around the fifth or sixth decade. In MS, accumulation might amplify neurodegenerative processes beyond those observed with age.

6.1.2.6.3 Cerebral Perfusion

Decreases in cerebral blood flow, thought to be mediated via release of vasoconstrictive peptides, such as endothelin-1 (ET-1) by activated astrocytes (D'haeseleer et al. 2015), has been reported in MS patients (Steen et al. 2013; Debernard et al. 2014; Narayana et al. 2014). Cerebral hypoperfusion might play a role in lesion formation (Lucchinetti et al. 2000), axonal and neuronal damage, and consequently, disability progression (Aviv et al. 2012; Francis et al. 2013). Our lab is currently evaluating the impact of therapies aimed at improving regional perfusion on disability accrual in MS.

Many well-conducted studies of postmortem tissue have shown that GM damage dominates the pathological process in progressive MS. These studies have demonstrated the clinical significance of the degenerative process occurring in the MS GM and underscored the need to understand its causes.

6.1.3 TREATMENT STRATEGIES IN MS

6.1.3.1 Overview of Treatments: Mechanisms of Action

There are now 13 therapies approved by the Food and Drug Administration (FDA) for the treatment of MS. Of these molecules, five belong to a class known as IFNs (either INF β 1a or 1b). IFNs are a group of cytokine products that perform fundamental physiologic functions. Two types of IFNs, α and β, have been evaluated in MS (Panitch et al. 2002, 2005; Kieseier 2011; Freedman 2014). While the exact mechanisms in the human have yet to be fully detailed, it is believed that β-IFNs modulate the interplay between pro-inflammatory and regulatory cells (Kieseier 2011). These compounds have been shown to increase anti-inflammatory cytokines such as IL10 and IL4, while decreasing the pro-inflammatory cytokines IL17, IFNγ, and TNF. In addition, IFNβ likely reduces cell trafficking across the BBB (Kieseier 2011). The second class of molecules is glatiramer acetate (GA), a proprietary mixture of four amino acids, tyrosine, glutamate, alanine, and lysine, in specific amounts that is believed, among other mechanisms, to enhance regulatory T cell function (Scott 2013). Arnon and Aharoni (2004) showed GA-specific Th2 cells as a key mechanism for the beneficial effects of GA in EAE. GA-specific Th2 cells isolated from treated EAE animals were shown to confer protection from EAE to untreated animals, in part by secreting anti-inflammatory cytokines such as IL4. These first two classes of molecules, typically referred to as platform agents, are the oldest therapies approved to treat MS and have been shown to be effective at reducing disease activity, as observed both clinically and radiologically (Panitch et al. 2005; Comi et al. 2012; Freedman 2014). Clinical relapse activity was shown to be reduced by about 30%, and radiologic activity by about 60%. Generally, the platform agents are well tolerated with minimal short- and long-term side effects (Freedman 2014).

There are currently three oral therapies to treat relapsing MS. Fingolimod is a sphingosine phosphate antagonist that binds sphingosine-1-phosphate (S1P) receptors, predominantly S1P1. S1P receptors are a group of cell surface molecules involved in the egress of naïve and central memory lymphocytes from lymph nodes. Activation of S1P1 results in reduced receptor expression, lymphocyte egress from the node, and circulating lymphocyte counts. In contrast to naïve and central memory cells, effector cells resident in tissue are less likely to migrate to lymph nodes and are less commonly reduced. Fingolimod may also have effects on cytokine signaling and cell activation (Xia and Wadham 2011). Fingolimod has been shown in a large 2-year randomized placebo-controlled study to reduce annualized clinical relapse rate by 54% (0.18 vs. 0.4) and radiologic activity, measured as gadolinium-enhancing lesion number, by 82% (0.2 vs. 1.1) (Kappos et al. 2010; Radue 2012). While the drug is generally well tolerated with minimal short-term safety concerns, the long-term safety has yet to be fully evaluated (Fonseca 2015; Dubey et al. 2016).

Teriflunomide is a dihydroorotate dehydrogenase (DHOHD) inhibitor purported to decrease activated lymphocyte numbers (Cherwinski et al. 1995; Rückemann et al. 1998). DHODH is a mitochondrial enzyme necessary for the *de novo* pyrimidine synthesis pathway. Rapidly dividing cells involved in MS pathology, such as lymphocytes and macrophages, require de novo synthesis of pyrimidine, as enough cannot be obtained from the salvage pathway. As such, teriflunomide purportedly preferentially decreases activity of cells involved in MS pathology (Gold and Wolinsky 2011). Teriflunomide has been shown in a large 2-year randomized placebo-controlled study to reduce the annualized clinical relapse rate by 31% (0.37 vs. 0.54) and radiologic activity by 80% (0.26 vs. 1.33) (O'Connor et al. 2011; Wolinsky et al. 2013). Teriflunomide is generally well tolerated

with minimal short-term side effects (Miller 2015). While this drug is relatively new, leflunomide, the parent molecule, has been approved for rheumatoid arthritis for more than a decade with few long-term side effects (Ishaq et al. 2011).

In preclinical models, dimethyl fumarate (DMF) has been shown to have beneficial effects on neuroinflammation and oxidative stress mediated through activation of the nuclear 1 factor (erythroid-derived 2)-like 2 (Nrf2) antioxidant pathway (Linker and Gold 2013). DMF is metabolized to monomethyl fumarate and exerts its effect in the cytoplasm. Nrf2 is typically upregulated in response to oxidative stress and translocated to the nucleus, where it activates several genes involved in cell survival (Albrecht et al. 2012). *In vitro* and *in vivo* studies suggest that fumaric acid esters shift cytokine production from a Th1 to a Th2 pattern (de Jong et al. 1996). DMF has been shown in a large 2-year randomized placebo-controlled study to reduce the annualized clinical relapse rate by 47% (0.17 vs. 0.36) and radiologic activity by 90% (0.1 vs. 1.8) (Gold et al. 2012). As with the other oral agents, DMF is generally well tolerated with few short-term side effects, predominantly gastrointestinal (Gold et al. 2012). However, the long-term safety profile of both fingolimod and DMF has yet to be fully evaluated.

Finally, there are two approved intravenous therapies. Natalizumab is an α-4 integrin antagonist purported to decrease cellular trafficking into tissues (Polman et al. 2006). Natalizumab inhibits α-4-mediated adhesion of leukocytes to associated receptors, such as VCAM-1, on the vascular endothelial surface. Receptor blockade results in reduced leukocyte extravasation through the BBB. In addition, within the brain, natalizumab might further inhibit recruitment and activity of various pro-inflammatory cells involved in lesion formation (Drews 2006). Natalizumab has been shown in a large 2-year randomized placebo-controlled study to reduce the annualized relapse rate by 67% (0.22 vs. 0.67) and radiologic activity by 92% (0.1 vs. 1.2) (Polman et al. 2006).

Alemtuzumab is a humanized CD52 antagonist that depletes circulating T and B lymphocytes through antibody-dependent cellular cytolysis (ADCC), leading to changes in the number, proportion, and function of lymphocyte subsets (Cox et al. 2005; Thompson et al. 2009). Repopulation of cells can take many months, with a potential for reduced myelin-specific lymphocyte subsets (Hill-Cawthorne et al. 2012). In addition to decreasing Th1 and cytotoxic T cells, the proportion of regulatory T cell subsets was shown to increase after treatment. Unlike the previously mentioned drugs that were compared against placebo, this drug has been compared against a thrice-weekly IFN (active comparator study) and shown to reduce the annualized relapse rate by 54% (0.18 vs. 0.39). The percentage of subjects in the study with gadolinium activity was reduced from 19% for thrice-weekly IFN to 7% for alemtuzumab (Cohen et al. 2012). Compared with platform and oral therapies, both intravenous therapies are generally considered to have potentially greater short-term and long-term side effects.

6.1.3.2 MS Phenotypes: Impact on Treatment Choice

As outlined before, previously defined clinical MS phenotypes have been revised, updating "active disease" to include clinical and/or radiographic change (Lublin and Reingold 1996; Lublin et al. 2014). In addition, the concept of no evidence of disease activity (NEDA) has been incorporated into recent clinical studies (Arnold et al. 2014; Nixon et al. 2014). While neither the revisions to clinical phenotypes nor the aforementioned studies recommend treatment change based solely on radiographic activity, several clinicians embraced a "zero-lesion" approach to patient management. Routine monitoring of subclinical disease activity with annual MRI, at least for patients early in disease, was recommended in recent consensus statements by both Lublin and Traboulsee (Lublin et al. 2014; Traboulsee et al. 2016). Patients with clinical activity, defined as relapse or rapidly worsening disability, or radiologic activity, defined as contrast-enhancing or new or unequivocally enlarged T2 lesions, should at least be counselled on alternate treatment strategies.

6.1.4 FUTURE GOALS

Inflammation and resultant demyelination are important pathologic processes in both WM and GM areas of the brain. Demyelinated axons are susceptible to focal membrane channel remodeling, resulting in calcium-mediated excitotoxicity and Wallerian degeneration. Once axons are demyelinated, remyelination and repair are often slow and ineffective (Crawford et al. 2013; Mahad et al. 2015).

As previously outlined, there are now numerous therapies approved to treat relapsing MS. In the aggregate, these therapies have been shown to reduce inflammation, acute clinical activity, and resultant short-term disability, usually measured as 3-month disability progression. Despite effectively reducing inflammatory activity, none of the therapies have been shown to reduce long-term disability or treat degeneration, the designated second stage of MS. It is therefore likely that therapies directed at and encouraging remyelination or decreasing axonal degeneration are needed to impact progressive disease.

There are several preclinical and clinical studies focusing on targeting neurodegenerative processes in MS. While a complete overview is beyond the scope of this text, two general pathways deserve further discussion.

6.1.4.1 Remyelinating Therapies

Several molecules and methods have been shown in preclinical MS models to encourage remyelination of the denuded axon (Pepinsky et al. 2011; Crawford et al. 2013; Deshmukh et al. 2013). Of these, antibodies against the leucine-rich repeat and immunoglobulin (Ig) domain–containing Nogo receptor interacting protein (LINGO) have recently been evaluated in phase II clinical trials. The RENEW study evaluated anti-LINGO + IFNβ 1a given intramuscularly once weekly to patients presenting within 28 days of acute optic neuritis. In this study, all patients were treated with IFNβ 1a weekly, a currently FDA-approved therapy for relapsing MS, and randomized to receive 100 mg/kg anti-LINGO (BIIB033) or placebo every 4 weeks for 24 weeks from enrollment. Optic nerve myelination was evaluated by full-field and multifocal visual evoked potentials (ffVEP and mfVEP, respectively), where distal latency of the action potential is correlated with the degree of demyelination. Results recently presented at the last European Committee for Treatment and Research in Multiple Sclerosis (ECTRIMS) meeting (ECTRIMS 2015, Barcelona) showed a significant improvement in latency for both ffVEP and mfVEP in favor of the treatment arm, suggesting that short-term therapy with BIIB033 improved remyelination after acute inflammatory injury.

6.1.4.2 Neuroprotection Strategies

Increases in expressed membrane channels, both Na^{2+} and K^+, are reported in demyelinated axons (Trapp and Stys 2009; Mahad et al. 2015). Increased sodium–potassium and sodium–calcium exchangers also increased energy utilization, leading to imbalances in supply and demand, causing the "virtual hypoxia" previously discussed. Several small molecules have been studied that might "stabilize" membranes and possibly be "neuroprotective." Of these, antiepileptic drugs such as lamotrigine and phenytoin have been evaluated in small clinical studies (Kapoor et al. 2010; Raftopoulos et al. 2016).

The clinical trial using lamotrigine failed to show benefit at reducing disability progression in SPMS patients and had mixed results for both clinical and imaging outcome measures. For example, lamotrigine treatment seemed to be associated with greater cerebral volume loss in the first year, suggesting a negative effect of treatment, while clinical measures of lower-extremity mobility (timed 25-foot walk) was improved for patients on lamotrigine, suggesting a positive effect of treatment (Kapoor et al. 2010).

A randomized placebo-controlled study using phenytoin as an adjunct therapy in acute optic neuritis, similar in design to the RENEW study previously described, showed a 30% reduction in the loss of retinal nerve fiber layer (RNFL) thickness, a measure reflecting axonal injury of ganglion cells, 6 months after acute injury (Raftopoulos et al. 2016). While the study was small, it supports proof-of-concept data that membrane-stabilizing therapies might function to "protect" the damaged axon and/or cell body from secondary degeneration.

It is unlikely that many preclinical and early-phase clinical studies will show similar benefits when evaluated in larger multicenter phase III studies. However, targeting mechanisms of neurodegeneration and remyelination will be necessary to decrease clinical disability progression, and likely is an important next step in expanding the MS treatment arsenal.

6.1.5 CONCLUSIONS

MS is a complex and devastating CNS disease. While early studies suggested immune mechanisms of focal injury, it seems more probable that both inflammatory and degenerative mechanisms are involved in disease pathology as either related, interdependent, or independent processes. While we now have many treatment options to suppress inflammation, the next wave of research and resultant therapies will focus on combating disability progression by targeting mechanisms involved in neurodegeneration and repair.

REFERENCES

Albrecht, P., I. Bouchachia, N. Goebels, N. Henke, H. H. Hofstetter, A. Issberner, Z. Kovacs et al. 2012. Effects of Dimethyl Fumarate on Neuroprotection and Immunomodulation. *Journal of Neuroinflammation* 9: 163.

Allen, I. V., and S. R. McKeown. 1979. A Histological, Histochemical and Biochemical Study of the Macroscopically Normal White Matter in Multiple Sclerosis. *Journal of the Neurological Sciences* 41 (1): 81–91.

Androdias, G., R. Reynolds, M. Chanal, C. Ritleng, C. Confavreux, and S. Nataf. 2010. Meningeal T Cells Associate with Diffuse Axonal Loss in Multiple Sclerosis Spinal Cords. *Annals of Neurology* 68 (4): 465–76.

Arnold, D. L., P. A. Calabresi, B. C. Kieseier, S. I. Sheikh, A. Deykin, Y. Zhu, S. Liu, X. You, B. Sperling, and S. Hung. 2014. Effect of Peginterferon Beta-1a on MRI Measures and Achieving No Evidence of Disease Activity: Results from a Randomized Controlled Trial in Relapsing-Remitting Multiple Sclerosis. *BMC Neurology* 14: 240.

Arnon, R., and R. Aharoni. 2004. Mechanism of Action of Glatiramer Acetate in Multiple Sclerosis and Its Potential for the Development of New Applications. *Proceedings of the National Academy of Sciences of the United States of America* 101 (Suppl 2): 14593–98.

Arstila, T. P., A. Casrouge, V. Baron, J. Even, J. Kanellopoulos, and P. Kourilsky. 1999. A Direct Estimate of the Human Alphabeta T Cell Receptor Diversity. *Science (New York, N.Y.)* 286 (5441): 958–61.

Aviv, R. I., P. L. Francis, R. Tenenbein, P. O'Connor, L. Zhang, A. Eilaghi, L. Lee, T. J. Carroll, J. Mouannes-Srour, and A. Feinstein. 2012. Decreased Frontal Lobe Gray Matter Perfusion in Cognitively Impaired Patients with Secondary-Progressive Multiple Sclerosis Detected by the Bookend Technique. *AJNR: American Journal of Neuroradiology* 33 (9): 1779–85.

Babbe, H., A. Roers, A. Waisman, H. Lassmann, N. Goebels, R. Hohlfeld, M. Friese et al. 2000. Clonal Expansions of CD8(+) T Cells Dominate the T Cell Infiltrate in Active Multiple Sclerosis Lesions as Shown by Micromanipulation and Single Cell Polymerase Chain Reaction. *Journal of Experimental Medicine* 192 (3): 393–404.

Barkhof, F., M. Filippi, D. H. Miller, P. Scheltens, A. Campi, C. H. Polman, G. Comi, H. J. Adèr, N. Losseff, and J. Valk. 1997. Comparison of MRI Criteria at First Presentation to Predict Conversion to Clinically Definite Multiple Sclerosis. *Brain: A Journal of Neurology* 120 (Pt 11): 2059–69.

Bashinskaya, V. V., O. G. Kulakova, A. N. Boyko, A. V. Favorov, and O. O. Favorova. 2015. A Review of Genome-Wide Association Studies for Multiple Sclerosis: Classical and Hypothesis-Driven Approaches. *Human Genetics* 134 (11–12): 1143–62.

Bergsland, N., M. M. Laganà, E. Tavazzi, M. Caffini, P. Tortorella, F. Baglio, G. Baselli, and M. Rovaris. 2015. Corticospinal Tract Integrity Is Related to Primary Motor Cortex Thinning in Relapsing-Remitting Multiple Sclerosis. *Multiple Sclerosis (Houndmills, Basingstoke, England)* 21 (14): 1771–80.

Bø, L., C. A. Vedeler, H. Nyland, B. D. Trapp, and S. J. Mørk. 2003. Intracortical Multiple Sclerosis Lesions Are Not Associated with Increased Lymphocyte Infiltration. *Multiple Sclerosis (Houndmills, Basingstoke, England)* 9 (4): 323–31.

Bodini, B., M. Veronese, D. García-Lorenzo, M. Battaglini, E. Poirion, A. Chardain, L. Freeman et al. 2016. Dynamic Imaging of Individual Remyelination Profiles in Multiple Sclerosis. *Annals of Neurology* 79 (5).

Bot, J. C. J., F. Barkhof, C. H. Polman, G. J. Lycklama à Nijeholt, V. de Groot, E. Bergers, H. J. Ader, and J. A. Castelijns. 2004. Spinal Cord Abnormalities in Recently Diagnosed MS Patients: Added Value of Spinal MRI Examination. *Neurology* 62 (2): 226–33.

Brosnan, C. F., L. Battistini, C. S. Raine, D. W. Dickson, A. Casadevall, and S. C. Lee. 1994. Reactive Nitrogen Intermediates in Human Neuropathology: An Overview. *Developmental Neuroscience* 16 (3–4): 152–61.

Brownell, B., and J. T. Hughes. 1962. The Distribution of Plaques in the Cerebrum in Multiple Sclerosis. *Journal of Neurology, Neurosurgery, and Psychiatry* 25: 315–20.

Calabrese, M., R. Magliozzi, O. Ciccarelli, J. J. G. Geurts, R. Reynolds, and R. Martin. 2015. Exploring the Origins of Grey Matter Damage in Multiple Sclerosis. *Nature Reviews: Neuroscience* 16 (3): 147–58.

Cepok, S., D. Zhou, R. Srivastava, S. Nessler, S. Stei, K. Büssow, N. Sommer, and B. Hemmer. 2005. Identification of Epstein-Barr Virus Proteins as Putative Targets of the Immune Response in Multiple Sclerosis. *Journal of Clinical Investigation* 115 (5): 1352–60.

Cherwinski, H. M., D. McCarley, R. Schatzman, B. Devens, and J. T. Ransom. 1995. The Immunosuppressant Leflunomide Inhibits Lymphocyte Progression through Cell Cycle by a Novel Mechanism. *Journal of Pharmacology and Experimental Therapeutics* 272 (1): 460–68.

Cheslow, L., and J. I. Alvarez. 2016. Glial-Endothelial Crosstalk Regulates Blood-Brain Barrier Function. *Current Opinion in Pharmacology* 26: 39–46.

Chhor, V., T. Le Charpentier, S. Lebon, M.-V. Oré, I. L. Celador, J. Josserand, V. Degos et al. 2013. Characterization of Phenotype Markers and Neuronotoxic Potential of Polarised Primary Microglia In Vitro. *Brain, Behavior, and Immunity* 32: 70–85.

Ciccarelli, O., F. Barkhof, B. Bodini, N. De Stefano, X. Golay, K. Nicolay, D. Pelletier et al. 2014. Pathogenesis of Multiple Sclerosis: Insights from Molecular and Metabolic Imaging. *Lancet: Neurology* 13 (8): 807–22.

Cohen, J. A., A. J. Coles, D. L. Arnold, C. Confavreux, E. J. Fox, H.-P. Hartung, E. Havrdova et al. 2012. Alemtuzumab versus Interferon Beta 1a as First-Line Treatment for Patients with Relapsing-Remitting Multiple Sclerosis: A Randomised Controlled Phase 3 Trial. *Lancet (London, England)* 380 (9856): 1819–28.

Comi, G., N. De Stefano, M. S. Freedman, F. Barkhof, C. H. Polman, B. M. J. Uitdehaag, F. Casset-Semanaz et al. 2012. Comparison of Two Dosing Frequencies of Subcutaneous Interferon Beta-1a in Patients with a First Clinical Demyelinating Event Suggestive of Multiple Sclerosis (REFLEX): A Phase 3 Randomised Controlled Trial. *Lancet: Neurology* 11 (1): 33–41.

Confavreux, C., and S. Vukusic. 2006. Natural History of Multiple Sclerosis: A Unifying Concept. *Brain: A Journal of Neurology* 129 (Pt 3): 606–16.

Confavreux, C., S. Vukusic, and P. Adeleine. 2003. Early Clinical Predictors and Progression of Irreversible Disability in Multiple Sclerosis: An Amnesic Process. *Brain: A Journal of Neurology* 126 (Pt 4): 770–82.

Confavreux, C., S. Vukusic, T. Moreau, and P. Adeleine. 2000. Relapses and Progression of Disability in Multiple Sclerosis. *New England Journal of Medicine* 343 (20): 1430–38.

Correale, J., and M. F. Farez. 2015. The Role of Astrocytes in Multiple Sclerosis Progression. *Frontiers in Neurology* 6 (180).

Cox, A. L., S. A. J. Thompson, J. L. Jones, V. H. Robertson, G. Hale, H. Waldmann, D. A. S. Compston, and A. J. Coles. 2005. Lymphocyte

Homeostasis Following Therapeutic Lymphocyte Depletion in Multiple Sclerosis. *European Journal of Immunology* 35 (11): 3332–42.

Crawford, A. H., C. Chambers, and R. J. M. Franklin. 2013. Remyelination: The True Regeneration of the Central Nervous System. *Journal of Comparative Pathology* 149 (2–3): 242–54.

Debernard, L., T. R. Melzer, S. Van Stockum, C. Graham, C. A. Wheeler-Kingshott, J. C. Dalrymple-Alford, D. H. Miller, and D. F. Mason. 2014. Reduced Grey Matter Perfusion without Volume Loss in Early Relapsing-Remitting Multiple Sclerosis. *Journal of Neurology, Neurosurgery, and Psychiatry* 85 (5): 544–51.

de Jong, R., A. C. Bezemer, T. P. Zomerdijk, T. van de Pouw-Kraan, T. H. Ottenhoff, and P. H. Nibbering. 1996. Selective Stimulation of T Helper 2 Cytokine Responses by the Anti-Psoriasis Agent Monomethylfumarate. *European Journal of Immunology* 26 (9): 2067–74.

Deshmukh, V. A, V. Tardif, C. A. Lyssiotis, C. C. Green, B. Kerman, H. J. Kim, K. Padmanabhan et al. 2013. A Regenerative Approach to the Treatment of Multiple Sclerosis. *Nature* 502 (7471): 327–32.

D'haeseleer, M., S. Hostenbach, I. Peeters, S. El Sankari, G. Nagels, J. De Keyser, and M. B. D'hooghe. 2015. Cerebral Hypoperfusion: A New Pathophysiologic Concept in Multiple Sclerosis? *Journal of Cerebral Blood Flow and Metabolism: Official Journal of the International Society of Cerebral Blood Flow and Metabolism* 35 (9): 1406–10.

Drews, J. 2006. Case Histories, Magic Bullets and the State of Drug Discovery. *Nature Reviews: Drug Discovery* 5 (8): 635–40.

Dubey, D., C. A. Cano, and O. Stüve. 2016. Update on Monitoring and Adverse Effects of Approved Second-Generation Disease-Modifying Therapies in Relapsing Forms of Multiple Sclerosis. *Current Opinion in Neurology* 29 (3): 278–85.

Dutta, R., A. Chang, M. K. Doud, G. J. Kidd, M. V. Ribaudo, E. A. Young, R. J. Fox, S. M. Staugaitis, and B. D. Trapp. 2011. Demyelination Causes Synaptic Alterations in Hippocampi from Multiple Sclerosis Patients. *Annals of Neurology* 69 (3): 445–54.

Dutta, R., A. M. Chomyk, A. Chang, M. V. Ribaudo, S. A. Deckard, M. K. Doud, D. D. Edberg et al. 2013. Hippocampal Demyelination and Memory Dysfunction Are Associated with Increased Levels of the Neuronal microRNA miR-124 and Reduced AMPA Receptors. *Annals of Neurology* 73 (5): 637–45.

Eisen, H. N. 2014. Affinity Enhancement of Antibodies: How Low-Affinity Antibodies Produced Early in Immune Responses Are Followed by High-Affinity Antibodies Later and in Memory B-Cell Responses. *Cancer Immunology Research* 2 (5): 381–92.

Feuillet, L., F. Reuter, B. Audoin, I. Malikova, K. Barrau, A. Ali Cherif, and J. Pelletier. 2007. Early Cognitive Impairment in Patients with Clinically Isolated Syndrome Suggestive of Multiple Sclerosis. *Multiple Sclerosis (Houndmills, Basingstoke, England)* 13 (1): 124–27.

Fischer, M. T., I. Wimmer, R. Höftberger, S. Gerlach, L. Haider, T. Zrzavy, S. Hametner et al. 2013. Disease-Specific Molecular Events in Cortical Multiple Sclerosis Lesions. *Brain: A Journal of Neurology* 136 (Pt 6): 1799–815.

Fonseca, J. 2015. Fingolimod Real World Experience: Efficacy and Safety in Clinical Practice. *Neuroscience Journal* 2015: 389360.

Francis, P. L., R. Jakubovic, P. O'Connor, L. Zhang, A. Eilaghi, L. Lee, T. J. Carroll, J. Mouannes-Srour, A. Feinstein, and R. I. Aviv. 2013. Robust Perfusion Deficits in Cognitively Impaired Patients with Secondary-Progressive Multiple Sclerosis. *AJNR: American Journal of Neuroradiology* 34 (1): 62–67.

Franklin, R. J. M., C. Ffrench-Constant, J. M. Edgar, and K. J. Smith. 2012. Neuroprotection and Repair in Multiple Sclerosis. *Nature Reviews: Neurology* 8 (11): 624–34.

Freedman, M. S. 2014. Efficacy and Safety of Subcutaneous Interferon-β-1a in Patients with a First Demyelinating Event and Early Multiple Sclerosis. *Expert Opinion on Biological Therapy* 14 (8): 1207–14.

Frischer, J. M., S. D. Weigand, Y. Guo, N. Kale, J. E. Parisi, I. Pirko, J. Mandrekar et al. 2015. Clinical and Pathological Insights into the Dynamic Nature of the White Matter Multiple Sclerosis Plaque. *Annals of Neurology* 78 (5): 710–21.

Gilmore, C. P., I. Donaldson, L. Bö, T. Owens, J. Lowe, and N. Evangelou. 2009. Regional Variations in the Extent and Pattern of Grey Matter Demyelination in Multiple Sclerosis: A Comparison between the Cerebral Cortex, Cerebellar Cortex, Deep Grey Matter Nuclei and the Spinal Cord. *Journal of Neurology, Neurosurgery, and Psychiatry* 80 (2): 182–87.

Giovannoni, G., and G. Ebers. 2007. Multiple Sclerosis: The Environment and Causation. *Current Opinion in Neurology* 20 (3): 261–68.

Gold, R., L. Kappos, D. L. Arnold, A. Bar-Or, G. Giovannoni, K. Selmaj, C. Tornatore et al. 2012. Placebo-Controlled Phase 3 Study of Oral BG-12 for Relapsing Multiple Sclerosis. *New England Journal of Medicine* 367 (12): 1098–107.

Gold, R., and J. S. Wolinsky. 2011. Pathophysiology of Multiple Sclerosis and the Place of Teriflunomide. *Acta Neurologica Scandinavica* 124 (2): 75–84.

Gran, B., B. Hemmer, M. Vergelli, H. F. McFarland, and R. Martin. 1999. Molecular Mimicry and Multiple Sclerosis: Degenerate T-Cell Recognition and the Induction of Autoimmunity. *Annals of Neurology* 45 (5): 559–67.

Gregory, A. P., C. A. Dendrou, K. E. Attfield, A. Haghikia, D. K. Xifara, F. Butter, G. Poschmann et al. 2012. TNF Receptor 1 Genetic Risk Mirrors Outcome of Anti-TNF Therapy in Multiple Sclerosis. *Nature* 488 (7412): 508–11.

Gregory, S. G., S. Schmidt, P. Seth, J. R. Oksenberg, J. Hart, A. Prokop, S. J. Caillier et al. 2007. Interleukin 7 Receptor α Chain (IL7R) Shows Allelic and Functional Association with Multiple Sclerosis. *Nature Genetics* 39 (9): 1083–91.

Haahr, S., N. Koch-Henriksen, A. Møller-Larsen, L. S. Eriksen, and H. M. Andersen. 1995. Increased Risk of Multiple Sclerosis after Late Epstein-Barr Virus Infection: A Historical Prospective Study. *Multiple Sclerosis (Houndmills, Basingstoke, England)* 1 (2): 73–77.

Hernán, M. A., S. S. Jick, G. Logroscino, M. J. Olek, A. Ascherio, and H. Jick. 2005. Cigarette Smoking and the Progression of Multiple Sclerosis. *Brain: A Journal of Neurology* 128 (Pt 6): 1461–65.

Hill-Cawthorne, G. A., T. Button, O. Tuohy, J. L. Jones, K. May, J. Somerfield, A. Green et al. 2012. Long Term Lymphocyte Reconstitution after Alemtuzumab Treatment of Multiple Sclerosis. *Journal of Neurology, Neurosurgery, and Psychiatry* 83 (3): 298–304.

Holick, M. F. 2007. Vitamin D Deficiency. *New England Journal of Medicine* 357 (3): 266–81.

Howell, O. W., C. A. Reeves, R. Nicholas, D. Carassiti, B. Radotra, S. M. Gentleman, B. Serafini et al. 2011. Meningeal Inflammation Is Widespread and Linked to Cortical Pathology in Multiple Sclerosis. *Brain: A Journal of Neurology* 134 (Pt 9): 2755–71.

Ishaq, M., J. S. Muhammad, K. Hameed, and A. I. Mirza. 2011. Leflunomide or Methotrexate? Comparison of Clinical Efficacy and Safety in Low Socio-Economic Rheumatoid Arthritis Patients. *Modern Rheumatology/The Japan Rheumatism Association* 21 (4): 375–80.

Kapoor, R., J. Furby, T. Hayton, K. J. Smith, D. R. Altmann, R. Brenner, J. Chataway, R. A. C. Hughes, and D. H. Miller. 2010. Lamotrigine for Neuroprotection in Secondary Progressive Multiple Sclerosis: A Randomised, Double-Blind, Placebo-Controlled, Parallel-Group Trial. *Lancet: Neurology* 9 (7): 681–88.

Kappos, L., E.-W. Radue, P. O'Connor, C. Polman, R. Hohlfeld, P. Calabresi, K. Selmaj et al. 2010. A Placebo-Controlled Trial of Oral Fingolimod in Relapsing Multiple Sclerosis. *New England Journal of Medicine* 362 (5): 387–401.

Kearney, H., D. H. Miller, and O. Ciccarelli. 2015. Spinal Cord MRI in Multiple Sclerosis—Diagnostic, Prognostic and Clinical Value. *Nature Reviews: Neurology* 11 (6): 327–38.

Kieseier, B. C. 2011. The Mechanism of Action of Interferon-β in Relapsing Multiple Sclerosis. *CNS Drugs* 25 (6): 491–502.

Klaver, R., V. Popescu, P. Voorn, Y. Galis-de Graaf, P. van der Valk, H. E. de Vries, G. J. Schenk, and J. J. G. Geurts. 2015. Neuronal and Axonal Loss in Normal-Appearing Gray Matter and Subpial Lesions in Multiple Sclerosis. *Journal of Neuropathology and Experimental Neurology* 74 (5): 453–58.

Kolasa, K. 2013. How Much Is the Cost of Multiple Sclerosis—Systematic Literature Review. *Przegląd Epidemiologiczny* 67 (1): 75–79, 157–60.

Kolasinski, J., C. J. Stagg, S. A. Chance, G. C. Deluca, M. M. Esiri, E.-H. Chang, J. A. Palace et al. 2012. A Combined Post-Mortem Magnetic Resonance Imaging and Quantitative Histological Study of Multiple Sclerosis Pathology. *Brain: A Journal of Neurology* 135 (Pt 10): 2938–51.

Kremenchutzky, M., G. P. A. Rice, J. Baskerville, D. M. Wingerchuk, and G. C. Ebers. 2006. The Natural History of Multiple Sclerosis: A Geographically Based Study 9: Observations on the Progressive Phase of the Disease. *Brain: A Journal of Neurology* 129 (Pt 3): 584–94.

Kuhlmann, T., G. Lingfeld, A. Bitsch, J. Schuchardt, and W. Brück. 2002. Acute Axonal Damage in Multiple Sclerosis Is Most Extensive in Early Disease Stages and Decreases over Time. *Brain: A Journal of Neurology* 125 (Pt 10): 2202–12.

Lassmann, H. 2014a. Mechanisms of White Matter Damage in Multiple Sclerosis. *Glia* 62 (11): 1816–30.

Lassmann, H. 2014b. Multiple Sclerosis: Lessons from Molecular Neuropathology. *Experimental Neurology* 262 (Pt A): 2–7.

Leray, E., S. Vukusic, M. Debouverie, M. Clanet, B. Brochet, J. de Sèze, H. Zéphir et al. 2015. Excess Mortality in Patients with Multiple Sclerosis Starts at 20 Years from Clinical Onset: Data from a Large-Scale French Observational Study. *PLoS One* 10 (7): e0132033.

Leray, E., J. Yaouanq, E. Le Page, M. Coustans, D. Laplaud, J. Oger, and G. Edan. 2010. Evidence for a Two-Stage Disability Progression in Multiple Sclerosis. *Brain: A Journal of Neurology* 133 (Pt 7): 1900–913.

Levin, L. I., K. L. Munger, M. V. Rubertone, C. A. Peck, E. T. Lennette, D. Spiegelman, and A. Ascherio. 2005. Temporal Relationship between Elevation of Epstein-Barr Virus Antibody Titers and Initial Onset of Neurological Symptoms in Multiple Sclerosis. *JAMA* 293 (20): 2496–500.

Lin, R., J. Charlesworth, I. van der Mei, and B. V. Taylor. 2012. The Genetics of Multiple Sclerosis. *Practical Neurology* 12 (5): 279–88.

Lincoln, J. A., and S. D. Cook. 2009. An Overview of Gene-Epigenetic Environmental Contributions to MS Causation. *Journal of the Neurological Sciences* 286 (54–57).

Lincoln, J. A., K. Hankiewicz, and S. D. Cook. 2008. Could Epstein-Barr Virus or Canine Distemper Virus Cause Multiple Sclerosis? *Neurologic Clinics* 26 (3): 699–715, viii.

Linker, R. A., and R. Gold. 2013. Dimethyl Fumarate for Treatment of Multiple Sclerosis: Mechanism of Action, Effectiveness, and Side Effects. *Current Neurology and Neuroscience Reports* 13 (11): 394.

Liu, J. S., M. L. Zhao, C. F. Brosnan, and S. C. Lee. 2001. Expression of Inducible Nitric Oxide Synthase and Nitrotyrosine in Multiple Sclerosis Lesions. *American Journal of Pathology* 158 (6): 2057–66.

Lublin, F. D., and S. C. Reingold. 1996. Defining the Clinical Course of Multiple Sclerosis: Results of an International Survey. National Multiple Sclerosis Society (USA) Advisory Committee on Clinical Trials of New Agents in Multiple Sclerosis. *Neurology* 46 (4): 907–11.

Lublin, F. D., S. C. Reingold, J. A. Cohen, G. R. Cutter, P. S. Sørensen, A. J. Thompson, J. S. Wolinsky et al. 2014. Defining the Clinical Course of Multiple Sclerosis: The 2013 Revisions. *Neurology* 83 (3): 278–86.

Lucchinetti, C., W. Brück, J. Parisi, B. Scheithauer, M. Rodriguez, and H. Lassmann. 2000. Heterogeneity of Multiple Sclerosis Lesions: Implications for the Pathogenesis of Demyelination. *Annals of Neurology* 47 (6): 707–17.

Lucchinetti, C. F., B. F. G. Popescu, R. F. Bunyan, N. M. Moll, S. F. Roemer, H. Lassmann, W. Brück et al. 2011. Inflammatory Cortical Demyelination in Early Multiple Sclerosis. *New England Journal of Medicine* 365 (23): 2188–97.

Lundgaard, I., M. J. Osório, B. T. Kress, S. Sanggaard, and M. Nedergaard. 2014. White Matter Astrocytes in Health and Disease. *Neuroscience* 276: 161–73.

Lünemann, J. D., and C. Münz. 2007. Epstein-Barr Virus and Multiple Sclerosis. *Current Neurology and Neuroscience Reports* 7 (3): 253–58.

Magliozzi, R., O. Howell, A. Vora, B. Serafini, R. Nicholas, M. Puopolo, R. Reynolds, and F. Aloisi. 2007. Meningeal B-Cell Follicles in Secondary Progressive Multiple Sclerosis Associate with Early Onset of Disease and Severe Cortical Pathology. *Brain: A Journal of Neurology* 130 (Pt 4): 1089–104.

Mahad, D. H., B. D. Trapp, and H. Lassmann. 2015. Pathological Mechanisms in Progressive Multiple Sclerosis. *Lancet: Neurology* 14 (2): 183–93.

Marrie, R. A., L. Elliott, J. Marriott, M. Cossoy, J. Blanchard, S. Leung, and N. Yu. 2015. Effect of Comorbidity on Mortality in Multiple Sclerosis. *Neurology* 85 (3): 240–47.

Marrie, R. A., C. Wolfson, M. C. Sturkenboom, O. Gout, O. Heinzlef, E. Roullet, and L. Abenhaim. 2000. Multiple Sclerosis and Antecedent Infections: A Case-Control Study. *Neurology* 54 (12): 2307–10.

Martinelli, V., G. D. Costa, B. Colombo, D. D. Libera, A. Rubinacci, M. Filippi, R. Furlan, and G. Comi. 2014. Vitamin D Levels and Risk of Multiple Sclerosis in Patients with Clinically Isolated Syndromes. *Multiple Sclerosis (Houndmills, Basingstoke, England)* 20 (2): 147–55.

McDonald, W. I., A. Compston, G. Edan, D. Goodkin, H. P. Hartung, F. D. Lublin, H. F. McFarland et al. 2001. Recommended Diagnostic Criteria for Multiple Sclerosis: Guidelines from the International Panel on the Diagnosis of Multiple Sclerosis. *Annals of Neurology* 50 (1): 121–27.

Mehta, V., W. Pei, G. Yang, S. Li, E. Swamy, A. Boster, P. Schmalbrock, and D. Pitt. 2013. Iron Is a Sensitive Biomarker for Inflammation in Multiple Sclerosis Lesions. *PloS One* 8 (3): e57573.

Melcon, M. O., J. Correale, and C. M. Melcon. 2014. Is It Time for a New Global Classification of Multiple Sclerosis? *Journal of the Neurological Sciences* 344 (1–2): 171–81.

Miller, A. E. 2015. Teriflunomide for the Treatment of Relapsing-Remitting Multiple Sclerosis. *Expert Review of Clinical Immunology* 11 (2): 181–94.

Mowry, E. M., M. Pesic, B. Grimes, S. R. Deen, P. Bacchetti, and E. Waubant. 2009. Clinical Predictors of Early Second Event in Patients with Clinically Isolated Syndrome. *Journal of Neurology* 256 (7): 1061–66.

Munger, K. L., J. Åivo, K. Hongell, M. Soilu-Hänninen, H.-M. Surcel, and A. Ascherio. 2016. Vitamin D Status during Pregnancy and Risk of Multiple Sclerosis in Offspring of Women in the Finnish Maternity Cohort. *JAMA Neurology* 73 (5): 515–19.

Najafian, N., T. Chitnis, A. D. Salama, B. Zhu, C. Benou, X. Yuan, M. R. Clarkson, M. H. Sayegh, and S. J. Khoury. 2003. Regulatory Functions of CD8+CD28− T Cells in an Autoimmune Disease Model. *Journal of Clinical Investigation* 112 (7): 1037–48.

Narayana, P. A., Y. Zhou, K. M. Hasan, S. Datta, X. Sun, and J. S. Wolinsky. 2014. Hypoperfusion and T1-Hypointense Lesions in White Matter in Multiple Sclerosis. *Multiple Sclerosis (Houndmills, Basingstoke, England)* 20 (3): 365–73.

Nixon, R., N. Bergvall, D. Tomic, N. Sfikas, G. Cutter, and G. Giovannoni. 2014. No Evidence of Disease Activity: Indirect Comparisons of Oral Therapies for the Treatment of Relapsing-Remitting Multiple Sclerosis. *Advances in Therapy* 31 (11): 1134–54.

Noonan, C. W., D. M. Williamson, J. P. Henry, R. Indian, S. G. Lynch, J. S. Neuberger, R. Schiffer, J. Trottier, L. Wagner, and R. A. Marrie. 2010. The Prevalence of Multiple Sclerosis in 3 US Communities. *Preventing Chronic Disease* 7 (1): A12.

Noseworthy, J. H., C. Lucchinetti, M. Rodriguez, and B. G. Weinshenker. 2000. Multiple Sclerosis. *New England Journal of Medicine* 343 (13): 938–52.

O'Connor, P., J. S. Wolinsky, C. Confavreux, G. Comi, L. Kappos, T. P. Olsson, H. Benzerdjeb et al. 2011. Randomized Trial of Oral Teriflunomide for Relapsing Multiple Sclerosis. *New England Journal of Medicine* 365 (14): 1293–303.

Oksenberg, J. R. 2013. Decoding Multiple Sclerosis: An Update on Genomics and Future Directions. *Expert Review of Neurotherapeutics* 13 (Suppl 2): 11–19.

Okuda, D. T., A. Siva, O. Kantarci, M. Inglese, I. Katz, M. Tutuncu, B. M. Keegan et al. 2014. Radiologically Isolated Syndrome: 5-Year Risk for an Initial Clinical Event. *PloS One* 9 (3): e90509.

Panitch, H., D. Goodin, G. Francis, P. Chang, P. Coyle, P. O'Connor, D. Li, B. Weinshenker, and EVIDENCE (EVidence of Interferon Dose-response: European North American Comparative Efficacy) Study Group and the University of British Columbia MS/MRI Research Group. 2005. Benefits of High-Dose, High-Frequency Interferon Beta-1a in Relapsing-Remitting Multiple Sclerosis Are Sustained to 16 Months: Final Comparative Results of the EVIDENCE Trial. *Journal of the Neurological Sciences* 239 (1): 67–74.

Panitch, H., D. S. Goodin, G. Francis, P. Chang, P. K. Coyle, P. O'Connor, E. Monaghan et al. 2002. Randomized, Comparative Study of Interferon Beta-1a Treatment Regimens in MS: The EVIDENCE Trial. *Neurology* 59 (10): 1496–506.

Pankratz, S., T. Ruck, S. G. Meuth, and H. Wiendl. 2016. CD4+HLA-G+ Regulatory T Cells: Molecular Signature and Pathophysiological Relevance. *Human Immunology* 77 (9): 727–33.

Papadopoulos, D., S. Dukes, R. Patel, R. Nicholas, A. Vora, and R. Reynolds. 2009. Substantial Archaeocortical Atrophy and Neuronal Loss in Multiple Sclerosis. *Brain Pathology (Zurich, Switzerland)* 19 (2): 238–53.

Patani, R., M. Balaratnam, A. Vora, and R. Reynolds. 2007. Remyelination Can Be Extensive in Multiple Sclerosis Despite a Long Disease Course. *Neuropathology and Applied Neurobiology* 33 (3): 277–87.

Pepinsky, R. B., Z. Shao, B. Ji, Q. Wang, G. Meng, L. Walus, X. Lee et al. 2011. Exposure Levels of Anti-LINGO-1 Li81 Antibody in the Central Nervous System and Dose-Efficacy Relationships in Rat Spinal Cord Remyelination Models after Systemic Administration. *Journal of Pharmacology and Experimental Therapeutics* 339 (2): 519–29.

Peterson, J. W., L. Bö, S. Mörk, A. Chang, and B. D. Trapp. 2001. Transected Neurites, Apoptotic Neurons, and Reduced Inflammation in Cortical Multiple Sclerosis Lesions. *Annals of Neurology* 50 (3): 389–400.

Pohl, D., B. Krone, K. Rostasy, E. Kahler, E. Brunner, M. Lehnert, H.-J. Wagner, J. Gärtner, and F. Hanefeld. 2006. High Seroprevalence of Epstein-Barr Virus in Children with Multiple Sclerosis. *Neurology* 67 (11): 2063–65.

Polman, C. H., P. W. O'Connor, E. Havrdova, M. Hutchinson, L. Kappos, D. H. Miller, J. T. Phillips et al. 2006. A Randomized, Placebo-Controlled Trial of Natalizumab for Relapsing Multiple Sclerosis. *New England Journal of Medicine* 354 (9): 899–910.

Polman, C. H., S. C. Reingold, B. Banwell, M. Clanet, J. A. Cohen, M. Filippi, K. Fujihara et al. 2011. Diagnostic Criteria for Multiple Sclerosis: 2010 Revisions to the McDonald Criteria. *Annals of Neurology* 69 (2): 292–302.

Polman, C. H., S. C. Reingold, G. Edan, M. Filippi, H.-P. Hartung, L. Kappos, F. D. Lublin et al. 2005. Diagnostic Criteria for Multiple Sclerosis: 2005 Revisions to the 'McDonald Criteria.' *Annals of Neurology* 58 (6): 840–46.

Popescu, V., R. Klaver, P. Voorn, Y. Galis-de Graaf, D. L. Knol, J. W. R. Twisk, A. Versteeg et al. 2015. What Drives MRI-Measured Cortical Atrophy in Multiple Sclerosis? *Multiple Sclerosis (Houndmills, Basingstoke, England)* 21 (10): 1280–90.

Radue, E.-W. 2012. Impact of Fingolimod Therapy on Magnetic Resonance Imaging Outcomes in Patients with Multiple Sclerosis. *Archives of Neurology* 69 (10): 1259.

Raftopoulos, R., S. J. Hickman, A. Toosy, B. Sharrack, S. Mallik, D. Paling, D. R. Altmann et al. 2016. Phenytoin for Neuroprotection in Patients with Acute Optic Neuritis: A Randomised, Placebo-Controlled, Phase 2 Trial. *Lancet: Neurology* 15 (3): 259–69.

Ramagopalan, S. V., and G. C. Ebers. 2009. Multiple Sclerosis: Major Histocompatibility Complexity and Antigen Presentation. *Genome Medicine* 1 (11): 105.

Reingold, S. C. 2002. Prevalence Estimates for MS in the United States and Evidence of an Increasing Trend for Women. *Neurology* 59 (2): 294; author reply 294–295.

Rovira, À., M. P. Wattjes, M. Tintoré, C. Tur, T. A. Yousry, M. P. Sormani, N. De Stefano et al. 2015. Evidence-Based Guidelines: MAGNIMS Consensus Guidelines on the Use of MRI in Multiple Sclerosis— Clinical Implementation in the Diagnostic Process. *Nature Reviews: Neurology* 11 (8): 471–82.

Rückemann, K., L. D. Fairbanks, E. A. Carrey, C. M. Hawrylowicz, D. F. Richards, B. Kirschbaum, and H. A. Simmonds. 1998. Leflunomide Inhibits Pyrimidine De Novo Synthesis in Mitogen-Stimulated T-Lymphocytes from Healthy Humans. *Journal of Biological Chemistry* 273 (34): 21682–91.

Scalfari, A., A. Neuhaus, M. Daumer, P. A. Muraro, and G. C. Ebers. 2014. Onset of Secondary Progressive Phase and Long-Term Evolution of Multiple Sclerosis. *Journal of Neurology, Neurosurgery, and Psychiatry* 85 (1): 67–75.

Scalfari, A., A. Neuhaus, A. Degenhardt, G. P. Rice, P. A. Muraro, M. Daumer, and G. C. Ebers. 2010. The Natural History of Multiple Sclerosis: A Geographically Based Study 10: Relapses and Long-Term Disability. *Brain: A Journal of Neurology* 133 (Pt 7): 1914–29.

Scott, L. J. 2013. Glatiramer Acetate: A Review of Its Use in Patients with Relapsing-Remitting Multiple Sclerosis and in Delaying the Onset of Clinically Definite Multiple Sclerosis. *CNS Drugs* 27 (11): 971–88.

Seewann, A., H. Vrenken, P. van der Valk, E. L. A. Blezer, D. L. Knol, J. A. Castelijns, C. H. Polman, P. J. W. Pouwels, F. Barkhof, and J. J. G. Geurts. 2009.

Diffusely Abnormal White Matter in Chronic Multiple Sclerosis: Imaging and Histopathologic Analysis. *Archives of Neurology* 66 (5): 601–9.

Sormani, M. P., L. Bonzano, L. Roccatagliata, G. L. Mancardi, A. Uccelli, and P. Bruzzi. 2010. Surrogate Endpoints for EDSS Worsening in Multiple Sclerosis. A Meta-Analytic Approach. *Neurology* 75 (4): 302–9.

Stankoff, B., S. Mrejen, A. Tourbah, B. Fontaine, O. Lyon-Caen, C. Lubetzki, and M. Rosenheim. 2007. Age at Onset Determines the Occurrence of the Progressive Phase of Multiple Sclerosis. *Neurology* 68 (10): 779–81.

Steen, C., M. D'haeseleer, J. M. Hoogduin, Y. Fierens, M. Cambron, J. P. Mostert, D. J. Heersema, M. W. Koch, and J. De Keyser. 2013. Cerebral White Matter Blood Flow and Energy Metabolism in Multiple Sclerosis. *Multiple Sclerosis (Houndmills, Basingstoke, England)* 19 (10): 1282–89.

Stojanovic, I. R., M. Kostic, and S. Ljubisavljevic. 2014. The Role of Glutamate and Its Receptors in Multiple Sclerosis. *Journal of Neural Transmission (Vienna, Austria: 1996)* 121 (8): 945–55.

Stys, P. K. 2004. Axonal Degeneration in Multiple Sclerosis: Is It Time for Neuroprotective Strategies? *Annals of Neurology* 55 (5): 601–3.

Su, K. G., G. Banker, D. Bourdette, and M. Forte. 2009. Axonal Degeneration in Multiple Sclerosis: The Mitochondrial Hypothesis. *Current Neurology and Neuroscience Reports* 9 (5): 411–17.

Swanton, J. K., K. Fernando, C. M. Dalton, K. A. Miszkiel, A. J. Thompson, G. T. Plant, and D. H. Miller. 2006. Modification of MRI Criteria for Multiple Sclerosis in Patients with Clinically Isolated Syndromes. *Journal of Neurology, Neurosurgery, and Psychiatry* 77 (7): 830–33.

Thompson, S. A. J., J. L. Jones, A. L. Cox, D. A. S. Compston, and A. J. Coles. 2009. B-Cell Reconstitution and BAFF after Alemtuzumab (Campath-1H) Treatment of Multiple Sclerosis. *Journal of Clinical Immunology* 30 (1): 99–105.

Tintoré, M. 2008. Rationale for Early Intervention with Immunomodulatory Treatments. *Journal of Neurology* 255 (Suppl 1): 37–43.

Tintore, M., À. Rovira, J. Río, S. Otero-Romero, G. Arrambide, C. Tur, M. Comabella et al. 2015. Defining High, Medium and Low Impact Prognostic Factors for Developing Multiple Sclerosis. *Brain: A Journal of Neurology* 138 (Pt 7): 1863–74.

Traboulsee, A., J. H. Simon, L. Stone, E. Fisher, D. E. Jones, A. Malhotra, S. D. Newsome et al. 2016. Revised Recommendations of the Consortium of MS Centers Task Force for a Standardized

MRI Protocol and Clinical Guidelines for the Diagnosis and Follow-Up of Multiple Sclerosis. *AJNR: American Journal of Neuroradiology* 37 (3): 394–401.

Trapp, B. D., J. Peterson, R. M. Ransohoff, R. Rudick, S. Mörk, and L. Bö. 1998. Axonal Transection in the Lesions of Multiple Sclerosis. *New England Journal of Medicine* 338 (5): 278–85.

Trapp, B. D., and P. K. Stys. 2009. Virtual Hypoxia and Chronic Necrosis of Demyelinated Axons in Multiple Sclerosis. *Lancet: Neurology* 8 (3): 280–91.

Vercellino, M., S. Masera, M. Lorenzatti, C. Condello, A. Merola, A. Mattioda, A. Tribolo et al. 2009. Demyelination, Inflammation, and Neurodegeneration in Multiple Sclerosis Deep Gray Matter. *Journal of Neuropathology and Experimental Neurology* 68 (5): 489–502.

Vercellino, M., F. Plano, B. Votta, R. Mutani, M. T. Giordana, and P. Cavalla. 2005. Grey Matter Pathology in Multiple Sclerosis. *Journal of Neuropathology and Experimental Neurology* 64 (12): 1101–7.

von Büdingen, H.-C., A. Palanichamy, K. Lehmann-Horn, B. A. Michel, and S. S. Zamvil. 2015. Update on the Autoimmune Pathology of Multiple Sclerosis: B-Cells as Disease-Drivers and Therapeutic Targets. *European Neurology* 73 (3–4): 238–46.

Vukusic, S., and C. Confavreux. 2003. Prognostic Factors for Progression of Disability in the Secondary Progressive Phase of Multiple Sclerosis. *Journal of the Neurological Sciences* 206 (2): 135–37.

Wegner, C., M. M. Esiri, S. A. Chance, J. Palace, and P. M. Matthews. 2006. Neocortical Neuronal, Synaptic, and Glial Loss in Multiple Sclerosis. *Neurology* 67 (6): 960–67.

Witte, M. E., D. J. Mahad, H. Lassmann, and J. van Horssen. 2014. Mitochondrial Dysfunction Contributes to Neurodegeneration in Multiple Sclerosis. *Trends in Molecular Medicine* 20 (3): 179–87.

Wolinsky, J. S., P. A. Narayana, F. Nelson, S. Datta, P. O'Connor, C. Confavreux, G. Comi et al. 2013. Magnetic Resonance Imaging Outcomes from a Phase III Trial of Teriflunomide. *Multiple Sclerosis Journal* 19 (10): 1310–19.

Xia, P., and C. Wadham. 2011. Sphingosine 1-Phosphate, a Key Mediator of the Cytokine Network: Juxtacrine Signaling. *Cytokine & Growth Factor Reviews* 22 (1): 45–53.

Zhang, Q., C.-Y. Lin, Q. Dong, J. Wang, and W. Wang. 2011. Relationship between HLA-DRB1 Polymorphism and Susceptibility or Resistance to Multiple Sclerosis in Caucasians: A Meta-Analysis of Non-Family-Based Studies. *Autoimmunity Reviews* 10 (8): 474–81.

Nanotherapeutics for Multiple Sclerosis

Yonghao Cao, Joyce J. Pan, Inna Tabansky, Souhel Najjar, Paul Wright, and Joel N. H. Stern

CONTENTS

6.2.1 INTRODUCTION

Multiple sclerosis (MS) is a complex neurodegenerative autoimmune disease characterized by demyelination of neurons and progressive destruction of the central nervous system (CNS) (Ransohoff et al. 2015). In MS, autoreactive immune cells permeate the blood–brain barrier (BBB) and catalyze an inflammatory process that causes perivenous demyelinating lesions, which results in multiple discrete plaques primarily manifesting in white matter (Goldenberg 2012).

The multifaceted pathogenesis of MS is reflected in the patients' clinical presentations and difficult diagnosis. MS typically begins with an acute neurological episode, also coined a "clinically isolated syndrome," which will then be succeeded by a period of relapses interspersed with remissions, and eventually will progress (on average over the course of 10–15 years) to a period of intensifying disability without relapses. One in five patients have no relapses, and their disease steadily progresses from the initial episode. As initial symptoms vary significantly between patients with MS, clinical tools such as magnetic resonance imaging (MRI) and lumbar puncture (LP) are used for diagnosis (Mahmoudi et al. 2011a).

There are various aspects and approaches of treating MS. According to the National MS Society, there are five modes of care provided for MS patients: (1) modifying disease course, (2) treating exacerbations, (3) managing symptoms, (4) promoting function through rehabilitation, and (5) providing emotional support. Comprehensive care includes all five of these aspects, but for the purpose of this section, we focus on current treatments that modify disease course (Tabansky et al. 2015). Although the exact mechanisms of the pathogenesis of MS remain unknown, substantial scientific advances in the treatment of this disease have been made (Loma and Heyman 2011). Treatments for MS can be broken down into two distinct categories: symptomatic therapies and disease-modifying therapies (DMTs). Symptomatic therapies are predicated on the management of the myriad symptoms and comorbidities that can afflict MS patients (Tabansky et al. 2015). DMTs are all therapies that modulate the pathogenesis of disease. There are currently more than 12 Food and Drug Administration (FDA)–approved DMTs for the treatment of MS, which differ in several respects, such as the efficacy of therapy, the ease of administration, and potential adverse effects (Table 6.1).

Breakthrough drug delivery technologies have the potential to reshape MS treatment not only by modifying the properties of current therapies, but also by enhancing targeting. This enhanced targeting would increase drug efficacy while mitigating potential adverse effects (Tabansky et al. 2015). This section focuses on one of the most important and rapidly emerging of these breakthrough technologies, nanotherapy. In order to understand the potentials of nanotherapy in MS, it is important to examine the characteristics of current DMTs.

6.2.2 NANOPARTICLES IN MEDICINE

Nanomedicine is a new and rapidly expanding field of nanotechnology that emerged at the interface between nanotechnology and biotechnology, the "nano-bio interface" (Nel et al. 2009; Mahmoudi et al. 2011b). The novel physicochemical properties of nanomaterials have offered many prospects in terms of clinical therapeutic possibilities (Mahmoudi et al. 2011b). Engineered nanomaterials have been studied across a broad array of biomedical applications, including biomedical imaging, transfection, gene delivery, tissue engineering, and stem cell tracking (Moghimi et al. 2005). For these reasons, it has been estimated that nanomaterials will experience rapid growth in the coming years. As of 2011, they were growing at a 17% compound annual growth rate (CAGR) and had produced a market worth of more than $50 billion (Mahmoudi et al. 2011b). This section examines the history and evolution of nanoparticles (NPs) in medicine, and focuses on how materials have been modified over time. We also discuss the discrepancy between the current and prior generation of formulations in terms of efficacy, targeting, and delivery.

For the purpose of this chapter, NPs and microparticles can be defined as small, physically concrete materials, 0.001–100 μm in size. They have diverse applications, including the ability to target drugs to specific tissues, tumors, and cells. They have also been shown to be involved in immunomodulation, vaccine delivery, and drug coating (Gharagozloo et al. 2015). The versatility of NP drug delivery systems is a primary feature behind their potential to improve the treatment of autoimmune diseases such as MS (Tabansky et al. 2015).

The development of synthetic polymers for controlled release of therapeutic agents was prompted by the discoveries of Folkman and Long in the 1960s, who demonstrated the potential use of silicone rubber as a carrier for prolonged drug delivery (Folkman and Long 1964; Folkman et al. 1966). In the decades that followed, several other polymeric materials and drug delivery devices were created, including films, tablets, gels, and microspheres (Folkman and Long 1964; Langer 2001; Richards Grayson et al. 2003; Kabanov and Gendelman 2007). Many of these materials have been implemented in drug formulations, such as controlled-release drug delivery systems for attention deficit hyperactivity disorder (ADHD) (Concerta). These drug delivery systems were also used for other clinical manifestations, such as polymeric implants, for the potential treatment of brain tumors and neurodegenerative diseases (Wu et al. 1994; Lesniak et al. 2001; Kabanov and Gendelman 2007). However, the efficacy of prior iterations of localized delivery systems and NPs was invasive, lead to an adverse response to implantation, and had insufficient diffusion of particles beyond the implantation site (Saltzman et al. 1999; Stroh et al. 2003; Siepmann et al. 2006; Kabanov and Gendelman 2007).

6.2.2.1 Nanomedicine in the Central Nervous System

In light of these flaws, the creation of polymer therapeutics and nanomedicines that can be delivered systemically and are able to penetrate the barriers to entry into the CNS would be a crucial development in the diagnosis and treatment of neurodegenerative diseases such as MS (Kabanov and Gendelman 2007). Between 2000 and 2005, a number of polymer therapeutics for cancer and other diseases either came on the market or underwent clinical evaluation for potential FDA approval. Among these, polyethylene glycol (PEG)–coated liposomal doxorubicin attained approval for treatment of hematological malignancies and AIDS-related Kaposi's sarcoma (Sharpe et al. 2002; Gabizon et al. 2003). Another

TABLE 6.1
Current treatments of MS.

	Treatment (chemical name)	Dosage	Type of MS	Mechanism of action (MOA)	Common side effects
Injectable	Avonex (interferon beta-1a), Plegridy, Rebif Biogen, EMD Serono, Inc./ Pfizer, Inc.	30 μg (into a large muscle) once weekly	Relapse–remitting MS	Immunosuppressive and anti-inflammatory through inhibiting transcription factors involved in inflammatory response	Headache, flu-like symptoms (chills, fever, muscle pain, fatigue, weakness), injection site pain and inflammation
	Betaseron (interferon beta-1b) Bayer Healthcare Pharmaceuticals, Inc.	0.25 mg every other day	Relapse–remitting MS	Same as Avonex	Flu-like symptoms (chills, fever, muscle pain, fatigue, weakness) following injection, headache, injection site reactions (swelling, redness, pain), injection site skin breakdown, low white blood cell count
	Copaxone, Glatopa (glatiramer acetate) Novartis Pharmaceuticals, Sandoz, a Novartis Company	20 mg every day, or 40 mg three times per week	Relapse–remitting MS	Induction of suppressor T cells	Injection site reactions (redness, pain, swelling), flushing, shortness of breath, rash, chest pain
	Extavia (interferon beta-1b) Novartis Pharmaceuticals	0.25 mg every other day	Relapse–remitting MS	Same as Avonex	Flu-like symptoms (chills, fever, muscle pain, fatigue, weakness) following injection, headache, injection site reactions (swelling, redness, pain)
Oral	Aubagio (teriflunomide) Sanofi Genzyme	7 or 14 mg pill once daily	Relapse–remitting MS	Limits proliferation of rapidly dividing T and B cells by inhibiting pyrimidine synthesis	Headache, hair thinning, diarrhea, nausea, abnormal liver tests
	Gilenya (fingolimod) Novartis Pharmaceuticals	0.5 mg capsule once daily	Relapse–remitting MS		Headache, flu, diarrhea, back pain, liver enzyme elevations, sinusitis, abdominal pain, pain in extremities, cough

(Continued)

TABLE 6.1 (CONTINUED)
Current treatments of MS.

	Treatment (chemical name)	Dosage	Type of MS	Mechanism of action (MOA)	Common side effects
	Tecfidera (dimethyl fumarate) Biogen	120 mg capsule taken twice daily for 1 week, then 240 mg capsule taken twice daily thereafter	Relapse–remitting MS	Modulates expression of cytokines and induction of antioxidant response; reduces frequency of attacks	Flushing (sensation of heat or itching and a blush on the skin), gastrointestinal issues (nausea, diarrhea, abdominal pain)
Intravenous infusion	Lamtrada (alemtuzumab) Sanofi Genzyme	12 mg/day for 5 consecutive days, followed by 12 mg/day for 3 consecutive days 1 year later	Relapse–remitting MS	Induces depletion of B and T cells; changes the profile of lymphocyte subsets	Rash, headache, fever, nasal congestion, nausea, urinary tract infection, fatigue, insomnia, upper respiratory tract infection, herpes viral infections, hives, itching, thyroid gland disorders, fungal infection, diarrhea, vomiting, flushing, and pain in joints, extremities, and back; infusion reactions (including nausea, hives, itching, insomnia, chills, flushing, fatigue, shortness of breath, changes in the sense of taste, indigestion, dizziness, and pain) are also common while the medication is being administered and for 24 hours or more after the infusion is over
	Mitoxanthone Only available generically	12 mg/m^2 IV every 3 months	Relapse–remitting MS Progressive–relapsing, secondary progressive	Disrupts DNA synthesis and repair; inhibits macrophage, B cell, and T cell proliferation; impairs antigen presentation	Nausea, hair loss, menstrual change, upper respiratory infection, urinary tract infection, mouth sores, irregular heartbeat, diarrhea, constipation, back pain, sinusitis, headache, blue–green urine
	Tysabri (natalizumab) Biogen	300 mg IV every 28 days	Relapse–remitting MS	MOA is not completely identified; it is believed to prevent leukocytes from crossing the BBB by interacting with integrins	Headache, fatigue, joint pain, chest discomfort, urinary tract infection, lower respiratory tract infection, gastroenteritis, vaginitis, depression, pain in extremity, abdominal discomfort, diarrhea, rash

SOURCE: Adapted from National MS Society and Tabansky, I. et al., Immunol. Res., 63 (1–3), 58–69, 2015.

example of a prior-generation nanomaterial that garnered approval for treatment was albumin-bound paclitaxel, a treatment for metastatic breast cancer that used 130 nm albumin-bound technology to circumvent solvent requirements and transport it across the endothelial cells lining the blood vessels, resulting in higher concentrations of the treatments in the area of the tumor (Gradishar 2006). These treatments demonstrated the capacity of polymer therapeutics and nanomedical applications to augment the delivery of drugs and vaccines to the specific regions of the body, in order to combat disease (Feng et al. 2004; Kabanov and Gendelman 2007).

Over the past decade, there have been several examples in the literature of NP or microparticle approaches being implemented in mouse models of MS with varying efficacies. As axonal degeneration is a key indicator of neurological impairment in MS, certain therapeutic approaches using NP and microparticles have focused on increasing neuroprotection (Nowacek et al. 2009). In 2008, an NP comprised of a water-soluble fullerene that was functionalized with an N-methyl-D-asparate (NMDA) receptor showed encouraging results in hindering neurodegeneration caused by MS. The NP achieved these results by reducing oxidative injury, CD11b infiltration, and CCL2 expression without modifying T cell responses (Basso et al. 2008). It was a complex NP that combined multiple functions into a single potent particle. Another NP formulation has been proposed based on experimental results indicating that infection with *Helicobacter pylori* is protective against MS. The authors proposed that *H. pylori* could potentially be packaged into NP, functionalized for neurospecific targets, and then delivered by cells to the CNS (Pezeshki et al. 2008).

A subsequent generation of nanomedicines has now emerged, whose novelty lies in new self-assembled nanomaterials for drug and gene delivery (Salem et al. 2003; Missirlis et al. 2005; Nayak and Lyon 2005; Trentin et al. 2005; Kabanov and Gendelman 2007). Subtypes of these include polymeric micelles (Kwon 2003; Savic et al. 2003; Kabanov et al. 2005; Torchilin 2007), DNA–polycation complexes ("polyplexes") (Wu and Wu 1987; Ogris and Wagner 2002; Read et al. 2005), block ionomer complexes (Kakizawa and Kataoka 2002; van Nostrum 2004; Oh et al. 2006), and nanogels (McAllister et al. 2002; Vinogradov et al. 2004; Kazakov et al. 2006; Soni et al. 2006). Of

those, the greatest evidence of efficacy in clinical applications has been seen among the polymeric micelles, which have been shown to be effective in human trials for the delivery of anticancer therapeutics (Kabanov and Gendelman 2007). Additional progress in polymer chemistry precipitated the creation of novel nanomaterials with unique spatial orientations, including dendrimers (Helms and Meijer 2006), star polymers (Tao and Uhrich 2006), and cross-linked polymer micelles (Pochan et al. 2004; Bronich et al. 2005; Wang et al. 2005; O'Reilly et al. 2006). Although these materials are fairly recent in terms of their development, their unique structural and mechanical properties offer great potential for drug delivery and bioimaging applications.

Over the past two decades, nanomedicines have emerged and demonstrated material improvements to drug delivery, targeting, and triggered release. They are able to go beyond targeting just one organ or one set of tissues to targeting individual cells or intracellular compartments. Many nanomedicines offered multiple specific properties that were conducive to the delivery and imaging of therapeutic agents to the CNS. Discussed below are some of these types of nanomedicines that have been considered for use in the delivery of molecules to the CNS (Kabanov and Gendelman 2007).

6.2.3 NANOMEDICINE AND MULTIPLE SCLEROSIS

Over the past 5 years, growing evidence suggests that nanotherapies are able to modulate immune responses within the CNS. Thus, direct delivery of drugs into the CNS is a feasible avenue for the treatment of MS and is likely to increase their efficacy and mitigate side effects (Gendelman et al. 2015). In a study of experimental autoimmune encephalomyelitis (EAE) (the most commonly used model of MS in rodents), Kizelsztein and colleagues (2009) showed that encapsulation of tempamine—a stable radical with antioxidant and proapoptotic activity—in liposomes could potentially inhibit EAE in mice. Other scholars have demonstrated the efficacy of using gold NPs to induce antispecific regulatory T cells by delivering a combination of a tolerogenic compound with an oligodendrite antigen. These NPs increased the T-reg population and inhibited the disease course of MS (Yeste et al. 2012). Using the relapsing–remitting EAE model, Hunter et al. (2014) developed a biodegradeable poly(lactic-co-glycololic acid) (PLGA) NP as a myelin antigen carrier and demonstrated its effectiveness in hindering MS progression.

Another recent advance in nanomolecule technology involves the use of dendrimers, which are large molecules with repetitive branching organized around a central core. They are usually spherical and approximate conventional NPs in size (Gendelman et al. 2015). The inherent capacity of many dendrimers to localize around activated microglia and astrocytes renders them good candidates for targeted delivery of immunosuppressive drugs (Gendelman et al. 2015). For example, Wang et al. (2009) have reported clinical evidence indicating that N-acetyl cysteine (NAC) showed enhanced antioxidant and anti-inflammatory properties when carried by polyamidoamine (PAMAM) dendrimers, when compared with free NAC. In two separate studies, this dendrimer-based strategy reduced neuroinflammation. In one case, the dendrimers mitigated cerebral palsy symptoms, and in the other, they suppressed neuroinflammatory biomarkers in a model of retinal degeneration (Iezzi et al. 2012; Kannan et al. 2012). Additionally, a methylprednisolone-loaded carboxylmethylchitisan dendrimer was reported by Cerqueira et al. (2013) to be internalized by astrocytes, microglia, and oligodendrocytes in the spinal cord, resulting in the regulation of growth factors while hindering the titer of pro-inflammatory molecules. Local delivery of dendrimer NPs to the spinal cords of Wistar rats after a hemisection lesion led to improved locomotor recovery after injury (Gendelman et al. 2015). Many of the applications thus far suggested for dendrimer NPs have been for neurological disorders characterized by the accumulation of neuroinflammation, especially when attributable to repetitive head trauma, such as chronic traumatic encephalopathy (CTE) (Gendelman et al. 2015). There is an increasing appreciation of a potential connection between neurodegenerative disease and the persistent neuroinflammation coupled with accumulation of tau plaques that is the hallmark of CTE (Gendelman et al. 2015). In light of this connection, nanoformulations could prove to have especially important applications in the treatment of CTE (Lin et al. 2012; Das et al. 2014; Samuel et al. 2014). For instance, in one recent study,

cerebrolysin-loaded PLGA NPs limited brain edema and possibly the degree of BBB permeability encountered after a traumatic brain injury, such as concussions (Ruozi et al. 2014). It is possible that limiting BBB permeability would prevent accumulation of autoreactive immune cells in the brain.

A particularly promising nanotherapy for the administration and delivery of MS treatments is the use of superparamagnetic iron oxide nanoparticles (SPIONs). Biocompatible SPIONs are composed of a surface coating such as gold, silica, dextran, or PEG. They have been used across a wide variety of biomedical applications, including contrast enhancement (Anderson et al. 2000; Cunningham et al. 2005), site-specific drug release (Polyak and Friedman 2009), and biomedical imaging (Amiri et al. 2011). Feridex—a dextran-coated SPION with a core size of 3–6 nm—has been approved for use in MRI with patients (Bartolozzi et al. 1999; Mahmoudi et al. 2011b).

6.2.4 CATEGORIES OF NANOFORMULATIONS IN THE CNS

The therapeutic effects of nanomedicines in MS are largely dependent on their ability to cross the BBB. The BBB selectively transports small molecules, polypeptides, and cells into the CNS and prevents the entry of potentially harmful compounds into the brain. In particular, nutrients and endogenous compounds required by the CNS, such as amino acids, glucose, essential fatty acids, vitamins, minerals, and electrolytes, are effectively carried into the brain by numerous saturable transport systems expressed at the BBB. Many of the nanoformulations that are currently in use are also able to cross the BBB. These include NPs, polymeric micelles, nanogels, SPIONs, and other nanomaterials. The following sections discuss the characteristics of several of the aforementioned nanocarriers that are able to cross into the CNS (Kabanov and Gendelman 2007).

It is important to note that there are few papers on the clinically proven use of nanotherapies specific to MS. The nanoformulations discussed in the following sections, with the exception of SPIONs, are merely inferred to have potential for treatment of MS in the future, as they have shown either the ability to cross the BBB or an effect on neurodegenerative diseases of the CNS in studies.

6.2.4.1 Liposomes

Liposomes are vesicular structures composed of lipid bilayers with an internal aqueous compartment (Kabanov and Gendelman 2007). Liposomes are often also modified or coated with PEG, often called PEGylated liposomes (Huwyler et al. 1996; Kozubek et al. 2000; Voinea and Simionescu 2002). PEGylation reduces the likelihood that the liposomes will be phagocytosed (also known as "opsonized") in the plasma and decreases the ability of the body to recognize them as a foreign object. This allows for liposomes to have a circulation half-life of up to 50 hours in the human body (Gabizon et al. 1994). PEGylated liposomes are already in clinical use. For instance, Doxil® is a lipsosome-encapsulated doxorubicin, which is used to primarily treat ovarian cancer and metastatic breast cancer (Gabizon et al. 1994; Papaldo et al. 2006). Liposomes can be effective in prolonging the circulation time of drugs in the bloodstream and mitigating drug side effects in clinical settings, making liposomes a prime candidate for drug delivery to the CNS (Umezawa and Eto 1988; Rousseau et al. 1999; Shi et al. 2001).

In animals with EAE, an experimentally induced disease often used to model MS, it is observed that PEGylated liposomes accumulate quickly near the areas of damage to the BBB (Rousseau et al. 1999). These liposomes can also be taken up by macrophages, microglia, and astrocytes within the CNS (Schmidt et al. 2003). Although liposomes are not able to cross the intact BBB, their ability to cross a compromised BBB makes them good candidates for the treatment of MS and other CNS diseases. For instance, immunoliposomes have been successfully used to treat glial brain tumors expressing glial fibrillary acidic protein (GFAP) (Chekhonin et al. 2005). The mechanism by which immunoliposomes and liposomes cross the BBB is not yet fully characterized, but crossing has been postulated to be mediated by fusions involving vesicular pits (Cornford and Cornford 2002). Alternatively, liposomes may be taken up by mononuclear phagocytes (MPs), which can cross the BBB.

6.2.4.2 Nanoparticles

NPs composed of insoluble polymers have demonstrated their potential efficacy for drug and nucleic acid delivery (Liu and Chen 2005). As the NP is formed in solution with a drug, the drug is caught by the precipitating polymer, to be released once the polymer disintegrates naturally in a biological environment (Kabanov and Gendelman 2007). Creators of NPs often utilize organic solvents that may cause some immobilized drug agents—such as biomacromolecules—to degrade. To optimize uptake into the cell, these particles should not exceed 200 nm in diameter. The surface of the NP is often coated with PEG to increase its dispersion stability and the amount of time that it can circulate within the body (Peracchia et al. 1998; Torchilin 1998; Calvo et al. 2002). Poly(butylcyanoacrylate) NPs coated with PEG have been evaluated for CNS delivery of several drugs (Calvo et al. 2002; Kreuter et al. 2003; Steiniger et al. 2004). Caution must be exercised in using this approach, as enhanced brain delivery with surfactant-coated poly-NPs may be positively associated with increased permeability of the BBB and subsequent toxicity (Olivier et al. 1999).

Despite the above caveat, these NPs have already been used in clinical applications to deliver several drugs, including analgesics (dalargin and loperamide), anticancer agents (doxorubicin), anticonvulsants (NMDA receptor antagonist), and peptides (dalargin and kyotorphin), with benefits over conventional drug administration (Kreuter et al. 2003; Steiniger et al. 2004). For example, NPs have been proposed to extend the anticonvulsive activities of MRZ 2/576 compared with the free drug; increase survival rates in rats with an aggressive form of glioblastoma, and the capability to cross the BBB; chelate metals; and then exit through the BBB without compromising their complexed metal ions (Steiniger et al. 2004; Cui et al. 2005; Liu and Chen 2005; Liu et al. 2005). The last application, in particular, may prove useful in hindering the negative consequences resulting from oxidative damage in Alzheimer's and other neurodegenerative diseases (Kabanov and Gendelman 2007).

Another subcategory of NPs is nanospheres, which are hollow substances created either by using microemulsion polymerization or by covering template with a thin polymer layer, followed by removal of the template (Hyuk Im et al. 2005; Kabanov and Gendelman 2007). Nanospheres have the potential for CNS delivery of both drugs and imaging agents, especially in conditions that disrupt the normal permeability of the BBB (Kabanov and Gendelman 2007). For instance, carboxylated polystyrene nanonspheres (20 nm) were unable to cross the BBB under normal conditions, but capable of doing so—at least partially—under ischemia-induced stress (Kreuter 2001; Kabanov and Gendelman 2007).

Another category of NPs that are potentially clinically useful is drug nanosuspensions, defined as crystalline drug particles stabilized by either nonionic surfactants containing PEG or a combination of lipids (Jacobs et al. 2000; Rabinow 2004; Kabanov and Gendelman 2007). Methods from their manufacture vary, and they include media milling, high-pressure homogenization, and templating with emulsions and/or microemulsions (Friedrich and Muller-Goymann 2003; Friedrich et al. 2005). As a result of their irregular, small, polydisperse material structure, they have a couple of major advantages over the other categories of NPs, including simplicity, comparatively large drug loading capacity, and diverse applicability, especially to highly hydrophobic compounds (Muller et al. 2001; Friedrich et al. 2005).

6.2.4.3 Polymerics and Polymeric Micelles

Polymeric micelles, also known as micellar nanocontainers, have been used as carriers for drugs and imaging dyes. Polymeric micelles, similar to liposomes, are formed spontaneously in aqueous solutions from amphiphilic block copolymers. Polymeric micelles have a core–shell architecture, with the core consisting of hydrophobic polymer blocks. The hydrophobic core can comprise up to 20%–30% water-insoluble drugs by mass, thereby preventing premature release and degradation of these drugs. These micelles can also contain hydrophilic polymer blocks (such as PEG). Polymeric micelles can be 10–100 nm in diameter. Once they are able to reach a target cell, the drug is released by diffusion (Danson et al. 2004; Kim et al. 2004; Matsumura et al. 2004).

6.2.4.4 SPIONs

SPIONs can be defined as magnetic (Fe_3O_4) or maghemite (γ-Fe_2O_3) cores that are stabilized with some kind of a hydrophilic surface coating, such as a polysaccharide, a synthetic polymer, or small molecules (Laurent et al. 2008, 2010; Mahmoudi et al. 2011b,c). In terms of their strengths vis-à-vis other kinds of NPs, SPIONs are generally well regarded due to both their biocompatibility and physiochemical properties, including effects of T_1, T_2, and T_2^* relaxation (Mahmoudi et al. 2011a,b). Additionally, they have several properties that are conducive to their implementation in biomedical applications (Mahmoudi et al. 2011b). Ideally, these particles have an average diameter of 5–10 nm and can demonstrate superparamagnetism, thus preventing the embolization in capillary vessels (Mahmoudi et al. 2011a). SPIONs can be used as MRI contrast agents (Kohler et al. 2004; Mahmoudi et al. 2011a), targeted drug delivery (Mahmoudi et al. 2011b), gene therapy (Laurent et al. 2008), and tissue repair, among other things (Gupta and Gupta 2005; Gupta et al. 2007; Mahmoudi et al. 2011b).

Of all the types of NPs, SPIONs are perhaps the most promising in terms of their applicability to diagnosis and treatment of MS. Certain SPIONs (such as those with core sizes of 3–6 nm and coated with dextran, including Feridex) have already attained approval for patient use in MRI (Liu and Chen 2005). Simultaneously, experiments have demonstrated that drug-loaded SPIONs with stimuli-sensitive polymeric shells can be directed to a target site, by using either an external magnetic field or other existing targeting methods. In fact, these experiments underscore the possibility of SPIONs having "theragnostic" value for MS patients, enabling both testing and treatment of the disease (Liu and Chen 2005; Mahmoudi et al. 2011b).

Imaging experiments have further underscored the potential for SPIONs in the diagnosis of MS. For example, an earlier MRI investigation showed that a new contrast agent (USPION) that has the capacity to accumulate in phagocytic cells enabled detection of macrophage brain infiltration *in vivo* (Vellinga et al. 2008). Vellinga et al. (2008) also showed that SPIONs could augment MRI signals in the brains of patients with MS, in a manner that is distinctive from those with Gd enhancement. The visualization of macrophage activity *in vivo* with SPIONs allows better monitoring of the dynamic process of lesion formation in MS. The macrophage activity information derived from the SPIONs is separate from and complementary to the increased BBB permeability visualized by using gadolinium (Mahmoudi et al. 2011b). Enhancing the power of this finding regarding macrophages is the fact that, recently, a method has been developed to label SPIONs with radiotracers (Jalilian et al. 2008), potentially presenting an unprecedented opportunity to develop multimodal (e.g., MRI and positron emission tomography [PET], and MRI and single-photon emission computer tomography [SPECT]) testing protocols for MS.

Thus, SPIONs have proven themselves especially useful in tracking neuroinflammatory processes. In the case of MS, SPIONs conjugated to targeting moieties and used with a magnetic field could enable simultaneous imaging and drug delivery to areas of inflammation. This method would maintain sufficient levels of the drug while mitigating the amount of drug needed and the side effects (Mahmoudi et al. 2011b). The paradigm of MS theragnosis that offers the greatest promise in terms of leveraging SPIONs is multimodal applications. For example, radiotracers would be a promising candidate for multimodal monitoring of immune system trafficking into the CNS.

In addition to the diagnostic applications discussed above, the same SPIONs could be deployed to deliver treatments to the CNS. In terms of treatment, it would be important to use "smart" stimuli-responsive polymers on the surface of SPIONs. These SPIONs are particularly adaptive in terms of their capacity to modulate themselves based on their environment, due to their ability to regulate the transportation of ions and molecules. A suitable drug (e.g., interferon B-1A or B-1B) could be utilized for aqueous colloidal dispersion of NPs using brushed polymer chains on the surface of SPIONS; in this case, the smart SPIONs would then have the capacity to release drug only in an area of inflammation, and not normal tissue (Mahmoudi et al. 2011b). Table 6.2 summarizes the differences between and functions of each of the nanocarriers mentioned above.

6.2.5 CONCLUSIONS

In this chapter, we have reviewed the applications of nanocarriers as potentially helpful therapeutic strategies in overcoming the biological obstacles posed by the complex pathoetiological mechanisms of MS. We also discussed current treatment options and the gravity of their adverse effects; nanotherapeutics offered a conduit to mitigate these negative effects by reducing the dosage and targeting inflammatory sites within tissues. We then articulated the long and often circuitous history around the creation and development of nanotherapeutics. We next discussed four key classes of nanomedicines—liposomes, NPs, polymeric micelles, and SPIONs—and elaborated on how each one could potentially be used to combat, treat, or diagnose MS, along with a discussion of the current state of the field. SPIONs have

TABLE 6.2
Summary of nanocarriers with potential for MS therapy.

Type of nanocarriers	Structure	Function and future potential
Liposomes	• Vesicular structures composed of lipid bilayer with internal aqueous compartment • Varies in size • Generally in PEGylated form	Liposomes are able to cross a compromised BBB. Prolongs circulation of delivered drugs while mitigating drug side effects.
Nanoparticles	• Insoluble polymers • ~200 nm in length for optimal cell intake • ~20 nm for nanospheres	Nanoparticles are versatile in applicability and show promising results for CNS delivery of drugs and imaging agents.
Polymeric micelles	• Amphiphilic block copolymers and a core–shell architecture with a core of hydrophobic polymer blocks • 10–100 nm in size	Shows promising ability to cross BBB; serves similar function as liposomes.
SPIONs	• Magnetic (Fe_3O_4) or maghemite (γ-Fe_2O_3) cores that are stabilized with some kind of a hydrophilic surface coating • Core size 3–6 nm, entire particle 5–10 nm	Most promising nanocarrier for MS diagnosis and treatment. Have shown efficacy in EAE models and great potential for multimodal development in imaging (e.g., MRI and PET, MRI and SPECT).

offered clinically important evidence attesting to the power of nanomedicine to reshape the treatment and diagnosis of MS. Although certain therapies, especially SPIONs, have the potential to vastly improve the theragnostic process, there are still significant obstacles to overcome—especially in terms of BBB permeability—for nanotherapeutics to emerge as a cornerstone of MS treatment.

ACKNOWLEDGMENTS

We are grateful to Adrienne M. Stoller, MA, for critically reading our chapter and providing insightful suggestions.

REFERENCES

Amiri, H., M. Mahmoudi, and A. Lascialfari. 2011. Superparamagnetic colloidal nanocrystal clusters coated with polyethylene glycol fumarate: A possible novel theranostic agent. *Nanoscale* 3 (3):1022–30.

Anderson, S. A., R. K. Rader, W. F. Westlin, C. Null, D. Jackson, G. M. Lanza, S. A. Wickline, and J. J. Kotyk. 2000. Magnetic resonance contrast enhancement of neovasculature with alpha(v)beta(3)-targeted nanoparticles. *Magn Reson Med* 44 (3):433–9.

Bartolozzi, C., R. Lencioni, F. Donati, and D. Cioni. 1999. Abdominal MR: Liver and pancreas. *Eur Radiol* 9 (8):1496–512.

Basso, A. S., D. Frenkel, F. J. Quintana, F. A. Costa-Pinto, S. Petrovic-Stojkovic, L. Puckett, A. Monsonego, A. Bar-Shir, Y. Engel, M. Gozin, and H. L. Weiner. 2008. Reversal of axonal loss and disability in a mouse model of progressive multiple sclerosis. *J Clin Invest* 118 (4):1532–43.

Bronich, T. K., P. A. Keifer, L. S. Shlyakhtenko, and A. V. Kabanov. 2005. Polymer micelle with cross-linked ionic core. *J Am Chem Soc* 127 (23):8236–7.

Calvo, P., B. Gouritin, H. Villarroya, F. Eclancher, C. Giannavola, C. Klein, J. P. Andreux, and P. Couvreur. 2002. Quantification and localization of PEGylated polycyanoacrylate nanoparticles in brain and spinal cord during experimental allergic encephalomyelitis in the rat. *Eur J Neurosci* 15 (8):1317–26.

Cerqueira, S. R., J. M. Oliveira, N. A. Silva, H. Leite-Almeida, S. Ribeiro-Samy, A. Almeida, J. F. Mano, N. Sousa, A. J. Salgado, and R. L. Reis. 2013. Microglia response and in vivo therapeutic potential of methylprednisolone-loaded dendrimer nanoparticles in spinal cord injury. *Small* 9 (5):738–49.

Chekhonin, V. P., Y. A. Zhirkov, O. I. Gurina, I. A. Ryabukhin, S. V. Lebedev, I. A. Kashparov, and T. B. Dmitriyeva. 2005. PEGylated immunoliposomes directed against brain astrocytes. *Drug Deliv* 12 (1):1–6.

Cornford, E. M., and M. E. Cornford. 2002. New systems for delivery of drugs to the brain in neurological disease. *Lancet Neurol* 1 (5):306–15.

Cui, Z., P. R. Lockman, C. S. Atwood, C. H. Hsu, A. Gupte, D. D. Allen, and R. J. Mumper. 2005. Novel D-penicillamine carrying nanoparticles for metal chelation therapy in Alzheimer's and other CNS diseases. *Eur J Pharm Biopharm* 59 (2): 263–72.

Cunningham, C. H., T. Arai, P. C. Yang, M. V. McConnell, J. M. Pauly, and S. M. Conolly. 2005. Positive contrast magnetic resonance imaging of cells labeled with magnetic nanoparticles. *Magn Reson Med* 53 (5):999–1005.

Danson, S., D. Ferry, V. Alakhov, J. Margison, D. Kerr, D. Jowle, M. Brampton, G. Halbert, and M. Ranson. 2004. Phase I dose escalation and pharmacokinetic study of pluronic polymer-bound doxorubicin (SP1049C) in patients with advanced cancer. *Br J Cancer* 90 (11):2085–91.

Das, M., C. Wang, R. Bedi, S. S. Mohapatra, and S. Mohapatra. 2014. Magnetic micelles for DNA delivery to rat brains after mild traumatic brain injury. *Nanomedicine* 10 (7):1539–48.

Feng, S. S., L. Mu, K. Y. Win, and G. Huang. 2004. Nanoparticles of biodegradable polymers for clinical administration of paclitaxel. *Curr Med Chem* 11 (4):413–24.

Folkman, J., and D. M. Long. 1964. The use of silicone rubber as a carrier for prolonged drug therapy. *J Surg Res* 4 (3):139–42.

Folkman, J., D. M. Long Jr., and R. Rosenbaum. 1966. Silicone rubber: A new diffusion property useful for general anesthesia. *Science* 154 (3745):148–9.

Friedrich, I., and C. C. Muller-Goymann. 2003. Characterization of solidified reverse micellar solutions (SRMS) and production development of SRMS-based nanosuspensions. *Eur J Pharm Biopharm* 56 (1):111–9.

Friedrich, I., S. Reichl, and C. C. Muller-Goymann. 2005. Drug release and permeation studies of nanosuspensions based on solidified reverse micellar solutions (SRMS). *Int J Pharm* 305 (1–2):167–75.

Gabizon, A., R. Catane, B. Uziely, B. Kaufman, T. Safra, R. Cohen, F. Martin, A. Huang, and Y. Barenholz. 1994. Prolonged circulation time and enhanced accumulation in malignant exudates of doxorubicin encapsulated in polyethylene-glycol coated liposomes. *Cancer Res* 54 (4):987–92.

Gabizon, A., H. Shmeeda, and Y. Barenholz. 2003. Pharmacokinetics of pegylated liposomal doxorubicin: Review of animal and human studies. *Clin Pharm* 42 (5):419–36.

Gendelman, H. E., V. Anantharam, T. Bronich, S. Ghaisas, H. Jin, A. G. Kanthasamy, X. Liu, J. McMillan, R. L. Mosley, B. Narasimhan, and S. K. Mallapragada. 2015. Nanoneuromedicines for degenerative, inflammatory, and infectious nervous system diseases. *Nanomedicine* 11 (3):751–67.

Gharagozloo, M., S. Majewski, and M. Foldvari. 2015. Therapeutic applications of nanomedicine in autoimmune diseases: From immunosuppression to tolerance induction. *Nanomedicine* 11 (4):1003–18.

Goldenberg, M. M. 2012. Multiple sclerosis review. *Pharm Ther* 37 (3):175–84.

Gradishar, W. J. 2006. Albumin-bound paclitaxel: A next-generation taxane. *Expert Opin Pharmacother* 7 (8):1041–53.

Gupta, A. K., and M. Gupta. 2005. Synthesis and surface engineering of iron oxide nanoparticles for biomedical applications. *Biomaterials* 26 (18):3995–4021.

Gupta, A. K., R. R. Naregalkar, V. D. Vaidya, and M. Gupta. 2007. Recent advances on surface engineering of magnetic iron oxide nanoparticles and their biomedical applications. *Nanomedicine (Lond)* 2 (1):23–39.

Helms, B., and E. W. Meijer. 2006. Chemistry. Dendrimers at work. *Science* 313 (5789):929–30.

Hunter, Z., D. P. McCarthy, W. T. Yap, C. T. Harp, D. R. Getts, L. D. Shea, and S. D. Miller. 2014. A biodegradable nanoparticle platform for the induction of antigen-specific immune tolerance for treatment of autoimmune disease. *ACS Nano* 8 (3):2148–60.

Huwyler, J., D. Wu, and W. M. Pardridge. 1996. Brain drug delivery of small molecules using immunoliposomes. *Proc Natl Acad Sci USA* 93 (24):14164–9.

Hyuk Im, S., U. Jeong, and Y. Xia. 2005. Polymer hollow particles with controllable holes in their surfaces. *Nat Mater* 4 (9):671–5.

Iezzi, R., B. R. Guru, I. V. Glybina, M. K. Mishra, A. Kennedy, and R. M. Kannan. 2012. Dendrimer-based targeted intravitreal therapy for sustained attenuation of neuroinflammation in retinal degeneration. *Biomaterials* 33 (3):979–88.

Jacobs, C., O. Kayser, and R. H. Muller. 2000. Nanosuspensions as a new approach for the formulation for the poorly soluble drug tarazepide. *Int J Pharm* 196 (2):161–4.

Jalilian, A. R., M. Nikzad, H. Zandi, A. N. Kharat, P. Rowshanfarzad, M. Akhlaghi, and F. Bolourinovin. 2008. Preparation and evaluation of [(61)Cu]-thiophene-2-aldehyde thiosemicarbazone for PET studies. *Nucl Med Rev Cent East Eur* 11 (2):41–7.

Kabanov, A. V., and H. E. Gendelman. 2007. Nanomedicine in the diagnosis and therapy of neurodegenerative disorders. *Progr Polym Sci* 32 (8–9):1054–82.

Kabanov, A., J. Zhu, and V. Alakhov. 2005. Pluronic block copolymers for gene delivery. *Adv Genet* 53:231–61.

Kakizawa, Y., and K. Kataoka. 2002. Block copolymer micelles for delivery of gene and related compounds. *Adv Drug Deliv Rev* 54 (2):203–22.

Kannan, S., H. Dai, R. S. Navath, B. Balakrishnan, A. Jyoti, J. Janisse, R. Romero, and R. M. Kannan. 2012. Dendrimer-based postnatal therapy for neuroinflammation and cerebral palsy in a rabbit model. *Sci Transl Med* 4 (130):130ra46.

Kazakov, S., M. Kaholek, I. Gazaryan, B. Krasnikov, K. Miller, and K. Levon. 2006. Ion concentration of external solution as a characteristic of micro- and nanogel ionic reservoirs. *J Phys Chem B* 110 (31):15107–16.

Kim, T. Y., D. W. Kim, J. Y. Chung, S. G. Shin, S. C. Kim, D. S. Heo, N. K. Kim, and Y. J. Bang. 2004. Phase I and pharmacokinetic study of Genexol-PM, a cremophor-free, polymeric micelle-formulated paclitaxel, in patients with advanced malignancies. *Clin Cancer Res* 10 (11):3708–16.

Kizelsztein, P., H. Ovadia, O. Garbuzenko, A. Sigal, and Y. Barenholz. 2009. Pegylated nanoliposomes remote-loaded with the antioxidant tempamine ameliorate experimental autoimmune encephalomyelitis. *J Neuroimmunol* 213 (1–2):20–25.

Kohler, N., G. E. Fryxell, and M. Zhang. 2004. A bifunctional poly(ethylene glycol) silane immobilized on metallic oxide-based nanoparticles for conjugation with cell targeting agents. *J Am Chem Soc* 126 (23):7206–11.

Kozubek, A., J. Gubernator, E. Przeworska, and M. Stasiuk. 2000. Liposomal drug delivery, a novel approach: PLARosomes. *Acta Biochim Pol* 47 (3): 639–49.

Kreuter, J. 2001. Nanoparticulate systems for brain delivery of drugs. *Adv Drug Deliv Rev* 47 (1):65–81.

Kreuter, J., P. Ramge, V. Petrov, S. Hamm, S. E. Gelperina, B. Engelhardt, R. Alyautdin, H. von Briesen, and D. J. Begley. 2003. Direct evidence that polysorbate-80-coated poly(butylcyanoacrylate) nanoparticles deliver drugs to the CNS via specific mechanisms requiring prior binding of drug to the nanoparticles. *Pharm Res* 20 (3):409–16.

Kwon, G. S. 2003. Polymeric micelles for delivery of poorly water-soluble compounds. *Crit Rev Ther Drug Carrier Syst* 20 (5):357–403.

Langer, R. 2001. Drugs on target. *Science* 293 (5527): 58–9.

Laurent, S., J. L. Bridot, L. V. Elst, and R. N. Muller. 2010. Magnetic iron oxide nanoparticles for biomedical applications. *Future Med Chem* 2 (3):427–49.

Laurent, S., D. Forge, M. Port, A. Roch, C. Robic, L. Vander Elst, and R. N. Muller. 2008. Magnetic iron oxide nanoparticles: Synthesis, stabilization, vectorization, physicochemical characterizations, and biological applications. *Chem Rev* 108 (6):2064–110.

Lesniak, M. S., R. Langer, and H. Brem. 2001. Drug delivery to tumors of the central nervous system. *Curr Neurol Neurosci Rep* 1 (3):210–16.

Lin, Y., Y. Pan, Y. Shi, X. Huang, N. Jia, and J. Y. Jiang. 2012. Delivery of large molecules via poly(butyl cyanoacrylate) nanoparticles into the injured rat brain. *Nanotechnology* 23 (16):165101.

Liu, G., M. R. Garrett, P. Men, X. Zhu, G. Perry, and M. A. Smith. 2005. Nanoparticle and other metal chelation therapeutics in Alzheimer disease. *Biochim Biophys Acta* 1741 (3):246–52.

Liu, X., and C. Chen. 2005. Strategies to optimize brain penetration in drug discovery. *Curr Opin Drug Discov Devel* 8 (4):505–12.

Loma, I., and R. Heyman. 2011. Multiple sclerosis: Pathogenesis and treatment. *Curr Neuropharmacol* 9 (3):409–16.

Mahmoudi, M., H. Hosseinkhani, M. Hosseinkhani, S. Boutry, A. Simchi, W. S. Journeay, K. Subramani, and S. Laurent. 2011a. Magnetic resonance imaging tracking of stem cells in vivo using iron oxide nanoparticles as a tool for the advancement of clinical regenerative medicine. *Chem Rev* 111 (2):253–80.

Mahmoudi, M., M. A. Sahraian, M. A. Shokrgozar, and S. Laurent. 2011b. Superparamagnetic iron oxide nanoparticles: Promises for diagnosis and treatment of multiple sclerosis. *ACS Chem Neurosci* 2 (3):118–40.

Mahmoudi, M., S. Sant, B. Wang, S. Laurent, and T. Sen. 2011c. Superparamagnetic iron oxide nanoparticles (SPIONs): Development, surface modification and applications in chemotherapy. *Adv Drug Deliv Rev* 63 (1–2):24–46.

Matsumura, Y., T. Hamaguchi, T. Ura, K. Muro, Y. Yamada, Y. Shimada, K. Shirao, T. Okusaka, H. Ueno, M. Ikeda, and N. Watanabe. 2004. Phase I

clinical trial and pharmacokinetic evaluation of NK911, a micelle-encapsulated doxorubicin. *Br J Cancer* 91 (10):1775–81.

McAllister, K., P. Sazani, M. Adam, M. J. Cho, M. Rubinstein, R. J. Samulski, and J. M. DeSimone. 2002. Polymeric nanogels produced via inverse microemulsion polymerization as potential gene and antisense delivery agents. *J Am Chem Soc* 124 (51):15198–207.

Missirlis, D., N. Tirelli, and J. A. Hubbell. 2005. Amphiphilic hydrogel nanoparticles. Preparation, characterization, and preliminary assessment as new colloidal drug carriers. *Langmuir* 21 (6):2605–13.

Moghimi, S. M., A. C. Hunter, and J. C. Murray. 2005. Nanomedicine: Current status and future prospects. *FASEB J* 19 (3):311–30.

Muller, R. H., C. Jacobs, and O. Kayser. 2001. Nanosuspensions as particulate drug formulations in therapy. Rationale for development and what we can expect for the future. *Adv Drug Deliv Rev* 47 (1):3–19.

Nayak, S., and L. A. Lyon. 2005. Soft nanotechnology with soft nanoparticles. *Angew Chem Int Ed* 44 (47):7686–708.

Nel, A. E., L. Madler, D. Velegol, T. Xia, E. M. Hoek, P. Somasundaran, F. Klaessig, V. Castranova, and M. Thompson. 2009. Understanding biophysicochemical interactions at the nano-bio interface. *Nat Mater* 8 (7):543–57.

Nowacek, A., L. M. Kosloski, and H. E. Gendelman. 2009. Neurodegenerative disorders and nanoformulated drug development. *Nanomedicine (Lond)* 4 (5):541–55.

O'Reilly, R. K., C. J. Hawker, and K. L. Wooley. 2006. Cross-linked block copolymer micelles: Functional nanostructures of great potential and versatility. *Chem Soc Rev* 35 (11):1068–83.

Ogris, M., and E. Wagner. 2002. Tumor-targeted gene transfer with DNA polyplexes. *Somat Cell Mol Genet* 27 (1–6):85–95.

Oh, K. T., T. K. Bronich, L. Bromberg, T. A. Hatton, and A. V. Kabanov. 2006. Block ionomer complexes as prospective nanocontainers for drug delivery. *J Control Release* 115 (1):9–17.

Olivier, J. C., L. Fenart, R. Chauvet, C. Pariat, R. Cecchelli, and W. Couet. 1999. Indirect evidence that drug brain targeting using polysorbate 80-coated polybutylcyanoacrylate nanoparticles is related to toxicity. *Pharm Res* 16 (12):1836–42.

Papaldo, P., A. Fabi, G. Ferretti, M. Mottolese, A. M. Cianciulli, B. Di Cocco, M. S. Pino et al. 2006. A phase II study on metastatic breast cancer patients treated with weekly vinorelbine with or without trastuzumab according to HER2 expression: Changing the natural history of HER2-positive disease. *Ann Oncol* 17 (4):630–6.

Peracchia, M. T., C. Vauthier, D. Desmaele, A. Gulik, J. C. Dedieu, M. Demoy, J. d'Angelo, and P. Couvreur. 1998. Pegylated nanoparticles from a novel methoxypolyethylene glycol cyanoacrylate-hexadecyl cyanoacrylate amphiphilic copolymer. *Pharm Res* 15 (4):550–6.

Pezeshki, M. Z., S. Zarrintan, and M. H. Zarrintan. 2008. *Helicobacter pylori* nanoparticles as a potential treatment of conventional multiple sclerosis. *Med Hypotheses* 70 (6):1223.

Pochan, D. J., Z. Chen, H. Cui, K. Hales, K. Qi, and K. L. Wooley. 2004. Toroidal triblock copolymer assemblies. *Science* 306 (5693):94–7.

Polyak, B., and G. Friedman. 2009. Magnetic targeting for site-specific drug delivery: Applications and clinical potential. *Expert Opin Drug Deliv* 6 (1):53–70.

Rabinow, B. E. 2004. Nanosuspensions in drug delivery. *Nat Rev Drug Discov* 3 (9):785–96.

Ransohoff, R. M., Hafler, D. A., Lucchinetti, C. F. 2015. Multiple sclerosis—A quiet revolution. *Nat Rev Neurol* 11 (3):134–42.

Read, M. L., A. Logan, and L. W. Seymour. 2005. Barriers to gene delivery using synthetic vectors. *Adv Genet* 53:19–46.

Richards Grayson, A. C., I. S. Choi, B. M. Tyler, P. P. Wang, H. Brem, M. J. Cima, and R. Langer. 2003. Multi-pulse drug delivery from a resorbable polymeric microchip device. *Nat Mater* 2 (11): 767–72.

Rousseau, V., B. Denizot, J. J. Le Jeune, and P. Jallet. 1999. Early detection of liposome brain localization in rat experimental allergic encephalomyelitis. *Exp Brain Res* 125 (3):255–64.

Ruozi, B., D. Belletti, F. Forni, A. Sharma, D. Muresanu, H. Mossler, M. A. Vandelli, G. Tosi, and H. S. Sharma. 2014. Poly (D,L-lactide-co-glycolide) nanoparticles loaded with cerebrolysin display neuroprotective activity in a rat model of concussive head injury. *CNS Neurol Disord Drug Targets* 13 (8):1475–82.

Salem, A. K., P. C. Searson, and K. W. Leong. 2003. Multifunctional nanorods for gene delivery. *Nat Mater* 2 (10):668–71.

Saltzman, W. M., M. W. Mak, M. J. Mahoney, E. T. Duenas, and J. L. Cleland. 1999. Intracranial delivery of recombinant nerve growth factor: Release kinetics and protein distribution for three delivery systems. *Pharm Res* 16 (2):232–40.

Samuel, E. L., M. T. Duong, B. R. Bitner, D. C. Marcano, J. M. Tour, and T. A. Kent. 2014. Hydrophilic carbon clusters as therapeutic, high-capacity antioxidants. *Trends Biotechnol* 32 (10):501–5.

Savic, R., L. Luo, A. Eisenberg, and D. Maysinger. 2003. Micellar nanocontainers distribute to defined cytoplasmic organelles. *Science* 300 (5619):615–8.

Schmidt, J., J. M. Metselaar, M. H. Wauben, K. V. Toyka, G. Storm, and R. Gold. 2003. Drug targeting by long-circulating liposomal glucocorticosteroids increases therapeutic efficacy in a model of multiple sclerosis. *Brain* 126 (Pt 8):1895–904.

Sharpe, M., S. E. Easthope, G. M. Keating, and H. M. Lamb. 2002. Polyethylene glycol-liposomal doxorubicin: A review of its use in the management of solid and haematological malignancies and AIDS-related Kaposi's sarcoma. *Drugs* 62 (14):2089–126.

Shi, N., Y. Zhang, C. Zhu, R. J. Boado, and W. M. Pardridge. 2001. Brain-specific expression of an exogenous gene after i.v. administration. *Proc Natl Acad Sci USA* 98 (22):12754–9.

Siepmann, J., F. Siepmann, and A. T. Florence. 2006. Local controlled drug delivery to the brain: Mathematical modeling of the underlying mass transport mechanisms. *Int J Pharm* 314 (2):101–19.

Soni, S., A. K. Babbar, R. K. Sharma, and A. Maitra. 2006. Delivery of hydrophobised 5-fluorouracil derivative to brain tissue through intravenous route using surface modified nanogels. *J Drug Target* 14 (2):87–95.

Steiniger, S. C., J. Kreuter, A. S. Khalansky, I. N. Skidan, A. I. Bobruskin, Z. S. Smirnova, S. E. Severin, R. Uhl, M. Kock, K. D. Geiger, and S. E. Gelperina. 2004. Chemotherapy of glioblastoma in rats using doxorubicin-loaded nanoparticles. *Int J Cancer* 109 (5):759–67.

Stroh, M., W. R. Zipfel, R. M. Williams, W. W. Webb, and W. M. Saltzman. 2003. Diffusion of nerve growth factor in rat striatum as determined by multiphoton microscopy. *Biophys J* 85 (1):581–8.

Tabansky, I., M. D. Messina, C. Bangeranye, J. Goldstein, K. M. Blitz-Shabbir, S. Machado, V. Jeganathan et al. 2015. Advancing drug delivery systems for the treatment of multiple sclerosis. *Immunol Res* 63 (1–3):58–69.

Tao, L., and K. E. Uhrich. 2006. Novel amphiphilic macromolecules and their in vitro characterization as stabilized micellar drug delivery systems. *J Colloid Interface Sci* 298 (1):102–10.

Torchilin, V. P. 1998. Polymer-coated long-circulating microparticulate pharmaceuticals. *J Microencapsul* 15 (1):1–19.

Torchilin, V. P. 2007. Micellar nanocarriers: Pharmaceutical perspectives. *Pharm Res* 24 (1):1–16.

Trentin, D., J. Hubbell, and H. Hall. 2005. Non-viral gene delivery for local and controlled DNA release. *J Control Release* 102 (1):263–75.

Umezawa, F., and Y. Eto. 1988. Liposome targeting to mouse brain: Mannose as a recognition marker. *Biochem Biophys Res Commun* 153 (3):1038–44.

van Nostrum, C. F. 2004. Polymeric micelles to deliver photosensitizers for photodynamic therapy. *Adv Drug Deliv Rev* 56 (1):9–16.

Vellinga, M. M., R. D. Oude Engberink, A. Seewann, P. J. Pouwels, M. P. Wattjes, S. M. van der Pol, C. Pering, C. H. Polman, H. E. de Vries, J. J. Geurts, and F. Barkhof. 2008. Pluriformity of inflammation in multiple sclerosis shown by ultra-small iron oxide particle enhancement. *Brain* 131 (Pt 3):800–7.

Vinogradov, S. V., E. V. Batrakova, and A. V. Kabanov. 2004. Nanogels for oligonucleotide delivery to the brain. *Bioconjug Chem* 15 (1):50–60.

Voinea, M., and M. Simionescu. 2002. Designing of "intelligent" liposomes for efficient delivery of drugs. *J Cell Mol Med* 6 (4):465–74.

Wang, B., R. S. Navath, R. Romero, S. Kannan, and R. Kannan. 2009. Anti-inflammatory and antioxidant activity of anionic dendrimer-N-acetyl cysteine conjugates in activated microglial cells. *Int J Pharm* 377 (1–2):159–68.

Wang, F., T. K. Bronich, A. V. Kabanov, R. D. Rauh, and J. Roovers. 2005. Synthesis and evaluation of a star amphiphilic block copolymer from poly(epsilon-caprolactone) and poly(ethylene glycol) as a potential drug delivery carrier. *Bioconjug Chem* 16 (2):397–405.

Wu, G. Y., and C. H. Wu. 1987. Receptor-mediated in vitro gene transformation by a soluble DNA carrier system. *J Biol Chem* 262 (10):4429–32.

Wu, M. P., J. A. Tamada, H. Brem, and R. Langer. 1994. In vivo versus in vitro degradation of controlled release polymers for intracranial surgical therapy. *J Biomed Mater Res* 28 (3):387–95.

Yeste, A., M. Nadeau, E. J. Burns, H. L. Weiner, and F. J. Quintana. 2012. Nanoparticle-mediated codelivery of myelin antigen and a tolerogenic small molecule suppresses experimental autoimmune encephalomyelitis. *Proc Natl Acad Sci USA* 109 (28):11270–5.

Bridging the Gap between the Bench and the Clinic

Yonghao Cao, Inna Tabansky, Joyce J. Pan, Mark Messina,
Maya Shabbir, Souhel Najjar, Paul Wright, and Joel N. H. Stern

CONTENTS

6.3.1 INTRODUCTION

Multiple sclerosis (MS) is a genetically mediated, inflammatory, demyelinating, and neurodegenerative disorder characterized by infiltration of immune cells into the central nervous system (CNS). Demyelination results from the recognition of myelin antigens by immune cells and subsequent attack on these antigens (McFarland and Martin 2007; Goverman 2009; Ransohoff et al. 2015). The etiology of MS is still unclear, but intensive research is being conducted into the causes of the disease. Ongoing studies on genetics, epidemiology, and immunology have revealed that the cause of MS is complex. The origin of the disease is rooted in the interactions between genetic factors and environmental factors, resulting in immunological imbalances and dysfunctions, ultimately leading to autoimmune responses (Ransohoff et al. 2015).

On the treatment front, there are currently about 10 first-line disease-modifying therapies (DMTs) and therapies alleviating symptoms, but the translation of basic research findings into the clinic has been slow in many areas (Wingerchuk and Carter 2014; Tabansky et al. 2016). As of now, there is still only one approved nanomedicine treatment for MS. In this section, we discuss some of the epidemiology that was elaborated on in Chapter 6.1 and discuss the causes and potential solutions of the gap between the bench and the clinic in MS, and how this relates to nanomedicine.

6.3.2 GENETIC FACTORS

6.3.2.1 Genetic Variation and MS

Identification and characterization of common variants in MS requires unbiased, whole-genome approaches, as contributions of individual loci to disease risk are quite small. Genome-wide association studies (GWASs) have identified more than 100 genetic variants that are associated with an increased susceptibility to MS, located in human leukocyte antigen (HLA), cytotoxic T lymphocyte–associated protein 4 (CTLA-4), interleukin (IL)-2/IL-7R, tumor necrosis factor (TNF) or nuclear factor-κB (NFκB), and other genes. These GWASs also provided exceptionally clear evidence that MS is an autoimmune disease (International Multiple Sclerosis Genetics Consortium et al. 2007, 2011). The majority of causal variants are noncoding and map to immune cell–specific enhancers (Farh et al. 2015).

The mechanisms that cause heritable differences among individuals remain largely unclear. Recently, an algorithm for fine-mapping single-nucleotide polymorphisms (SNPs) associated with many autoimmune diseases was generated to identify and characterize the causal variants driving autoimmune disease risk. Candidate causal variants tend to coincide with nucleosome-depleted regions bound by master regulators of immune differentiation and stimulus-dependent gene activation, including IRF4, PU.1, NFKB, and AP-1 family transcription factors (Farh et al. 2015). Identifying specific sites where a single, noncoding nucleotide variant is responsible for disease risk may pinpoint specific disruptions of consensus transcription factor binding sites that ultimately define disease risk, as related to environmental factors. These data clearly demonstrate that the genetic variants associated with MS are primarily related to immune genes. MS clusters with other autoimmune disorders, finally answering the question as to whether MS is an immunologic disease.

6.3.2.2 Genetic Locus to Cellular Pathways

In the post-GWAS era, researchers are moving from genetics to functional immunology, correlating the phenotypic differences with variants in disease states, and elucidating how genetic variations influence immune cell function to drive disease development and progression. Allelic variants in genes for cytokine receptors and costimulatory molecules have been associated with T cell function. For example, the MS risk allele in the CD6 locus affects *Cd6* mRNA alternative splicing and alters CD4+ T cell proliferation (Kofler et al. 2011). Moreover, blocking CD6 signaling has shown promising effects in autoimmune diseases (Kofler et al. 2016). Other genetic variants associated with risk of MS alter NFκB signaling pathways, resulting in enhanced NFκB activation and a lower CD4+ T cell activation threshold to inflammatory stimuli (Housley et al. 2015a). Despite these advances in the genetic understanding of MS pathogenesis, the contribution of genetics to the MS clinical course has not yet been articulated.

An integrative approach is needed to leverage genetic and cellular immunology data in predicting disease course and patient response to treatment. The better understanding of the genetic basis of autoimmunity may lead to more sophisticated models of underlying cellular phenotypes and, eventually, novel diagnostics and targeted therapies, such as precision and personal medicine (Marson et al. 2015).

6.3.3 ENVIRONMENTAL FACTORS

Although the role of genetics in MS has been extensively established through monozygotic twin studies and multiple GWASs, there is also ample evidence supporting a role for environmental factors in the etiology of the disease (Wingerchuk 2011; O'Gorman et al. 2012). Furthermore, as stated on the National MS Society webpage,

> Migration from one geographic area to another seems to alter a person's risk of developing MS. Studies indicate that immigrants and their descendants tend to take on the risk level—either higher or lower—of the area to which they move. Those who move in early childhood tend to take on the new risk themselves. For those who move later in life, the change in risk level may not appear until the next generation.

There is considerable evidence of interplay between environmental factors and genetics in the onset of MS, and the present section of this chapter delineates the known environmental factors that affect the onset of MS.

Some of the most prevalent environmental factors that have been associated with MS include vitamin D and sun exposure, dietary salt intake, Epstein–Barr virus (EBV), John Cunningham virus (JCV), and smoking cigarettes. The first three of these factors will be discussed at length in this section, with a focus on exploring how they might help translate scientific research into clinical applications.

6.3.3.1 Vitamin D and Sun Exposure

Low levels of vitamin D and sun exposure have previously been linked to a host of autoimmune diseases, including lupus, type 1 diabetes, inflammatory bowel diseases, and rheumatoid arthritis (Arnson et al. 2007; Agmon-Levin et al. 2015; Azrielant and Shoenfeld 2016; Watad et al. 2016b). These findings have been reflected in clinical studies of MS. Most notably, the BENEFIT trial demonstrated an inverse relationship between vitamin D levels and several metrics of MS severity, including disease activity, progression, lesion volume, and brain volume (Watad et al. 2016a). Similar findings, with respect to the correlation between low vitamin D levels and increased likelihood of MS, were made in another study (Ascherio et al. 2014). A study of the potential causes of the comparatively reduced incidence of MS in Norway, as opposed to countries with comparable climate, pointed to the effect of vitamin D in modulating MS frequency based on latitude (Kampman and Brustad 2008; Holick et al. 2011; Watad et al. 2016a). Taken collectively, the results of these studies have been sufficiently compelling that clinical trials have been undertaken to better understand the safety and efficacy of vitamin D therapy for relapsing–remitting MS patients (Ascherio et al. 2014; Bhargava et al. 2014; Watad et al. 2016a).

Although the preponderance of evidence suggests a vital role for vitamin D and latitude-dependent sun exposure in the prevalence of MS, there is still a dearth of literature on the precise role of vitamin D in MS and autoimmunity.

The uncertainty of the exact role of Ultraviolet B (UVB)-derived vitamin D in MS and the multifaceted nature of the causes of the disease make the translation of findings on vitamin D into the clinical setting all the more difficult.

6.3.3.2 Virus and EBV

Although widely recognized as an autoimmune disease, MS can in some ways mimic the progression of infectious and viral diseases (Olson et al. 2001). This is especially evident in animal models of MS, where experimental autoimmune encephalomyelitis (EAE) and Theiler's murine encephalomyelitis virus (TMEV) are used to study various aspects of the disease, such as progression, drug treatments, and diagnosis (Oleszak et al. 2004; Constantinescu et al. 2011). A recent study shows that EBV-related lymphocryptovirus (LCV) B cell infection causes the conversion of destructive processing of the myelin antigen into cross-presentation to strongly autoaggressive CTLs (Jagessar et al. 2016).

Many infections have been proposed to play a role in MS pathogenesis, but it is EBV that has presented the strongest evidence of a connection. Late infection with EBV, and a past history of infectious mononucleosis (IM), is particularly associated with MS risk (Haahr et al. 1995, 2004). Both case-control and cohort studies have repeatedly reported an association between a past history of IM—or with higher levels of EBV-specific antibodies—and susceptibility to MS (Lucas et al. 2011; O'Gorman et al. 2012). In a recent meta-analysis of 18 case-control and cohort studies, a history of IM was associated with a twofold increase in the risk of developing MS (Handel et al. 2010). The increased risk of MS posed by exposure to viruses has been specifically attributed to higher titers of immunoglobulin (Ig) G antibodies specific to Epstein–Barr nuclear antigens (EBNAs), with far less statistically meaningful links to other EBV antigens or other viral infections, such as measles, herpes simplex virus (HSV), JCV, varicella zoster virus (VZV), LCV, or cytomegalovirus (CMV) (Ascherio et al. 2001; Sundstrom et al. 2004; Levin et al. 2005; O'Gorman et al. 2012).

6.3.3.3 Salt

Over the past decade, significant evidence has accumulated that attests to a potential role of dietary salt in autoimmunity. Dietary salt is believed to possess pro-inflammatory properties, but more research needs to be undertaken to better understand its effect on the pathogenesis and susceptibility to MS. For example, an observational study of relapsing–remitting MS revealed an increased rate of clinical flare-ups and MRI activity in patients with higher dietary salt intake versus those who consumed a lower-salt diet (Farez et al. 2015). Over the past 2 years, at least two other studies have found a link between high-salt diets and earlier onset or progression of MS in the EAE animal models (Kleinewietfeld et al. 2013; Wu et al. 2013). These studies similarly suggested that dietary salt was associated with pro-inflammatory changes. Recent studies have suggested that high salt promotes autoimmunity by inducing pro-inflammatory responses in effector T cells and inhibiting the suppressive function of Foxp3+ regulatory T cells (Tregs) (Hernandez et al. 2015). Moreover, high salt induces macrophage activation, which may thus lead to an overall imbalance in immune homeostasis (Binger et al. 2015; Zhang et al. 2015). However, other studies cast doubt on that finding. A multicenter case-control study showed no evidence that dietary salt increased MS susceptibility in children (McDonald et al. 2016). More studies are therefore needed to determine whether salt contributes to the pathogenesis of human autoimmune diseases.

6.3.4 INTERACTIONS BETWEEN GENETICS AND ENVIRONMENTAL FACTORS

The current concept is that the interactions between genetics and environmental factors lead to immune dysregulation, and eventually cause human autoimmune diseases. Thus, particular genetic variants will only cause deleterious effects in specific environmental conditions, and may be neutral or advantageous in other conditions.

Genetic variants that increase risk of MS cause different cell types of the adaptive and innate immune system—including Th17 and B cells and macrophages—to become more easily activated (Nylander and Hafler 2012). Genetic and epigenetic fine mapping shows the loss of immune regulation with environmental factors that link to genetic loci (Farh et al. 2015). As mentioned in Chapter 6.1, there are many useful biomarkers for disease susceptibility (Housley et al. 2015b). How the interactions between genetics and environmental factors contributed to immune dysregulation is the next step toward understanding immune function in the post-GWAS era.

6.3.5 IMMUNOLOGICAL FACTORS

The idea that MS is an autoimmune disease originated in studies of EAE, which is induced by myelin antigen–induced activation or adoptive transfer of myelin protein–specific self-reactive

T cells (Rangachari and Kuchroo 2013). MS occurs in the context of breaks in tolerance to self-antigens driven by both genetic and environmental factors. Ultimately, self-reactive immune cells in the cerebrospinal fluid (CSF) and in the circulation play a critical role in the pathogenesis of this disease. New therapies for relapsing–remitting MS have been found in the last two decades, but there are still no effective therapies that can alter disease course for progressive MS. Most drugs used to treat MS are immunologic modulators that try to reduce the duration and frequency of the inflammatory phase of the disease.

6.3.5.1 Self-Antigens

Myelin protein–derived antigens have been extensively studied and considered the main autoreactive targets in patients with MS. As demyelination is the key feature of MS neuropathology, the well-characterized autoantigens are myelin proteins, including myelin oligodendrocyte glycoprotein (MOG), myelin basic protein (MBP), proteolipid protein dendrocyte (PLP), and myelin-associated glycoprotein (MAG) (Siglec-4). Autoimmune attack against these antigens is thought to be involved in both MS and EAE (Dendrou et al. 2015).

Myelin-reactive T cells and B cells are well characterized in the EAE model and MS, and a significant percentage of MS patients are also seropositive for antibodies against myelin antigens (Reindl et al. 1999). A recent study has reported that the potassium channel KIR4.1 is a putative self-antigen for MS (Srivastava et al. 2012), but this finding was challenged by other groups (Brickshawana et al. 2014; Chastre et al. 2016; Pröbstel et al. 2016). Thus, whether other antigens besides myelin are involved in MS remains unknown.

6.3.5.2 CD4+ T Cells

Investigations of effector CD4+ T cells in the EAE model have demonstrated that the minimal requirement for inducing an inflammatory autoimmune demyelinating disease is the activation of interferon (IFN)-γ-producing T helper (Th) 1 and IL-17-producing Th17 cells (Kroenke et al. 2008; Ghoreschi et al. 2010). Confounding these observations, mice treated with either anti-IFN-γ- or anti-IL-17-blocking antibodies show an exacerbation of EAE (Haak et al. 2009; Axtell et al. 2010). Recent studies have shown that granulocyte–macrophage colony-stimulating factor (GM-CSF)-producing Th cells displayed a district transcriptional profile (Codarri et al. 2011; El-Behi et al. 2011), and may represent a new subset of T cells that play a pathogenic role in EAE (Noster et al. 2014; Sheng et al. 2014). Collectively, these studies demonstrate that Th1-, Th17-, and GM-CSF-producing T cells participate in MS pathogenesis, but their role might be complicated.

Autoreactive T cells are thought to play a critical role in MS pathogenesis through the recognition of myelin antigens. Antigen-presenting cells (APCs) process and present microbial or self-antigen—such as myelin antigens—to T cells. A certain cytokine condition drives these T cells to differentiate into Th1 or Th17 cells to become putative autoreactive T cells (Nylander and Hafler 2012). In healthy subjects, these potentially pathogenic T cells are not activated due to suppression by Tregs, such as FoxP3$^+$ Tregs and type 1 regulatory (Tr1) T cells. However, patients with MS have defects in peripheral immune regulation, such as downregulated expression of CTLA-4 on Foxp3$^+$ Treg cells, and lower IL-10 production in Tr1 cells (Kleinewietfeld and Hafler 2014). Additionally, MS patients possess a population of PI3K/AKT/Foxo1/3 pathway-dependent, dysfunctional IFN-γ-secreting Tregs that do not suppress autoimmunity properly (Dominguez-Villar et al. 2011; Kitz et al. 2016). Thus, the activation barrier for autoreactive T cells is lowered in MS patients. Activated myelin-reactive T cells can then cross the blood–brain barrier (BBB) into the CNS, where they will encounter myelin antigens that will induce them to undergo clonal expansion and activation. They will then release pro-inflammatory cytokines, mediators, and proteases, leading to the destruction of myelin and axonal damage.

Myelin-reactive T cells have been identified in the circulation of patients with MS and healthy subjects at comparable frequencies over the last quarter century, which raised questions as to the relevance of these cells in patients (Ota et al. 1990; Martin et al. 1991; Raddassi et al. 2011). Over this period, it has been very difficult to identify the particular T cells involved in the disease (Raddassi et al. 2011). Most recently, we have been able to identify, for the first time, the functional and phenotypic properties that distinguish pathogenic T cells in patients with MS. Myelin-reactive T cells from patients with MS produce more pro-inflammatory cytokines IFN-γ, IL-17, and GM-CSF,

but less IL-10 in comparison with healthy subjects. Moreover, myelin-specific T cells from MS patients demonstrated a transcriptome homology with encephalitogenic T cells isolated from mice with EAE and revealed a transcriptional profile that distinguishes MS-derived myelin-reactive T cells. For instance, myelin-specific T cells from healthy subjects showed considerable enrichment of transcripts associated with the CTLA-4 inhibitory pathway (Cao et al. 2015). Correlatively, inflammatory CNS demyelination and enhanced myelin-reactive T cell responses were observed in melanoma patients after treatment with ipilimumab, a monoclonal antibody against CTLA-4 (Cao et al. 2016). Targeting these cytokines or pathways holds promise as potential treatments for MS. Taken together, these findings enhance our understanding of MS pathogenesis, and they can guide the development of focused therapies that can circumvent the need to impact the immune system more broadly, thereby optimizing the risk–benefit profile of future interventions.

6.3.5.3 CD8$^+$ T Cells

The role of CD8$^+$ T cells in MS pathology remains unclear, as EAE models have shown it to be a primarily CD4$^+$ T cell–mediated disease. Studies on the effect of genetic deletion or antibody-mediated depletion of CD8$^+$ T cells in EAE suggest a regulatory role for CD8$^+$ T cells (Koh et al. 1992). CD8$^+$ T cells predominate in the CSF and brain tissue sections of MS patients with clonal expansion of immune cells (Friese and Fugger 2009; Goverman et al. 2009). Major histocompatibility complex (MHC) class I–restricted MBP-specific CD8$^+$ T cells are pathogenic and cause lesions distinct from those observed in CD4$^+$ T cell–mediated conventional EAE, but similar to some lesions in MS patients. These observations suggest that CD8$^+$ T cells play a different role than CD4$^+$ T cells in disease induction and development (Huseby et al. 2001; Ji et al. 2010). It is important to identify the specific effects of CD8$^+$ T cells in order to design targeted therapies to ameliorate their potentially negative effects.

6.3.5.4 B Cells

The identification of specific autoantibodies associated with the disease would certainly benefit the diagnosis and treatment. Peripheral myelin antigen destruction and exposure result in B cell activation, affinity maturation, clonal expansion, and transmigration across the BBB into the CNS. Self-antigen primed B cells produce autoantibodies, and contribute to the pathogenesis of MS. Related B cell clones that populate the CNS of patients with MS have been found in extraparenchymal lymphoid tissue and parenchymal infiltrates (Lovato et al. 2011). Moreover, recent studies by two independent groups showed that antigen-experienced B cells mature in peripheral draining cervical lymph nodes prior to transmigration into the CNS (Palanichamy et al. 2014; Stern et al. 2014), providing evidence that peripheral deletion or modulation of specific B cell subsets could have a therapeutic benefit in MS, and supporting the anti-CD20 monoclonal therapies for relapsing MS (Hauser et al. 2008; Kappos et al. 2011).

In addition to producing antibodies, activated B cells upregulate MHC expression and serve as professional APCs. B cells can also produce pro-inflammatory cytokines. A recent study described a GM-CSF-producing memory B cell subset in patients with MS. These pro-inflammatory B cells were diminished after B cell depletion therapy, even as new B cells repopulated. Instead, the new B cells secreted more anti-inflammatory cytokine IL-10, suggesting a rationale for selective targeting of distinct B cell subsets in MS and other autoimmune diseases, with great potential for antigen-specific therapy in the clinic (Li et al. 2015).

6.3.6 GAP BETWEEN THE BENCH AND CLINICAL APPLICATIONS

Suppression of autoimmunity is a cornerstone of treatment for autoimmune diseases such as MS. The issue with treatment to date is that it generally lacks specificity; systemically administered immunosuppressants and agents that interfere with immune cell activation and migration are associated with myriad adverse effects.

Direct targeting of diseased tissue allows for decreased effective doses, and therefore reduced toxicity to the patient. Nanotechnology has made direct targeting feasible, by allowing the creation of nanocarriers that can deliver drugs to specific targets. For example, disruption of the BBB is an integral aspect of the pathophysiology of MS, and therapeutics have been developed with the goal of promoting repair of the BBB, or of preventing passage of peripheral immune cells into the CNS. This BBB disruption is an opportunity for the field of nanomedicine. In rats with EAE, liposomes (spherical vesicles made of phospholipid membranes in which water-soluble and lipid-soluble compounds can be packaged) accumulate to a greater extent in diseased brains than in those that are healthy (Ozbakir et al. 2014). In light of these findings, it is possible that in the clinic liposomes can be used to facilitate the entrance of therapeutics already in use in clinical practice into the CNS.

Due to their immunomodulatory capabilities, corticosteroids have become a mainstay of treatment of MS, not only for the treatment of active relapses, but also to prevent future relapses and halt disease progression. Polyethylene glycol (PEG)–coated liposomes containing corticosteroids have been shown to achieve higher concentrations in the brains of rats with EAE and, critically, to significantly reduce BBB disruption compared with equal doses of free corticosteroid (Schmidt et al. 2003). PEGylation is also known to increase circulating time of lyposomes, by preventing phagocytosis by macrophages of the reticuloendothelial system, allowing for a longer bioavailability, and potentially reducing the therapeutic dosage. Additionally, a 2-year phase III multicenter, double-blind study of patients with relapsing–remitting MS indicated that subcutaneous PEGylated IFN-β decreased the annualized relapse rate, disability progression, and new lesion formation (as measured by MRI), without increasing the rate of adverse effects, compared with traditional IFN-β therapy. One of the greatest benefits of nanotechnology use in this case was a significant decrease in the minimum dosage required to achieve a therapeutic effect. The traditional dosing frequency of IFN-β therapy comprises one to three injections per week at 30–44 μg, while the two different regimens used in this study were 125 μg injected once either every 2 or 4 weeks (Kieseier et al. 2015).

Aptamers (synthetic molecules that are engineered to specifically bind to other molecules) have also shown promise as nanotherapeutic tools for MS. These peptide or oligonucleotide molecules are essentially artificial antibodies; they can be made to target specific nucleic acids or proteins in any host. Aptamers made against IL-17A, a cytokine known to have pro-inflammatory properties, have been shown to reduce neuritis in EAE models (Gharagozloo et al. 2015).

Nanomedicine is already being used in clinical practice for the treatment of MS. In 1996, glatiramer acetate (GA), a synthetic polymer consisting of four amino acids (L-alanine, L-lysine, L-glutamic acid, and L-tyrosine), was approved by the Food and Drug Administration (FDA). This amino acid sequence is found within the MBP antigen, and it can successfully bind to autoreactive immune cells as a lone polymer. The interaction of GA with immune cells prevents them from binding to MBP, producing self-tolerance in previously autoreactive cells. Additionally, GA has been shown to stimulate Th2 and regulatory T cell responses, causing suppression of autoimmunity and reducing the rate of relapses in patients with relapsing–remitting MS (Lalive et al. 2011).

As previously discussed, disruption of regulatory T cells is believed to be at the root of autoimmunity. Nanomedicine is currently being investigated as a potential means of addressing this issue directly. A multicenter phase II study is planned that will elaborate on a prior 2008 study of 27 MS patients who were administered a trivalent T cell receptor peptide with the goal of inducing regulatory T cell development. Conjugation of this peptide antigen to a nanoparticle is believed to facilitate its delivery to APCs, leading to the induction of Tregs and suppression of autoimmunity (Gharagozloo et al. 2015).

6.3.7 FUTURE DIRECTIONS

The increasing incidence of autoimmune diseases, along with the lack of target specificity in current therapeutics, poses a great area of difficulty for present-day healthcare providers. It is possible that nanomedicine can bridge the gap between drug development and patient populations by providing safer and more effective administration of therapeutics. Through improved targeting, decreased dosages can be achieved, allowing for less toxicity and greater patient compliance. Nanomedicine has been used in the management of MS for more than a decade, but recent advances in the field give hope that groundbreaking changes may be coming in the near future.

REFERENCES

Agmon-Levin, N., Kopilov, R., Selmi, C., Nussinovitch, U., Sanchez-Castanon, M., Lopez-Hoyos, M., Amital, H., Kivity, S., Gershwin, E. M., and Shoenfeld, Y.
2015. Vitamin D in primary biliary cirrhosis, a plausible marker of advanced disease. *Immunol Res* 61 (1–2): 141–146.

Arnson, Y., Amital, H., and Shoenfeld, Y. 2007. Vitamin D and autoimmunity: New aetiological and therapeutic considerations. *Ann Rheum Dis* 66 (9): 1137–1142.

Ascherio, A., Munger, K. L., Lennette, E. T., Spiegelman, D., Hernan, M. A., Olek, M. J., Hankinson, S. E., and Hunter, D. J. 2001. Epstein-Barr virus antibodies and risk of multiple sclerosis: A prospective study. *JAMA* 286 (24): 3083–3088.

Ascherio, A., Munger, K. L., White, R., Kochert, K., Simon, K. C., Polman, C. H., Freedman, M. S. et al. 2014. Vitamin D as an early predictor of multiple sclerosis activity and progression. *JAMA Neurol* 71 (3): 306–314.

Axtell, R. C., de Jong, B. A., Boniface, K., van der Voort, L. F., Bhat, R., De Sarno, P., Naves, R. et al. 2010. T helper 1 and 17 cells determine efficacy of interferon-β in multiple sclerosis and experimental encephalomyelitis. *Nat Med* 16 (4): 406–412.

Azrielant, S., and Shoenfeld, Y. 2016. Eppur Si Muove: Vitamin D is essential in preventing and modulating SLE. *Lupus* 25 (6): 563–572.

Bhargava, P., Cassard, S., Steele, S. U., Azevedo, C., Pelletier, D., Sugar, E. A., Waubant, E., and Mowry, E. M. 2014. The Vitamin D to Ameliorate Multiple Sclerosis (VIDAMS) trial: Study design for a multicenter, randomized, double-blind controlled trial of vitamin D in multiple sclerosis. *Contemp Clin Trials* 39 (2): 288–293.

Binger, K. J., Gebhardt, M., Heinig, M., Rintisch, C., Schroeder, A., Neuhofer, W., Hilgers, K. et al. 2015. High salt reduces the activation of IL-4- and IL-13-stimulated macrophages. *J Clin Invest* 125 (11): 4223–4238.

Brickshawana, A., Hinson, S. R., Romero, M. F., Lucchinetti, C. F., Guo, Y., Buttmann, M., McKeon, A. et al. 2014. Investigation of the KIR4.1 potassium channel as a putative antigen in patients with multiple sclerosis: A comparative study. *Lancet Neurol* 13: 795–806.

Cao, Y., Goods, B. A., Raddassi, K., Nepom, G. T., Kwok, W. W., Love, J. C., and Hafler, D. A. 2015. Functional inflammatory profiles distinguish myelin-reactive T cells from patients with multiple sclerosis. *Sci Transl Med* 7 (287): 287ra74.

Cao, Y., Nylander, A., Ramanan, S., Goods, B. A., Ponath, G., Zabad, R., Chiang, V. L., Vortmeyer, A. O., Hafler, D. A., and Pitt, D. 2016. CNS Demyelination and enhanced myelin-reactive responses after ipilimumab treatment. *Neurology* 86 (16): 1553–1556.

Chastre, A., Hafler, D. A., and O'Connor, K. C. 2016. Evaluation of KIR4.1 as an immune target in multiple sclerosis. N Engl J Med 374: 1495–1496.

Codarri, L., Gyülvészi, G., Tosevski, V., Hesske, L., Fontana, A., Magnenat, L., Suter, T., and Becher, B. 2011. RORγt drives production of the cytokine GM-CSF in helper T cells, which is essential for the effector phase of autoimmune neuroinflammation. Nat Immunol 12 (6): 560–567.

Constantinescu, C. S., Farooqi, N., O'Brien, K., and Gran, B. 2011. Experimental autoimmune encephalomyelitis (EAE) as a model for multiple sclerosis (MS). Br J Pharmacol 164 (4): 1079–1106.

Dendrou, C., Fugger, L. F., and Friese, M. A. 2015. Immunopathology of multiple sclerosis. Nat Rev Immunol 15: 545–558.

Dominguez-Villar, M., Baecher-Allan, C. M., and Hafler, D. A. 2011. Identification of T helper type 1-like, Foxp3+ regulatory T cells in human autoimmune disease. Nat Med 17 (6): 673–675.

El-Behi, M., Ciric, B., Dai, H., Yan, Y., Cullimore, M., Safavi, F., Zhang, G. X., Dittel, B. N., and Rostami, A. 2010. The encephalitogenicity of T_H17 cells is dependent on IL-1- and IL-23-induced production of the cytokine GM-CSF. Nat Immunol 12 (6): 568–575.

Farez, M. F., Fiol, M. P., Gaitán, M. I., Quintana, F. J., and Correale, J. 2015. Sodium intake is associated with increased disease activity in multiple sclerosis. J Neurol Neurosurg Psychiatry 86 (1): 26–31.

Farh, K. K., Marson, A., Zhu, J., Kleinewietfeld, M., Housley, W. J., Beik, S., Shoresh, N. et al. 2015. Genetic and epigenetic fine mapping of causal autoimmune disease variants. Nature 518 (7539): 337–343.

Friese, M. A., and Fugger, L. 2009. Pathogenic CD8(+) T cells in multiple sclerosis. Ann Neurol 66 (2): 132–141.

Gharagozloo, M., Majewski, S., and Foldvari, M. 2015. Therapeutic applications of nanomedicine in autoimmune diseases: From immunosuppression to tolerance induction. Nanomedicine 11 (4): 1003–1018.

Ghoreschi, K., Laurence, A., Yang, X. P., Tato, C. M., McGeachy, M. J., Konkel, J. E., Ramos, H. L. et al. 2010. Generation of pathogenic T_H17 cells in the absence of TGF-β signalling. Nature 467 (7318): 967–971.

Goverman, J. 2009. Autoimmune T cell responses in the central nervous system. Nat Rev Immunol 9 (6): 393–407.

Haahr, S., Koch-Henriksen, N., Moller-Larsen, A., Eriksen, L. S., and Andersen, H. M. 1995. Increased risk of multiple sclerosis after late Epstein-Barr virus infection: A historical prospective study. Mult Scler 1 (2): 73–77.

Haahr, S., Plesner, A. M., Vestergaard, B. F., and Hollsberg, P. 2004. A role of late Epstein-Barr virus infection in multiple sclerosis. Acta Neurol Scand 109 (4): 270–275.

Haak, S., Croxford, A. L., Kreymborg, K., Heppner, F. L., Pouly, S., Becher, B., and Waisman, A. 2009. IL-17A and IL-17F do not contribute vitally to autoimmune neuro-inflammation in mice. J Clin Invest 119 (1): 61–69.

Handel, A. E., Williamson, A. J., Disanto, G., Handunnetthi, L., Giovannoni, G., and Ramagopalan, S. V. 2010. An updated meta-analysis of risk of multiple sclerosis following infectious mononucleosis. PLoS One 5 (9): e12496.

Hauser, S. L., Waubant, E., Arnold, D. L., Vollmer, T., Antel, J., Fox, R. J., Bar-Or, A. et al. 2008. B-cell depletion with rituximab in relapsing-remitting multiple sclerosis. N Engl J Med 358 (7): 676–688.

Hernandez, A. L., Kitz, A., Wu, C., Lowther, D. E., Rodriguez, D. M., Vudattu, N., Deng, S., Herold, S. C., Kuchroo, V. K., Kleinwietfeld, M., and Hafler, D. A. 2015. Sodium chloride inhibits the suppressive function of Foxp3+ regulatory T cells. J Clin Invest 125 (11): 4212–4222.

Holick, M. F., Binkley, N. C., Bischoff-Ferrari, H. A., Gordon, C. M., Hanley, D. A., Heaney, R. P., Murad, M. H., and Weaver, C. M. 2011. Evaluation, treatment, and prevention of vitamin D deficiency: An Endocrine Society clinical practice guideline. J Clin Endocrinol Metab 96 (7): 1911–1930.

Housley, W. J., Fernandez, S. D., Vera, K., Murikinati, S. R., Grutzendler, J., Cuerdon, N., Glick, L., De Jager, P. L., Mitrovic, M., Cotsapas, C., and Hafler, D. A. 2015a. Genetic variants associated with autoimmunity drive NFκB signaling and responses to inflammatory stimuli. Sci Transl Med 7 (291): 291ra93.

Housley, W. J., Pitt, D., and Hafler D. A. 2015b. Biomarkers in multiple sclerosis. Clin Immunol 161 (1): 51–58.

Huseby, E. S., Liggitt, D., Brabb, T., Schnabel, B., Ohlen, C., and Goverman, J. 2001. A pathogenic role for myelin-specific CD8+ T cells in a model for multiple sclerosis. J Exp Med 194 (5): 669–676.

International Multiple Sclerosis Genetics Consortium, Hafler, D. A., Compston, A., Sawcer, S., Lander, E. S., Daly, M. J., De Bakker, P. I. et al. 2007. Risk alleles for multiple sclerosis identified by a genomewide study. N Engl J Med 357 (9): 851–862.

International Multiple Sclerosis Genetics Consortium, Wellcome Trust Case Control Consortium 2, Sawcer, S., Hellenthal, G., Pirinen, M., Spencer, C. C., Patsopoulos, N. A. et al. 2011. Genetic risk and a primary role for cell-mediated immune mechanisms in multiple sclerosis. *Nature* 476 (7359): 214–219.

Jagessar, S. A., Holtman, I. R., Hofman, S., Morandi, E., Heijmans, N., Laman, J. D., Gran, B., Faber, B. W., van Kasteren, S. I., Eggen, B. J., and 't Hart, B. A. 2016. Lymphocryptovirus infection of nonhuman primate B cells converts destructive into productive processing of the pathogenic CD8 T cell epitope in myelin oligodendrocyte glycoprotein. *J Immunol* 197 (4): 1074–1088.

Ji, Q., Perchellet, A., and Goverman, J. M. 2010. Viral infection triggers central nervous system autoimmunity via activation of CD8+ T cells expressing dual TCRs. *Nat Immunol* 11 (7): 628–634.

Kampman, M. T., and Brustad, M. 2008. Vitamin D: A candidate for the environmental effect in multiple sclerosis—Observations from Norway. *Neuroepidemiology* 30 (3): 140–146.

Kappos, L., Li, D., Calabresi, P. A., O'Connor, P., Bar-Or, A., Barkhof, F., Yin, M., Leppert, D., Glanzman, R., Tinbergen, J., and Hauser, S. L. 2011. Ocrelizamab in relapsing-remitting multiple sclerosis: A phase 2, randomised, placebo-controlled, multicentre trial. *Lancet* 378 (9805): 1779–1787.

Kieseier, B. C., Arnold, D. L., Balcer, L. J., Boyko, A. A., Pelletier, J., Liu, S., Zhu, Y. et al. 2015. Peginterferon beta-1a in multiple sclerosis: 2-Year results from ADVANCE. *Mult Scler* 21 (8): 1025–1035.

Kitz, A., de Marchen, M., Gautron, A. S., Mitrovic, M., Hafler, D. A., and Dominguez-Villar, M. 2016. AKT isoforms modulate Th1-like Treg generation and function in human autoimmune disease. *EMBO Rep* 17 (8): 1169–1183.

Kleinewietfeld, M., and Hafler, D. A. 2014. Regualtory T cells in autoimmune neuroinflammation. *Immunol Rev* 259: 231–244.

Kleinewietfeld, M., Manzel, A., Titze, J., Kvakan, H., Yosef, N., Linker, R. A., Muller, D. N., and Hafler, D. A. 2013. Sodium chloride drives autoimmune disease by the induction of pathogenic TH 17 cells. *Nature* 496 (7446): 518–522.

Kofler, D. M., Farkas, A., von Bergwelt-Baildon, M., and Hafler, D. A. 2016. The link between CD6 and autoimmunity: Genetic and cellular associations. *Curr Drug Targets* 17 (6): 651–665.

Kofler, D. M., Severson, C. A., Mousissian, N., De Jager, P. L., and Hafler, D. A. 2011. The CD6 multiple sclerosis susceptibility allele is associated with alterations in CD4+ T cell proliferation. *J Immunol* 187 (6): 3286–3291.

Koh, D. R., Fung-Leung, W. P., Ho, A., Gray, D., Acha-Orbea, H., and Mak, T. W. 1992. Less mortality but more relapses in experimental allergic encephalomyelitis in CD8−/− mice. *Science* 256 (5060): 1210–1213.

Kroenke, M. A., Carlson, T. J., Andjelkovic, A. V., and Segal, B. M. 2008. IL-12- and IL-23-modulated T cells induce distinct types of EAE based on histology, CNS chemokine profile, and response to cytokine inhibition. *J Exp Med* 205 (7): 1535–1541.

Lalive, P. H., Neuhaus, O., Benkhoucha, M., Burger, D., Hohlfeld, R., Zamvil, S. S., and Weber, M. S. 2011. Glatiramer acetate in the treatment of multiple sclerosis: Emerging concepts regarding its mechanism of action. *CNS Drugs* 25 (5): 401–414.

Levin, L. I., Munger, K. L., Rubertone, M. V., Peck, C. A., Lennette, E. T., Spiegelman, D., and Ascherio, A. 2005. Temporal relationship between elevation of Epstein-Barr virus antibody titers and initial onset of neurological symptoms in multiple sclerosis. *JAMA* 293 (20): 2496–2500.

Li, R., Rezk, A., Miyazaki, Y., Hilgenberg, E., Touil, H., Shen, P., Moore, C. S. et al. 2015. Proinflammatory GM-CSF-producing B cells in multiple sclerosis and B cell depletion therapy. *Sci Transl Med* 7 (310): 310ra166.

Lovato, L., Willis, S. N., Rodig, S. J., Caron, T., Almendinger, S. E., Howell, O. W., Reynolds, R., O'Connor, K. C., and Hafler, D. A. 2011. Related B cell clones populate the meninges and parenchyma of patients with multiple sclerosis. *Brain* 134 (Pt 2): 534–541.

Lucas, R. M., Hughes, A. M., Lay, M. L., Ponsonby, A. L., Dwyer, D. E., Taylor, B. V., and Pender, M. P. 2011. Epstein-Barr virus and multiple sclerosis. *J Neurol Neurosurg Psychiatry* 82 (10): 1142–1148.

McDonald, J., Graves, J., Waldman, A., Lotze, T., Schreiner, T., Belman, A., Greenberg, B. et al. 2016. A case-control study of dietary salt intake in pediatric-onset multiple sclerosis. *Mult Scler Relat Disord* 6: 87–92.

McFarland, H. F., and Martin, R. 2007. Multiple sclerosis: A complicated picture of autoimmunity. *Nat Immunol* 8: 913–919.

Marson, A., Housley, W. J., and Hafler, D. A. 2015. Genetic basis of autoimmunity. *J Clin Invest* 125 (6): 2234–2241.

Martin, R., Howell, M. D., Jaraquemada, D., Flerlage, M., Richert, J., Brostoff, S., Long, E. O., McFarlin, D. E., and McFarland, H. F. 1991. A myelin basic protein peptide is recognized by cytotoxic T cells in the context of four HLA-DR types associated with multiple scleoris. *J Exp Med* 173 (1): 19–24.

Noster, R., Riedel, R., Mashreghi, M. F., Radbruch, H., Harms, L., Haftmann, C., Chang, H. D., Radbruch, A., and Zielinski, C. E. 2014. IL-17 and GM-CSF expression are antagonistically regulated by human T helper cells. *Sci Transl Med* 6 (241): 241ra280.

Nylander, A., and Hafler, D. A. 2012. Multiple sclerosis. *J Clin Invest* 122: 1180–1188.

O'Gorman, C., Lucas, R., and Taylor, B. 2012. Environmental risk factors for multiple sclerosis: A review with a focus on molecular mechanisms. *Int J Mol Sci* 13 (9): 11718–11752.

Oleszak, E. L., Chang J. R., Friedman, H., Katsetos, C. D., and Platsoucas, C. D. 2004. Theiler's virus infection: A model for multiple sclerosis. *Clin Microbiol Rev* 17 (1): 174–207.

Olson, J. K., Croxford, J. L., Calenoff, M. A., Dal Canto, M. C., and Miller, S. D. 2001. A virus-induced molecular mimicry model of multiple scleorsis. *J Clin Invest* 108 (2): 311–318.

Ota, K., Matsui, M., Milford, E. L., Mackin, G. A., Weiner, H. L., and Hafler, D. A. 1990. T-cell recognition of an immuno-dominant myelin basic protein epitope in multiple sclerosis. *Nature* 346 (6280): 183–187.

Ozbakir, B., Crielaard, B. J., Metselaar, J. M., Storm, G., and Lammers, T. 2014. Liposomal corticosteroids for the treatment of inflammatory disorders and cancer. *J Control Release* 190: 624–636.

Palanichamy, A., Apeltsin, L., Kuo, T. C., Sirota, M., Wang, S., Pitts, S. J., Sundar, P. D. et al. 2014. Immunoglobulin class-switched B cells form an active immune axis between CNS and periphery in multiple sclerosis. *Sci Transl Med* 6 (248): 248ra106.

Pröbstel, A. K., Kuhle, J., Lecourt, A. C., Vock, I., Sanderson, N. S., Kappos, L., and Derfuss T. 2016. Multiple sclerosis and antibodies against KIR4.1. *N Engl J Med* 374: 1496–1498.

Raddassi, K., Kent, S. C., Yang, L., Bourcier, K., Bradshaw, E. M., Seyfert-Margolis, V., Nepom, G. T., Kwok, W. W., and Hafler, D. A. 2011. Increased frequencies of myelin oligodendrocyte glycoprotein/MHC class II-binding CD4 cells in patients with multiple sclerosis. *J Immunol* 187: 1039–1046.

Rangachari, M., and Kuchroo, V. K. 2013. Using EAE to better understand principles of immune function and autoimmune pathology. *J Autoimmun* 45: 31–39.

Ransohoff, R. M., Hafler, D. A., and Lucchinetti, C. F. 2015. Multiple sclerosis—A quiet revolution. *Nat Rev Neurol* 11 (3): 134–142.

Reindl, M., Linington, C., Brehm, U., Egg, R., Dilitz, E., Deisenhammer, F., Poewe, W., and Berger, T. 1999. Antibodies against the myelin oligodendrocyte glycoprotein and the myelin basic protein in multiple sclerosis and other neurological diseases: A comparative study. *Brain* 122 (Pt 11): 2047–2056.

Schmidt, J., Metselaar, J. M., Wauben, M. H., Toyka, K. V., Storm, G., and Gold, R. 2003. Drug targeting by long-circulating liposomal glucocorticosteroids increases therapeutic efficacy in a model of multiple sclerosis. *Brain* 126 (8): 1895–1904.

Sheng, W., Yang, F., Zhou, Y., Yang, H., Low, P. Y., Kemeny, D. M., Tan, P., Moh, A., Kaplan, M. H., Zhang, Y., and Fu, X. Y. 2014. STAT5 programs a distinct subset of GM-CSF-producing T helper cells that is essential for autoimmune neuroinflammation. *Cell Res* 24 (12): 1387–1402.

Srivastava, R., Aslam, M., Kalluri, S. R., Schirmer, L., Buck, D., Tackenberg, B., Rothhammer, V. et al. 2012. Potassium channel KIR4.1 as an immune target in multiple sclerosis. *N Engl J Med* 367: 115–123.

Stern, J. N., Yaari, G., Vander Heiden, J. A., Church, G., Donahue, W. F., Hintzen, R. Q., Huttner, A. J., et al. 2014. B cells populating the multiple sclerosis brain mature in the draining cervical lymph nodes. *Sci Transl Med* 6(248): 248ra107.

Sundstrom, P., Juto, P., Wadell, G., Hallmans, G., Svenningsson, A., Nystrom, L., Dillner, J., and Forsgren, L. 2004. An altered immune response to Epstein-Barr virus in multiple sclerosis: A prospective study. *Neurology* 62 (12): 2277–2282.

Tabansky, I., Messina, M. D., Bangeranye, C., Goldstein, J., Blitz-Shabbir, K. M., Machado, S., Jeganathan, V. et al. 2016. *Immunol Res* 64 (2): 640.

Watad, A., Azrielant, S., Soriano, A., Bracco, D., Abu Much, A., and Amital, H. 2016a. Association between seasonal factors and multiple sclerosis. *Eur J Epidemiol* 31 (11): 1081–1089.

Watad, A., Neumann, S. G., Amital, H., and Shoenfeld, Y. 2016b. Vitamin D and systemic lupus erythematosus: Myth or reality? *Isr Med Assoc J* 18 (3–4): 177–182.

Wingerchuk, D. M. 2011. Environmental factors in multiple sclerosis: Epstein-Barr virus, vitamin D, and cigarette smoking. *Mt Sinai J Med* 78 (2): 221–230.

Wingerchuk, D. M., and Carter, J. L. 2014. Multiple sclerosis: Current and emerging disease-modifying therapies and treatment strategies. *Mayo Clin Proc* 89 (2): 225–240.

Wu, C., Yosef, N., Thalhamer, T., Zhu, C., Xiao, S., Kishi, K., Regev, A., and Kuchroo, V. K. 2013. Induction of pathogenic TH 17 cells by inducible salt-sensing kinase SGK1. *Nature* 496 (7446): 513–517.

Zhang, W. C., Zheng, X. J., Du, L. J., Sun, J. Y., Sheng, Z. X., Shi, C., Sun, S. et al. 2015. High salt primers a specific activation state of macrophages, M(Na). *Cell Res* 25 (8): 893–910.

The Biology and Clinical Treatment of Asthma

Rima Kandil, Jon R. Felt, Prashant Mahajan, and Olivia M. Merkel

CONTENTS

7.1.1 INTRODUCTION

Asthma is among the most common chronic diseases globally, and represents one of the most common inflammatory conditions of both children and adults. It is a condition characterized by a chronic inflammatory state of the lungs and lower respiratory tract with acute episodes of increased inflammation and airflow obstruction (National Institutes of Health 2007). Acute obstruction is usually reversible and therefore manageable, but the chronic inflammatory state results in physiologic changes within the lungs that can lead to more severe and possibly life-threatening exacerbations, permanently decreased lung function, and poor quality of life. The high social and economic burdens associated with both the management and sequelae of asthma have fostered a significant effort in researching the underlying mechanisms and effective management of this disease.

7.1.2 EPIDEMIOLOGY

Asthma affects an estimated 330 million people worldwide, including 15% of children and adolescents (Global Asthma Network 2014). It is found in all parts of the world, but appears to have the highest prevalence in Australia, northern and western Europe, and Brazil (Lai et al. 2009). In childhood, the disease is more common in males, but as the disease progresses into adulthood, slightly more women are affected. Although asthma can be diagnosed at any age, generally symptoms present in childhood, with 80% of asthmatics diagnosed by age 6 (Center for Disease Control and Prevention 2012b). Many childhood asthmatics do have significant improvement of symptoms and even full remission in adolescence and young adulthood, but some have persistent or a full return of symptoms as adults. Children who experience multiple wheezing episodes before age 3 and who continue to wheeze after age 6 have been shown to experience a decline in lung function growth and often have persistent symptoms through adolescence and into adulthood (Martinez et al. 1995; Covar et al. 2004; Morgan et al. 2005). Other factors associated with persistence of asthma into adulthood are parental asthma, low birth weight or prematurity, male gender, atopic dermatitis (eczema), allergic rhinitis, and food allergies (Covar et al. 2010).

The development and severity of asthma is most closely linked to atopic conditions such as eczema and allergic rhinitis. The overall prevalence in the developing world is less than that of the more metropolitan and affluent nations, but is increasing steadily with greater urbanization by approximately 50% per decade (Robinson et al. 2011; Rodriguez et al. 2011). In contrast, the prevalence in the developed world remains relatively unchanged. Death from asthma is fairly uncommon in the developed world (1.1 per 100,000 population in the United States), although the risk is much greater for the elderly, those with severe and persistent symptoms, and those in the developing world (Akinbami et al. 2012).

The economic and social burden of asthma is significant. It is a disease that disproportionately affects those of lower income and education, and those of minority status. In the Global Asthma Report (GAR), the burden of asthma on society, as measured by years of life lost due to premature death and years of life living with disability, is greatest in the young and the elderly and leads among chronic diseases in these categories (Murray et al. 2012). In the United States, more than 50% of asthmatics report having an exacerbation each year, accounting for more than 14 million doctor's visits, 2 million emergency room trips, and close to half a million hospital admissions (Centers for Disease Control and Prevention 2012a). The economic cost in the United States alone approaches US$50 billion in direct healthcare expenditure, and more than $6 billion in lost productivity (Barnett and Nurmagambetov 2011).

7.1.3 ETIOLOGY

While the exact causes and mechanisms for the development of asthma remain unclear, it is understood to be greatly influenced by both genetics and environmental exposures. There is a high degree of concordance for asthma in identical twins, and those individuals with severe asthma are more likely to have children who develop asthma. More than 100 genetic loci have been linked to asthma in multiple different studies, but no single gene or family of genes has been identified that can explain the development or severity of the disease (Melen and Pershagen 2012). Most of these genes have been identified as related to other allergic or atopic conditions, as well as pro-inflammatory states, suggesting that the development of asthma is polygenic and likely requires significant environmental interaction.

The high incidence of allergy or atopy in those with asthma, as well as the striking lack of asthma and atopy among those born and raised in rural farming communities with early exposure to domestic animals, suggests the role of epigenetics, or environmental influence (von Mutius and Vercelli 2010; Wells et al. 2014). This is also supported by the increasing prevalence of the disease in developing countries with the spread of urbanization. Additionally, recurrent episodes of wheezing in infants from acute viral respiratory infections due to influenza virus, respiratory syncytial virus, and human metapneumovirus, among others, have been shown to be directly related to later development of airway hyperreactivity and asthma (Juntti et al. 2003; Holt and Sly 2012). All these highlight the significant role that environmental exposures, in combination with genetic predilections, have in the development and natural progression of asthma. Understanding these interactions and modifying environmental exposures have long been a focus for management of the disease (Bateman et al. 2008a).

7.1.4 PATHOPHYSIOLOGY

The underlying mechanisms behind the development of asthma in those who are genetically susceptible are not completely understood. It is known that asthmatics have both structurally and functionally different lower airways than non-asthmatics, although it is unclear whether these changes are what cause the disease or are simply a result of frequent exacerbations. Overall, asthmatics have a nearly constant level of underlying inflammation of the lower respiratory tract and a hyperresponsiveness to triggers. The underlying reasons for this persistent inflammation are not fully understood (Martinez 2006).

The pathophysiology of an acute asthma attack is mediated by a multitude of different inflammatory cells, structural cells, and inflammatory mediators in a complex interaction (Robinson 2004; Pelaia et al. 2008; Nakagome and Nagata 2011). Generally, the underlying cause of airflow obstruction and the patient's distress is twofold: smooth muscle surrounding the airways contracts with resultant bronchospasm, and increased inflammation causes edema and mucus production, both resulting in sometimes severe airway narrowing. Figure 7.1b compares the appearance of an asthmatic airway with a normal one.

7.1.4.1 Bronchospasm

Bronchospasm of the airways is a reflexive measure designed to protect the lungs from exposure to harm. In healthy individuals, this is commonly seen as a response to extreme cold, heat, and environmental pollution (smoke, dust, and heavy fog). In asthmatics, the lungs become hyperresponsive to triggers, and therefore the response is exaggerated and often persistent, causing a significant narrowing of the airways and decreased airflow (Robinson 2004). Multiple factors play a role in this mechanism, and they are discussed here.

Physical disruption of the epithelium from trauma, invading pathogens, extremes in temperature, or changes in osmolality expose sensory nerves, which directly activate smooth muscle cells to contract. Indirectly, damage or loss of epithelial cells also causes decreases of various enzymes, inflammatory mediators, and relaxants that are secreted by the epithelial cells. Each of these acts either directly to modify the reaction of smooth muscle cells (epithelial-derived relaxant factor) or indirectly to degrade inflammatory mediators, which also promote muscle contraction (Pelaia et al. 2008).

Multiple inflammatory mediators (histamine, prostaglandin D_2, and cysteinyl-leukotriene), released by mast cells, act directly on smooth muscle to induce bronchospasm. Histamine acts directly on peripheral nerves innervating smooth muscle, specifically causing bronchial smooth muscle to contract via H_1 receptors. Mast cells are found lining the epithelium, as well as the airway smooth muscle layer. While mast cells have important roles in wound healing and coagulation, they are best known for their release of histamine and its role in allergic reactions (Robinson 2004).

They are activated by two mechanisms, namely, binding of immunoglobulin E (IgE) and changes in osmolality. Mast cells express a high-affinity receptor that is highly specific to IgE, resulting in the cell being coated with IgE antibodies. When an allergen attaches itself across two matching IgE molecules (cross-linking), the high-affinity receptor is activated, beginning a tyrosine-dependent cascade, resulting in activation of the mast cell and release of its granule stores. Cross-linking by certain antigens, rather than full activation, causes the mast cell to become sensitized to environmental changes in the airway, most commonly changes in osmolality (Koskela 2007). Changes in air temperature or humidity cause a shift in the osmolality of the thin fluid lining the mucosal layer of the airways, which stimulates degranulation of the mast cells and subsequent bronchospasm (Robinson 2004).

7.1.4.2 Airway Inflammation

Inflammation of the airways is characterized by mucosal edema and hypersecretion of mucus. In addition to mast cells and epithelial cells continuing to play a role in these events, dendritic cells, T-lymphocytes, and eosinophils also play a critical role. Figure 7.2 illustrates the many cells involved in the pathophysiology of inflammation in the lungs and their interactions. Dendritic cells, specialized macrophages in the airways, are the major antigen-presenting cells that activate the

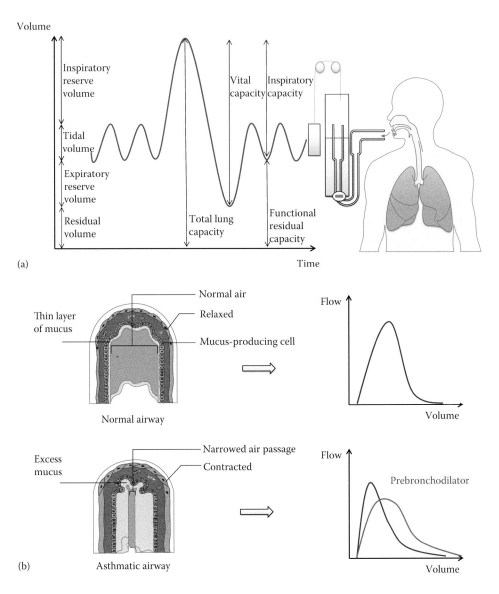

Figure 7.1. (a) Exemplary spirometry curve with its component parts. (b) How obstruction of asthmatic airways affects the resulting flow diagram.

cell-mediated immune response. In healthy lungs, T-helper 1 cells predominate. However, for less understood reasons, in asthmatic lungs, the dendritic cells are mostly immature and promote differentiation and response of T-helper 2 (T_H2) cells instead. This is partially accomplished through the release of thymic stromal lymphopoietin (TSLP) from epithelial cells, which causes the immature dendrites to release chemokines attracting T_H2 cells to the airways (Cianferoni and Spergel 2014).

T_H2 cells are critical to the coordination of the inflammatory response through the release of interleukin-4 (IL-4), IL-5, and IL-13. Release of these cytokines results in recruitment of eosinophils (IL-5), the maintenance and survival of both eosinophils and mast cells within the airways, and increased IgE formation (IL-4 and IL-13). IL-13 also induces hypersecretion of mucus from goblet cells lining the airways. Eosinophils increase airway hyperreactivity by releasing major basic protein and free radicals, which cause injury to the mucosa. They also release a small amount of histamine, although much less than mast cells. Overall, the importance of eosinophils in asthma

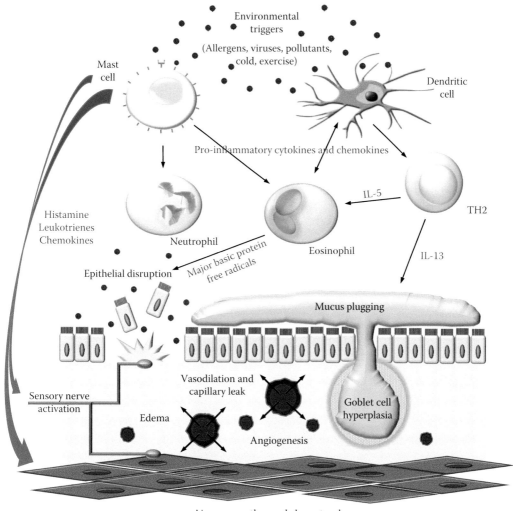

Figure 7.2. Cell-mediated and molecular pathogenesis of bronchospasm and airway inflammation during an acute asthma exacerbation.

is unclear, as they appear to play dual roles in both a dysregulated allergic response and tissue homeostasis and repair (Furuta et al. 2014). In some patients, particularly those with adult-onset asthma, neutrophils are attracted into the airways in larger numbers than eosinophils, but their role and how they affect the course of the disease is largely unknown (Bruijnzeel et al. 2015).

The disruption of the epithelium, already mentioned, also releases several additional inflammatory mediators, including vascular endothelial growth factor (VEGF), which promotes angiogenesis in the lungs. The significant increase in vasculature surrounding the airways results in more edema during periods of high inflammation, due to leakage of plasma into the interstitium (Hosoki et al. 2016). Many additional inflammatory cytokines and chemokines, including tumor necrosis factor (TNF)-α and IL-1β, which amplify the overall inflammatory response (Tillie-Leblond et al. 2005), are also involved but cannot be described here in detail.

Over time, due to constant inflammation, a physical remodeling of the lungs occurs. The mucosa of the lower airways is characterized by infiltration of a larger number of eosinophils, T-lymphocytes, and mast cells when compared with healthy lungs. There is a greater susceptibility for viral and bacterial infections and an exaggerated inflammatory response to such infections (Holt and Sly 2012). The basement membrane is thickened due to subepithelial collagen deposition from fibrogenic mediators released by the increased numbers of eosinophils (Furuta et al. 2014). The smooth muscle surrounding the airways becomes hypertrophic, in part due to the increased vascularity, but also from growth factor stimulation and frequent episodes of bronchospasm (Pelaia et al. 2008).

These changes are most often found in the bronchi; however, they can be present anywhere throughout the respiratory tract, from the trachea to the terminal bronchioles, in either a continuous or patchy distribution (Elliot et al. 2015). Unfortunately, the severity and/or extensiveness of these changes does not correlate with the severity of patient symptoms or disease.

7.1.5 DIAGNOSIS

Chronic symptoms of asthma include more frequent dry cough, often worse at night, occasional chest tightness and shortness of breath, and wheezing. Acute exacerbations lead to significant airflow obstruction and air trapping. Patients present with wheezing, dyspnea, and worsening cough. They often complain of significant chest tightness and pain. There is increased mucus production that is difficult to clear and contributes to air trapping (difficulty exhaling) and dyspnea (Tillie-Leblond et al. 2005). If suffering from severe exacerbations, patients present with a significant increase in their work of breathing, with increased respiratory rate, accessory respiratory muscle use, and nasal flaring. Although many will have audible wheezing, auscultation of the lungs reveals an expiratory wheeze (sometimes also inspiratory wheezing in severe attacks) and a prolongation of the expiratory phase due to air trapping. There will often be asymmetry of breath sounds due to segmental atelectasis from mucus plugging. This mucus production and inflammatory exudate can also cause crackles to be heard. In severe exacerbations, airflow may be so limited that no wheezing may be heard and is an ominous finding (National Institutes of Health 2007).

Diagnosis of asthma is based primarily on symptoms and response to therapy. Testing of lung function can confirm the diagnosis, but is difficult to perform on very young patients, and therefore has limited utility in the initial diagnosis of patients under the age of 6 years. While full pulmonary function testing can confirm the diagnosis, simple spirometry and a flow–volume loop are sufficient. A reduction in forced expiratory volume in 1 second (FEV1), but no reduction in forced vital capacity (FVC) (reduced FEV1/FVC ratio <70% of predicted) is suggestive of the disease (National Institutes of Health 2007). This, combined with a reduction in peak expiratory flow (PEF), as seen by a convex shape on the flow–volume loop, is diagnostic of asthma. The measurements obtained by lung function tests are depicted in Figure 7.1, which also demonstrates the characteristic convex shape of the flow–volume loop obtained by spirometry seen in asthmatic versus normal lungs. Often, spirometry is repeated after administration of a bronchodilator, and a subsequent improvement of FEV1 of greater than 10% confirms the diagnosis (National Institutes of Health 2007).

Generally, a diagnosis of asthma can be made with observation of multiple episodes of airway obstruction and wheezing over time. However, not all wheezing is from asthma, and careful examination and review of the history may indicate a different etiology. Table 7.1 lists the other causes of wheezing categorized by age when onset of symptoms is often observed. Overall, failure to correctly diagnose any of these other conditions will, at a minimum, result in improper use of medication with subsequent risk of side effects. In the case of inhaled foreign bodies, delay in correct diagnosis and prompt removal can result in permanent, sometimes life-threatening, injury and scarring (Goussard et al. 2014).

Once an asthma diagnosis is made, the severity of the disease is classified to guide initial management. Severity is based on patient symptoms, functional impairment, and the patient's risk for future adverse events. Symptom severity can range from mild infrequent episodes of wheeze or cough without limitation to daily activities, to frequent episodes of severe exacerbation with daily symptoms causing considerable limitation in daily activities.

Asthma is classified as either intermittent or persistent based on the number of days with symptoms, frequency of nighttime symptoms, use of short-acting rescue medications, interference with normal activities, objective measurements of lung function, and the frequency of

TABLE 7.1
Nonasthmatic causes of wheezing by age group.

Infant	Toddler	School-aged	Adolescent	Adult
Bronchiolitis	Bronchiolitis	Foreign body	Vocal cord dysfunction	Interstitial lung disease
Bronchopulmonary dysplasia	Foreign body	Cardiomegaly	Cardiomegaly	Chronic obstructive pulmonary disease
Tracheobronchomalacia	Cystic fibrosis	Pulmonary edema	Pulmonary edema	Cardiomegaly
Vascular ring	Bacterial tracheitis	Tumor	Tumor	Pulmonary edema
Tracheal stenosis	Cardiomegaly			Tumor
Laryngeal web	Pulmonary edema			
Cystic lesion/tumor	Tumor			
Cardiomegaly				
Pulmonary edema				

exacerbations requiring systemic corticosteroids (National Institutes of Health 2007). Patients with persistent symptoms (occurring at least biweekly, with some limitation of daily activity) are then classified as mild, moderate, or severe. Care should be taken to provide equal consideration to both symptoms and objective data when classifying asthma. This is especially true in pediatric patients, as they may demonstrate normal lung function with objective testing but still suffer from severe or frequent symptoms (Bacharier et al. 2004; Spahn et al. 2004; Paull et al. 2005). In addition, it should be recognized that even those classified with mild or intermittent disease can still present with severe, life-threatening exacerbations. Table 7.2 lists the criteria used to classify asthma severity and stratify risk.

Assessing risk of future adverse events helps clinicians identify those patients that require more careful monitoring and also helps to influence more aggressive management strategies to prevent complications. Patients are categorized as low, medium, and high risk. Those factors that are predictive of increased exacerbations, severe exacerbations, or death include two or more emergency department visits or hospitalizations in the past year for asthma, previous intensive care unit admission, previous endotracheal intubation from asthma, tobacco use or secondhand exposure, severe and persistent airflow obstruction (seen with objective lung function testing), history of medication noncompliance, and low socioeconomic status (National Institutes of Health 2007; Lai et al. 2009; Akinbami et al. 2012).

7.1.6 MANAGEMENT

The goals of management are to reduce the frequency and severity of exacerbations and use of rescue medication, prevent chronic symptoms (cough, exercise intolerance, shortness of breath, etc.), maintain normal activity, prevent loss in pulmonary function, and reduce the side effects of medication. Although pharmacologic treatment is the mainstay in asthma management, it is only a part of successful treatment of the disease. Once the severity of asthma has been classified, successful management comes from identification and avoidance of triggers, identification and management of comorbid conditions that can worsen the disease, education and empowerment of the patient and family to develop skills in self-management, and frequent assessment and monitoring (National Institutes of Health 2007).

Comorbidities that directly affect the ability to successfully manage asthma are important to identify and treat. Foremost are other allergic or inflammatory conditions, such as sinusitis, rhinitis, eczema, and gastroesophageal reflux disease, which add to the severity of inflammation in the body and can worsen asthma symptoms. Obesity and obstructive sleep apnea alter pulmonary mechanics and add to airflow obstruction, particularly at night. Additionally, obesity and metabolic syndrome have been linked to an increase in inflammatory markers, which would impact the chronic inflammatory state associated with asthma (Fitzpatrick et al. 2012). Finally, multiple psychosocial factors, including stress and

TABLE 7.2

Determination of asthma severity and risk.

Components of severity		Intermittent			Persistent — Mild			Persistent — Moderate			Persistent — Severe		
Age (years)		0–4	5–11	>12	0–4	5–11	>12	0–4	5–11	>12	0–4	5–11	>12
Impairment	Symptoms	≤2 days/week			≤2 days/week, but not daily			Daily			Throughout the day		
	Nocturnal symptoms	0	≤2×/month	≤2×/month	1–2×/month	3–4×/month	3–4×/month	3–4×/month	≥1×/week	≥1×/week	≥2×/week	Often 7×/week	Often 7×/week
	SABA use, excluding exercise-induced treatment	≤2 days/week			≥2 days/week			Daily			Several times/day		
	Limitation of normal activity	None			Minor			Some			Constant		
	Lung function — FEV1	—	>80%	>80%	—	>80%	>80%	—	60%–80%	60%–80%	—	<60%	<60%
	Lung function — FEV1/FVC	—	>85%	Normal	—	>80%	Normal	—	75%–80%	Reduced >5%	—	<75%	Reduced >5%
Risk	Exacerbations requiring oral steroids	0–1×/year			≥2×/6 months or >4×/year + risk factors			≥2×/6 months or >2×/year + risk factors					

SOURCE: Adapted from National Institutes of Health, International consensus report on diagnosis and treatment of asthma, National Heart, Lung, and Blood Institute, National Institutes of Health, Bethesda, MD, 2007. http://www.nhlbi.nih.gov/files/docs/guidelines/asthgdln.pdf.

NOTE: SABA, short-acting β_2 adrenergic agonist.

depression, have been shown to worsen asthma symptoms and outcome. While the exact cause of this correlation has yet to be identified, there is evidence to suggest that increased stress is associated with more pro-inflammatory cytokines released throughout the body (Jiang et al. 2014).

Patients with asthma should be monitored frequently to assess whether symptoms are being well-controlled. Clinicians are encouraged to use judgment in the frequency of clinical assessment, but according to the U.S. Department of Health and Human Services guidelines for management of asthma, the interval should not exceed 6 months in the well-controlled patient and should be much more frequent in the poorly controlled or those in the high-risk category (National Institutes of Health 2007). Assessments should include both clinical assessment via history and focused physical exam and patient self-assessment of their own symptoms and understanding of their disease control and compliance. Objective measurements with spirometry should be obtained every 1–2 years.

Pharmacologic management of asthma is divided into acute and chronic therapies. Chronic management is aimed at reducing underlying lung inflammation and chronic symptoms while preventing exacerbations. Acute management consists of providing rapid relief of symptoms and airflow obstruction during exacerbation and returning inflammation to its baseline as quickly as possible. To minimize adverse effects, pharmacologic agents are preferably administered directly into the respiratory tree via inhalation, with systemic agents reserved either for acute exacerbations or to escalate chronic management in the poorly controlled or moderate to severe cases (Newhouse and Dolovich 1986).

7.1.6.1 Chronic Management

Pharmacologic agents designed for chronic management generally fall within two main categories, β receptor agonists and corticosteroids. Inhaled short-acting β_2 adrenergic agonists (SABAs), albuterol and levalbuterol, act directly to relax the smooth muscles of the respiratory tree and decrease bronchospasm. Inhaled, long-acting β_2 adrenergic agonists (LABA), salmeterol and formeterol, also act directly to relax smooth muscle in the lungs but are more β_2 selective, and therefore have less effect on the heart rate. When used alone, they have been shown to increase the risk of asthma-related death, and therefore are only recommended for use when combined with an inhaled corticosteroid (ICS) and when a patient's symptoms are difficult to control (Nelson et al. 2006; Bateman et al. 2008b; McMahon et al. 2011).

Corticosteroids work via direct immune suppression of leukocytes and inhibition of inflammatory products. They inhibit the genes that encode for a host of inflammatory cytokines that activate and direct the immune response of leukocytes. Specifically, the suppression of IL-2 prevents T-lymphocyte activation and suppression of phospholipase A2, which is a critical factor in lymphocyte adhesion, emigration, and chemotaxis. The inhibition of prostaglandins and leukotrienes also helps to reduce mast cell activation, bronchial vasodilation, and smooth muscle contraction (Rowe et al. 2004). Inhaled formulations (budesonide, mometasone, and fluticasone) are preferred in order to decrease systemic side effects, but are ineffective in treating the increased inflammation during asthma exacerbations. In these instances, the more potent systemic corticosteroids (prednisone, methylprednisolone, and dexamethasone) are preferred (Kravitz et al. 2011).

Additional agents are used that either target other steps in the inflammatory pathway or work to decrease bronchospasm and hyperreactivity. Montelukast and zafirlukast are leukotriene receptor antagonists, and zileuton is a 5-lipoxygenase inhibitor that blocks the formation of the leukotrienes (Montuschi and Peters-Golden 2010). Several anticholinergics, such as ipratropium and tiotropium, help to counter bronchospasm and are generally used in conjunction with a SABA. The cromoglycates cromolyn and nedocromil (available only outside the United States) act as mast cell stabilizers, but have additional anti-inflammatory effects on eosinophils, monocytes, and T-lymphocytes (Netzer et al. 2012). Theophylline is a methylxanthine and acts to inhibit leukotriene production as well as relax bronchial smooth muscle. Historically, it was the agent of choice in management of acute exacerbations, but due to its narrow therapeutic window and many drug interactions, it is only recommended as adjunctive treatment in severe asthma (National Institutes of Health 2007). Finally, several monoclonal antibodies (mAbs) have been developed that specifically target IgE (omalizumab) and IL-5 (mepolizumab) and have been shown to be effective in reducing asthma symptoms in the poorly controlled. Although both drugs have been shown to be effective in treatment of younger children with severe asthma, neither drug is yet approved for those under age 12 in the United States (Bel et al. 2014; O'Quinn et al. 2014).

The choice of which medication to use focuses on a stepwise approach, which escalates therapy (adding new medication or increasing doses) based on a patient's symptoms and control, and de-escalates therapy as symptoms improve and control is maintained. The main goal is to use the least number and lowest doses of medications to achieve symptom control and thereby avoid unnecessary side effects. Figure 7.3 illustrates the stepwise approach to pharmacologic treatment of asthma. Patients with intermittent asthma are generally treated with as-needed use of a SABA. For those with exercise-induced symptoms, often pretreatment with a SABA is all that is needed to decrease exacerbations. As symptoms progress, a daily controller medication is added, generally a low-dose ICS. The next step up is to increase the dose of ICS or add an adjunct, such as a leukotriene modifier (LM), cromoglycate, or theophylline

(in adolescents or adults). If symptoms remain uncontrolled, a LABA (in adolescents and adults) or LM is added, if it has not been done previously. If therapy is escalated, high-dose ICS is used, plus introduction of a mAb in adolescents or adults. The final step in the worst-controlled asthmatics is the addition of a low-dose oral corticosteroid to the previous regimen of high-dose ICS, LABA, LM, and mAb (in older patients) (National Institutes of Health 2007). Patients should be frequently assessed and stepped down in therapy if their symptoms have been controlled for at least 3 months.

7.1.6.2 Acute Management

The management of an acute asthma exacerbation should focus on relief of airflow obstruction, correction of hypoxemia if present, and reduction in acute inflammation. The majority of the medications used to control asthma previously mentioned have a longer onset of action, and therefore have no utility in the acute exacerbation (National Institutes of Health 2007). For most exacerbations, multiple doses of a SABA and initiation of a short course of systemic corticosteroids is sufficient to achieve improvement of symptoms and return to

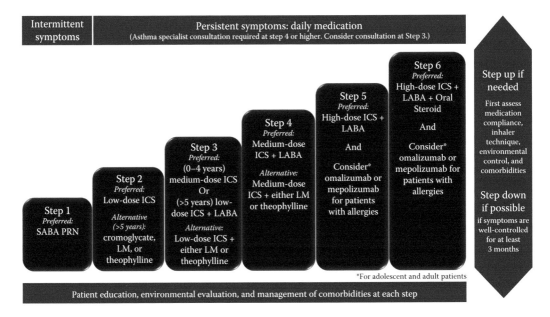

Figure 7.3. Treatment algorithm for the chronic management of asthma demonstrating a stepwise approach to pharmacotherapy. (Adapted from National Institutes of Health, International consensus report on diagnosis and treatment of asthma, National Heart, Lung, and Blood Institute, National Institutes of Health, Bethesda, MD, 2007. http://www.nhlbi.nih.gov /files/docs/guidelines/asthgdln.pdf.)

baseline. Many scoring systems have been developed and validated to determine the severity of an acute attack and may be useful for clinicians to guide management. In general, a patient should be assessed for his or her work of breathing, subjective feeling of dyspnea, degree of air exchange, and presence of hypoxemia (National Institutes of Health 2007).

Mild exacerbations are characterized by dyspnea with exertion, subjective feeling of tightness in the chest or shortness of breath, increased cough (particularly at night), and occasional expiratory wheeze or prolonged expiratory phase on auscultation. These patients can be managed at home with intermittent use of a SABA, generally with prompt relief. If symptoms have persisted for more than 48 hours or are worsening, consideration can be given to start a short (3–5 days) course of oral corticosteroids (Kravitz et al. 2011). Moderate exacerbations are generally characterized by dyspnea interfering with normal activity, persistent expiratory wheeze, slight decrease in air exchange, and increased cough. These patients have no increased work of breathing or accessory muscle use. Those in moderate exacerbation should be evaluated promptly by a healthcare provider. Several doses of a SABA, either repeated every 20 minutes or given continuously via aerosolization, should be administered and the patient started on a short course of oral corticosteroids. Often, these patients will require scheduled use of their SABA (frequently every 4–6 hours) for the first 24–48 hours of treatment until the peak effect of the steroids has been obtained (National Institutes of Health 2007).

Severe exacerbations present with dyspnea at rest, difficulty speaking, and increased work of breathing with accessory muscle use. Air exchange is usually diminished, and wheezing may occur during both inspiration and expiration. Often, these patients are hypoxemic and require urgent attention. They should be helped into a position of comfort and provided supplemental oxygen if needed. A SABA should be given continuously over the first hour of evaluation and systemic corticosteroids initiated. The addition of ipratropium in combination with a SABA early in the course of treatment has been shown to reduce hospitalization and is recommended (Plotnick and Ducharme 2000; Rodrigo and Castro-Rodriguez 2005). Patients should be reassessed frequently and care escalated quickly if little or no improvement is observed. Being unable to speak at all, appearing panicked, little air movement, no wheezing, or altered mental status are all ominous signs and indicate the patient is rapidly progressing toward respiratory arrest and subsequent cardiovascular collapse (National Institutes of Health 2007).

The use of epinephrine or terbutaline (both potent β_2 adrenergic agonists), given intramuscularly, can provide temporary bronchodilation, allowing the inhaled particles of the SABA to penetrate deeper into the lungs, and should be considered very early in those presenting with severe or life-threatening symptoms. Intravenous access should be established as quickly as possible and isotonic fluid administration begun to support cardiac function and correct dehydration if present. Magnesium sulfate, a potent bronchodilator, can be given intravenously in severe exacerbations not responding to SABAs, and can improve pulmonary function and possibly prevent hospitalization (Rowe et al. 2000; Silverman et al. 2002). Additional adjunctive therapies that have shown some benefit in severe asthma attacks include heliox (a helium and oxygen mixture), noninvasive positive-pressure ventilation (NIPPV), intravenous terbutaline, and ketamine sedation.

Heliox (with the oxygen component not exceeding 40%) has been shown to increase laminar flow into and throughout the respiratory tree. It is felt that this improved flow allows better penetration of inhaled medications and improved ventilation (Rodrigo et al. 2003; Kim et al. 2005). Intravenous β_2 agonists (terbutaline) can augment the action of inhaled SABAs when air exchange is poor, although care must be taken with these formulations due to their more prominent cardiac side effects (Travers et al. 2002). Ketamine, an N-methyl-D-aspartate receptor antagonist and dissociative anesthetic, causes bronchodilation and can be used in refractory cases. It can also be useful in decreasing agitation and anxiety in those suffering severe exacerbations and is a preferred induction agent if endotracheal intubation is required (Howton et al. 1996; Allen and Macias 2005). It is effective both in small titrated doses and as a continuous drip.

NIPPV can ease dyspnea, improve work of breathing, and in some cases, prevent endotracheal intubation. Previously, the addition of positive pressure was contraindicated in asthmatics out of concern for worsening air trapping and increasing the risk of barotrauma to the lungs (Ram et al. 2005). However, research now suggests that the positive pressure can actually reduce air trapping by stenting open the airways and allowing better ventilation. This has been shown in several small studies in adult patients, and has yet to be validated in the pediatric population, but provides an alternative to avoid endotracheal intubation (Ram et al. 2005).

If necessary, endotracheal intubation of an asthmatic patient should not be delayed; however, extreme care should be taken to avoid complication. Manipulation of the airway can cause severe bronchospasm due to airway hyperreactivity, often resulting in rapid decline and progression to cardiac arrest from loss of the airway and no ventilation. In addition, there is a significant increase in the risk of barotrauma from increased hyperinflation, air trapping, and airway pressure, all of which are already elevated within the lungs during exacerbation (Zimmerman et al. 1993). Indications for endotracheal intubation include persistent hypoxemia despite optimal delivery of supplemental oxygen or NIPPV, persistent respiratory distress and respiratory muscle fatigue, altered mental status, and respiratory or cardiac arrest. Endotracheal intubation should be performed by the most experienced provider to avoid repeated attempts (Ono et al. 2015).

Once intubated, great care must be taken in choosing correct ventilator settings to avoid barotrauma or cardiac compromise from hyperinflation. Generally, in intubated patients, the goal of ventilation is to reduce hypercarbia, but this is not always possible or advisable in those with asthma. Due to air trapping and inflammation, asthmatics often have prolonged expiration, which poses a significant problem with ventilator management (Zimmerman et al. 1993). If the expiratory time during the respiratory cycle is not long enough, there will be a steady increase in functional residual capacity (known as breath stacking), which results in increasingly higher airway pressures, which can lead to alveolar rupture, pneumothorax, and increased intrathoracic pressure that reduces pulmonary blood flow and venous return to the heart. This leads to both reduction in preload and outflow obstruction and results in eventual cardiovascular collapse (Zimmerman et al. 1993). To avoid these complications, often respiratory rate and tidal volumes must be decreased, and inspiratory-to-expiratory ratios adjusted. In most cases, this does not allow for adequate removal of carbon dioxide from the lungs, and therefore a certain degree of hypercarbia must be allowed. Overall, patients who require endotracheal intubation generally suffer higher morbidity and mortality and longer hospital stays (Zimmerman et al. 1993; Ono et al. 2015).

In intubated patients, a previously unavailable therapeutic option can be considered. Inhalational anesthetics (halothane, isoflurane, and sevoflurane) are all potent bronchodilators and can be administered to improve airflow obstruction (Carrie and Anderson 2015). Other rescue medications should continue to be used, either intravenously or via inhalation through the anesthesia circuit. The final option of rescue in the severe asthmatic, in whom all other therapies have failed, is extracorporeal membrane oxygenation, although this should be reserved as a last resort in the direst of circumstances (Alzeer et al. 2015).

Overall, most patients in acute exacerbation will respond completely to repeated use of a SABA and initiation of a short course of oral steroids. Generally, SABAs act as a rescue medication to reverse acute airway obstruction and buy time until steroids can decrease the flare-up in inflammation and subsequent edema. In the most severe cases, steroids are still the mainstay of treatment and the only thing known to decrease the acute inflammation and eventually end the acute exacerbation. All other treatments mentioned here, including SABAs, are simply short-term rescue measures that provide relief from the symptoms of acute airflow obstruction and cannot treat the underlying inflammation. This highlights the importance of optimizing the long-term chronic management of asthma, with the most important goal being minimization of acute exacerbations and the degree of chronic inflammation in the patient's lungs.

7.1.7 BARRIERS AND HURDLES FOR EFFICIENT TREATMENT

Due to the detailed understanding of asthma and its underlying pathologic conditions, as well as the great range of available treatment options, the disease and its symptoms should nowadays be expected to be controllable in a satisfactory way. A Canadian study, however, revealed that only in 47% of patients are asthma symptoms appropriately controlled according to symptom-based guideline criteria, while 53% of patients are considered to have "poorly controlled" asthma. Nevertheless, 97% of the questioned patients would describe their asthma as "adequately controlled" themselves (FitzGerald et al. 2006).

According to the GAR 2014 (Global Asthma Network 2014), asthma is not only a cause of substantial burden to patients regarding their reduced quality of life due to physical, psychological, and social effects, but also a tremendously high global economic burden. Thus, the indirect costs of the disease, with its negative impact on productivity leading the way, are at least as high as its direct costs, making it a problem of worldwide concern. Given the fact that there are essential asthma medicines with proven benefit available for most patients, a major global focus should be put on the improvement of access to care and adherence to these evidence-based treatments. In both developing and developed countries, this especially requires education of both healthcare providers and patients about the correct use and respective long-term benefits of medications. Additional barriers to effective management occur in developing countries, including lack of affordability of quality-assured medicines and poor infrastructure, indicating the need for political commitment for better asthma care. The key intention from the public health perspective should be the systemic implementation of the best standards of care in everyday practice in order to reduce both human suffering and the associated societal costs. Healthcare professionals and asthma experts are responsible for collaborating with national public health authorities and international organizations to develop national strategies and action plans. However, in 2013, approximately only one in four countries had national asthma strategies in place for children and/or adults. Therefore, the GAR's key recommendation to health authorities in all countries is to ensure the availability of nationally appropriate asthma management guidelines and provide access for everyone to the quality-assured, affordable essential medicines these guidelines suggest. Although such guidelines, first created in the 1980s, were in a great measure commercially sponsored consensus statements, today they are most commonly evidence based and independent of support from the pharmaceutical industry. They play a crucial role in standardizing both timely and correct assessment of asthma symptoms and severity and effective case management, hence lessening the overall burden of the disease. The World Health Organization (WHO) recently published respective asthma management guidelines in its report "Prevention and Control of Noncommunicable Diseases: Guidelines for Primary Health Care in Low Resource Settings" (WHO 2012).

7.1.7.1 Detection and Diagnosis Problems

One hurdle in the very beginning in the way of treating asthma efficiently is the correct detection and diagnosis of the disease. Since the before-described symptoms are manifold and can vary strongly between patients, it can be challenging to detect and treat asthma early on without missing other possible disorders. Particular emphasis has to be put on a distinct and thorough differential diagnosis, as most symptoms are also observed in a number of other pulmonary and airway diseases that can mimic asthma clinically. Hence, it is necessary to maintain a rather broad differential diagnosis in order not to misdiagnose other lung diseases with airflow obstruction as asthma, especially for patients believed to have severe asthma that do not respond adequately to standard asthma therapy.

Although a clear and precise official definition is important to distinguish similar diseases, the guidelines concerning asthma have become rather ambiguous and vague over the last years. While former directives still highlighted eosinophils and mast cells as predominant cells in asthma development (NHLBI 1991), the emphasis in the latest specifications by the Global Initiative for Asthma (GINA) is put only on variability in lung function and airflow limitation (GINA 2015). These symptoms, however, are also present in chronic obstructive pulmonary disease (COPD) and other related lung diseases. Particularly in atopic patients with a long smoking history, the differentiation between asthma and COPD is a demanding task, especially since it was noted that some COPD patients do in fact benefit from high doses of bronchodilators in terms of an improved FEV1 (Virchow 2016). Before, it had been assumed that COPD-related airway obstruction could be regarded as irreversible in response to this treatment, a traditional way to demarcate it from asthmatic origins. Since it became obvious that a considerable group of asthma patients also have a clinically relevant history of smoking, and therefore respective severe airway symptoms due to the two different kinds of airway inflammation, this condition is described as so-called asthma–COPD overlap syndrome (ACOS). Although recent studies suggest that subgroups of asthma and COPD patients do have overlapping immune responses (Gelb et al. 2016), critics question the benefit of this new affiliation of two diseases with different underlying pathological processes.

During viral infections caused, for example, by respiratory syncytial virus or adenoviruses, especially children can present with airflow obstruction, wheezing, and other asthma-like symptoms (Gern et al. 2005), which can be easily misdiagnosed as childhood asthma. Furthermore, dysfunctional breathing disorders like the vocal cord dysfunction (VCD) syndrome can show clinical pictures very similar to asthma, and often even coexist in patients with actual bronchial asthma. Although in many cases these patients receive high-dose antiasthmatic therapy, treatment of the underlying disorders, such as gastroesophageal reflux and postnasal drip, is really needed. Pulmonary sarcoidosis can, likewise, not satisfactorily be treated with bronchodilators or corticosteroids and has to be distinguished from asthma by the lack of seasonal symptom variation and wheezing on auscultation.

Because of the described circumstances, it is strongly suspected that some diagnoses of asthma have been made overhastily and haphazardly in the past. In order to prevent these misjudgments, the first measure for patients with alleged asthma should be a chest x-ray to preclude parenchymal disease, tumors, pneumothorax, and other thoracic conditions. A particular mistrust of asthma diagnosis is applicable in patients without eosinophilia and in those who do not respond to adequate antiasthmatic, anti-inflammatory treatment with bronchodilators and ICSs with an expected normalization of lung function. Here, the differential diagnosis has to be broadened and the diagnostic accuracy has to be improved by considering ancillary examinations, such as high-resolution computed tomography (CT), bronchoalveolar lavage, and sputum cytology or even open lung biopsy (Virchow 2016).

Besides the dissociation from other possible diseases, in modern asthma therapy, it is crucial to identify and differentiate clinical subphenotypes within patients, which is emphasized more precisely in the last segment of this chapter.

7.1.7.2 Corticosteroid Resistance

Another major hurdle in the effective treatment of asthma, especially in smokers and patients with more severe forms of the disease, is the reduced responsiveness to the anti-inflammatory effects of corticosteroids. Although the regular application of low doses of ICSs can control most patients' symptoms appropriately nowadays, approximately 10% need maximal doses and in 1%, even oral corticosteroids are required in order to maintain optimal control. In contrast to this so-called steroid-insensitive asthma, complete resistance to corticosteroids is very rare, but still presents a serious problem in asthma care. The general definition of this resistance is no clinical improvement after high doses of an oral corticosteroid, meaning 40 mg/day prednisone or prednisolone for 2 weeks. As there is no well-defined procedure to quantify clinical steroid responsiveness, it remains a difficult task to measure the degree of resistance in individual patients. However, trying an oral application or a single injection of a depot corticosteroid such as triamcinolone acetonide can be helpful to identify complete resistance (Barnes 2013). Resistant patients were observed to clinically differ from responsive asthmatics by showing a longer duration of symptoms, a greater degree of airway hyperresponsiveness, and a more frequent family history of asthma (Carmichael et al. 1981). The lower responsiveness in patients with severe asthma implies that the mechanisms of steroid resistance are possibly contributing to the grade of disease severity (Moore et al. 2007). In contrast to general familial glucocorticoid resistance, steroid-resistant asthma patients are not cortisol deficient, nor do they have any abnormalities in sex hormones (Lamberts 2001). Furthermore, plasma cortisol and adrenal suppression in response to exogenous corticosteroids are normal, and usually the typical side effects of systemic corticosteroids can be observed. Nevertheless, bronchial biopsies of resistant patients showed increased eosinophil and lymphocyte counts, as well as a missing suppression of the T_H2 cytokines IL-4 and IL-5, compared with sensitive patients, despite the treatment with high doses of corticosteroids (Leung et al. 1995).

As circulating cells from patients with steroid resistance asthma also show reduced responses in vitro, it is feasible to investigate the underlying molecular mechanisms experimentally.

Hence, it could be shown that proliferation of peripheral blood mononuclear cells (PBMCs) from steroid-resistant patients and complement receptors on monocytes from the latter were not inhibited by steroids, indicating that circulating T-lymphocytes and monocytes are resistant in these patients (Poznansky et al. 1984). Subsequently, it was observed that IL-2 and interferon (IFN)-γ secretion are not inhibited (Corrigan et al. 1991) and secretion of cytokines and chemokines of peripheral monocytes and alveolar macrophages is less restricted than in patients with corticosteroid-sensitive asthma (Hew et al. 2006; Bhavsar et al. 2008).

Several mechanisms resulting in a reduced responsiveness to steroids have already been identified, indicating that individual therapeutic approaches may be needed to overcome this pitfall in asthma treatment. As this phenomenon is apparently more common within families, it can be expected that genetic factors play a crucial role in its occurrence. Eleven genes have been found to differ between patients with corticosteroid-resistant and -sensitive asthma in microarray studies of PBMCs (Hakonarson et al. 2005), giving reason to assume that it might be possible to develop a genetic test for steroid resistance. A large proportion of patients with steroid-insensitive asthma show a reduced nuclear translocation of glucocorticoid receptor (GR) α after binding of corticosteroids, which might be explained by modification of the GR via phosphorylation (Matthews et al. 2004). This can be the result of activation of several kinases, such as p38 mitogen-activated protein kinase α or γ, which in turn might be due to reduced activity and expression of phosphatases (Barnes 2013). A further related mechanism is the increased expression of GR β, which inhibits activated GR α. Besides that, increased secretion of macrophage migration inhibitory factor, competition with the transcription factor activator protein 1, and reduced expression of histone deacetylase 2 (HDAC 2) were proposed to be possible causes. The decreased activity of HDAC 2 in severe asthma patients is caused by oxidative stress via activation of phosphoinositide 3-kinase δ (Osoata et al. 2009) and appears to interfere with the action of steroids to switch off activated inflammatory genes. Therefore, a novel approach to reverse steroid resistance is the enhancement of HDAC 2 expression by theophylline or other phosphoinositide 3-kinase δ inhibitors. Respective long-term studies with low-dose theophylline combined with oral corticosteroids and ICSs are under way (Barnes 2013). Besides that, common strategies to manage steroid resistance include the use of alternative broad-spectrum anti-inflammatory drugs, such as phosphodiesterase-4 (PDE4) and p38MAPK inhibitors. Unfortunately, when given systematically, these treatments show side effects that limit the oral dose and efficacy; however, developing inhalable versions has proven to be difficult. Therefore, the well-tolerated ICSs would be the preferred anti-inflammatory therapy, in combination with other drugs for increased responsiveness. This effect can, for example, be achieved by LABAs via the reversion of GR α phosphorylation (Barnes 2013).

7.1.7.3 Failure to Treat Vascular Problems

In a recent review, Harkness et al. (2015) identified the failure of current asthma medication to regularize the vascular remodeling in affected airways as a further barrier in efficient treatment. Numerous studies have shown the abnormal expansion and morphological dysregulation of the bronchial vascular network in asthmatic lungs by reporting an increased number, size, and density of blood vessels, vascular leakage, and plasma engorgement. A more intense blood flow to the airway tissue is suggested to promote chronic influx of inflammatory mediators and pathological cell proliferation (Bergeron et al. 2010).

Nevertheless, appropriate treatment attempts and respective trials are scarce. However, some novel findings give reason to further explore the potential of antiangiogenic therapies as a new drug class for future asthma management. Repurposing antiangiogenic tumor therapy agents for asthma as a genomically and phenotypically more stable target is claimed to be relatively straightforward (Harkness et al. 2015), and some of those agents have indeed already been tested successfully in asthma mouse models. As VEGF-A is a key driver of microvascular remodeling in asthma (Meyer et al. 2012), several treatment approaches have tried to use it as a target. It was shown that the multikinase inhibitor sunitinib significantly inhibits eosinophilic airway inflammation and remodeling in chronic experimental asthma in response to either ovalbumin (OVA) or toluene diisocyanate (Lee et al. 2002; Huang et al. 2009). Likewise, the VEGF-neutralizing mAb bevucizamab reduced epithelial, airway smooth muscle, and basement membrane thickness compared with untreated OVA-challenged mice (Yuksel et al. 2013). In another study, coadministration of the collagen XVIII fragment endostatin with OVA challenge in mice inhibited airway hyperresponsiveness, as well as pulmonary inflammation, and subepithelial angiogenesis was greatly reduced (Suzaki et al. 2005). Further trials targeted circulating endothelial precursor cells, vasculogenesis, or lymphangiogenesis. Nonetheless, the authors concede that there are still substantial hurdles to therapeutic implementation of antiangiogenic medication in asthma treatment, most notably limiting side effects and the lack of good biomarkers to predict therapy responsiveness (Harkness et al. 2015).

7.1.8 PATIENT CARE

7.1.8.1 Patient Monitoring and Assessment Tools

The National Institutes of Health and expert panels currently agree that achieving and maintaining asthma control are essential goals of therapy, and the periodic assessment of this control should be part of patient monitoring once treatment is established. Therefore, asthma control is also increasingly being used as an outcome measure in respective research studies. Due to the complexity and individuality of the disease, precise determination of the control level is difficult, and commonly used assessment methods, such as lung function tests, have their limitations. As patients can show normal lung function between exacerbations but may still not have adequate control over their symptoms, the degree of control is often overestimated by both clinicians and the patients themselves. In order to measure the level of asthma control in a quantitative and comparable way, several composite assessment tools have been developed.

In a recent work, studies were identified by a comprehensive literature search that are attempting to develop and/or test composite score instruments for asthma control (Cloutier et al. 2012). Seventeen score instruments with published validation information were identified, all of which have comparable content and assess nocturnal symptoms or interference with sleep. Symptom frequency, either of specific symptoms, such as cough, wheeze, or dyspnea, or of any asthma symptom, is captured in all but one instrument,

and most of them also detect use of SABAs. All but one evaluate some form of activity limitation, such as interference with daily activities and school or work attendance, whereas only two (Asthma Control Questionnaire [ACQ] and Asthma Control Scoring System [ACSS]) include pulmonary function parameters, and one (ACSS) assesses sputum eosinophilia. While more than half of the evaluated instruments assess exacerbations, only the Test for Respiratory and Asthma Control in Kids (TRACK) includes the "risk of exacerbations" domain, although this is recommended by the National Asthma Education and Prevention Program Expert Panel Report 3 guidelines (NAEPP 2007). In most cases, the score tools are designed to reflect the disease activity over a 1- to 4-week time period and are not validated to be used during asthma exacerbations.

The most widely used tool in patients older than 12 years is the Asthma Control Test (ACT) (Revicki and Weiss 2006), which quantifies asthma control as a continuous variable and offers a numeric value to differentiate between controlled and uncontrolled asthma, similar to most available instruments. It is a multidimensional, standardized, and validated patient-centered questionnaire that inquires about the patient's experience of nocturnal and daytime asthma symptoms, the use of rescue medications, the effect of asthma on daily functioning, and the patient's perception of asthma control over the previous 4 weeks (Nathan et al. 2004). For children aged 4–11 years, the Childhood Asthma Control Test (cACT) was developed in 2006 (Liu et al. 2007), a self-administered tool that involves the child's, as well as his or her caregiver's, perspectives. Another option is a parent-completed version of the Asthma Therapy Assessment Questionnaire (ATAQ), designed to identify children and adolescents of 5–17 years with current problems in asthma control (Skinner et al. 2004).

To ensure that the selected instrument will really measure the wanted outcome, profound knowledge of the psychometric properties of the tool is needed. Furthermore, it is essential to choose tools that have been evaluated with a comparable population and in a similar setting as their intended utilization (Alzahrani and Becker 2016).

In order to facilitate implementation of asthma guidelines to primary care practices, the Asthma APGAR scheme was developed, documenting activity limitation, persistence of symptoms, triggers, asthma medications, and response to therapy. This tool provides a patient survey to collect information based on control scores and contains a management algorithm to incorporate recommendations for education, therapy adherence, adequate inhaler techniques, and follow-up visits. A study evaluating the benefits of implementing the Asthma APGAR in practice demonstrated that it resulted in both enhanced medical record documentation and significantly increased care processes. Involved patients and clinical staff described the tools to be easy to use and stated that they "made sense" and led to "improved care" (Yawn et al. 2008).

7.1.8.2 Pharmaceutical Care and Patient Training

Beyond the care in medical practices, pharmacists play an equally important role in achieving optimal therapy and maximum improvement in the patient's quality of life. Although the scope of pharmaceutical care is often primarily focused on pharmacotherapy and drug-related outcomes, it is encroaching upon other areas of asthma patient care, too, beginning from the very early step of diagnosis. As a recent study reported, pharmacists are qualified to take part in this procedure by performing quality spirometry testing (Cawley and Warning 2015). Spirometry is the most commonly used means to detect and quantify the degree of airflow obstruction in patients and the lung function test of choice for both diagnosing asthma and assessing asthma control in response to treatment. This method is easy and safe to perform; however, for the patient, it is physically demanding, as it requires maximal effort and may cause breathlessness, syncope, and cough, and can even induce bronchospasm in poorly controlled patients. Two important values obtained from the spirometry measurement are the FEV1, a criterion for airway caliber, and the FVC, the maximum volume of air that can be expired during the test. An exemplary spirometer curve with its components is depicted in Figure 7.1. As the spirometer measures timed expired and inspired volumes, it can be calculated from these how effectively and quickly the patient's lungs can be emptied or filled, making the spirogram a volume–time curve. Conventional volume-displacement spirometers provide a direct measurement of the respired volume from the displacement of a water-sealed bell, a rolling-sealed piston, or bellows. New forms of advanced spirometers, however, utilize a sensor detecting flow as the primary signal, for example, from the pressure drop across a resistance or the cooling of a heated wire. The sensed flow is then converted into volume by electronic or numerical integration of the signal. These portable devices are more suitable for personal use and generally easier to clean and disinfect (Johns and Pierce 2008).

A further pulmonary function test used to detect and measure a patient's variability of lung capacity is the PEF measurement. It assesses the maximum expiratory flow occurring just after the start of a forced expiration from the point of maximum inspiration. In contrast to a spirometer, the peak flow meter has significant limitations, as its measurements are effort dependent and results vary considerably between different instruments. Furthermore, the range of values regarded as normal and healthy is rather wide. Nevertheless, PEF monitoring by patients can be useful when they suffer from intermittent symptoms and find it difficult to gauge asthma severity based on those (NACA 2016).

Besides lung function measurements, pharmacists are a crucial factor in the patients' education. As a successful treatment outcome in asthmatic patients depends on continuing compliance with their therapy, even in times they do not suffer from any obvious symptoms, the importance of regular intake of medication has to be emphasized. In addition to a personalized drug regimen, every patient should get an individually tailored care plan in order to cope with variations in asthma severity. In this way, asthmatics are able to slightly adjust their therapy themselves without having to consult their physician every time. When they use PEF measurements to monitor their disease, as with inhaler devices, their technique should be regularly inspected and reinforced. In general, patients have to be familiarized with clinical symptoms of toxicity or undertreatment, such as shortness of breath, wheeze, tremor, or change in O_2 saturation. Patients receiving corticosteroids, in particular, should be educated and monitored concerning adverse effects of this drug class and respective measures they can take to avoid or reduce them. While local side effects of ICSs include oral thrush, hoarseness, and dysphonia, patients on oral therapy should be monitored for hyperglycemia, gastrointestinal and neuropsychiatric effects, adrenal suppression, osteoporosis, and infections (Boyter et al. 2000).

Oftentimes, medication adherence rates decline for children becoming teenagers. In a study assessing adolescent asthmatic needs and preferences regarding medication counseling and self-management, it was shown that effective adherence-enhancing interventions for this patient group are missing. Although lack of perceived need or beneficial effects were also mentioned as reasons for not taking the medication as prescribed, forgetting was identified as being the major cause for nonadherence. Participating adolescents revealed that their parents mainly still play a role in reminding and collecting refills and suggested smartphone applications with a reminder function and easy access to online information as favorable means for successful self-management (Koster et al. 2015). It has been reported that inhaler reminders also offer an effective strategy to improve adherence in adult patients. A 6-month randomized trial compared three patient groups receiving asthma controller treatment: while one just obtained the usual care, personalized adherence discussions were performed with the second group. In the third group, patients received twice-daily SmartTrack reminders for missed doses and automated e-mails about their daily inhaler use were sent to their practitioner. Results demonstrated that the electronic inhaler reminders, including adherence feedback, were able to improve the compliance even more than the patient-specific behavioral interventions (Foster et al. 2014).

Besides poor adherence, inadequate inhaler techniques can also be a substantial cause for persisting asthma symptoms. The correct use of devices is central to effectively deliver the contained medication; however, especially for dry powder inhalers, serious technique errors are not uncommon. The failure to exhale before inhalation, insufficient breath-hold at the end of the inhalation process, and inhalation not forceful from the start are among the most frequent mistakes that are often connected to not only poor asthma control, but also severe consequences, such as asthma-related hospitalizations (Westerik et al. 2016). This demonstrates the great need of comprehensive education and training measures, whose success has already been proven in numerous studies. Among others, Plaza et al. (2015) showed the effectiveness of repeated educational interventions, including a written personalized action plan and training on inhaler techniques

in improving control over asthma symptoms, as well as future risk and quality of life. Moreover, it was reported how efficient symptom management training increases self-efficacy for children and adolescents (Cevik Guner and Celebioglu 2015). Especially in younger children, particular attention has to be drawn to a thorough training of correct inhalation techniques involving encompassing education of the parents.

Although asthma was and is often regarded as being mainly a childhood disease, it is also a relevant origin of morbidity and mortality in the older generation, leading to hospitalization, medical costs, and most importantly, a significant decrease in health-related quality of life, being an even greater burden for seniors. Recent findings suggest that asthma in the elderly phenotypically differs from that in younger patients (Yanez et al. 2014), showing the need for new diagnostic and therapeutic strategies in this population, especially considering the fact that life spans are rising and the proportion of individuals aged 65 and older is evermore increasing worldwide. The high occurrence of various comorbidities, aging-related lung and immune alterations, and epigenetic factors can lead to complex interactions and a diverse pathophysiology. As most former studies were based on allergic or T_H2-mediated asthma, which is not a predominant characteristic of asthma in the elderly (Hanania et al. 2011), and respective clinical studies often exclude older patients (Bauer et al. 1997), the knowledge in this area is still incomplete. Examinations have to consider numerous age-related changes, for example, decreasing respiratory mechanical properties, a shift in immune cells from naïve to memory lymphocytes, reduction of total serum IgE, loss of lung volume, and decline in FEV1, to name only a few. Regarding pharmacotherapy, special attention has to be paid to the higher sensitivity to side effects of medications (Stupka and deShazo 2009), the frequent polypharmacy, and thereby increasing risk of drug interactions, as well as factors like misunderstanding of the disease and treatment regimen, poor compliance, and memory problems (Baptist et al. 2010; Reed 2010).

Furthermore, as aging is often associated with a sedentary everyday life and weight gain, recommendations for regular exercise and promotion of a healthy lifestyle are particularly advisable (Yanez et al. 2014); however, this is not less true for all other age groups. A recent study evaluating the association of severity and control of asthma with factors like body mass index, insulin resistance, levels of adipokines, and inflammatory markers in asthmatic children and adolescents found that asthma was associated with insulin resistance and a systemic inflammatory response possibly mediated by adipokines, with leptin levels standing out among the subjects with excess weight (Morishita et al. 2016).

Assessing the effect of exercise therapy on overweight women with chronic inflammatory diseases led to the observation that neutrophil counts can be reduced (Michishita et al. 2010), and in asthma patients in particular, aerobic exercise training was proven to lower the number of eosinophils in induced sputum, as well as the levels of the fraction of exhaled nitric oxide (FeNO) (Mendes et al. 2011). In children with persistent allergic asthma, a physical training program decreased their total and allergen-specific IgE levels (Moreira et al. 2008). The molecular background of this phenomenon was investigated in a study determining the effect of aerobic exercise in an OVA asthma mouse model. It was shown that exercise is able to reverse the OVA-induced reduction of GR and consequently induce an increased expression of the anti-inflammatory cytokines IL-10 and IL-1ra, while inflammatory mediators like nuclear factor (NF)-κB and transforming growth factor (TGF)-β, as well as airway inflammation and remodeling, were reduced (Silva et al. 2016). Besides that, aerobic exercise was proven to show an anti-inflammatory effect in mice exposed to air pollution (Vieira et al. 2012), and a single session of moderate aerobic exercise can downregulate inflammatory mediators' genes expression and T_H2-derived cytokine production (Hewitt et al. 2009).

Nevertheless, it has to be minded that vigorous activity can also provoke asthma symptoms like cough, wheeze, or dyspnea. While sports with physical efforts of rather short durations and low ventilary levels have a small risk for the development of asthma symptoms, team sports in general can be regarded as medium-risk sports. Participating in swimming, endurance, and winter sports, however, entails a higher risk for triggering exercise-induced asthma (EIA) and bronchoconstriction (EIB) due to the long duration of exertion and the low air temperature, respectively. Other factors influencing how beneficial or detrimental different sports can be for asthmatic patients are,

for example, the humid air inhaled during swimming versus chlorine-based irritants in the pool or the exposure to environmental pollutants and allergens like pollen and molds in outdoor sports (Del Giacco et al. 2015).

7.1.9 RECENT DEVELOPMENTS AND NEW TREATMENT OPTIONS

Although in most patients asthma symptoms can nowadays effectively be managed by guideline-directed conventional medications, some disease forms are more severe and complex, and therefore difficult to control. This may only apply to a relatively small group of the overall patient population; however, they account for more than 50% of asthma-related healthcare utilization (Chung et al. 2014) and are at increased risk of asthma-caused death (Desai and Brightling 2009), reasons enough to make them a top priority in respective research.

One approach to find new therapies is to modulate and improve currently successful drugs, for example, by advancing delivery systems and prolonging their duration of action. Among others, indacaterol, a 24-hour ultra-LABA, was shown to be safe and effective in clinical trials, and dissociated corticosteroids, a new class of glucocorticoids still in preclinical development, are expected to maintain efficacy with reduced side effects (Olin and Wechsler 2014). However, there are also some completely new drug classes in development. Overall, the main interest currently concentrates on finding and targeting novel specific pathways in order to optimize treatments for individual patients.

7.1.9.1 Pharmacogenetics

More than 100 genes are considered conducive to asthma manifestations, including primary disease–conferring genes, asthma severity–varying genes, and treatment-modifying genes. A well-investigated example that has been associated with a more severe form of asthma and a decreased response to β agonists is the substitution of arginine with glycine at position 16 of the $β_2$ adrenergic receptor (Skloot 2016). Recent genome-wide association studies, combined with replication in additional cohorts and in vitro cell-based models, identified novel pathway-related pharmacogenetic variations. As these have the potential to influence the efficacy of therapeutic measures, a more detailed understanding of the underlying genetic mechanisms may lead to the development of biomarkers to determine the most suitable therapy for individual patients (Miller and Ortega 2013).

7.1.9.2 Asthma Phenotyping and Personalized Treatment Approaches

For a long time, asthma has been viewed as a single disease characterized by chronic airway inflammation and remodeling, with anti-inflammatory therapy as the major approach to treatment. But more recently, the disease is recognized to be a multidimensional syndrome involving clinical, physiologic, and pathologic domains, which may coexist, but are not necessarily related (Desai and Brightling 2009). As the understanding of the heterogeneity of asthma is increasing, more and more different phenotypes and endotypes are being identified, incorporating observable clinical characteristics and specific biologic mechanisms in a more complex way (Skloot 2016). A phenotype can be "defined as the composite observable characteristics or traits of an organism that result from genetic as well as environmental influences" (Chung 2016). These kinds of phenotypes, including dependence on high-dose corticosteroid treatment, severe airflow obstruction, and recurrent exacerbations concomitant with an allergic background and late onset of disease, have been revealed by analytical clustering methods, among others in the Severe Asthma Research Program (SARP) (Moore et al. 2010). This investigation identified five clusters of patients differing in age, sex, age of onset, presence of atopy and/or obesity, degree of lung dysfunction, and reversibility of airflow obstruction. In another approach using sputum analysis, four phenotypes based on the predominant inflammatory cell type, such as eosinophilic or neutrophilic, were found (Simpson et al. 2006). In these groupings, the most evident differentiation was between patients with mild to moderate, early-onset asthma with eosinophilic or paucigranulocytic predominant sputum patterns and patients with a more severe form of asthma and greater sputum neutrophilia. Furthermore, to not only compose the clinical picture, but also find subtypes based on the underlying biologic mechanisms, different endotypes were defined, including biomarkers, lung physiology, genetics,

histopathology, epidemiology, and respective treatment response. For example, the aspirin-exacerbated respiratory disease endotype features leukotriene-related genetic polymorphisms leading to an upregulated leukotriene synthesis. As 5-lipoxygenase inhibitors block the synthesis pathway upstream, they are recognized as the superior treatment for this patient group (Fanta and Long 2012). The allergic bronchopulmonary mycosis endotype, in contrast, is associated with colonization of the airways by mold and might therefore benefit from antifungal agents (Skloot 2016). Nevertheless, the major part of novel individual treatment approaches concentrates on biologic therapies.

7.1.9.3 Biologics

The most common grouping of asthma patients divides their asthma into allergic and nonallergic forms. While allergic asthma is present in all groups of asthma severity, with 50%–80% it is of particular importance in patients with severe asthma (Dolan et al. 2004); therefore, finding new treatments for this population is an interesting area of research. Besides the well-known function of IgE, it is also assumed that several different cytokines and chemokines play a crucial role in the pathogenesis of the disease. A number of these have already been identified as suitable targets for therapy, demonstrating the potential of biological drugs such as mAbs and small-molecule inhibitors, especially as reasonable add-on treatments for those patients whose severe asthma forms do not respond to conventional therapies in a satisfactory way. While the anti-inflammatory effect of corticosteroids interferes with several pathways that are involved in asthma pathogenesis, cytokine-based therapy usually only targets a restricted cascade. As experimental anticytokine therapies have also been shown to induce variable responses in individual patients, the need to accurately characterize the patient's phenotypic pattern becomes even more evident. A certain biological drug addressing the particular molecular targets relevant for each subgroup has to be found in order to achieve the best-possible outcome for individual patients (Pelaia et al. 2012).

As the first biologic in asthma treatment, omalizumab (Xolair®, Novartis) was approved in the United States in 2003. The murine mAb is produced by the somatic cells hybridization method and contains a paratope able to bind to high- and low-affinity IgE receptors on basophils and mast cells, inhibiting both degranulation and activation of respective cellular mediators. Although several clinical trials have already proven the effectiveness of omalizumab with a significant decline in asthma exacerbations, improvement of quality of life, and steroid-sparing effects, it still has some limitations. To overcome those, new mAbs are currently under investigation showing a greater avidity for IgE, such as RG7449, which targets B-lymphocytes before they are activated to produce IgE (Menzella et al. 2015).

After more than a decade without any new appearances on the market, the anti-IL-5 humanized mAb mepolizumab (Nucala®) by GlaxoSmithKline was approved at the end of 2015 as an add-on treatment for patients with severe asthma and eosinophilic inflammation. IL-5 was identified as a useful target as it promotes eosinophil growth and activation, and two additional mAbs, reslizumab (Cinqair®) and benralizumab, are already in the pipeline. While Nucala and Cinqair both bind IL-5 directly to hinder it from tacking to its receptors on eosinophils, benralizumab targets the receptor α subunit in order to mediate the death of eosinophils by enhanced antibody-dependent cell-mediated cytotoxicity. As IL-5 is not the only cytokine promoting eosinophil growth and survival, this active cell depletion approach can potentially be even more efficient (Azvolinsky 2016).

Besides that, several different biological targets are currently explored in asthma research. In addition to the IL-5 mAbs, atopic patients with T_H2-driven eosinophilic asthma could also benefit from inhibition of IL-4 and IL-13. A blockade of IL-17 that also contributes to steroid resistance, instead, would most likely be useful for patients with severe neutrophilic asthma. Combinations of biologics targeting different types of cytokines could be applied for the mixed neutrophilic–eosinophilic phenotypes, a group represented quite frequently among the exacerbation-prone form of severe asthma. Another promising approach is the neutralization of the effects of innate cytokines IL-25, IL-33, and TSLP, all of which play a crucial role in the initial priming of T_H2-mediated airway inflammation. This strategy could have the potential to disconnect the link between adaptive and innate immune responses that might be responsible for the development of severe subtypes of asthma that are difficult to treat (Pelaia et al. 2012).

The two major hurdles yet to be overcome on the way to more efficient and easily accessible biological therapies are the lack of reliable biomarkers to characterize specific phenotypes and predict medication responsiveness and the need to design reasonable clinical trials to evaluate the long-term safety of these immunomodulatory agents. Furthermore, the cost factor is not to be underestimated. As Nucala, for example, has a wholesale acquisition cost of $2500 per single-use vial (Azvolinsky 2016), it remains a demanding task to define which patients really need expensive new biologics and which are able to cope with their symptoms by using standard therapies.

GLOSSARY

airway hyperresponsiveness: an exaggerated response of the airways to nonspecific stimuli that results in airway obstruction and can be assessed with bronchial challenge tests, for example, with methacholine or histamine.

barotrauma: injury caused by changes in air pressure, most often increased pressure in the lungs.

cell-mediated immune response: immune response independent of antibodies, involving direct activation of immune cells, including macrophages, phagocytes, and lymphocytes, and subsequent release of cytokines in response to an antigen.

endotracheal intubation: insertion of breathing tube directly into the trachea via the esophagus to allow for mechanical ventilation.

exercise-induced asthma (EIA): a condition of respiratory difficulty related to histamine release, triggered by aerobic exercise.

exercise-induced bronchoconstriction (EIB): airway narrowing occurring during or after exercise or physical exertion, most common in cold weather or indoor sports.

extracorporeal membrane oxygenation: medical procedure allowing for redirection of a patient's blood volume from the venous system through an artificial lung and directly back into circulation, bypassing the heart and lungs to provide oxygenation.

forced expiratory volume in 1 second (FEV1): the volume of air that is forcefully exhaled within the first second.

forced vital capacity (FVC): the greatest volume of air that can be expelled when performing a rapid, forced exhalation; includes the FEV1. An FEV1/FVC ratio of <80% is an indicator for an obstructive defect.

fraction of exhaled nitric oxide (FeNO): biomarker for diagnosis, follow-up, and as a therapy guide for asthma patients, as nitric oxide is produced by certain cell types in an inflammatory response.

free radicals: uncharged molecules with an unpaired electron, highly reactive and short-lived, which are capable of damaging cell walls and proteins and inducing changes in DNA.

hypoxemia: abnormally low blood oxygen concentration.

major basic protein: cationic protein stored in eosinophilic granules that is toxic to epithelial tissue.

n-methyl-D-aspartate receptor: glutamate receptor and ion protein channel found in nerve cells important for memory formation and synaptic plasticity.

noninvasive positive-pressure ventilation: a form of mechanical ventilation allowing for increased air pressure delivery to the lungs without the need for insertion of an endotracheal tube.

ovalbumin (OVA): the main protein in egg white, an established model allergen for airway hyperresponsiveness.

peak expiratory flow (PEF): the maximum flow generated during expiration performed with maximal force and started after a full inspiration.

segmental atelectasis: partial collapse of a discrete segment of alveoli in the lung due to proximal obstruction of the bronchial tree.

RECOMMENDED READING

Barnes, P. J. 2015. Asthma. In *Harrison's Principles of Internal Medicine*, ed. Kasper, D., Fauci, A., Hauser, S., Longe, D., Jameson, J., and Loscalzo, J., Chapter 39,

19th ed. New York: McGraw-Hill. http://access medicine.mhmedical.com/content.aspx?bookid=1130&sectionid=63653136.

Global Asthma Network. 2014. The global asthma report 2014. Auckland: Global Asthma Network. http://www.globalasthmareport.org/resources/Global_Asthma_Report_2014.pdf (accessed March 28, 2016).

Liu, A. H., Covar, R. A., Spahn, J. D., and Sicherer, S. H. 2016. Childhood asthma. In *Nelson Textbook of Pediatrics*, ed. Kliegman, R., Stanton, B., St. Geme, S., and Schor, N., 1095–115.e1. 20th ed. Philadephia: Elsevier.

Mendis, S., Chestnov, O., and Bettcher, D. 2012. WHO prevention and control of noncommunicable diseases: Guidelines for primary health care in low-resource settings. Part II. Management of asthma and chronic obstructive pulmonary disease in primary health care in low-resource settings. Geneva: World Health Organization.

National Institutes of Health. 2007. International consensus report on diagnosis and treatment of asthma. Bethesda, MD: National Heart, Lung, and Blood Institute, National Institutes of Health. http://www .nhlbi.nih.gov/files/docs/guidelines/asthgdln.pdf (accessed March 28, 2016).

REFERENCES

Akinbami, L. J., J. E. Moorman, C. Bailey, H. S. Zahran, M. King, C. A. Johnson, and X. Liu. 2012. Trends in asthma prevalence, health care use, and mortality in the United States, 2001–2010. https://www .cdc.gov/nchs/data/databriefs/db94.htm (accessed March 25).

Allen, J. Y., and C. G. Macias. 2005. The efficacy of ketamine in pediatric emergency department patients who present with acute severe asthma. *Ann Emerg Med* 46 (1):43–50.

Alzahrani, Y. A., and E. A. Becker. 2016. Asthma control assessment tools. *Respir Care* 61 (1):106–16.

Alzeer, A. H., H. A. Al Otair, S. M. Khurshid, S. E. Badrawy, and B. M. Bakir. 2015. A case of near fatal asthma: The role of ECMO as rescue therapy. *Ann Thorac Med* 10 (2):143–5.

Azvolinsky, A. 2016. Severe asthma gets first biologic in decades. *Nat Biotechnol* 34 (1):10–1.

Bacharier, L. B., R. C. Strunk, D. Mauger, D. White, R. F. Lemanske Jr., and C. A. Sorkness. 2004. Classifying asthma severity in children: Mismatch between symptoms, medication use, and lung function. *Am J Respir Crit Care Med* 170 (4):426–32.

Baptist, A. P., B. B. Deol, R. C. Reddy, B. Nelson, and N. M. Clark. 2010. Age-specific factors influencing asthma management by older adults. *Qual Health Res* 20 (1):117–24.

Barnes, P. J. 2013. Corticosteroid resistance in patients with asthma and chronic obstructive pulmonary disease. *J Allergy Clin Immunol* 131 (3):636–45.

Barnett, S. B., and T. A. Nurmagambetov. 2011. Costs of asthma in the United States: 2002–2007. *J Allergy Clin Immunol* 127 (1):145–52.

Bateman, E., H. Nelson, J. Bousquet, K. Kral, L. Sutton, H. Ortega, and S. Yancey. 2008b. Meta-analysis: Effects of adding salmeterol to inhaled corticosteroids on serious asthma-related events. *Ann Intern Med* 149 (1):33–42.

Bateman, E. D., S. S. Hurd, P. J. Barnes, J. Bousquet, J. M. Drazen, M. FitzGerald, P. Gibson et al. 2008a. Global strategy for asthma management and prevention: GINA executive summary. *Eur Respir J* 31 (1):143–78.

Bauer, B. A., C. E. Reed, J. W. Yunginger, P. C. Wollan, and M. D. Silverstein. 1997. Incidence and outcomes of asthma in the elderly. A population-based study in Rochester, Minnesota. *Chest* 111 (2):303–10.

Bel, E. H., S. E. Wenzel, P. J. Thompson, C. M. Prazma, O. N. Keene, S. W. Yancey, H. G. Ortega, I. D. Pavord, and Sirius Investigators. 2014. Oral glucocorticoid-sparing effect of mepolizumab in eosinophilic asthma. *N Engl J Med* 371 (13):1189–97.

Bergeron, C., M. K. Tulic, and Q. Hamid. 2010. Airway remodelling in asthma: From benchside to clinical practice. *Can Respir J* 17 (4):e85–93.

Bhavsar, P., M. Hew, N. Khorasani, A. Torrego, P. J. Barnes, I. Adcock, and K. F. Chung. 2008. Relative corticosteroid insensitivity of alveolar macrophages in severe asthma compared with non-severe asthma. *Thorax* 63 (9):784–90.

Boyter, A., J. Currie, K. Dagg, F. Groundland, and S. Hudson. 2000. Pharmaceutical care (8) asthma. *Pharm J* 264 (7091):546–56.

Bruijnzeel, P. L., M. Uddin, and L. Koenderman. 2015. Targeting neutrophilic inflammation in severe neutrophilic asthma: Can we target the disease-relevant neutrophil phenotype? *J Leukoc Biol* 98 (4):549–56.

Carmichael, J., I. C. Paterson, P. Diaz, G. K. Crompton, A. B. Kay, and I. W. Grant. 1981. Corticosteroid resistance in chronic asthma. *Br Med J (Clin Res Ed)* 282 (6274):1419–22.

Carrie, S., and T. A. Anderson. 2015. Volatile anesthetics for status asthmaticus in pediatric patients: A comprehensive review and case series. *Paediatr Anaesth* 25 (5):460–7.

Cawley, M. J., and W. J. Warning. 2015. Pharmacists performing quality spirometry testing: An evidence based review. *Int J Clin Pharm* 37 (5):726–33.

Centers for Disease Control and Prevention. 2012a. National surveillance of asthma: United States, 2001–2010. http://www.cdc.gov/nchs/data/series/sr_03/sr03_035.pdf (accessed March 25).

Center for Disease Control and Prevention, National Center for Health Statistics. 2012b. Health data interactive summary health statistics for U.S. children. http://www.cdc.gov/nchs/data/series/sr_10/sr10_258.pdf (accessed March 25).

Cevik Guner, U., and A. Celebioglu. 2015. Impact of symptom management training among asthmatic children and adolescents on self-efficacy and disease course. *J Asthma* 52 (8):858–65.

Chung, K. F. 2016. Asthma phenotyping: A necessity for improved therapeutic precision and new targeted therapies. *J Intern Med* 279 (2):192–204.

Chung, K. F., S. E. Wenzel, J. L. Brozek, A. Bush, M. Castro, P. J. Sterk, I. M. Adcock et al. 2014. International ERS/ATS guidelines on definition, evaluation and treatment of severe asthma. *Eur Respir J* 43 (2):343–73.

Cianferoni, A., and J. Spergel. 2014. The importance of TSLP in allergic disease and its role as a potential therapeutic target. *Expert Rev Clin Immunol* 10 (11):1463–74.

Cloutier, M. M., M. Schatz, M. Castro, N. Clark, H. W. Kelly, R. Mangione-Smith, J. Sheller, C. Sorkness, S. Stoloff, and P. Gergen. 2012. Asthma outcomes: Composite scores of asthma control. *J Allergy Clin Immunol* 129 (3 Suppl):S24–33.

Corrigan, C. J., P. H. Brown, N. C. Barnes, J. J. Tsai, A. J. Frew, and A. B. Kay. 1991. Glucocorticoid resistance in chronic asthma. Peripheral blood T lymphocyte activation and comparison of the T lymphocyte inhibitory effects of glucocorticoids and cyclosporin A. *Am Rev Respir Dis* 144 (5):1026–32.

Covar, R. A., J. D. Spahn, J. R. Murphy, S. J. Szefler, and Group Childhood Asthma Management Program Research. 2004. Progression of asthma measured by lung function in the Childhood Asthma Management Program. *Am J Respir Crit Care Med* 170 (3):234–41.

Covar, R. A., R. Strunk, R. S. Zeiger, L. A. Wilson, A. H. Liu, S. Weiss, J. Tonascia, J. D. Spahn, S. J. Szefler, and Group Childhood Asthma Management Program Research. 2010. Predictors of remitting, periodic, and persistent childhood asthma. *J Allergy Clin Immunol* 125 (2):359–366.e3.

Del Giacco, S. R., D. Firinu, L. Bjermer, and K. H. Carlsen. 2015. Exercise and asthma: An overview. *Eur Clin Respir J* 2:27984.

Desai, D., and C. Brightling. 2009. Cytokine and anticytokine therapy in asthma: Ready for the clinic? *Clin Exp Immunol* 158 (1):10–9.

Dolan, C. M., K. E. Fraher, E. R. Bleecker, L. Borish, B. Chipps, M. L. Hayden, S. Weiss, B. Zheng, C. Johnson, S. Wenzel, and Tenor Study Group. 2004. Design and baseline characteristics of the epidemiology and natural history of asthma: Outcomes and Treatment Regimens (TENOR) study: A large cohort of patients with severe or difficult-to-treat asthma. *Ann Allergy Asthma Immunol* 92 (1):32–9.

Elliot, J. G., R. L. Jones, M. J. Abramson, F. H. Green, T. Mauad, K. O. McKay, T. R. Bai, and A. L. James. 2015. Distribution of airway smooth muscle remodelling in asthma: Relation to airway inflammation. *Respirology* 20 (1):66–72.

Fanta, C. H., and A. A. Long. 2012. Difficult asthma: Assessment and management. Part 2. *Allergy Asthma Proc* 33 (4):313–23.

FitzGerald, J. M., L. P. Boulet, R. A. McIvor, S. Zimmerman, and K. R. Chapman. 2006. Asthma control in Canada remains suboptimal: The Reality of Asthma Control (TRAC) study. *Can Respir J* 13 (5):253–9.

Fitzpatrick, S., R. Joks, and J. I. Silverberg. 2012. Obesity is associated with increased asthma severity and exacerbations, and increased serum immunoglobulin E in inner-city adults. *Clin Exp Allergy* 42 (5):747–59.

Foster, J. M., T. Usherwood, L. Smith, S. M. Sawyer, W. Xuan, C. S. Rand, and H. K. Reddel. 2014. Inhaler reminders improve adherence with controller treatment in primary care patients with asthma. *J Allergy Clin Immunol* 134 (6):1260–68.e3.

Furuta, G. T., F. D. Atkins, N. A. Lee, and J. J. Lee. 2014. Changing roles of eosinophils in health and disease. *Ann Allergy Asthma Immunol* 113 (1):3–8.

Gelb, A. F., S. A. Christenson, and J. A. Nadel. 2016. Understanding the pathophysiology of the asthma-chronic obstructive pulmonary disease overlap syndrome. *Curr Opin Pulm Med* 22 (2):100–5.

Gern, J. E., L. A. Rosenthal, R. L. Sorkness, and R. F. Lemanske Jr. 2005. Effects of viral respiratory infections on lung development and childhood asthma. *J Allergy Clin Immunol* 115 (4):668–74; quiz 675.

GINA (Global Initiative for Asthma). 2015. Global strategy for asthma management and prevention. http://www.ginasthma.org (accessed December 14).

Global Asthma Network. 2014. The global asthma report. http://www.globalasthmareport.org/resources/Global_Asthma_Report_2014.pdf (accessed March 28).

Goussard, P., R. Gie, S. Andronikou, and J. L. Morrison. 2014. Organic foreign body causing lung collapse and bronchopleural fistula with empyema. *BMJ Case Rep* 2014.

Hakonarson, H., U. S. Bjornsdottir, E. Halapi, J. Bradfield, F. Zink, M. Mouy, H. Helgadottir et al. 2005. Profiling of genes expressed in peripheral blood mononuclear cells predicts glucocorticoid sensitivity in asthma patients. *Proc Natl Acad Sci USA* 102 (41):14789–94.

Hanania, N. A., M. J. King, S. S. Braman, C. Saltoun, R. A. Wise, P. Enright, A. R. Falsey et al. 2011. Asthma in the elderly: Current understanding and future research needs—A report of a National Institute on Aging (NIA) workshop. *J Allergy Clin Immunol* 128 (3 Suppl):S4–24.

Harkness, L. M., A. W. Ashton, and J. K. Burgess. 2015. Asthma is not only an airway disease, but also a vascular disease. *Pharmacol Ther* 148:17–33.

Hew, M., P. Bhavsar, A. Torrego, S. Meah, N. Khorasani, P. J. Barnes, I. Adcock, and K. F. Chung. 2006. Relative corticosteroid insensitivity of peripheral blood mononuclear cells in severe asthma. *Am J Respir Crit Care Med* 174 (2):134–41.

Hewitt, M., A. Creel, K. Estell, I. C. Davis, and L. M. Schwiebert. 2009. Acute exercise decreases airway inflammation, but not responsiveness, in an allergic asthma model. *Am J Respir Cell Mol Biol* 40 (1):83–9.

Holt, P. G., and P. D. Sly. 2012. Viral infections and atopy in asthma pathogenesis: New rationales for asthma prevention and treatment. *Nat Med* 18 (5):726–35.

Hosoki, K., T. Itazawa, I. Boldogh, and S. Sur. 2016. Neutrophil recruitment by allergens contribute to allergic sensitization and allergic inflammation. *Curr Opin Allergy Clin Immunol* 16 (1):45–50.

Howton, J. C., J. Rose, S. Duffy, T. Zoltanski, and M. A. Levitt. 1996. Randomized, double-blind, placebo-controlled trial of intravenous ketamine in acute asthma. *Ann Emerg Med* 27 (2):170–5.

Huang, M., X. Liu, Q. Du, X. Yao, and K. S. Yin. 2009. Inhibitory effects of sunitinib on ovalbumin-induced chronic experimental asthma in mice. *Chin Med J (Engl)* 122 (9):1061–6.

Jiang, M., P. Qin, and X. Yang. 2014. Comorbidity between depression and asthma via immune-inflammatory pathways: A meta-analysis. *J Affect Disord* 166:22–9.

Johns, D. P., and R. Pierce. 2008. Spirometry—The measurement and interpretation of ventilatory function in clinical practice. http://www.nationalasthma.org.au/uploads/content/211-spirometer_handbook_naca.pdf (accessed March 19).

Juntti, H., J. Kokkonen, T. Dunder, M. Renko, A. Niinimaki, and M. Uhari. 2003. Association of an early respiratory syncytial virus infection and atopic allergy. *Allergy* 58 (9):878–84.

Kim, I. K., E. Phrampus, S. Venkataraman, R. Pitetti, A. Saville, T. Corcoran, E. Gracely, N. Funt, and A. Thompson. 2005. Helium/oxygen-driven albuterol nebulization in the treatment of children with moderate to severe asthma exacerbations: A randomized, controlled trial. *Pediatrics* 116 (5):1127–33.

Koskela, H. O. 2007. Cold air-provoked respiratory symptoms: The mechanisms and management. *Int J Circumpolar Health* 66 (2):91–100.

Koster, E. S., D. Philbert, T. W. de Vries, L. van Dijk, and M. L. Bouvy. 2015. "I just forget to take it": Asthma self-management needs and preferences in adolescents. *J Asthma* 52 (8):831–7.

Kravitz, J., P. Dominici, J. Ufberg, J. Fisher, and P. Giraldo. 2011. Two days of dexamethasone versus 5 days of prednisone in the treatment of acute asthma: A randomized controlled trial. *Ann Emerg Med* 58 (2):200–4.

Lai, C. K., R. Beasley, J. Crane, S. Foliaki, J. Shah, S. Weiland, and ISAAC Phase Three Study Group. 2009. Global variation in the prevalence and severity of asthma symptoms: Phase three of the International Study of Asthma and Allergies in Childhood (ISAAC). *Thorax* 64 (6):476–83.

Lamberts, S. W. 2001. Hereditary glucocorticoid resistance. *Ann Endocrinol (Paris)* 62 (2):164–7.

Lee, Y. C., Y. G. Kwak, and C. H. Song. 2002. Contribution of vascular endothelial growth factor to airway hyperresponsiveness and inflammation in a murine model of toluene diisocyanate-induced asthma. *J Immunol* 168 (7):3595–600.

Leung, D. Y., R. J. Martin, S. J. Szefler, E. R. Sher, S. Ying, A. B. Kay, and Q. Hamid. 1995. Dysregulation of interleukin 4, interleukin 5, and interferon gamma gene expression in steroid-resistant asthma. *J Exp Med* 181 (1):33–40.

Liu, A. H., R. Zeiger, C. Sorkness, T. Mahr, N. Ostrom, S. Burgess, J. C. Rosenzweig, and R. Manjunath. 2007. Development and cross-sectional validation of the Childhood Asthma Control Test. *J Allergy Clin Immunol* 119 (4):817–25.

Martinez, F. D. 2006. Inhaled corticosteroids and asthma prevention. *Lancet* 368 (9537):708–10.

Martinez, F. D., A. L. Wright, L. M. Taussig, C. J. Holberg, M. Halonen, and W. J. Morgan. 1995. Asthma and wheezing in the first six years of life. The Group Health Medical Associates. *N Engl J Med* 332 (3):133–8.

Matthews, J. G., K. Ito, P. J. Barnes, and I. M. Adcock. 2004. Defective glucocorticoid receptor nuclear translocation and altered histone acetylation patterns in glucocorticoid-resistant patients. *J Allergy Clin Immunol* 113 (6):1100–8.

McMahon, A. W., M. S. Levenson, B. W. McEvoy, A. D. Mosholder, and D. Murphy. 2011. Age and risks of FDA-approved long-acting beta(2)-adrenergic receptor agonists. *Pediatrics* 128 (5):e1147–54.

Melen, E., and G. Pershagen. 2012. Pathophysiology of asthma: Lessons from genetic research with particular focus on severe asthma. *J Intern Med* 272 (2):108–20.

Mendes, F. A., F. M. Almeida, A. Cukier, R. Stelmach, W. Jacob-Filho, M. A. Martins, and C. R. Carvalho. 2011. Effects of aerobic training on airway inflammation in asthmatic patients. *Med Sci Sports Exerc* 43 (2):197–203.

Menzella, F., M. Lusuardi, C. Galeone, and L. Zucchi. 2015. Tailored therapy for severe asthma. *Multidiscip Respir Med* 10 (1):1.

Meyer, N., J. Christoph, H. Makrinioti, P. Indermitte, C. Rhyner, M. Soyka, T. Eiwegger et al. 2012. Inhibition of angiogenesis by IL-32: Possible role in asthma. *J Allergy Clin Immunol* 129 (4):964–73.e7.

Michishita, R., N. Shono, T. Inoue, T. Tsuruta, and K. Node. 2010. Effect of exercise therapy on monocyte and neutrophil counts in overweight women. *Am J Med Sci* 339 (2):152–6.

Miller, S. M., and V. E. Ortega. 2013. Pharmacogenetics and the development of personalized approaches for combination therapy in asthma. *Curr Allergy Asthma Rep* 13 (5):443–52.

Montuschi, P., and M. L. Peters-Golden. 2010. Leukotriene modifiers for asthma treatment. *Clin Exp Allergy* 40 (12):1732–41.

Moore, W. C., E. R. Bleecker, D. Curran-Everett, S. C. Erzurum, B. T. Ameredes, L. Bacharier, W. J. Calhoun et al. 2007. Characterization of the severe asthma phenotype by the National Heart, Lung, and Blood Institute's Severe Asthma Research Program. *J Allergy Clin Immunol* 119 (2):405–13.

Moore, W. C., D. A. Meyers, S. E. Wenzel, W. G. Teague, H. Li, X. Li, R. D'Agostino Jr. et al. 2010. Identification of asthma phenotypes using cluster analysis in the Severe Asthma Research Program. *Am J Respir Crit Care Med* 181 (4):315–23.

Moreira, A., L. Delgado, T. Haahtela, J. Fonseca, P. Moreira, C. Lopes, J. Mota, P. Santos, P. Rytila, and M. G. Castel-Branco. 2008. Physical training does not increase allergic inflammation in asthmatic children. *Eur Respir J* 32 (6):1570–5.

Morgan, W. J., D. A. Stern, D. L. Sherrill, S. Guerra, C. J. Holberg, T. W. Guilbert, L. M. Taussig, A. L. Wright, and F. D. Martinez. 2005. Outcome of asthma and wheezing in the first 6 years of life: Follow-up through adolescence. *Am J Respir Crit Care Med* 172 (10):1253–8.

Morishita, R., M. D. Franco, F. I. Suano-Souza, D. Sole, R. F. Puccini, and M. W. Strufaldi. 2016. Body mass index, adipokines and insulin resistance in asthmatic children and adolescents. *J Asthma* 53 (5):478–84.

Murray, C. J., T. Vos, R. Lozano, M. Naghavi, A. D. Flaxman, C. Michaud, M. Ezzati et al. 2012. Disability-adjusted life years (DALYs) for 291 diseases and injuries in 21 regions, 1990–2010: A systematic analysis for the Global Burden of Disease Study 2010. *Lancet* 380 (9859):2197–223.

NACA (National Asthma Council Australia). 2016. Asthma & lung function tests. http://www.nationalasthma.org.au/uploads/publication/asthma-lung-function-tests-hp.pdf (accessed March 12).

NAEPP (National Asthma Education and Prevention Program). 2007. Expert Panel Report 3 (EPR-3): Guidelines for the diagnosis and management of asthma—Summary report 2007. *J Allergy Clin Immunol* 120 (5 Suppl):S94–138.

Nakagome, K., and M. Nagata. 2011. Pathogenesis of airway inflammation in bronchial asthma. *Auris Nasus Larynx* 38 (5):555–63.

Nathan, R. A., C. A. Sorkness, M. Kosinski, M. Schatz, J. T. Li, P. Marcus, J. J. Murray, and T. B. Pendergraft. 2004. Development of the asthma control test: A survey for assessing asthma control. *J Allergy Clin Immunol* 113 (1):59–65.

National Institutes of Health. 2007. International consensus report on diagnosis and treatment of asthma. Bethesda, MD: National Heart, Lung, and Blood Institute. http://www.nhlbi.nih.gov/files/docs/guidelines/asthgdln.pdf (accessed March 28).

Nelson, H. S., S. T. Weiss, E. R. Bleecker, S. W. Yancey, P. M. Dorinsky, and Smart Study Group. 2006. The Salmeterol Multicenter Asthma Research Trial: A comparison of usual pharmacotherapy for asthma or usual pharmacotherapy plus salmeterol. *Chest* 129 (1):15–26.

Netzer, N. C., T. Kupper, H. W. Voss, and A. H. Eliasson. 2012. The actual role of sodium cromoglycate in the treatment of asthma—A critical review. *Sleep Breath* 16 (4):1027–32.

Newhouse, M. T., and M. B. Dolovich. 1986. Control of asthma by aerosols. *N Engl J Med* 315 (14):870–4.

NHLBI (National Heart, Lung, and Blood Institute). 1991. Guidelines for the diagnosis and management of asthma. National Heart, Lung, and Blood Institute. National Asthma Education Program. Expert Panel Report. *J Allergy Clin Immunol* 88 (3 Pt 2):425–534.

Olin, J. T., and M. E. Wechsler. 2014. Asthma: Pathogenesis and novel drugs for treatment. *BMJ* 349:g5517.

Ono, Y., H. Kikuchi, K. Hashimoto, T. Sasaki, J. Ishii, C. Tase, and K. Shinohara. 2015. Emergency endotracheal intubation-related adverse events in bronchial asthma exacerbation: Can anesthesiologists attenuate the risk? *J Anesth* 29 (5):678–85.

O'Quinn, J., S. Santucci, D. Pham, Z. Chad, I. MacLusky, J. Reisman, and W. Yang. 2014. Omalizumab treatment of moderate to severe asthma in the adolescent and pediatric population. *Allergy Asthma Clin Immunol* 10 (Suppl 2):A34.

Osoata, G. O., T. Hanazawa, C. Brindicci, M. Ito, P. J. Barnes, S. Kharitonov, and K. Ito. 2009. Peroxynitrite elevation in exhaled breath condensate of COPD and its inhibition by fudosteine. *Chest* 135 (6):1513–20.

Paull, K., R. Covar, N. Jain, E. W. Gelfand, and J. D. Spahn. 2005. Do NHLBI lung function criteria apply to children? A cross-sectional evaluation of childhood asthma at National Jewish Medical and Research Center, 1999–2002. *Pediatr Pulmonol* 39 (4):311–7.

Pelaia, G., T. Renda, L. Gallelli, A. Vatrella, M. T. Busceti, S. Agati, M. Caputi, M. Cazzola, R. Maselli, and S. A. Marsico. 2008. Molecular mechanisms underlying airway smooth muscle contraction and proliferation: Implications for asthma. *Respir Med* 102 (8):1173–81.

Pelaia, G., A. Vatrella, and R. Maselli. 2012. The potential of biologics for the treatment of asthma. *Nat Rev Drug Discov* 11 (12):958–72.

Plaza, V., M. Peiro, M. Torrejon, M. Fletcher, A. Lopez-Vina, J. M. Ignacio, J. A. Quintano, S. Bardagi, I. Gich, and Prometheus Study Group. 2015. A repeated short educational intervention improves asthma control and quality of life. *Eur Respir J* 46 (5):1298–307.

Plotnick, L. H., and F. M. Ducharme. 2000. Combined inhaled anticholinergics and beta2-agonists for initial treatment of acute asthma in children. *Cochrane Database Syst Rev* (4):CD000060.

Poznansky, M. C., A. C. Gordon, J. G. Douglas, A. S. Krajewski, A. H. Wyllie, and I. W. Grant. 1984. Resistance to methylprednisolone in cultures of blood mononuclear cells from glucocorticoid-resistant asthmatic patients. *Clin Sci (Lond)* 67 (6):639–45.

Ram, F. S., S. Wellington, B. Rowe, and J. A. Wedzicha. 2005. Non-invasive positive pressure ventilation for treatment of respiratory failure due to severe acute exacerbations of asthma. *Cochrane Database Syst Rev* (3):CD004360.

Reed, C. E. 2010. Asthma in the elderly: Diagnosis and management. *J Allergy Clin Immunol* 126 (4):681–7; quiz 688–9.

Revicki, D., and K. B. Weiss. 2006. Clinical assessment of asthma symptom control: Review of current assessment instruments. *J Asthma* 43 (7):481–7.

Robinson, C. L., L. M. Baumann, K. Romero, J. M. Combe, A. Gomez, R. H. Gilman, L. Cabrera et al. 2011. Effect of urbanisation on asthma, allergy and airways inflammation in a developing country setting. *Thorax* 66 (12):1051–7.

Robinson, D. S. 2004. The role of the mast cell in asthma: Induction of airway hyperresponsiveness by interaction with smooth muscle? *J Allergy Clin Immunol* 114 (1):58–65.

Rodrigo, G. J., and J. A. Castro-Rodriguez. 2005. Anticholinergics in the treatment of children and adults with acute asthma: A systematic review with meta-analysis. *Thorax* 60 (9):740–6.

Rodrigo, G. J., C. Rodrigo, C. V. Pollack, and B. Rowe. 2003. Use of helium-oxygen mixtures in the treatment of acute asthma: A systematic review. *Chest* 123 (3):891–6.

Rodriguez, A., M. Vaca, G. Oviedo, S. Erazo, M. E. Chico, C. Teles, M. L. Barreto, L. C. Rodrigues, and P. J. Cooper. 2011. Urbanisation is associated with prevalence of childhood asthma in diverse, small rural communities in Ecuador. *Thorax* 66 (12):1043–50.

Rowe, B. H., J. A. Bretzlaff, C. Bourdon, G. W. Bota, and C. A. Camargo Jr. 2000. Magnesium sulfate for treating exacerbations of acute asthma in the emergency department. *Cochrane Database Syst Rev* (2):CD001490.

Rowe, B. H., M. L. Edmonds, C. H. Spooner, B. Diner, and C. A. Camargo Jr. 2004. Corticosteroid therapy for acute asthma. *Respir Med* 98 (4):275–84.

Silva, R. A., F. M. Almeida, C. R. Olivo, B. M. Saraiva-Romanholo, M. A. Martins, and C. R. Carvalho. 2016. Exercise reverses OVA-induced inhibition of glucocorticoid receptor and increases anti-inflammatory cytokines in asthma. *Scand J Med Sci Sports* 26 (1):82–92.

Silverman, R. A., H. Osborn, J. Runge, E. J. Gallagher, W. Chiang, J. Feldman, T. Gaeta et al. 2002. IV magnesium sulfate in the treatment of acute severe asthma: A multicenter randomized controlled trial. *Chest* 122 (2):489–97.

Simpson, J. L., R. Scott, M. J. Boyle, and P. G. Gibson. 2006. Inflammatory subtypes in asthma: Assessment and identification using induced sputum. *Respirology* 11 (1):54–61.

Skinner, E. A., G. B. Diette, P. J. Algatt-Bergstrom, T. T. Nguyen, R. D. Clark, L. E. Markson, and A. W. Wu. 2004. The Asthma Therapy Assessment Questionnaire (ATAQ) for children and adolescents. *Dis Manag* 7 (4):305–13.

Skloot, G. S. 2016. Asthma phenotypes and endotypes: A personalized approach to treatment. *Curr Opin Pulm Med* 22 (1):3–9.

Spahn, J. D., R. Cherniack, K. Paull, and E. W. Gelfand. 2004. Is forced expiratory volume in one second the best measure of severity in childhood asthma? *Am J Respir Crit Care Med* 169 (7):784–6.

Stupka, E., and R. deShazo. 2009. Asthma in seniors. Part 1. Evidence for underdiagnosis, undertreatment, and increasing morbidity and mortality. *Am J Med* 122 (1):6–11.

Suzaki, Y., K. Hamada, M. Sho, T. Ito, K. Miyamoto, S. Akashi, H. Kashizuka et al. 2005. A potent anti-angiogenic factor, endostatin prevents the development of asthma in a murine model. *J Allergy Clin Immunol* 116 (6):1220–7.

Tillie-Leblond, I., P. Gosset, and A. B. Tonnel. 2005. Inflammatory events in severe acute asthma. *Allergy* 60 (1):23–9.

Travers, A. H., B. H. Rowe, S. Barker, A. Jones, and C. A. Camargo Jr. 2002. The effectiveness of IV beta-agonists in treating patients with acute asthma in the emergency department: A meta-analysis. *Chest* 122 (4):1200–7.

Vieira, R. P., A. C. Toledo, L. B. Silva, F. M. Almeida, N. R. Damaceno-Rodrigues, E. G. Caldini, A. B. Santos et al. 2012. Anti-inflammatory effects of aerobic exercise in mice exposed to air pollution. *Med Sci Sports Exerc* 44 (7):1227–34.

Virchow, J. C. 2016. Diagnostic challenges of adult asthma. *Curr Opin Pulm Med* 22 (1):38–45.

von Mutius, E., and D. Vercelli. 2010. Farm living: Effects on childhood asthma and allergy. *Nat Rev Immunol* 10 (12):861–8.

Wells, A. D., J. A. Poole, and D. J. Romberger. 2014. Influence of farming exposure on the development of asthma and asthma-like symptoms. *Int Immunopharmacol* 23 (1):356–63.

Westerik, J. A., V. Carter, H. Chrystyn, A. Burden, S. L. Thompson, D. Ryan, K. Gruffydd-Jones et al. 2016. Characteristics of patients making serious inhaler errors with a dry powder inhaler and association with asthma-related events in a primary care setting. *J Asthma* 1–9.

WHO (World Health Organization). 2012. Prevention and control of noncommunicable diseases: Guidelines for primary health care in low resource settings. Geneva: WHO.

Yanez, A., S. H. Cho, J. B. Soriano, L. J. Rosenwasser, G. J. Rodrigo, K. F. Rabe, S. Peters et al. 2014. Asthma in the elderly: What we know and what we have yet to know. *World Allergy Organ J* 7 (1):8.

Yawn, B. P., S. Bertram, and P. Wollan. 2008. Introduction of Asthma APGAR tools improve asthma management in primary care practices. *J Asthma Allergy* 1:1–10.

Yuksel, H., O. Yilmaz, M. Karaman, H. A. Bagriyanik, F. Firinci, M. Kiray, A. Turkeli, and O. Karaman. 2013. Role of vascular endothelial growth factor antagonism on airway remodeling in asthma. *Ann Allergy Asthma Immunol* 110 (3):150–5.

Zimmerman, J. L., R. P. Dellinger, A. N. Shah, and R. W. Taylor. 1993. Endotracheal intubation and mechanical ventilation in severe asthma. *Crit Care Med* 21 (11):1727–30.

NANOMEDICINE FOR INFLAMMATORY DISEASES

Nanotherapeutics for Asthma

Adriana Lopes da Silva, Fernanda Ferreira Cruz, and Patricia Rieken Macedo Rocco

CONTENTS

7.2.1 ASTHMA PHYSIOPATHOLOGY AND TREATMENTS

Allergic asthma is a major global health problem. Approximately 334 million people worldwide are believed to be affected by asthma (GINA, 2015), with negative consequences such as poor health-related quality of life, impaired work productivity, and limitations of the activities of daily living (Peters et al., 2006).

Classically, CD4[+] T helper (Th) type 2 cells have been considered the primary regulators of the allergic response through production of Th2 cytokines, which lead to airway inflammation and hyperresponsiveness (Holloway et al., 2010). Patients with asthma may also display airway hyperresponsiveness with no early- or late-phase inflammation present, which is likely due to long-term changes in the structure of the airway, such as hypertrophy and/or hyperplasia of the smooth muscle layer (Murdoch and Lloyd, 2010). The most widely employed therapies for asthma, including inhaled corticosteroids and bronchodilators (short- or long-lasting β-adrenergic agonists or muscarinic antagonists), reduce disease symptoms (Kim et al., 2011; Raissy et al., 2013). However, corticosteroids cause long-lasting side effects and may not provide instant relief, thus decreasing treatment adherence by patients (Cooper et al., 2015) and accelerating disease progress and lung remodeling, which leads to progressive loss of lung function (Pascual and Peters, 2005). Moreover, long-term use of corticosteroids cannot revert lung remodeling and may result in immunosuppression, predisposing patients to lung infection with opportunistic microorganisms (Barnes, 2012). Finally, not all asthmatic patients benefit from inhaled corticosteroid treatment; those with a more neutrophilic lung inflammatory status are particularly resistant to standard glucocorticoid anti-inflammatory therapy (Barnes, 2013). Thus, alternative therapeutic approaches that can both attenuate inflammatory and remodeling processes in different asthma phenotypes and reduce airway resistance without leading to immunosuppression are needed.

In this context, nanotechnology has emerged as a new paradigm for the treatment of diseases refractory to conventional therapeutics, including asthma. The use of nanoparticles (NPs) facilitates the delivery of biological materials and drugs directly to target tissues while minimizing adverse effects.

A variety of nanoparticulate systems (nano-complexes) have been evaluated in the lung. However, several physiological barriers, for example, mucus in the airway lumen, make application of this therapeutic strategy to the lung very challenging. NPs encounter different barriers in the central and peripheral airways because the lung has evolved innate mechanisms of defense to limit access of foreign particles (Figure 7.4). NPs must cross the airway and the pulmonary epithelium to reach the underlying tissues or the systemic circulation, by overcoming the mucus layer, mucociliary clearance in the airways, and surfactant, as well as the phagocytosis mediated by alveolar macrophages in the peripheral lung. Neutrophils represent additional barriers and sentinels for the innate and adaptive immune systems in the airways, including cytokines, chemokines, and the complement system (Di Gioia et al., 2015).

Therefore, nanometer size, the material used to manufacture NPs, and their sustained activity must be evaluated in order to avoid clearance by macrophages and allow transcytosis and treatment of the submucosa, while crossing of the endothelium allows systemic treatment.

7.2.2 NANOPARTICLES FOR ASTHMA TREATMENT

7.2.2.1 Nanoparticles and Drug Delivery

Successful management of asthmatic patients depends on achieving adequate delivery of inhaled drugs to the lung. This assumes particular importance for inhaled corticosteroids, the therapeutic goal for which should be to achieve a high rate of airway anti-inflammatory efficacy while avoiding systemic side effects (Jackson and Lipworth, 1995). However, the availability of user-friendly inhaler NPs requires critical appraisal of their effectiveness and evaluation of whether improved lung deposition of antiasthma drugs translates into improved clinical efficacy with sustained therapeutic effects.

NPs have been manufactured from various materials—including polymers, ceramics, metals, and biologics—with unique architectures to serve as possible drug vehicles to treat particular diseases. Generally, the choice of material and format depends on the particular therapy to be employed, and is selected with reduction of toxicity in mind (Da Silva et al., 2013). Nevertheless,

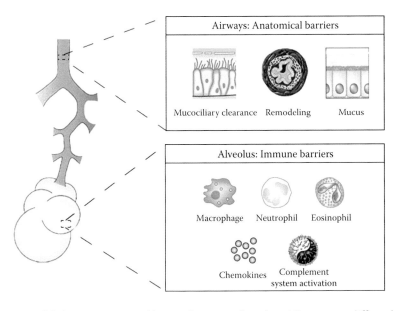

Figure 7.4. Overview of the barriers encountered by nanotherapeutics for asthma. NPs encounter different barriers in the central and peripheral airways because the lung has evolved innate defense mechanisms to limit access to foreign particles. NPs need to overcome the anatomical barriers created by asthma, including mucociliary clearance, lung remodeling, and mucus, as well as native immune barriers such as phagocytosis mediated by alveolar macrophages. Neutrophils represent additional barriers for the innate and adaptive immune systems in the airways, as do eosinophils, chemokines, and complement system activation.

pharmaceutical nanocarriers must be manufactured from biocompatible materials, and their quality, safety, and efficacy have to be demonstrated by appropriate preclinical and clinical studies (Yih and Al-Fandi, 2006).

In this context, many studies have been developing and testing new NPs for asthma drug delivery. A recent study showed that stealth steroids compacted with NPs produce prolonged and greater benefits at the site of airway inflammation compared with free steroids (Matsuo et al., 2009). Furthermore, NPs compacted with salbutamol interact more with the lung membrane, causing greater and more sustained relief of bronchospasm due to drug concentration in the target area (Bhavna et al., 2009). Studies have also demonstrated that liposomes could increase the concentration and retention time of salbutamol sulfate in the lungs, thus prolonging its therapeutic effect for up to 10 hours more than uncapsulated drugs (Chen et al., 2012).

Solid lipid nanoparticles (SLNs) have attracted increasing attention as a potential drug delivery carrier because of their unique structure and properties, such as good biocompatibility, protection of the incorporated compound against degradation, and controlled release (Wang et al., 2012). Wang et al. (2012) reported enhanced bioavailability and efficiency of curcumin, a potent anti-inflammatory supplement that could be used as an add-on therapy in patients with bronchial asthma, by its formulation in SLNs.

Another compound widely used to prevent the wheezing, exercise-induced bronchospasm, chest tightness, and coughing caused by asthma is montelukast, one of a class of medications known as leukotriene receptor antagonists (LTRAs). Patil-Gadhe and colleagues (2014a, 2014b) demonstrated that a montelukast-loaded nanostructured lipid carrier improved the systemic bioavailability and performance of the drug, bypassing hepatic metabolism and thus reducing hepatocellular toxicity.

Furthermore, a well-defined nontoxic telodendrimer has also been reported to be an efficient nanocarrier that has greater loading capacity and superior stability (longer than 6 months) than micelles reported in the literature (Jackson et al., 2004). This nanocarrier allows delivery of slow-release formulations of hydrophobic drugs, such as dexamethasone, directly to the lung, decreasing allergic lung inflammation and reducing the number of eosinophils and inflammatory cytokines, thus improving airway hyperresponsiveness to a greater degree than equivalent doses of dexamethasone alone (Kenyon et al., 2013).

According to the literature, the expansion of novel NPs for drug delivery is an interesting and challenging research field, particularly for the delivery of emerging asthma therapies. Development of NP formulations has already greatly improved the stability and therapeutic effectiveness of several agents.

7.2.2.2 Nanoparticles and Gene Therapy

Gene therapy has emerged as an alternative for the treatment of diseases refractory to conventional therapeutics, including asthma, with the aim of inhibiting Th2 transcription factors, cytokines, and function or overexpressing Th2 antagonists. According to a recent study, synthetic NP-based gene delivery systems offer highly tunable platforms for delivery of therapeutic genes, but the inability to achieve sustained, high-level transgene expression in vivo presents a significant hurdle (Mastorakos et al., 2015). The respiratory system, although readily accessible, remains a challenging target, since effective gene therapy mandates colloidal stability in physiological fluids and the ability to overcome biological barriers found in the lung; as nucleic acids are prone to degradation by nucleases, delivery vectors must mediate protection against degradation (Sanders et al., 2009).

A number of lipid- and polymer-based nonviral vectors have been developed to formulate nucleic acids into nanosized particles for pulmonary delivery (Di Gioia et al., 2015). One of the most prominent polymeric gene delivery vectors is poly(ethylene imine) (PEI) (Merdan et al., 2003). Use of the polysaccharide chitosan (Koping-Hoggard et al., 2001), dendrimers (Rudolph et al., 2000), and poly(lactic-co-glycolic acid) (PLGA)–based polymers (Bivas-Benita et al., 2009) for pulmonary delivery of nucleic acids has also been described.

In this line, Kumar and colleagues (2003) showed that chitosan IFN-γ-pDNA nanoparticle (CIN) treatment significantly lowered airway hyperresponsiveness to methacholine and reduced lung histopathology in a BALB/c mouse model of ovalbumin (OVA)-induced allergic asthma. In a following study aimed to understand the reason for these effects, the same group demonstrated that CIN treatment was able to reduce cytokine production by a population of OVA-specific pro-inflammatory CD8[+] T lymphocytes in the lung, leading to decreased activation of dendritic cells. Because of the reciprocal regulation of Th cells, it was anticipated that increasing IFN-γ levels would promote a Th1 response by blocking Th2 cytokine production (Kong et al., 2008).

More recently, researchers used highly compacted DNA NPs, composed of pDNA compacted with block copolymers of poly-L-lysine and polyethylene glycol linked by a cysteine residue (CK30PEG), which have been shown to be nontoxic and nonimmunogenic in the lungs of mice and humans. Indeed, Da Silva et al. (2014) employed this NP system to deliver thymulin, a nonapeptide known for its anti-inflammatory and antifibrotic effects in the lung, to OVA-challenged allergic asthmatic BALB/c mice. A single intratracheal instillation of DNA NPs carrying thymulin plasmids prevented lung inflammation, collagen deposition, and smooth muscle hypertrophy in the murine lungs up to 27 days after administration, leading to improved lung mechanics (Da Silva et al., 2014).

On the other hand, due to its high cationic charge density, PEI is able to efficiently condense negatively charged DNA into nanosized complexes, thereby protecting it from degradation by nucleases. However, this positive charge causes toxicity, and makes it a poor vehicle for asthma gene therapy. PEGylated PEI can also activate the complement system and induce expression of apoptosis-related genes (Merkel et al., 2011a, 2011b). Therefore, the need for biodegradable vectors seems obvious. Mastorakos et al. (2015) presented a highly compacted, biodegradable DNA NP capable of overcoming the mucus barrier for inhaled lung gene therapy with a favorable safety profile and no signs of toxicity following intratracheal administration. In addition, a promising cytosine–phosphate–guanine (CpG) adjuvant-loaded biodegradable NP-based vaccine for the treatment of dust mite allergies has been recently developed and used as a potent adjuvant for shifting immune responses to the Th1 type, suppressing Th2-triggered asthma responses (Salem, 2014).

In summary, it is well known that asthma is a very complex inflammatory disease, in which numerous cells and cellular factors are implicated. Thus, there are many potential molecular targets for therapy, including cytokines and chemokines, transcription factors, tyrosine kinases and their receptors, and costimulatory molecules for gene silencing or overexpression of specific targets, together with drug delivery by NPs. Its many pros and cons notwithstanding (Figure 7.5), development of nanotherapeutics is paving the way for new therapeutic strategies.

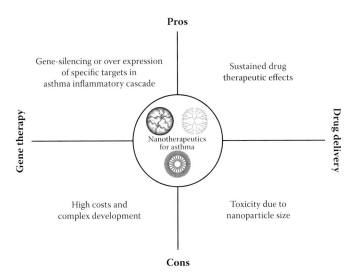

Figure 7.5. Major pros and cons of nanotherapeutics for asthma. Due to the complex inflammatory cascade involved in the pathogenesis of asthma, there are many advantages and disadvantages of these new therapeutic approaches, both for drug delivery and for gene therapy.

7.2.3 NANOPARTICLE THERAPY

NPs, including SLNs, polymeric NPs, and liposomes, have been extensively studied as potential vehicles for pulmonary drug delivery (Paranjpe and Muller-Goymann, 2014). Recently, however, nanocompounds themselves have been described to promote beneficial effects in preclinical models of asthma. Therapies using metallic NPs and extracellular vesicles (EVs) derived from stem cells have proved to be efficient for the treatment of allergic airway inflammation.

7.2.3.1 Metallic Nanoparticle Therapy

The nanotechnology revolution offers enormous societal and economic benefits for innovation in the fields of engineering, electronics, and medicine. Nevertheless, evidence from rodent studies shows that biopersistent engineered nanomaterials can stimulate immune, inflammatory, and fibroproliferative responses in the lung, suggesting possible risks for lung diseases or systemic immune disorders as a consequence of occupational, environmental, or consumer exposure. Due to their nanoscale dimensions, nanomaterials have a much greater potential to reach the distal regions of the lung and generate oxidative stress. Some nanomaterials, for example, nanofibers and nanotubes, can activate inflammasomes in macrophages, triggering interleukin (IL) 1β release and neutrophilic infiltration into the lungs. Moreover, some NPs can alter allergen-induced eosinophilic inflammation by immunostimulation, immunosuppression, or modulation of the balance between Th1, Th2, and Th17 cells, thereby influencing the nature of the inflammatory response. Some major metallic NPs produced include zinc oxide (ZnO), titanium dioxide (TiO$_2$), silver (Ag), and gold (Au), which present the greatest likelihood for human exposure (Thompson et al., 2014). Thus, some groups decided to study the impact of these metallic nanomaterials in the lungs. Nickel (Ni) and cobalt (Co) NPs, despite being produced in lower quantities, have potent effects on the immune system, and could present significant health risks (Wan et al., 2011; Glista-Baker et al., 2014; Thompson et al., 2014).

Exposure to ambient NPs has been associated with increased risk of childhood and adult asthma. Scuri et al. (2010) showed for the first time that exposure to titanium dioxide nanoparticles (TiO-NPs) upregulates the expression of mediators associated with airway hyperresponsiveness and inflammation, such as lung neurotrophins, in an age-dependent fashion. These results suggest the presence of a critical window of vulnerability in the earlier stages of lung development, which may lead to a higher risk of developing asthma (Scuri et al., 2010). Rossi et al. (2010) showed that exposure of healthy mice to TiO-NPs elicited lung inflammation, but on the other hand, surprisingly, levels of leukocytes, cytokines, chemokines, and antibodies characteristic of allergic asthma were dramatically reduced in asthmatic mice exposed to these particles, showing that allergic pulmonary inflammation can be substantially suppressed in allergic mice, and thus suggesting that repeated airway exposure to TiO$_2$ particles modulates airway inflammation depending on the immunological status of the exposed mice (Rossi et al., 2010).

Silver nanoparticles (Ag-NPs) have been classically associated with exacerbation of airway hyperresponsiveness. Chuang et al. (2013) showed that continuous exposure to Ag-NP (33 nm Ag-NP at 3.3 mg/m^3) evoked inflammatory infiltration of the airways and elevated levels of allergic markers, the Th2 cytokine IL-13, and indicators of oxidative stress in healthy and allergic mice (Seiffert et al., 2015). Indeed, over the past few years, nanotechnology has provided a way of producing pure Ag-NPs that exhibit cytoprotective and pro-healing properties. The anti-inflammatory and antioxidant properties of Ag-NP in a murine model of asthma have been reported (Park et al., 2010). Park et al. (2010) showed that increased levels of inflammatory cells, airway hyperresponsiveness, IL-4, IL-5, IL-13, and nuclear factor-κB (NF-κB) in lungs after OVA inhalation were reduced after administration of Ag-NP (24 nm, 20 ppm, 40 mg/kg). In addition, they also found that the increased intracellular reactive oxygen species levels in bronchoalveolar lavage fluid observed after OVA inhalation are decreased after Ag-NP administration. The exact mechanisms of the anti-inflammatory effects of these particles are not fully understood, and their ability to reduce reactive oxygen species has been attributed to the capacity of noble metal NPs to act as catalysts for reduction reactions. Furthermore, Ag-NPs substantially suppressed mucus hypersecretion and vascular permeability through the PI3K/HIF-1α/VEGF signaling pathway in an allergic airway inflammation model (Jang et al., 2012).

Gold NPs have been shown to exhibit a range of beneficial biological properties, including anti-inflammatory and antioxidant effects (Leibfarth and Persellin, 1981; Pedersen et al., 2009), but their putative impact on allergic asthma has only recently been addressed. Barreto et al. (2015) evaluated the potential of nasally instilled gold NPs (6 nm, 6 and 60 μg/kg) to prevent allergen-induced asthma in distinct murine models of this disease. Gold NPs clearly inhibited (70%–100%) allergen-induced accumulation of inflammatory cells, as well as production of both pro-inflammatory cytokines and reactive oxygen species. In addition, instilled gold NPs clearly prevented the mucus production, peribronchiolar fibrosis, and airway hyperreactivity triggered by allergen provocation. These findings demonstrate that gold NPs are able to prevent pivotal features of asthma, including airway hyperreactivity, inflammation, and lung remodeling. Such protective effects are accounted for by a reduction in generation of pro-inflammatory cytokines and chemokines in lung tissue, in a mechanism probably related to downregulation of oxidative stress levels (Barreto et al., 2015).

7.2.3.2 Extracellular Vesicles

Since the beginning of the last decade, the scientific community has shown increasing interest in exosomes. Cells release several distinct types of membrane vesicles of endosomal and plasma membrane origin into the extracellular environment. These vesicles are known as exosomes (40–100 nm) and microvesicles (100–1000 nm), respectively (Raposo and Stoorvogel, 2013). Such EVs represent an important mode of intercellular communication by serving as vehicles for intercell transfer of physiologically active membrane and cytosolic proteins, lipids, mRNA, and microRNA (Chlebowski et al., 2013; Aryani and Denecke, 2014).

A growing number of studies demonstrate that EVs released by mesenchymal stromal cells (MSCs) derived from bone marrow and other sources are as effective as MSCs themselves in mitigating inflammation and injury. Cruz et al. (2015) investigated whether the administration of EVs from human or mouse bone marrow–derived MSCs would be effective in a murine model of mixed Th2/Th17, neutrophilic-mediated allergic airway inflammation, reflective of severe refractory asthma, induced by repeated mucosal exposure to *Aspergillus* hyphal extract (AHE) in immunocompetent C57Bl/6 mice. Systemic administration of EVs isolated from human and murine MSCs at the onset of antigen challenge in previously sensitized mice significantly ameliorated AHE-associated increases in airway hyperreactivity, lung inflammation, and the antigen-specific CD4 T-cell Th2 and Th17 phenotypes. Notably, EVs from human MSCs were generally more potent than those from mouse MSCs in most of the outcome measures. These results demonstrate potent effects of EVs from MSCs in an immunocompetent mouse model of allergic airway inflammation (Cruz et al., 2015).

One alternative option to the use of endogenous exosomes is the design of artificial exosomes with a clearly defined therapeutic active cargo and surface markers that ensure specific targeting to recipient cells. This has been proposed as a promising approach for the treatment of lung diseases such as asthma (Aryani and Denecke, 2014).

7.2.4 CONCLUSIONS AND FUTURE DIRECTIONS

This chapter summarizes the potential beneficial effects of NP therapy in lung diseases, including acute respiratory distress syndrome (ARDS), chronic obstructive pulmonary disease (COPD), and asthma. Nanotechnology has generated much enthusiasm as a potentially beneficial approach to the treatment of lung diseases and other critical illnesses. Preclinical studies using NPs showed many advancements, and current trials show that many NPs are safe for administration, with few adverse effects. Nevertheless, substantial challenges still have to be overcome before nanotherapeutic approaches can be used for clinical practice in asthma. As such, further studies focusing on understanding the mechanisms of action of NPs are warranted to continue development of rational approaches for clinical trials. NP-based therapies offer potential hope for asthma (Figure 7.6) through the possibility of focusing on specific targets, both for gene silencing or overexpression and to accomplish drug delivery, in order to reduce airway inflammation and remodeling, improving lung repair, without adverse effects.

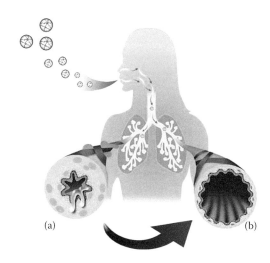

(a)   (b)

Figure 7.6. Benefits of nanotechnology for asthma. Nanotechnology-based approaches act on airway inflammation and remodeling (a), improving lung repair (b).

REFERENCES

Aryani A, Denecke B (2014). Exosomes as a nanodelivery system: A key to the future of neuromedicine? *Mol Neurobiol* 53: 818–834.

Barnes PJ (2012). Severe asthma: Advances in current management and future therapy. *J Allergy Clin Immunol* 129(1): 48–59.

Barnes PJ (2013). New anti-inflammatory targets for chronic obstructive pulmonary disease. *Nat Rev Drug Discov* 12(7): 543–559.

Barreto E, Serra MF, Dos Santos RV, Dos Santos CE, Hickmann J, Cotias AC et al. (2015). Local administration of gold nanoparticles prevents pivotal pathological changes in murine models of atopic asthma. *J Biomed Nanotechnol* 11(6): 1038–1050.

Bhavna, Ahmad FJ, Mittal G, Jain GK, Malhotra G, Khar RK et al. (2009). Nano-salbutamol dry powder inhalation: A new approach for treating bronchoconstrictive conditions. *Eur J Pharm Biopharm* 71(2): 282–291.

Bivas-Benita M, Lin MY, Bal SM, van Meijgaarden KE, Franken KL, Friggen AH et al. (2009). Pulmonary delivery of DNA encoding *Mycobacterium tuberculosis* latency antigen Rv1733c associated to PLGA-PEI nanoparticles enhances T cell responses in a DNA prime/protein boost vaccination regimen in mice. *Vaccine* 27(30): 4010–4017.

Chen X, Huang W, Wong BC, Yin L, Wong YF, Xu M et al. (2012). Liposomes prolong the therapeutic effect of anti-asthmatic medication via pulmonary delivery. *Int J Nanomed* 7: 1139–1148.

Chlebowski A, Lubas M, Jensen TH, Dziembowski A (2013). RNA decay machines: The exosome. *Biochim Biophys Acta* 1829(6–7): 552–560.

Chuang HC, Hsiao TC, Wu CK, Chang HH, Lee CH, Chang CC et al. (2013). Allergenicity and toxicology of inhaled silver nanoparticles in allergen-provocation mice models. *Int J Nanomed* 8: 4495–4506.

Cooper V, Metcalf L, Versnel J, Upton J, Walker S, Horne R (2015). Patient-reported side effects, concerns and adherence to corticosteroid treatment for asthma, and comparison with physician estimates of side-effect prevalence: A UK-wide, cross-sectional study. *NPJ Prim Care Respir Med* 25: 15026.

Cruz FF, Borg ZD, Goodwin M, Sokocevic D, Wagner DE, Coffey A et al. (2015). Systemic administration of human bone marrow-derived mesenchymal stromal cell extracellular vesicles ameliorates *Aspergillus* hyphal extract-induced allergic airway inflammation in immunocompetent mice. *Stem Cells Transl Med* 4(11): 1302–1316.

Da Silva AL, Martini SV, Abreu SC, Samary C, Diaz BL, Fernezlian S et al. (2014). DNA nanoparticle-mediated thymulin gene therapy prevents airway remodeling in experimental allergic asthma. *J Control Release* 180: 125–133.

Da Silva AL, Santos RS, Xisto DG, Alonso Sdel V, Morales MM, Rocco PR (2013). Nanoparticle-based therapy for respiratory diseases. *An Acad Bras Cienc* 85(1): 137–146.

Di Gioia S, Trapani A, Castellani S, Carbone A, Belgiovine G, Craparo EF et al. (2015). Nanocomplexes for gene therapy of respiratory diseases: Targeting and overcoming the mucus barrier. *Pulm Pharmacol Ther* 34: 8–24.

Glista-Baker EE, Taylor AJ, Sayers BC, Thompson EA, Bonner JC (2014). Nickel nanoparticles cause exaggerated lung and airway remodeling in mice lacking the T-box transcription factor, TBX21 (T-bet). *Part Fibre Toxicol* 11: 7.

Global Initiative for Asthma (GINA) (2015). http://ginasthma.org/

Holloway JW, Yang IA, Holgate ST (2010). Genetics of allergic disease. *J Allergy Clin Immunol* 125(2 Suppl 2): S81–94.

Jackson C, Lipworth B (1995). Optimizing inhaled drug delivery in patients with asthma. *Br J Gen Pract* 45(401): 683–687.

Jackson JK, Zhang X, Llewellen S, Hunter WL, Burt HM (2004). The characterization of novel polymeric paste formulations for intratumoral delivery. *Int J Pharm* 270(1–2): 185–198.

Jang S, Park JW, Cha HR, Jung SY, Lee JE, Jung SS et al. (2012). Silver nanoparticles modify VEGF signaling pathway and mucus hypersecretion in allergic airway inflammation. *Int J Nanomed* 7: 1329–1343.

Kenyon NJ, Bratt JM, Lee J, Luo J, Franzi LM, Zeki AA et al. (2013). Self-assembling nanoparticles containing dexamethasone as a novel therapy in allergic airways inflammation. *PLoS One* 8(10): e77730.

Kim SH, Ye YM, Lee HY, Sin HJ, Park HS (2011). Combined pharmacogenetic effect of ADCY9 and ADRB2 gene polymorphisms on the bronchodilator response to inhaled combination therapy. *J Clin Pharm Ther* 36(3): 399–405.

Kong X, Hellermann GR, Zhang W, Jena P, Kumar M, Behera A et al. (2008). Chitosan interferon-gamma nanogene therapy for lung disease: Modulation of T-cell and dendritic cell immune responses. *Allergy Asthma Clin Immunol* 4(3): 95–105.

Koping-Hoggard M, Tubulekas I, Guan H, Edwards K, Nilsson M, Varum KM et al. (2001). Chitosan as a nonviral gene delivery system. Structure-property relationships and characteristics compared with polyethylenimine in vitro and after lung administration in vivo. *Gene Ther* 8(14): 1108–1121.

Kumar M, Kong X, Behera AK, Hellermann GR, Lockey RF, Mohapatra SS (2003). Chitosan IFN-gamma-pDNA nanoparticle (CIN) therapy for allergic asthma. *Genet Vaccines Ther* 1(1): 3.

Leibfarth JH, Persellin RH (1981). Mechanisms of action of gold. *Agents Actions* 11(5): 458–472.

Mastorakos P, da Silva AL, Chisholm J, Song E, Choi WK, Boyle MP et al. (2015). Highly compacted biodegradable DNA nanoparticles capable of overcoming the mucus barrier for inhaled lung gene therapy. *Proc Natl Acad Sci USA* 112(28): 8720–8725.

Matsuo Y, Ishihara T, Ishizaki J, Miyamoto K, Higaki M, Yamashita N (2009). Effect of betamethasone phosphate loaded polymeric nanoparticles on a murine asthma model. *Cell Immunol* 260(1): 33–38.

Merdan T, Callahan J, Petersen H, Kunath K, Bakowsky U, Kopeckova P et al. (2003). Pegylated polyethylenimine-Fab′ antibody fragment conjugates for targeted gene delivery to human ovarian carcinoma cells. *Bioconjug Chem* 14(5): 989–996.

Merkel OM, Beyerle A, Beckmann BM, Zheng M, Hartmann RK, Stoger T et al. (2011a). Polymer-related off-target effects in non-viral siRNA delivery. *Biomaterials* 32(9): 2388–2398.

Merkel OM, Urbanics R, Bedocs P, Rozsnyay Z, Rosivall L, Toth M et al. (2011b). In vitro and in vivo complement activation and related anaphylactic effects associated with polyethylenimine and polyethylenimine-graft-poly(ethylene glycol) block copolymers. *Biomaterials* 32(21): 4936–4942.

Murdoch JR, Lloyd CM (2010). Chronic inflammation and asthma. *Mutat Res* 690(1–2): 24–39.

Paranjpe M, Muller-Goymann CC (2014). Nanoparticle-mediated pulmonary drug delivery: A review. *Int J Mol Sci* 15(4): 5852–5873.

Park HS, Kim KH, Jang S, Park JW, Cha HR, Lee JE et al. (2010). Attenuation of allergic airway inflammation and hyperresponsiveness in a murine model of asthma by silver nanoparticles. *Int J Nanomed* 5: 505–515.

Pascual RM, Peters SP (2005). Airway remodeling contributes to the progressive loss of lung function in asthma: An overview. *J Allergy Clin Immunol* 116(3): 477–486; quiz 487.

Patil-Gadhe A, Kyadarkunte A, Patole M, Pokharkar V (2014a). Montelukast-loaded nanostructured lipid carriers. Part II. Pulmonary drug delivery and in vitro-in vivo aerosol performance. *Eur J Pharm Biopharm* 88(1): 169–177.

Patil-Gadhe A, Pokharkar V (2014b). Montelukast-loaded nanostructured lipid carriers. Part I. Oral bioavailability improvement. *Eur J Pharm Biopharm* 88(1): 160–168.

Pedersen MO, Larsen A, Pedersen DS, Stoltenberg M, Penkowa M (2009). Metallic gold reduces TNFalpha expression, oxidative DNA damage and pro-apoptotic signals after experimental brain injury. *Brain Res* 1271: 103–113.

Peters SP, Ferguson G, Deniz Y, Reisner C (2006). Uncontrolled asthma: A review of the prevalence, disease burden and options for treatment. *Respir Med* 100(7): 1139–1151.

Raissy HH, Kelly HW, Harkins M, Szefler SJ (2013). Inhaled corticosteroids in lung diseases. *Am J Respir Crit Care Med* 187(8): 798–803.

Raposo G, Stoorvogel W (2013). Extracellular vesicles: Exosomes, microvesicles, and friends. *J Cell Biol* 200(4): 373–383.

Reddel HK, Bateman ED, Becker A, Boulet LP, Cruz AA, Drazen JM et al (2015). A summary of the new GINA strategy: A roadmap to asthma control. *Eur Respir J* 46(3):622–39.

Rossi EM, Pylkkanen L, Koivisto AJ, Nykasenoja H, Wolff H, Savolainen K et al. (2010). Inhalation exposure to nanosized and fine TiO2 particles inhibits features of allergic asthma in a murine model. *Part Fibre Toxicol* 7: 35.

Rudolph C, Lausier J, Naundorf S, Muller RH, Rosenecker J (2000). In vivo gene delivery to the lung using polyethylenimine and fractured polyamidoamine dendrimers. *J Gene Med* 2(4): 269–278.

Salem AK (2014). A promising CpG adjuvant-loaded nanoparticle-based vaccine for treatment of dust mite allergies. *Immunotherapy* 6(11): 1161–1163.

Sanders N, Rudolph C, Braeckmans K, De Smedt SC, Demeester J (2009). Extracellular barriers in respiratory gene therapy. *Adv Drug Deliv Rev* 61(2): 115–127.

Scuri M, Chen BT, Castranova V, Reynolds JS, Johnson VJ, Samsell L et al. (2010). Effects of titanium dioxide nanoparticle exposure on neuroimmune responses in rat airways. *J Toxicol Environ Health A* 73(20): 1353–1369.

Seiffert J, Hussain F, Wiegman C, Li F, Bey L, Baker W et al. (2015). Pulmonary toxicity of instilled silver nanoparticles: Influence of size, coating and rat strain. *PLoS One* 10(3): e0119726.

Thompson EA, Sayers BC, Glista-Baker EE, Shipkowski KA, Taylor AJ, Bonner JC (2014). Innate immune responses to nanoparticle exposure in the lung. *J Environ Immunol Toxicol* 1(3): 150–156.

Wan R, Mo Y, Chien S, Li Y, Tollerud DJ, Zhang Q (2011). The role of hypoxia inducible factor-1alpha in the increased MMP-2 and MMP-9 production by human monocytes exposed to nickel nanoparticles. *Nanotoxicology* 5(4): 568–582.

Wang W, Zhu R, Xie Q, Li A, Xiao Y, Li K et al. (2012). Enhanced bioavailability and efficiency of curcumin for the treatment of asthma by its formulation in solid lipid nanoparticles. *Int J Nanomed* 7: 3667–3677.

Yih TC, Al-Fandi M (2006). Engineered nanoparticles as precise drug delivery systems. *J Cell Biochem* 97(6): 1184–1190.

Bridging the Gap between the Bench and the Clinic

ASTHMA

Yuran Xie, Rima Kandil, and Olivia M. Merkel

CONTENTS

7.3.1 BACKGROUND

Asthma is a disease characterized by chronic inflammation of the respiratory airways. Asthma remains a public health problem, particularly in developing countries (Global Asthma Network 2014). A large global market for asthma drugs (BCC Research 2012) fuels research and development by pharmaceutical companies to develop new therapies to treat asthma. A stepwise approach for controlling asthma symptoms is recommended by the Global Initiative for Asthma (GINA). For this stepwise approach, inhaled rescue medications (e.g., short-acting β_2-adrenoceptor agonist

[SABA]) are prescribed and taken as needed to relieve asthma symptoms in all asthmatic patients, and inhaled corticosteroids (ICSs) with or without long-acting β_2-adrenoceptor agonist (LABA) are recommended for patients with moderate persistent asthma. Additionally, oral corticosteroids (OCSs) and anti–immunoglobulin E (IgE) (omalizumab) may be applied in patients who have severe asthma (GINA 2015). Although current treatment strategies can control asthma symptoms in most patients, there is still a need to develop alternative therapies for asthma, primarily because of concerns of low patient compliance (Gamble et al. 2009). Patient compliance remains low for asthma therapies, especially because of the necessary frequent dosing (inhalation) and a fear for undesirable side effects. Nonadherence, on the other hand, results in the overuse of rescue medications and a high frequency of uncontrolled asthma symptoms. Moreover, 5%–10% of the patients whose asthma symptoms remain uncontrolled, despite the maintenance therapies, have a higher risk of exacerbation and hospitalization due to asthma (Barnes 2010).

Extensive research has been conducted to understand the detailed pathologic mechanisms of asthma and to develop new therapies for asthma. Numerous drugs have entered clinical trials but failed to demonstrate satisfactory therapeutic efficacy and safety. Nanomedicine may be a way to solve the poor safety and efficacy problems encountered by previously tested new drug candidates. Nanomedicine was defined by the Medical Standing Committee of the Europe Science Foundation in 2004 as "the science and technology of diagnosing, treating, and preventing disease and traumatic injury, of relieving pain, and of preserving and improving human health, using molecular tools and molecular knowledge of the human body" (Webster 2006). The advantages of nanomedicine include increased solubility, sustained release, and enhanced specificity of drug treatments. For example, a doxorubicin-encapsulating liposome (Doxil®), the first nanodrug approved by the U.S. Food and Drug Administration (USFDA) in 1995 to treat cancers, demonstrates increased circulation time, consequently enhanced amounts of drug in cancer cells, and fewer side effects compared with free doxorubicin (Pillai 2014). Nanomedicine is also suitable for the delivery of biopharmaceutics (Mansour et al. 2009). Despite the promise of nanomedicine for the development of new asthma treatments, no asthma treatments utilizing nanotechnology have undergone clinical trials or been brought to market so far. Here, we first discuss several new asthma drugs (currently in clinical trials or on the market) and how the development process of these drugs may be utilized in developing a nanomedicine for asthma therapies. Then, we elaborate on how studies need to be planned for inhaled nanomedicines, and what the regulatory hurdles are, before we give an outlook (Figure 7.7).

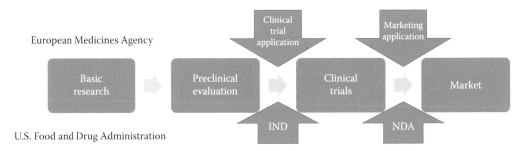

Figure 7.7. Process of developing new nanomedicine from laboratory to market.

NANOMEDICINE FOR INFLAMMATORY DISEASES

7.3.2 DRUGS FOR ASTHMA

Drugs available for asthma therapy can be categorized into small-molecule drugs and biopharmaceutics. The majority of asthma drugs are small molecules formulated for either oral administration or inhaled administration. Theophylline, leukotriene (LT) receptor antagonists (e.g., montelukast), cromones, and corticosteroid are usually formulated for oral delivery. Alternatively, corticosteroids, SABAs, LABAs, and anticholinergic and long-acting muscarinic antagonists (LAMAs) are preferably formulated for pulmonary delivery. Currently, there are only two biopharmaceutical treatments for asthma on the market, both of which are monoclonal antibodies, formulated for subcutaneous (s.c.) administration. The first antibody, omalizumab, targets a high-affinity receptor binding site on IgE and is approved for the treatment of patients with severe persistent allergic asthma. The other antibody, mepolizumab, is raised against the antigen interleukin (IL)-5, and was approved by the USFDA in 2015 to treat patients with severe asthma.

7.3.2.1 Small Molecules

Most small molecules developed for the treatment of asthma are formulated for either oral or pulmonary delivery. New drugs for anti-LT therapies have demonstrated oral activity, such as a dual antagonist of cysteinyl-leukotriene receptor 1 (CysLT1) and CysLT2 (gemilukast) and a 5-lipoxygenase-activating protein (FLAP) inhibitor (GSK2190915). Furthermore, chemokine receptor antagonists (e.g., GW766944), phosphodiesterase (PDE) inhibitors (e.g., revamilast), and mast cell inhibitors (e.g., Masitinib) are also administered orally in clinical trials.

Another typical example of orally administered drugs is the group of antagonists of chemoattractant receptor-homologous molecule expressed on T helper 2 (Th2) cells (CRTH2), also called D prostanoid receptor 2 (DP$_2$) (Pettipher and Whittaker 2012). Selective antagonists of CRTH2 alleviate allergic inflammation by inhibiting the activation and recruitment of pro-inflammatory cells, such as Th2 cells, eosinophils, and basophils (Hirai et al. 2001). At least 13 different CRTH2 antagonists have been brought to clinical trials for the treatment of asthma, allergic rhinitis, and chronic obstructive pulmonary disease (COPD) (Pettipher and Whittaker 2012). Most of the new CRTH2 antagonists are delivered orally. OC000459 is a promising CRTH2 antagonist that is currently in clinical trials. In preclinical studies, the ability of OC000459 to displace prostaglandin D$_2$ (PGD$_2$) from human DP$_2$ was first evaluated using recombinant DP$_2$ expressed on the membrane of CHO cells and native DP$_2$ expressed on Th2 cells. Both of these studies demonstrated the high potency of OC000459 for DP$_2$ (K(i) = 0.013 μM and K(i) = 0.004 μM, respectively). OC000459 had minimal inhibitory activity when assessed using a library of 69 nonrelated receptors and 19 enzymes, suggesting OC000459 was highly specific for DP$_2$. OC000459 treatment inhibited PGD$_2$-induced chemotaxis of Th2 cells (IC$_{50}$ = 0.028 μM) and IL-13 production of Th2 cells (IC$_{50}$= 0.019 μM), as well as competitively antagonized eosinophil shape change in response to PGD$_2$ (pK$_B$ = 7.9). The pharmacokinetics of OC000459 oral administration has been investigated in rats, and the plasma concentration of OC000459 at t$_{1/2}$ was found to be much higher than its *in vitro* IC$_{50}$ evaluated on cells. Additionally, the distribution volume (V_d) was estimated to be 0.5 L/kg, indicating high oral activity of OC000459. The anti-inflammatory effect of OC000459 was evaluated in rats, and oral administration of drug inhibited blood eosinophilia induced by systemic administration of 13,14-dihydro-15-keto-PGD$_2$ (DK-PGD$_2$) in a dose-dependent manner. Furthermore, in guinea pigs, OC000459 treatment inhibited airway eosinophilia induced by inhalation of aerosolized DK-PGD$_2$ (Pettipher et al. 2012).

Promising preclinical results have led to the clinical investigation of OC000459 for the treatment of asthma. To determine the efficacy of OC000459, in a phase II clinical trial, patients with mild persistent asthma received 200 mg of OC000459 orally or placebo twice daily for 28 days and the lung function was assessed by measuring the forced expiratory volume in 1 second (FEV$_1$). Results revealed that treatment with OC000459, and not placebo, increased FEV$_1$ while also reducing serum IgE and the sputum eosinophil count. Therefore, it was concluded that OC000459 is pharmacologically active in asthma patients (Barnes et al. 2012). To investigate if OC000459 can reduce lung inflammation in response to an allergic stimulus, another clinical trial was conducted in corticosteroid-naïve asthmatic patients. It was reported that treatment with 200 mg of OC000459 twice daily for 16 days reduced allergen-induced late asthmatic response (LAR), determined by calculations of the area under the curve (AUC) of FEV_1. Furthermore, this treatment regimen inhibited the induction of sputum eosinophils 1 day following allergen challenge when compared with placebo. Together, OC000459 was shown to inhibit allergic asthma–related inflammation when administered orally to human patients (Singh et al. 2013). To determine the optimal dose of OC000459 for the treatment of asthma, patients with mild to moderate persistent asthma were treated with three doses of OC000459 (25 or 200 mg once daily or 100 mg twice daily) for 12 weeks. Results reflected that the lung function, as determined by change in FEV_1, was improved in all patients who received OC000459 when compared with placebo. Therefore, the lowest dose of 0C000459 tested (25 mg daily) was sufficient to provide therapeutic efficacy. To investigate whether subpopulations would respond to treatment differently, lung function data from asthmatic patients with eosinophilia (blood eosinophil count $\geq 250/\mu l$) and without eosinophilia (blood eosinophil count $<250/\mu l$) were compared. Patients with eosinophilia showed significant improvement of lung function with OC000459 treatment when compared with those given the placebo. Conversely, lung function was not improved in patients without eosinophilia following OC000459 treatment, suggesting OC000459 was more effective for the treatment of eosinophilic asthma (Pettipher et al. 2014). More studies are needed to explore the application of OC000459 in severe eosinophilic asthma or other diseases, such as eosinophilic esophagitis, atopic dermatitis, and allergic rhinitis. The safety of OC000459 is also under investigation in healthy participants because of a potential interaction between OC000459 and cytochrome P450 3A4.

Numerous small-molecule asthma drugs have been formulated for pulmonary delivery because inhalation of the drug has the advantage of rapid delivery directly to the affected tissue (i.e., the lung). Small-molecule drugs formulated for inhalation in clinical trials include LABAs (e.g., GW642444), LAMAs (e.g., Seebri), nonsteroidal selective glucocorticoid receptor agonists (e.g., AZD7594), PDE inhibitors (e.g., RPL554), and very late antigen (VLA)-4 inhibitors (e.g., GW559090X).

RPL554, a dual inhibitor of PDE3 and PDE4, is administered through inhalation to avoid adverse gastrointestinal (GI) side effects reported for oral formulations of PDE4 inhibitors (Lipworth 2005). In preclinical studies, RPL554 was shown to potently inhibit the activity of isolated human PDE3 and PDE4 ($IC_{50} = 0.4$ nM and $IC_{50} = 1479$ nM, respectively). Inhibition of PDE3 was thought to mediate human airway smooth muscle relaxation, and both an in vitro model and an isolated guinea pig tracheal tissue model were used to investigate the ability of RPL554 to inhibit the activity of PDE3. RPL554 can relax the contractility of tracheal smooth muscle elicited by electrical field stimulation in a concentration-dependent manner. The ability to inhibit stimulation of immune cells by RPL554 was evaluated in human primary cells. It was shown that RPL554 can inhibit the release of tumor necrosis factor (TNF)-α from lipopolysaccharide (LPS)-activated human mononuclear cells ($IC_{50} = 0.52$ μM) and inhibit the proliferation of mononuclear cells stimulated by phytohemagglutinin ($IC_{50} = 0.46$ μM). In guinea pigs, inhalation of RPL554 dry powder (3–5 mg, blended with lactose) can reduce the bronchoconstriction and airway edema in response to intravenous (i.v.) administration of histamine, as well as decrease the infiltration of eosinophils to the airway of ovalbumin (OVA)-sensitized asthmatic animals, indicating the feasibility to deliver RPL554 through inhalation for anti-inflammatory therapies (Boswell-Smith et al. 2006).

Based on the efficient activity of bronchodilation and anti-inflammation, RPL554 was evaluated in clinical trials for asthma therapy. The safety of a single administration of different doses of RPL554 (0.003, 0.009, and 0.018 mg/kg) was evaluated in both healthy and asthmatic participants. Inhalation of RPL554 was well tolerated at all doses tested based on adverse event reports, vital signs, and electrocardiograph (ECG) data. RPL554 treatment can improve lung function of asthmatic patients, especially at high doses, as determined by more than 1.5-fold of the concentration of methacholine needed to induce a 20% decrease in FEV_1 compared with the original dose. To further evaluate the efficacy as a bronchodilator and the safety of RPL554, asthmatic patients received multiple administrations of inhaled RPL554 at a dose of 0.018 mg/kg for 6 days. FEV_1 was determined for 6 h postinhalation of RPL554 or placebo on days 1, 3, and 6.

Results showed that RPL554 improved FEV_1 when compared with placebo. In another clinical trial, the anti-inflammatory effect of RPL554 was investigated in healthy volunteers challenged by inhaled LPS. Participants received a daily dose of RPL554 of 0.018mg/kg for 6 days before LPS challenge, and their sputum was collected 6 h and 24 h postchallenge. Results showed significantly less macrophages, lymphocytes, neutrophils, and eosinophils in the sputum from participants treated with RPL554 when compared with the placebo-treated group. Therefore, inhalation of RPL554 was reported to inhibit the inflammatory response (Franciosi et al. 2013). Currently, RPL554 is being investigated in a clinical trial to compare its efficacy with that of an active comparator, salbutamol, in asthmatic patients.

Small molecular drugs for asthma are rarely administered intravenously. However, the novel β_2-adrenoceptor agonist, MN-221, was evaluated in clinical trials as an adjunct to standard therapy in patients experiencing an acute exacerbation of asthma. Phase I and II studies demonstrated that MN-221 (dose, 5.25−1125 µg; rate, 0.35−60 µg/min) was well tolerated in patients with mild to moderate or moderate to severe asthma, according to the reported adverse events, laboratory tests, vital signs, and ECG. Moreover, lung function was improved, as determined by FEV_1 measurements, in a dose-dependent manner in patients treated with MN-221 (Matsuda et al. 2012). Another clinical trial was conducted to evaluate the efficacy of MN-221 in patients with an acute exacerbation of asthma. In this trial, patients admitted to the emergency room received the standard treatment for acute asthma exacerbation, and those who did not respond to standard therapy (FEV_1 was less than 50% of predicted) received an i.v. infusion of MN-221 (1200 µg) or placebo, followed by measurements of FEV_1 for 3 h. MN-221 treatment, in addition to standard therapy, failed to significantly improve the lung function determined by AUC_{0-3h} of %FEV_1 when compared with standard treatment (House et al. 2015). The apparent lack of improvement with MN-221 treatment could be explained by the fact that the treated patients were resistant to any asthma treatment, because they also did not respond sufficiently to standard treatment. To better determine the efficacy of MN-221, future clinical trials should include a more heterogeneous population, to avoid a disproportionate number of treatment-resistant patients.

7.3.2.2 Biopharmaceutics

Asthma is a complicated disease involving numerous inflammatory cells and cellular mediators. Biopharmaceutical drugs can be used to intervene in specific events, for example, blockage of interaction between cells or between receptors and cytokines and chemokines and inhibiting the expression of key pathological factors (e.g., cytokines, receptors, and transcription factors). Numerous biopharmaceutical drugs are currently in preclinical and clinical development for the treatment of asthma. There are two major kinds of biopharmaceutical drugs that have been in clinical trials for the treatment of asthma: (1) protein-based drugs, such as monoclonal antibodies and cytokines, and (2) nucleic acid–based drugs, including DNAzymes, antisense oligonucleotides, and small interference RNA (siRNA).

Protein-based drugs, specifically monoclonal antibodies, are particularly promising treatments for asthma. Monoclonal antibodies have been successfully brought to the market for the treatment of various diseases. Monoclonal antibodies can be raised against specific antigens (i.e., drug targets) and administered through i.v. or s.c. injection. Monoclonal antibodies have the benefit of being highly specific for a particular drug target, and with little off-target binding. Currently, there is a large number of monoclonal antibodies under clinical investigation. Their targets include (1) cytokines such as IL-4 (e.g., pascolizumab), IL-5 (e.g., mepolizumab), IL-9 (e.g., MEDI-528), IL-13 (e.g., QAX576 and CAT-354), and IL-17A (e.g., secukinumab); (2) receptors such as IL-13 receptors α1 and α2 (e.g., GSK679586), IL-4 receptor α (e.g., AMG 317), and IL-5 receptor α (e.g., MEDI-563); and (3) other cellular elements, such as IgE (e.g., MEDI-4212) and OX40L (e.g., huMAb OX40L).

Mepolizumab is a humanized monoclonal antibody raised against IL-5. Mepolizumab can bind to IL-5, inhibiting the interaction between IL-5 and its receptor (Tsukamoto et al. 2015). Because IL-5 regulates the activation and survival of eosinophils, a pro-inflammatory cell type involved in asthma, mepolizumab could serve as a potential therapy for asthma. Mepolizumab has been evaluated both preclinically and in clinical trials. Mepolizumab was determined to have high binding affinity for purified human IL-5 (K_d of 4.2 pM) as determined by surface plasmon resonance

(Biacore) and titration microcalorimetry. In human TF-1.28 cells, a human erythroleukemia cell line, mepolizumab treatment can inhibit IL-5-dependent cellular proliferation with an IC_{50} of <150 pM. Mepolizumab binding is highly specific for the human IL-5 protein and does not bind to IL-5 derived from mouse, rat, guinea pig, or dog (European Medicines Agency 2009). Therefore, pharmacokinetic and pharmacodynamic studies of mepolizumab were conducted in healthy cynomolgus monkeys. The mean terminal half-lives of mepolizumab after i.v. administration (13.0 ± 2.2 days) and s.c. administration (14.5 ± 3.8 days) were similar and relatively long, indicating mepolizumab was bioavailable following the s.c. route. A single dose (1 mg/kg) or repeated doses (10 mg/ml) of s.c. administration of mepolizumab can reduce the number of peripheral eosinophils in the monkeys, and the IC_{50} of mepolizumab was only 1–2 μg/ml, suggesting it could be a valuable drug candidate for diseases associated with eosinophilia, such as asthma (Zia-Amirhosseini et al. 1999). To characterize the pharmacologic activity and long-term safety profile, mepolizumab was evaluated in a monkey allergy model or healthy monkeys. In monkeys naturally sensitized to *Ascaris suum*, a single dose of mepolizumab (i.v., 10 mg/kg) inhibited the infiltration of eosinophils to the airways and blood eosinophilia induced by inhaled *A. suum* antigen. Healthy monkeys were treated (i.v.) twice with mepolizumab at different doses (0.05, 0.5, or 50 mg/kg) on days 1 and 29. Mepolizumab mediated a dose-dependent reduction of peripheral blood eosinophils and inhibition of blood eosinophilia induced by s.c. injection of rhIL-2 on day 30. Repeated i.v. (10 or 100 mg/kg) or s.c. (10 mg/kg) administration of mepolizumab to monkeys monthly for 6 months was conducted to investigate long-term toxicity and pharmacodynamics. All doses were well tolerated, and no adverse side effects were observed. More importantly, all doses effectively decreased circulating eosinophils but did not significantly affect immature or mature eosinophils in bone marrow. Therefore, mepolizumab treatment appears to be a safe and effective treatment for the inhibition of eosinophil-mediated inflammation (Hart et al. 2001).

Several clinical trials of mepolizumab have been conducted. The efficacy and safety of single i.v. administration of mepolizumab (2.5 and 10 mg/kg) were tested in patients with mild allergic asthma. Mepolizumab treatment can inhibit the increased blood eosinophil and sputum eosinophil counts induced by inhaled allergen. However, mepolizumab did not significantly change airway response to inhaled histamine as determined by histamine PC_{20} (mg/ml), nor did it attenuate early asthmatic response (EAR) and LAR. These results indicated that eosinophils may not play a significant role for airway hyperresponsiveness (AHR) and LAR, but may be more relevant to the pathogenesis of asthma (Leckie et al. 2000). Larger-scale clinical studies on the safety and efficacy of mepolizumab were conducted and provide more information about the role of eosinophils in asthma pathogenesis. In a placebo-controlled clinical trial, asthma patients with persistent symptoms despite ICS therapy received three i.v. infusions of mepolizumab (250 and 750 mg) at 4-week intervals. Results showed that mepolizumab treatment was well tolerated and significantly reduced the eosinophil counts in sputum and blood. Similar to the previous clinical study, mepolizumab did not improve lung function, as determined by changed morning peak expiratory flow (PEF) and FEV_1, when compared with placebo, indicating that mepolizumab treatment may not be effective in alleviating asthma symptoms and other clinical outcome measures. However, patients that received high-dose mepolizumab treatment (750 mg) had a 50% lower exacerbation rate (although not statistically significant) when compared with placebo, suggesting a potential benefit of mepolizumab treatment in preventing exacerbation (Flood-Page et al. 2007). Because mepolizumab may be most effective in reducing the exacerbation rate of asthmatics, two clinical trials were conducted with patients displaying eosinophilia to determine the efficacy of mepolizumab to control exacerbation. In one of the clinical trials, patients who had refractory eosinophilic asthma and a history of recurrent severe exacerbation received i.v. infusions of 750 mg of mepolizumab every month for 12 consecutive months (Haldar et al. 2009). In the other clinical trial, patients who had persistent sputum eosinophilia despite prednisone treatment received i.v. infusions of 750 mg of mepolizumab every 4 weeks for 24 weeks (Nair et al. 2009). Both studies confirmed that mepolizumab treatment significantly reduced the exacerbation rate in asthmatic patients with eosinophilia when compared with placebo, without significantly improving lung function. To optimize the administration routes and determine an efficient dose of mepolizumab, a phase III placebo-controlled clinical trial studying mepolizumab as adjunctive therapy in patients with severe asthma was conducted. The efficacy of s.c. administration of 100 mg of mepolizumab was compared with i.v. administration of 75 mg of mepolizumab for 32 weeks at a 4-week interval. Results showed that mepolizumab treatment reduced the frequency of clinically significant exacerbations when compared with placebo. More importantly, s.c. administration achieved slightly better efficacy than i.v. administration. The mepolizumab treatment group showed a trend of increased lung function, as determined by FEV_1 (although not statistically significant) and an increased quality of life (QoL) according to the score of the Asthma Control Questionnaire (ACQ) and St. George's Respiratory Questionnaire (SGRQ) (Ortega et al. 2014). Based on the outcome from these clinical trials, in 2015, the s.c. administration of 100 mg of mepolizumab was approved by the USFDA for the treatment of severe asthma.

Another class of biopharmaceutical drugs in clinical development is nucleic acids (da Silva et al. 2014). Specific nucleic acid–based therapies include DNA vaccines to induce immune tolerance (Li et al. 2006), antisense oligonucleotides (Popescu and Popescu 2007), siRNA (Popescu and Popescu 2007; Xie and Merkel 2015), and deoxyribozymes (DNAzyme) for the selective silencing of asthma-related genes. For example, SB010 is a 10–23 DNAzyme that targets (i.e., cleaves) GATA-3, possessing a catalytic domain and two GATA-3 mRNA-specific binding sites (Santoro and Joyce 1997). GATA-3 is a key factor in asthma pathogenesis, and it is important for Th2 cell differentiation, the secretion of Th2 cytokines, and the production of IgE and airway recruitment of eosinophils. In preclinical evaluation, in a mouse model of acute asthma, 200 µg of intranasally (i.n.) applied SB010 (hgd40), which is more active in humans, significantly reduced infiltration of leukocytes, while the scrambled DNAzyme was inactive. In the same study, administration (i.n.) of another GATA-3 DNAzyme, Gd21, which is more active in mice, to asthmatic mice inhibited the secretion of IL-5 and infiltration of leukocytes and alleviated AHR and airway remodeling (e.g., goblet cell hyperplasia). These results indicate the therapeutic potential of GATA-3 DNAzymes in asthma (Sel et al. 2008). Potential off-target effects of SB010 in immune cells were investigated in TLR-9-transfected HEK293 cells, macrophage cell lines, and primary innate immune cells. Results showed that SB010 did not induce nonspecific innate cell stimulation, such as activation of neutrophils and degranulation of mast cells and basophils, suggesting that SB010 could be used as a safe treatment of allergic asthma (Dicke et al. 2012). In animal studies, potential toxicity of SB010 inhalation was evaluated in healthy rats and dogs because both rats and dogs share the same mRNA sequence as human GATA-3, which SB010 binds to. Results showed that prolonged exposure (28 days) to inhaled SB010 in dogs (14.6, 40.8, and 115.5 mg/m^3 with corresponding theoretical doses of 103.3, 250.4, and 739 µg of SB010/kg of body weight [BW]) and in Wistar rats at different doses (14, 38, and 113 mg/m^3 with corresponding theoretical doses of 112 and 302 µg of SB010/kg of BW) did not produce signs of systemic toxicity as measured by clinical chemistry, urinalysis, or gross pathology and histological analysis of the lung. Although minimal histopathological changes, including interstitial leukocyte infiltration, bronchus-associated lymphoid tissue hyperplasia, and compound-related lesions in the lung, were observed at a very high dose of SB010 (888.8 µg of SB010/kg of BW) in rats, these changes only occurred at the highest dose and were completely recovered after withdrawing SB010. Furthermore, long-term exposure in rats did not change the splenic cell population (e.g., Th cells and natural killer [NK] cells) or serum levels of IgG, IgA, and IgE antibodies, and there was no abnormal expression of cytokines in the bronchoalveolar lavage fluid (BALF) except a small increase of IL-10 and interferon (IFN)-γ. Dogs received i.v. infusion of SB010 (10 µg/kg) for 10 min, and ECG and blood pressure were monitored. The results suggested that SB010 treatment did not cause cardiac abnormalities. Together, these toxicity results suggested that SB010 treatment was well tolerated in dogs and rats and did not cause significant histopathological changes (e.g., respiratory or cardiac) (Fuhst et al. 2013). Biodistribution studies of inhaled SB010 were conducted in mice with OVA-induced airway inflammation, and in healthy rats and dogs. Fluorescently labeled SB010 (200 µg) was retained in the lungs of asthmatic mice up to 24 h after administration (i.n.) as observed by confocal microscopy. The long-term pharmacokinetics of intratracheal (i.t.) instillation of

[111]indium-labeled SB010 was monitored by single-photon emission computed tomography (SPECT) in asthmatic and healthy mice. Results showed that [111]indium-labeled SB010 rapidly accumulated in the lung and bladder and was detectable in the lungs of asthmatic mice for up to 150 h. In comparison, less [111]indium-labeled SB010 accumulation was observed in the lungs of healthy mice. Pharmacokinetics of inhaled different doses of SB010 were evaluated in rats (38.3 and 113.2 mg/m³ with corresponding theoretical doses of 303 and 889 μg of SB010/kg of BW) and dogs (40 and 115 mg/m³ with corresponding theoretical doses of 250 and 750 μg of SB010/kg of BW). SB010 exposure lasted for 28 days, serum samples were collected at different postexposure time points, and the concentration of SB010 was measured by enzyme-linked immunosorbent assay (ELISA). SB010 was detectable shortly after application in both species, indicating high systemic availability, while no plasma accumulation was observed after multiple administrations. These favorable pharmacokinetic characteristics support that inhalation of SB010 could be a safe and efficient administration route and also provide guidance for future dosing and regimens (Turowska et al. 2013). In human clinical trials, the safety, tolerability, and pharmacokinetics of inhaled SB010 were first evaluated in healthy participants. Inhalation of a single dose of SB010 (0.4, 2, 5, 10, 20, and 40 mg) and multiple doses (5, 10, and 20 mg) for 12 days did not cause serious adverse events. In the single-dose treatment, the drug plasma concentration was below the detection limit in patients receiving doses in the range from 0.4 to 5 mg. In contrast, SB010 was detected in plasma after inhalation of 20 and 40 mg of SB010 and reached the maximal concentration (C_{max}) at 0.5–2 h after administration. Next, the safety and pharmacokinetics of a single inhaled dose of SB010 (5, 10, and 20 mg) were investigated in asthmatic patients. The asthmatic patients overall had higher AUC and C_{max} at given doses than healthy participants since their airway epithelium may be damaged (Homburg et al. 2015). Since SB010 treatment was safe and well tolerated, its pharmacologic activity was further evaluated in asthmatic patients. Patients with mild asthma received 10 mg of SB010 or placebo via inhalation for 28 days and an allergen challenge after the 28-day treatment. Results showed that SB010 treatment improved the lung function and attenuated allergen-induced EAR and LAR, as

measured by FEV₁. SB010 treatment also resulted in reduced eosinophils in the sputum and lower IL-5 plasma levels when compared with controls, indicating that SB010 alleviated Th2-driven inflammatory responses and may consequently improve lung function (Krug et al. 2015). Despite promising results, clinical trials assessing the long-term safety and efficacy of SB010 still need to be conducted in a larger number of asthmatic patients to determine the usefulness of SB010 for the treatment of asthma.

7.3.2.3 Nanomedicine for Asthma

As described above, numerous drugs for the treatment of asthma are currently available or in clinical development; however, none of them can be considered nanomedicine. The use of nanomedicine for the treatment of asthma is the subject of numerous preclinical studies. Nanomedicine holds the promise of providing a superior drug formulation that may enhance drug delivery. Specifically, for asthma therapy, nanomedicine should focus on refining the formulation of potential small molecules and biopharmaceutical drugs for oral administration, inhalation, and s.c. injection.

Oral delivery of a drug is most desirable, as it can greatly increase patient compliance, especially in chronic diseases that require frequent administration, such as asthma. Nanomedicine allows for small-molecule drugs to be encapsulated in a polymer enhancing the drugs' solubility and stability in the GI tract, ultimately increasing bioavailability (Ensign et al. 2012). Only a few orally administered anti-inflammatory drugs are currently being developed using nanomedicine. For example, a nanoemulsion of vitamin D demonstrated increased bioavailability and attenuated inflammatory response in OVA-induced asthmatic mice when administered orally (Wei-hong et al. 2014). A nanoemulsion of curcumin was evaluated in transgenic mice with a firefly luciferase reporter gene driven by nuclear factor κB (NFκB) responsive elements, and may serve as a potential therapy for inflammatory diseases through suppression of the NFκB signaling pathway and macrophage migration (Young et al. 2014). However, results reported in these two studies, such as decreased production of IL-1β and TNF-α (Wei-hong et al. 2014) and the reduction of peripheral macrophages (Young et al. 2014), are insufficient to prove their clinical potential for asthma therapy. Additional investigations are needed to assess therapeutic effects in asthma animal models, such as the inhibition of the production of asthma-related cytokines (e.g., IL-13 and IL-5) in plasma or BALF, improved lung function, and reduced infiltration of leukocytes in BALF. Furthermore, the stability of the loaded drug and nanocarrier in the GI tract and potential systemic toxicity induced by drugs (e.g., β_2-adrenoceptor agonists and glucocorticoids) or nanocarriers also need to be considered. Regarding the peroral delivery of biologics in nanomedicines, we can learn from other examples where efforts have been made to develop an oral formulation. Polyester (poly(ε-caprolactone)) and a polycationic nonbiodegradable acrylic polymer were used to encapsulate insulin for oral administration to diabetic rats. Decreased glycemia was achieved by insulin nanoparticles, indicating the potential of using polymers for oral administration carriers for proteins (Damgé et al. 2007). Further evaluation needs to be performed to apply biopharmaceutical drugs orally for asthma therapy. Oral formulation of nanomedicine still requires several steps of development to reach the clinic.

Subcutaneous injections or i.v. infusions are the preferred means of administration of monoclonal antibodies in asthma therapy because they are degraded in the GI and generally do not readily cross biological barriers such as cell membranes. Compared with i.v. infusion, s.c. injection is a preferred route of administration for monoclonal antibodies because it is less invasive, less time-consuming, and more convenient, and could be self-administered at home (Misbah et al. 2009). Currently, there are several nanomedicines under development for s.c. administration; however, there are none for the treatment of asthma. Poly(lactic-co-glycolic acid) (PLGA) (DeYoung et al. 2011), cyclodextrins (Gaur et al. 2012), and a liposomal polymeric gel (Park et al. 2012) have been used to encapsulate proteins for s.c. or i.v. administration. Since there is only a limited amount of s.c. formulations that are in development, nanomedicines delivered through the s.c. route will most probably not enter the clinic any time soon.

The most common route of administration for nanomedicines in preclinical development for asthma therapy is inhalation. Inhalation is a non-invasive route, and therefore is thought to have better patient compliance. Additionally, pulmonary inhalation has the advantages of delivering the drug directly to the diseased organ, of delivering the drug to a large absorption surface area, and of generally lower dose requirements. Thus, this route can consequently increase bioavailability and reduce the possibility of systemic toxicity. Small-molecule drugs can be encapsulated in polymers (Oyarzun-Ampuero et al. 2009; Oh et al. 2011; Patel et al. 2014) or liposomes (Chen et al. 2012) to achieve increased solubility and sustained release after pulmonary administration. For example, a commercially available corticosteroid, budesonide, has been encapsulated in porous PLGA, a biodegradable polymer approved by the USFDA for its use in the clinic, and was evaluated in a murine asthma model. Encapsulated budesonide provided sustained release for 24 h as observed in vitro. In the murine model, an aerosol of encapsulated budesonide significantly decreased the infiltration of inflammatory cells in BALF, reduced airway thickness, and improved lung function when compared with an aerosol of free budesonide (Oh et al. 2011). To achieve a long-lasting effect, a β_2-adrenoceptor agonist, salbutamol sulfate, has also been formulated using nanomedicine. Specifically, the drug was incorporated into artificial lipid vesicles, termed liposomes. When administered (i.t.) to healthy rats, the aerosol of salbutamol sulfate–loaded liposomes resulted in an increased half-life of salbutamol when compared with the free drug. Additionally, larger AUCs of salbutamol sulfate in the lung and plasma were observed compared with those of free salbutamol sulfate, indicating that liposome encapsulation achieved sustained release of the drug. Salbutamol sulfate–loaded liposomes increased the time of bronchodilation in an asthmatic guinea pig model when compared with free drug, confirming that the liposome suspension could provide a longer-lasting therapeutic effect (Chen et al. 2012).

Currently, a limited number of biopharmaceutical drugs have been formulated using nanomedicine for the treatment of asthma. One example is an IFN-γ-plasmid DNA that was encapsulated in chitosan and administered (i.n.) to asthmatic mice. Chitosan-encapsulated IFN-γ-plasmid DNA attenuated AHR and reduced the infiltration of eosinophils. Additionally, the levels of IL-4 and IL-5 in BALF, as well as the infiltration of leukocytes in the airway, were decreased (Kumar et al. 2003). However, pulmonary delivery of bio-pharmaceutical drugs such as siRNA and proteins has been extensively studied for diseases other than asthma. For example, inhalation of nebu-lized insulin–loaded liposomes can continuously reduce blood glucose up to 6 h, while inhala-tion of nebulized insulin with empty liposomes decreased blood glucose temporarily, which, however, bumped back 2 h postinhalation (Huang and Wang 2006). Polyethylenimine (PEI) (Beyerle et al. 2011) and chitosan (Nielsen et al. 2010) are most commonly used for pulmonary delivery of nucleic acids, such as siRNA.

Pulmonary delivery requires the use of inha-lation devices, such as (1) pressurized metered dose inhalers (pMDIs), (2) nebulizers, and (3) dry powder inhalers (DPIs) (Ibrahim et al. 2015). A pMDI is a widely used and inexpensive device; however, many therapeutics do not readily dis-solve in the hydrofluoroalkane (HFA) propel-lants compatible with pMDIs. To administer nanomedicines using a pMDI, the properties of the nanomedicine, including physical stability in the pMDI formulation (e.g., aggregation), need to be carefully evaluated. As an example, Conti et al. reported in 2014 that polyamidoamine (PAMAM) dendrimer–siRNA nanoparticles for-mulated as a pMDI efficiently knocked down enhanced green fluorescent protein (eGFP) in eGFP-expressing A549 cells. Similar knockdown efficiency was achieved following a 2-month incubation of PAMAM-siRNA in HFA, indicating that the siRNA nanoparticles were sufficiently stable in HFA. For this formulation, the spray-dried nanoparticles were embedded in mannitol or chitosan–lactic acid and dispersed in HFA, and no large or irreversible aggregation was observed within 5 h, suggesting that the nanoparticles were stable in this pMDI formulation. The pMDI formulation was characterized by an eight-stage Andersen cascade impactor. Results showed that the aerosol contained a respirable fraction and fine particle fraction of approximately 77% and 50%, respectively, similar to commercially avail-able pMDIs. Taken together, the authors con-cluded that the nanoparticles demonstrated high potential to be used in a clinical setting as a pMDI (Conti et al. 2014).

A nebulizer is an inhalation device that requires less coordination of the patient, compared with pMDIs and DPIs, and therefore it is suitable for pediatric, elderly, and unconscious patients.

Nebulization is suitable for nanomedicines that are water soluble. However, special consideration needs to be taken regarding the stability of the drugs during the production of the aerosol. There are three different nebulizers available on the market: jet, ultrasonic, and mesh nebulizers. If the nanomedicine payload is temperature sensitive, as in case of proteins and DNA, nebulizers need to be chosen carefully because of heat generated inside the medication reservoir of the vibrating mesh (Hertel et al. 2014) or inside ultrasonic nebulizers (Ibrahim et al. 2015). For example, plasmid DNA (15–20 kb) is easily degraded during aerosol pro-duction because of the shear effects (Arulmuthu et al. 2007). The degradation of plasmid DNA can be prevented by encapsulation by both PEI (Gautam et al. 2000; Rudolph et al. 2002; Davies et al. 2008) and liposomes (Birchall et al. 2000). Another con-sideration when formulating a nanomedicine for nebulizers is the large air–liquid interface created during nebulization, which may induce protein unfolding and aggregation (Hertel et al. 2015). Many nanomedicine formulations tend to aggre-gate during nebulization (Dailey et al. 2003; Beck-Broichsitter et al. 2013) due to the concentration of nanoparticles in jet and ultrasonic nebulizers (Beck-Broichsitter et al. 2013). For example, Ewe and Aigner reported in 2014 that lipopolyplexes formulated from liposome–PEI loaded with pDNA or siRNA can mediate efficient transfection in SKOV-3 and in a luciferase-expressing SKOV-3 cell line (SKOV-3-LUC). The size of pDNA and siRNA lipopolyplexes after nebulization was generally larger than that prior to nebulization, suggesting that aggregation occurred. However, the apparent aggregation did not decrease the in vitro transfec-tion efficiency of lipopolyplexes. In fact, the trans-fection efficiency of pDNA lipopolyplexes right after nebulization or when stored at 4°C for 1 day after nebulization was enhanced compared with that of the corresponding pDNA lipopolyplexes that were not nebulized. This effect can be under-stood as the result of accelerated sedimentation of larger particles in cell culture. In vivo, however, the transfection efficacy is yet to be assessed. The transfection efficiency of siRNA lipopolyplexes was also retained after nebulization or following storage, suggesting that pDNA and siRNA were protected in the formulation during nebulization (Ewe and Aigner 2014).

DPIs are the most popular inhalation devices since they are portable, and no propellant is used. The dry powder formulation demonstrates better chemical stability than the liquid formulation (Ibrahim et al. 2015). Several key considerations should be made when formulating nanomedicines for DPIs, including the stability of the drug and nanocarrier during the production of the dry powder. Spray drying is a common method used to produce inhalable dry powders of drugs formulated with polymers (e.g., PLGA [Beck-Broichsitter et al. 2012; Ungaro et al. 2012], chitosan [Grenha et al. 2005; Al-Qadi et al. 2012], and polymer–lipid [Jensen et al. 2012]). However, during the spray-drying process, droplets of the formulation are rapidly dried by a hot gas, and therefore the use of thermolabile drugs or thermosensitive polymers is limited. Spray freeze drying (Lu and Hickey 2005) or freeze drying followed by milling (Lu and Hickey 2005) may be a better choice for heat-sensitive drugs, particularly proteins. Aqueous solubility, yield, and size change of reconstituted nanoparticles must also be carefully assessed. Specifically for the lung, deposition of inhaled dry powder highly depends on its size, with particles of 1–5 μm needed to achieve deep lung deposition. It is also important to assess the nanomedicine particle size once it is reconstituted in an aqueous environment (e.g., lung tissue). Once delivered to the aqueous environment of the target tissue, the nanomedicine must maintain or regain nanometer sizes to penetrate through the mucus layer and be taken up by cells. Spray-dried and spray-freeze-dried formulations of drugs (e.g., levofloxacin) loaded into polymer (e.g., PLGA) or lipid–polymer (lecithin–PLGA) have been characterized for pulmonary drug delivery. Spray-freeze-dried formulations of nanomedicines achieve better aerosol properties than spray-dried formulations, including approximately 26% fine particle fraction and a mass median aerodynamic diameter (MMAD) of 5.8 μm. To ensure the desired particle size in the aerosol requires optimization of the formulation methodology by carefully controlling the ratio of lipid, polymer, and excipients. However, in a study by Wang et al. (2012), the production yield of 33% (w/w) following optimization was relatively low, making further optimization necessary. In addition, reconstitution of nanoparticles from both formulations in an aqueous environment showed increased particle sizes, suggesting that aggregation occurred.

7.3.3 ANIMAL MODELS OF ASTHMA

Animal models are required to determine the pharmacokinetics, therapeutic effects, and biocompatibility of new asthma therapies. Specifically, asthma is a heterogeneous disease involving numerous types of immune cells and cellular mediators, rendering in vitro models of limited use. The European Federation of Pharmaceutical Industries and Associations (EFPIA 2014) and the USFDA (2004) agree that the combination of both new technologies and animal models is required for drug discovery. Several animal models of asthma are currently used, including mice, rats, guinea pigs, dogs, and sheep. In the next sections, we review the applicability of these animal models for the development of treatments of asthma.

7.3.3.1 Rodent Asthma Models

7.3.3.1.1 Mouse

Mouse asthma models are most commonly used to study potential asthma therapies. Mouse models of asthma have several advantages, including cost-effectiveness, availability of several transgenic strains, and a large number of commercially available mouse-specific probes and tools. Because mice do not naturally develop asthma, allergens, including OVA and house dust mite (HDM) extract, have been used to artificially induce airway inflammation. In both models, mice are first sensitized by several injections (intraperitoneal [i.p.]) of an allergen with or without adjuvant. Following sensitization (2–5 weeks later), mice are exposed to the inhaled allergen, which produces an inflammation response and AHR. For example, to test the therapeutic effects of the GATA-3-specific DNAzyme SB010, of which phase II clinical trials have been completed (Krug et al. 2015), Sel et al. (2008) established a murine asthma model with the allergen OVA and adjuvant $Al(OH)_3$ that can preferentially induce a Th2-biased response. Female BALB/c mice were sensitized by i.p. injection of 10 μg of OVA/1.5 mg of $Al(OH)_3$ suspended in 200 μl of PBS solution on days 0 and 14. The first challenge of 3 consecutive days, starting on day 24, sensitized mice that inhaled aerosolized 1% (w/v) OVA dissolved in PBS for 20 min daily. Before and during the second 3-day challenge, starting on day 36, animals received a 4-day i.n. treatment of GATA-3 DNAzyme (days 35–38).

The efficacy of DNAzyme was assessed on days 39 and 40. Treatment with GATA-3 DNAzyme relieved AHR, decreased mucus hypersecretion, and reduced the influx of eosinophils (Sel et al. 2008). This short-term (4–6 weeks) mouse model has been widely used and demonstrates high reproducibility. However, asthma in humans is a chronic airway inflammatory disease, and many patients show airway remodeling, such as airway wall fibrosis, goblet cell hyperplasia, smooth muscle thickening, and long-term AHR, which is absent in this acute allergic inflammation model. To address this shortcoming, a model of chronic airway inflammation has been developed in mice. This chronic model is similar to the acute mouse asthma model, with the exception that the challenge phase is extended. In the same study, Sel et al. (2008) established a chronic murine asthma model to further investigate the efficacy of GATA-3 DNAzyme. Mice were sensitized by three i.p injections of OVA/Al(OH)$_3$ on days 0, 14, and 21. Sensitized mice were challenged on 2 consecutive days each week for a total of 14 weeks (5th to 15th week and 19th to 21st week). This chronic asthma model more closely approximated human asthma indicated by eosinophil influx, AHR, goblet cell hyperplasia, mucus hypersecretion, and subepithelial collagen deposition (Sel et al. 2008). However, there are still some limitations to this chronic mouse asthma model. The inhalation of allergen causes an EAR in patients due to constriction of smooth airway muscles (Weersink et al. 1994). In some patients, 3–4 h after EAR, a second phase of decreased lung function, called LAR, often associated with AHR, may occur due to released cytokines and recruited eosinophils in EAR (Weersink et al. 1994). Mice can develop AHR to nonspecific bronchoconstriction agents such as methacholine, like humans, but there is a lack of evidence to prove EAR or LAR in mice after inhalation of the allergen.

7.3.3.1.2 Rat

Rat asthma models are also commonly used in asthma studies. The rat models of asthma, like the mouse models, are cheap and have many commercially available biological probes. In the rat asthma models, AHR and LAR can be induced by controlled exposure to allergens. Therefore, the rat models are better than the mouse models to study the effects of new therapies on EAR and LAR. Rats can also be sensitized by OVA (Belvisi et al. 2005; Liu et al; 2005, Lührmann et al. 2010) and HDM extracts (Singh et al. 2003). To establish the asthma models, rats are often sensitized by systemic injection of allergen with one (Al(OH)$_3$) or two adjuvants (Al(OH)$_3$ and heat-killed *Bordetella pertussis* bacilli), followed by challenge of inhaled allergen aerosol or i.t. instillation of allergen. A rat asthma model was established for testing the preclinical profile of ciclesonide. To this extent, brown Norway (BN) rats, a strain of rats commonly used for allergic models, were sensitized by i.p. injection of suspension of 100 μg of OVA/100 mg of Al(OH)$_3$ on days 0, 12, and 21 and challenged with inhalation of 1% OVA for 30 min daily for 4 consecutive days on days 27–30. Meanwhile, ciclesonide and control drug fluticasone were administered via i.t. instillation 1 and 24 h before each challenge. Rats were sacrificed 24 h after the final challenge, and the efficacy of ciclesonide to inhibit eosinophilia in BALF and lung tissue, a measure of airway inflammation, was assessed (Belvisi et al. 2005). A rat chronic asthma model has been reported, in which BN rats were sensitized by i.p. injection of 1 mg of OVA/200 μg of Al(OH)$_3$ on day 0. Sensitized rats were challenged with 1% OVA aerosol three times each week for 12 weeks, starting from day 14. Clear airway structure changes were observed, including goblet cell hyperplasia and subepithelial deposition of collagen and fibronectin; however, prolonged exposure of allergen led to reduced airway wall thickness and loss of AHR (Palmans et al. 2000). Therefore, further optimization of rat chronic asthma models is necessary.

7.3.3.1.3 Guinea Pig

As an asthma model, guinea pigs are less popular than mice and rats. However, guinea pig asthma models demonstrated well-defined EAR and LAR after allergen challenge (Hutson et al. 1988). The pharmacology and anatomy in the guinea pig model are more similar to the conditions in humans than rats and mice, and this model can be used for certain subtypes of asthma, such as the cough variant asthma (Nishitsuji et al. 2008). An acute guinea pig model of asthma has been reported where Dunkin–Hartley guinea pigs were sensitized by i.p. injection of 100 μg of OVA/100 mg of Al(OH)$_3$ on days 1 and 5 and challenged by inhalation of 0.01% OVA for 1 h on day 15. In the same study, a chronic asthma guinea pig model was established, where the sensitization and first challenge were the same as in the acute model, but the animals received eight additional challenges via inhalation of 0.1% OVA aerosol once every 2 days. To prevent fatal anaphylaxis, all additional challenges but the last one were performed under mepyramine cover (30 mg/kg, i.p.). Fluticasone propionate, roflumilast, and GW274150, an inducible nitric oxide synthase (iNOS) inhibitor, have been tested in clinical trials for patients with mild asthma, and were preclinically tested in both aforementioned guinea pig models. The drugs were administered three times at 24 h and 30 min before and 6 h after the final challenge either by aerosol (fluticasone, 0.51 mg/ml) or orally (roflumilast, 1 mg/kg; GW274150, 5 mg/kg as the phosphate). EAR, LAR, AHR, and influx of inflammatory cells were observed in both models. There was no airway remodeling observed in the acute model compared with the saline-challenged group, but in the chronic model, such airway remodeling was shown, including increased thickness of airway walls, bronchiolar collagen, or hyperplasia of goblet cells (Evans et al. 2012). GW274150 was effective in the acute model but not in the chronic model, which agreed with the results of its phase I clinical trial (Singh et al. 2007). Therefore, the authors proposed that this chronic model could be a very good animal model for asthma to predict results in humans.

7.3.3.2 Larger-Animal Models

Larger-animal asthma models have been developed in dogs, sheep, and monkeys. They are not as popular as rodent models because the costs are much higher. However, it is necessary to test therapeutic reagents in larger-animal models since they are more physiologically and immunologically relevant for humans.

7.3.3.2.1 Dog

Dog asthma models have been used for decades to study physiological and pathological mechanisms of asthma, as well as the pharmacological response of new therapeutic reagents. There are three ways to establish dog asthma models: allergen challenge, hyperventilation-induced, and ozone-induced asthma. Dogs naturally sensitized to *A. suum* (Woolley et al. 1994) can be challenged by

inhalation of an *A. suum* extract aerosol. Asthmatic response can be induced in dogs that are neonatally sensitized to ragweed by the inhaled antigen. Becker et al. (1989) described a ragweed-sensitized dog model of asthma where newborn Basenji–Greyhound dogs were immunized by i.p. injection of a suspension of 500 μg of short ragweed and 30 mg of $Al(OH)_3$ within 24 h of birth. Injection was repeated weekly for 8 weeks and biweekly for another 8 weeks. In this model, after inhalation of short ragweed (antigen E content 120 U/ml), all sensitized dogs demonstrated EAR and some individuals developed LAR. Furthermore, AHR in sensitized dogs was determined by nonallergic airway response to inhaled acetylcholine (Becker et al. 1989). The efficacy of GS-5759, a phosphodiesterase 4 (PDE4) inhibitor with long-acting β_2-adrenoceptor activity, was compared with that of indacaterol in the ragweed-sensitized dog asthma model. Different doses of micronized powder of GS-5759 and indacaterol were insufflated into the lung of dogs 1 h before ragweed challenge, and the change of lung resistance was monitored for the next 30 min. GS-5759 demonstrated a better ability to decrease the allergen-induced pulmonary resistance than indacaterol, which is in line with the results in different animal asthma models, including guinea pig, monkey, and rat (Salmon et al. 2014). Ozone-induced bronchoconstriction, airway inflammation, and AHR in dogs were used to study the therapeutic effect of MK-0591, an antagonist of FLAP, namely, the inhibitor of biosynthesis of leukotrienes (LTs). MK-0591 (2 mg/kg) was administered intravenously to dogs, and drug blood concentration was maintained by infusion of 8 μg/kg*min of MK-0591 throughout the whole experiment. Ozone was delivered through an endotracheal tube at a concentration of 3 ppm for 30 min to establish airway inflammation. MK-0591 treatment remarkably inhibited the production of LTB_4 in blood and in BALF cells, as well as the production of LTE_4 in urine. However, there was no effect on ozone-induced AHR, bronchoconstriction, and influx of neutrophils in the airway (Stevens et al. 1994). This result in the dog asthma model in fact predicts the result of the preclinical evaluation of MK-0591 in patients with mild asthma. Oral administration of MK-0591 for three times in asthmatic patients before allergen challenge can inhibit LTB_4 biosynthesis in blood and urinary LTE_4 excretion. Furthermore, EAR and LAR determined by change of FEV_1 were alleviated by MK-0591 treatment but not AHR, as measured by inhalation of histamine (Diamant et al. 1995). Reversible bronchoconstriction in patients with asthma can be induced by exercise and inhaled cold air, and this process can be mimicked by the hyperventilation-induced dog asthma model. Anesthetized dogs were intubated endotracheally and ventilated mechanically with room temperature dry air with 5% CO_2 at 200 ml/min. The flow rate was increased to 2000 ml/min for 2 min, and bronchoconstriction, AHR, and late-phase airway obstruction could be observed (Freed et al. 2000; Davis et al. 2002).

7.3.3.2.2 Sheep

Sheep are cheaper to purchase, maintain, and handle than dogs and nonhuman primates. Furthermore, there are many similarities between the lungs of sheep and humans, and sheep can be naturally sensitized by A. suum. These characteristics make sheep asthma models the most popular large-animal models for asthma. As an example, Abraham et al. (2000) selected sheep that were naturally sensitized to A. suum and which can develop EAR and LAR upon inhalation of A. suum to establish experimental asthma. Sheep were challenged by inhaled A. suum extract aerosol (82,000 protein nitrogen units/ml) at 20 breaths/min for 20 min. EAR and LAR were determined by specific lung resistance (SR_L). AHR continued for 9 days after antigen challenge, and an increased influx of inflammatory cells was observed in BALF and bronchial biopsies. The anti-inflammatory effects of multiple doses, a single dose, and different concentrations and formulations of BIO-1211, a small-molecule inhibitor of integrin $\alpha_4\beta_1$, also known as VLA-4, were tested in this model (Abraham et al. 2000). An experimental sheep asthma model where the animals were sensitized and challenged by HDM, a relevant human allergen, was developed by Bischof et al. (2003). Sheep were immunized subcutaneously with a 1 ml suspension of 50 μg of HDM mixed with 50 μg of $Al(OH)_3$ on days 0, 14, and 28. On day 42, tracheal instillation of 5 ml of HDM solution (0.2 mg/ml) was guided by a flexible fiber-optic bronchoscope deep into the left caudal lung lobe. For comparison, saline was instilled into the right caudal lung. HDM-specific antibodies IgE, IgG_1, and IgG_2 were increased on 7 days postthird

immunization, and increased eosinophil cell counts were observed in peripheral blood and BALF 48 h after HDM challenge in allergic sheep. In addition, infiltration of lymphocytes and eosinophil into the peribronchial region was shown in hematoxylin and eosin (H&E)–stained lung tissue (Bischof et al. 2003). A sheep model of chronic asthma was developed based on this protocol by the same group in which sheep were challenged twice a week for the first 3 months and once weekly for another 3 months. Airway remodeling, including hyperplasia of goblet cells and increased collagen deposition and thickness of smooth muscle, was observed; however, the total time frame of approximately 7 months to establish this model is very long and time-consuming (Snibson et al. 2005). Van Der Velden et al. (2013) investigated the effect of senicapoc (ICA-17073), an inhibitor of the $K_{Ca}3.1$ ion channel expressed widely in various cells involved in asthma, in a modified sheep model of chronic asthma. Sheep were sensitized with HDM as described by Bischof et al. (2003). Two weeks after the final immunization, sheep were challenged with nebulized HDM solution (1 mg/ml) at 20 breaths/min for 10 min once every 2 weeks for 14 weeks. Allergic sheep received 30 mg/kg senicapoc orally twice daily from 7 days before the first challenge and throughout the 14-week challenge phase. The lung resistance, R_L, in allergic sheep was increased, and EAR was observed but not LAR. However, increased collagen production and airway smooth muscle remodeling were not shown in this model, but hyperplasia of goblet cells and increased density of blood vessels in the airway wall were observed. Senicapoc treatment maintained normal R_L, reduced the EAR in allergic sheep, and inhibited the increased density of blood vessels. However, there was no significant difference in the influx of eosinophils or hyperplasia of goblet cells between the senicapoc and vehicle-treated groups (Van Der Velden et al. 2013). This result is in line with the result of a phase II clinical trial for allergic asthma. Senicapoc treatment reduced the allergen-induced increased airway resistance and exhaled nitric oxide, an inflammatory marker; however, it did not achieve improved lung function (Wulff and Castle 2010). An HDM challenge protocol to establish a sheep asthma model is more standardized than the A. suum challenge protocol. The HDM challenge sheep model of chronic asthma could be a relatively accurate model to predict clinical results, but further optimization is still needed to balance the time and development of physiological asthma hallmarks, such as airway remodeling.

There are several nanoscaled medicines currently being investigated for diseases concerning the respiratory system, including asthma, with the aim of overcoming the limitations of conventional drug therapy. Among other possible administration routes, the pulmonary delivery of particles is gaining special interest, as the lung offers a great diversity of advantages, such as its large surface area, distinct blood perfusion, and possibility to circumvent the first-pass effect, to name a few. However, there are still some hurdles to overcome in order to translate nano-based formulations for treating asthma into clinical routine. In the following, several important issues to address on this way from nanotechnology to nanomedicine will be discussed, with an insight into the current research situation and what obstacles still lie ahead.

7.3.4 STUDY PLANNING

7.3.4.1 Nanomedicine-Specific Considerations

A constitutive sector to examine when developing a new nanotherapy is the influence of the medication on the condition of the human body. Besides possible side effects that nanomedicines can induce, it is also important to monitor any changes they can provoke, for example, in pH or temperature, as those might decrease the valuable therapeutic effects by altering the drug characteristics. As the success of nanotherapy in asthmatic lungs hinges on several factors, such as administration route and characteristics of the particles, but also physiological aspects of the diseased organ, it is crucial to have an encompassing understanding of the respective anatomy, molecular biology, and cell physiology. In order to observe the distribution of nanomedicines within the intended target organ, as well as other parts of the body, different imaging techniques can be applied. Previous investigations focusing on nanoparticles as contrast enhancers (Harrington et al. 2000), intracellular trafficking of nanocarriers (Huang et al. 2013), real-time monitoring of pharmacokinetics and biodistribution (Taylor et al. 2001), and high-throughput pharmacokinetic

screening (Watt et al. 2000) have pointed out capabilities for high-quality and precise data collection (Da Silva et al. 2013).

7.3.4.1.1 Toxicity

Although the utilization of nanoscaled medications entails numerous advantages, including new material properties, an increased surface capability due to an enlarged surface–volume ratio, shorter transport times, the potential of selective targeting, and the minimized exposure of healthy tissue to the incorporated drug (Bhaskar et al. 2010), the miniaturization of systems always involves the danger of arousing toxicity. To determine a safe dose within the therapeutic window, as well as the lethal dose of drug-loaded nanoparticles, it is essential to perform toxicological testing *in vitro*, *ex vivo*, and *in vivo* in cell lines, tissue, and animal models before starting clinical trials with humans. Although it is not possible to convey the data gained from those experiments directly to the conditions of an actual patient, it is crucial to reduce the risk of toxic effects and possible adverse reaction as much as possible beforehand.

Several assays to ascertain the toxicity of nano-based drug formulations have already been established and can assess different cellular targets, such as mitochondria, lysosomal activity, cell membrane integrity, or DNA ladder assays, to determine cell death mechanisms. The standard testing method for cell viability after treatment with nanomaterials in a great range of cell lines is the colorimetric 3-(4,5-dimethylthiazol-2-yl)-2,5-diphenyltetrazolium bromide (MTT) assay. Other tests resting on the principle of the metabolic activity of mitochondria involve other tetrazolium salts or resazurin. To improve statistical validity and minimize the error chance, it is recommended to combine multiple experiments. For example, the simultaneous performance of MTT and resazurin assay can be rational, as they both utilize a similar principle of testing (Paranjpe and Muller-Goymann 2014). Studies revolving around the specific disease instances of asthmatic lungs should take into consideration the correspondent conditions, preeminently AHR, mucus hypersecretion, and the influx of inflammatory cells, as well as their cytokines (Xie and Merkel 2015). In order to detect possible immune-related and inflammatory responses, variations in activation levels of cytokines, for example, TNF-α, ILs, and prostaglandin, should be monitored.

Particular materials have to be chosen to reduce toxic effects contingent upon the desired target region. While PAMAM, for example, indeed exhibited favorable characteristics as nanocarriers in several studies, it was shown to foster acute lung injury by inducing autophagic cell death via the Akt-TSC2-mTOR signaling pathway (Bhavna et al. 2009). Card et al. (2008) reviewed different imaging, diagnostic, and therapeutic applications of engineered nanoparticles in the lung and identified whole groups of nanomaterials that can have negative repercussions on the pulmonary structure and function. According to their findings, nonbiodegradable substances such as carbon nanotubes, carbon black, fullerenes, silica, metals, and metal oxides can generate inflammation and/or fibrosis in the lung after inhalation, intranasal or oropharyngeal aspiration, or systemic administration (Card et al. 2008). Biodegradable nanoparticles, for example, made of polyethylene glycol–polylactide (PEG-PLA) (Da Silva et al. 2013) or PLGA (Dailey et al. 2006), on the contrary, have been proven to be useful pulmonary drug carriers.

Different studies in humans have been performed to investigate the deposition of inhaled nanoparticles in healthy and diseased lungs (Daigle et al. 2003; Moller et al. 2008). An increased pulmonary deposition and retention in constricted airways was predicted by computational models (Farkas et al. 2006) and demonstrated in obstructive lung disease (Anderson et al. 1990) and asthma (Chalupa et al. 2004) patients. Regarding the latter, the exposure of subjects with mild to moderate asthma to ultrafine carbon particles during spontaneous breathing led to an increased fraction of deposited particles compared with healthy individuals. Pietropaoli et al. (2004) nevertheless did not ascertain any differences in respiratory parameters between healthy and asthmatic subjects after inhalation of respective particles. No airway inflammation was observed in either group, but an exposure of healthy individuals to a higher concentration of particles resulted in a decreased midexpiratory flow rate and carbon monoxide diffusing capacity, indicating that nanoparticles may influence respiratory function and gas exchange (Pietropaoli et al. 2004). Moreover, several investigations have been conducted to test the translocation of nanoparticles

from the lung to the systemic circulation after inhalation, as cardiovascular effects similar to those from the impact of urban air pollution were apprehended. Most findings indicate that the tested [99m]techneticum-labeled carbon nanoparticles are not detected outside the lungs in appreciable concentrations (Mills et al. 2006; Wiebert et al. 2006; Moller et al. 2008). Nevertheless, it remains uncertain whether other nanoparticles behave similarly, and the possibility of particles influencing the vasculature is still not ruled out. Besides that, all studies used single-inhalation-exposure protocols. Further investigations on the repercussions of repeated exposure, stronger pulmonary accumulation, and therefore translocation of greater particle quantities are urgently needed (Card et al. 2008).

In a recent study, it was examined whether i.t. instillation studies can be used for evaluating any harmful effects of inhaled nanoparticles. Therefore, rats were exposed to nanoparticles composed of nickel oxide and titanium dioxide as high- and low-toxicity examples. Among others, increases in neutrophils in BALF and concentration of cytokine-induced neutrophil chemoattractants were compared after single i.t. installations and inhalations over 4 weeks, and results suggested that i.t. studies can be a useful tool in ranking adverse influences of nanoparticles (Morimoto et al. 2015).

7.3.4.1.2 In Vivo Pharmacokinetics, Administration, and Metabolism

The lung and its large surface area, with a high vascularization, as well as a thin air–blood barrier, on the one hand display an ideal location for the absorption of agents (Merkel et al. 2014). On the other hand, several physicochemical and biological barriers await the nanotherapeutics in the pulmonary system, making it essential to thoroughly track their routes and deposition in the body. Figure 7.8 illustrates the different defense mechanisms nanoscale particles have to encounter in the lung.

The most important parameter influencing the deposition of particles in the different areas of the lung is their size. Depending on it, three different mechanisms of allocation are possible: impaction, sedimentation, and Brownian diffusion. Particles with an MMAD greater than 5 μm pass through the oropharynx and upper respiratory passage

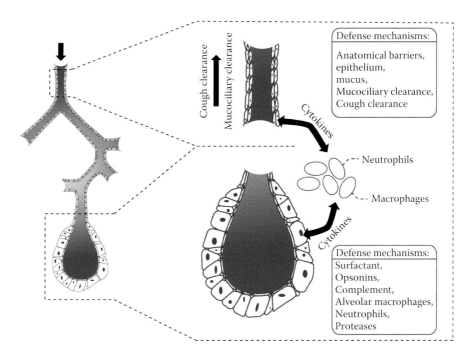

Figure 7.8. Lung-intrinsic barriers to efficient pulmonary siRNA delivery. (From Merkel, O. M., and Kissel, T., *Acc. Chem. Res.*, 45(7), 961–970, 2012. Copyright 2012 American Chemical Society. Reprinted with permission.)

with a higher pace, collide with the respiratory wall due to the centrifugal force, and are deposited in the mouth and pharyngeal regions (Merkel and Kissel 2012). This so-called impaction usually occurs with dry powder inhalation (DPI) and metered dose inhalators (MDIs). Deposition of drugs formulated as DPIs is especially dependent on the inspiratory effort of the patient: an insufficient force of inhalation leads to aggregations of the particles in the upper airways. Nevertheless, large and aggregated particles can also become subject to this process when MDIs are used, despite the higher speed of the generated aerosol. As gravitational forces preponderantly condition the sedimentation process, particles with an MMAD between 1 and 5 μm are slowly deposited in the smaller airways and bronchioles, whereas Brownian motion is the prevalent mechanism in the lower alveolar sections. The molecules surrounding the aqueous lung surfactant underlie the Brownian motion itself and induce a random moving of the particles. The dissolution of the therapeutic agents in the lung surfactant, depending on the concentration gradient, influences this process as well. Particles with a size smaller than 1 μm deposit in the alveolar region or can be exhaled. Therefore, sedimentation is the

preferably achieved process for therapeutic nanosystems in order for them to stay in the bronchiolar area for a long time and to result in the desired effects. In addition to these particle- and target surface–dependent characteristics, breathing patterns, the holding of breaths, and tidal volume, as well as air velocity and humidity, are factors influencing the deposition, and hence need to be considered (Yang et al. 2008).

The inside of the upper airways is covered with a film of mucus that is responsible for trapping and purging invading particles. Before they can reach lower sections of the lung, they are often cleared by mucociliary movements, through coughing or swallowing. Consequentially, nanosized drugs should be able to cross the mucus layer and reach the sol that covers the stratum below the gel coat (Bur et al. 2009). As PEG nanoparticles were repeatedly shown to be capable of permeating across mucus (Tang et al. 2009; Dunnhaupt et al. 2015; Inchaurraga et al. 2015), PEGylation is a possible approach to avoid bronchial clearance of nanomedicines. In the alveolar regions, the alveolar lining consists of various proteins and lipids. Additionally, the existing tight junctions hamper the transport of molecules. According to the structure of the nanomedicine, active transport or passive diffusion through those transporter proteins is possible. Larger particles are furthermore prone to being cleared through phagocytosis by alveolar macrophages (Patton et al. 2010).

Having arrived inside the peripheral lung, the particles have to dissolve, and for systemic effects, the incorporated drug has to diffuse through the epithelial barrier in order to reach the blood. There are still some deficiencies in the exact understanding of the process of cell uptake and how the particles are transported and reach the systemic circulation. Despite the existence of *in vitro* models for studying the uptake and permeation, the precise behavior of the cells under disease conditions has yet to be examined further (Paranjpe and Muller-Goymann 2014).

7.3.4.2 Disease-Specific Considerations: Asthma

Since asthma is a disease presenting itself in a very heterogeneous range of manifestations, it is of particular importance to define criteria both for categorizing the severity of the existing condition and for evaluating the success of any treatments. According to the current GINA report, "asthma severity" should be "assessed retrospectively from the level of treatment required to control symptoms and exacerbations" (GINA 2015). This severity is not necessarily a constant feature, but can change over time, so that it can only be evaluated after a patient has already been experiencing a controlling treatment for several months. Subsequently, the patient's status can be rated as mild, moderate, or severe asthma. The latter is thereby important to be distinguished from uncontrolled asthma due to incorrect inhaler technique or poor adherence (Ehmann et al. 2013).

Corresponding to this prior definition, the main goal in asthma treatment is to maintain "asthma control," which is described as "the extent to which the various manifestations of asthma have been reduced or removed by treatment" (GINA 2015). Among the current clinical situation of the patient, and therefore factors such as symptoms, lung function, the use of reliever medication, and night awakenings, this concept additionally comprises the patient's future risk of exacerbations and decline in lung function, as well as treatment-related side effects. According to the GINA guidelines, asthma can be regarded as controlled when daytime symptoms occur only twice per week or less, daily activities are not limited, the lung function is normal or near normal, there are no exacerbations and no symptoms at nighttime, and the reliever medications only need to be used twice per week or less. The obtained control level through therapy determines the necessity to go up or down to the adjacent treatment level in the stepped management approach for asthma therapy that both the respective U.S. and European guidelines recommend. Patients whose disease symptoms cannot be controlled in an adequate way by the highest stage of treatment, which means a reliever medication and two or more controller treatments, are considered to suffer from *difficult-to-treat* asthma (Ehmann et al. 2013).

7.3.4.2.1 Patient Selection

When choosing patients for clinical studies, it is important to use uniform clinical guidelines to diagnose and predefine the disease status based on clinical symptoms and disease history of the patient, as well as on assessment of airflow limitation. To examine this factor, the preferred approach is to perform spirometry under standardized conditions. In this way, both the FEV_1 and the forced vital capacity (FVC) can be measured to conclude on the patient's existing airflow limitation and its variability. In case these data are difficult to generate, for example, because the patient is treated with a controller medication, the PEF can also be ascertained. A determination of AHR can be of use for patients who show clinical symptoms, but possess a normal lung function (Ehmann et al. 2013).

For patients who already receive a treatment against asthma, it is especially important to assess and establish the minimum level of treatment necessary to maintain control over their symptoms and standardize this treatment as much as possible in order to ensure an appropriate baseline to interpret subsequent results. An accurate and detailed profile of every patient has to be compiled, including characterization of attributes such as lung function, day- or nighttime symptoms, and previous history of exacerbations, as well as a documentation of all comorbidities and concomitant treatments. Concerning the treatment of asthma in comorbid patients, particular attention has to be drawn to COPD, as these two conditions indeed have different pathways of causation, but can still coexist in one patient. Therefore, patients who suffer from COPD can distort respective results and should not be included in asthma studies. As opposed to this, smokers are able to take part, although their smoking history has to be accurately recorded, and it is advisable to conduct a subgroup analysis to discover any influence the smoking could have on study results. Other possible classifications for relevant subpopulations would be frequency of exacerbations, sensitivity to nonsteroidal anti-inflammatory drugs (NSAIDs), eosinophilia, or cosensitivity to other allergens. When investigating a specific immunotherapy, documentation and characterization of the patients' allergy history and inflammatory airway profile are required. Moreover, a crucial point to consider before starting any clinical asthma study is the equalization of clinical procedures. All patients have to be trained in an adequate way to use their drug administration devices, such as inhalers, as well as equipment for lung function testing and, where applicable, diary cards or other relevant evaluation tools (Ehmann et al. 2013).

The high incidence of asthma in children leads to them being a target group of special interest in the respective research. For them, as well as for the older generation, particular attention has to be drawn to the adequate utilization of inhalation devices. Studies in children should be conducted in the early development process, when the therapy holds promise to be a major advance for the pediatric population, of course implying that appropriate safety data are available. More detailed advice regarding studies for these special patient groups can be found in the respective International Conference on Harmonisation of Technical Requirements for Pharmaceuticals for Human Use (ICH) notes for guidance (Geriatrics: ICH Topic E7 [ICH 1993] and Pediatrics: CPMP/ICH/2711/99 [ICH 2000]) (Ehmann et al. 2013).

7.3.4.2.2 Efficacy Evaluation

To appraise the success of the examined treatment forms, there are several measurable parameters available for asthma studies.

For characterizing the effect of therapies on the lung function, the previously mentioned FEV_1 and PEF are suitable to detect and evaluate any airway obstruction. The most accurate factor represents the prebronchodilator FEV_1, being influenced even by short-term fluctuations. Nevertheless, it rather poorly correlates with the symptoms the patient experiences. To achieve a more relevant portraiture of the actual experienced lung function, the measurement of PEF would be a more appropriate approach. Further spirometric recordings, such as vital capacity (VC), the flow rates at 25% or 75% of VC above residual volume (RV), or postbronchodilator FEV_1 and FVC, are possible supplementary study endpoints.

In achieving asthma control, the avoidance of *exacerbations* is a crucial factor, which makes it a relevant endpoint to evaluate controller treatments in patients. According to the European Medicines Agency (EMA) notes for guidance on clinical investigation of medicinal products for treatment of asthma, severe exacerbations can be described as "a requirement for systemic corticosteroids or an increase from the maintenance dose of corticosteroids for at least three days and/or a need for an emergency visit, or hospitalization due to asthma" (Ehmann et al. 2013). Moderate exacerbations, however, arouse the need for a change in treatment in order to prevent a deterioration of asthma symptoms and, therefore, the occurrence of severe exacerbations. In order to really capture any changes in the number of exacerbation events in an appropriate way, the EMA guidelines advise a study length of at least 12 months, with a subsequent equally long follow-up period and the documentation of the respective seasons.

When studying specific immunotherapies, *challenge testing* with an applicable antigen can be of use to grade the tendency of airways to narrow after being stimulated in a way that has hardly any influence on healthy people. Therefore, direct factors, such as histamine, or indirect ones, such as mannitol or hypertonic saline, can be applied. Although the hyperresponsiveness shows only a weak correlation with symptoms, a respective increase can be a predictor of losing asthma control (Ehmann et al. 2013).

Symptom scores are another possible evaluation tool. Here, day- as well as nighttime symptoms have to be gathered, and predefined variables such as symptom-free days or number of night awakenings should be utilized. Besides that, there are special *composite scores* with categorical or numerical variables available for the measurement of asthma control, for example, the Asthma Therapy Assessment Questionnaire (ATAQ), the Asthma Control Test (ACT), or the Asthma Control Scoring System (ACSS) (Cloutier et al. 2012).

A different approach to capture the patients' symptoms is to take a look at the need of *reliever use*, as this correlates with the frequency and intensity of the symptoms and their tolerance, when being separated from the prophylactic use. Likewise, the *reduction of controller medication* can provide information about the therapy's influence on the patient.

The underlying airway inflammation can be determined by measuring certain biomarkers, such as eosinophil counts and fractional concentration of exhaled nitric oxide (FE_{NO}).

Since the patients' sensation of the disease status might be different from the way clinicians would evaluate it, it is also expedient to utilize health-related quality of life (HRQoL) questionnaires to integrate their point of view. One asthma-specific kind of these questionnaires is provided by the American Thoracic Society (Ehmann et al. 2013; ATS 2015).

7.3.4.2.3 Design of the Study

Before starting a clinical study, answers for the following questions need be found: How does the new therapy relate to current treatments? Is it a primary or add-on therapy? Is it a reliever or controller treatment? What is the intended mechanism of action? Specific immunotherapies do not belong to either reliever or controller medications and therefore have to be evaluated in an individual way.

The guideline ICH E-4, "Dose Response Information to Support Drug Registration," recommends examining the dose-related benefit and adverse effects of new pharmaceutical products in randomized, double-blind, placebo-controlled studies (Belvisi et al. 2005). While cumulative dose–response studies with FEV_1 or PEF as pharmacodynamic endpoints are applicable for β2 adrenergic agonists, anti-inflammatory therapies are advised to be tested in group comparative studies parallel to a control group. Both types of treatment can also be investigated with the bronchoprotection or bronchial reactivity model. More detailed information on this matter can be found in the CHMP guideline for orally inhaled products (CHMP 2009).

The duration of the trial is contingent on both the mechanism of action of the tested drug and a defined efficacy measurement. Studies of 6–12 weeks are recommended by the EMA for long-acting bronchodilators. Short-acting ones can,

on the other hand, also be acceptably examined in shorter periods of time. The exploration of an effect on inflammation or exacerbations, however, asks for a longer trial duration. Reliever medications are advised to be investigated in parallel group studies, where efficacy tests can be performed in short-term trials of 4 weeks that should show the maintenance of efficacy without any tolerance. The effects on controller medication, in contrast, can only be examined in an encompassing way in studies of at least 6 months, according to the selected endpoint, but possibly for even longer periods of time. Products for specific immunotherapy should usually be evaluated as an add-on treatment to required controller and/or reliever medication during a time span that comprises the period of high allergen exposure, for example, pollen season (Ehmann et al. 2013).

Regarding comparator products, the preferential approach for reliever medication would be a three-arm study in which the new product is compared with effects of both placebo and a short-acting β2 agonist. Medications that are expected to be utilized as a first-line controller therapy should be investigated in an active comparator trial, comparing them with a standard treatment for a specific step in the treatment scheme. Typically, an inhalable corticosteroid is included in all steps, which is supposed to be administered in appropriate doses and durations. To guarantee assay sensitivity, it is generally advisable to perform a three-arm study in patients with milder asthma with a placebo adjustment in at least one fundamental clinical trial. When ICSs are not planned to substitute the new medication, add-on study designs are necessary, in which the drug being tested is compared with a placebo and, as the circumstances require, also a standard comparator of the next-higher treatment scheme step.

Since asthma is a disease that can lead to severe medical conditions, it is crucial to secure the patients' safety and well-being in each instance, meaning that they receive appropriate treatment according to their severity level and that concomitant rescue therapy always has to be available. However, this proceeding ought to be standardized and facilitated as much as possible to ensure that the interpretation of generated results is not distorted. Every concomitant treatment with bronchodilators, corticosteroids, antibiotics, or mucolytic antioxidants has to be thoroughly documented and also balanced between different trial groups in order to establish a common base (Ehmann et al. 2013).

If applicable, it is advised to always carry out clinical trials for asthma in a double-blinded way. When this cannot be achieved, for example, for some inhalable products, a three-arm study with a blinded comparison with placebo and an unblinded comparison with an active comparator is possible. Here, the respective personnel measuring and documenting data should be left unaware of the received treatment option. It is always preferable that the most important results concerning safety and efficacy are collected by an independent adjusting committee.

When selecting suitable endpoints for asthma studies, it is essential to consider the diversity of this disease. Since it shows multiple manifestations that might not all correlate with each other, it is highly recommended to use a range of evaluation measures. The choice of the most convenient primary endpoint hinges on the type of the new medication (reliever or controller), its particular way of action, and the asthma severity level. Whereas the examination should be concentrated on the airway construction, and therefore FEV_1 assessments are suitable for reliever treatments, controller therapies are expected to control the asthma and diminish exacerbations. Hence, studying the latter has to include both lung function and symptom improvement as primary endpoints. For anti-inflammatory drugs, the focus particularly lies on the occurrence of exacerbations; however, in patients with milder asthma, this might not be adequate enough, and other symptom-based endpoints can be more suitable. Since specific immunotherapy is aimed at regulating immunological mechanisms, which need some onset time, their exploration should begin as an add-on treatment and can, for example, be studied during a stepwise reduction of controller therapy. Possible primary endpoints in this approach, next to lung function and exacerbation number, are composite scores or the reduction in need for controller medication. Irrespective of the chosen primary endpoint, it is crucial to determine the minimally important difference *a priori* according to the disease characteristics and severity, the control group, the duration of the study, and the tested conjecture. As a secondary endpoint, if not already chosen for the primary one, the lung function is recommended to be investigated in any case. Furthermore, variables such as symptom scores, hyperresponsiveness, biomarkers, or the need for rescue medication can be considered (Ehmann et al. 2013).

7.3.4.2.4 Safety

As asthma therapy is often started at a young age and has to be received over a great period of time, the safety of respective medications over a longer time is a crucial factor that needs to be addressed with appropriate attentiveness. Therefore, long-term safety studies of at least 1 year should be carried out, whereas the asthma severity determines the exact time span and intensity of treatment. Special attention has to be paid when investigating immunosuppressive therapies to detect any malignancy-evoking effects. Concerning inhalable products, specific security issues, such as oral infections, vocal cord myopathy, or cataract formation, might be necessary to address. Although the systemic exposition is lower here, its extent is advised to be evaluated regardless. Regarding ICSs, for example, the impacts on the hypothalamic–pituitary–adrenocortical (HPA) axis function, bone mineral density, and eyes should be monitored. In all cases, both pharmacokinetic and pharmacodynamic studies have to be used to examine the overall systemic safety (Ehmann et al. 2013).

7.3.5 REGULATORY AND ETHICAL HURDLES: THE DRUG APPROVAL PROCESS

Before a new drug can be introduced into the market, by law it has to be shown to be safe and effective before it can be approved by the appropriate regulatory agency governing the respective market. Apart from Japan, the United States and Europe are home to the two main regulatory agencies in the world. The USFDA and the EMA set the most demanding and stringent standards for approving new medicinal products, bringing out legislations and guidelines for developing, testing, trialing, and manufacturing drugs (Kashyap et al. 2013). Overall, both departments base their work on similar key principles to warrant public safety and health; however, the exact courses of actions display some minor differences.

The USFDA drug approval process generally consists of submitting two essential applications, an investigational new drug (IND) application, followed by a new drug application (NDA). As soon as a new drug is considered safe in preclinical trials, a firm or institution as a so-called sponsor is responsible for submitting the IND to the USFDA in order to start clinical trials in humans. In a pre-IND meeting, issues such as design of supporting animal research, clinical trial protocols, and the manufacturing and control of the novel drug can be deliberated with the USFDA. Subsequently, the Center for Drug Evaluation and Research (CDER) evaluates medicinal, chemical, pharmacological, toxicological, statistical, and safety aspects in a review. When the product is confirmed to be effective and not causing unreasonable risks in clinical trials, the manufacturer can next request to manufacture and sell the drug in the United States by submitting the NDA (Kashyap et al. 2013). In the following process, the USFDA reviewers decide whether the new drug is safe and effective in its proposed use and the benefits outweigh the risks, whether its labeling is appropriate, and whether manufacturing and control methods are adequate to preserve the drug's identity, strength, quality, and purity (USFDA 2015).

Comparably, the approval procedure in the European Union (EU) likewise involves two main steps, the clinical trial application and the marketing authorization application. While the former is approved at the level of one of the current 28 member states, marketing authorizations can be administered at both member state and centralized levels. Altogether, it can be distinguished between four different processes: the centralized, mutual recognition, nationalized, and decentralized procedures. The centralized procedure, which is compulsory for certain medicines, for example, deriving from biotechnological processes, intended for cancer, HIV/AIDS, or diabetes, and so-called orphan drugs, is mandatory in order to be allowed to obtain a marketing authorization valid throughout the EU. The respective application gets reviewed by an assigned rapporteur and submitted to the European Commission for final approval. To receive a marketing authorization in a member state (concerned member state) other than the one where the drug was originally approved (reference member state), the mutual recognition procedure is appropriate. The applicant submits dossiers to all desired states, from which one decides to take command and evaluates the drug and afterwards reports its findings to the other states. As opposed to this, the nationalized procedure only accomplishes an authorization for one member state and involves just the respective authority. Medicinal products that have not yet been approved in any EU country and do not come within the centralized process can undertake the decentralized procedure and be authorized simultaneously in several member states (Kashyap et al. 2013).

Focusing on nanoscience in particular, the USFDA implemented the 2013 Nanotechnology Regulatory Science Research Plan with the aim to lay out a framework and implementation plan to provide coordinated leadership and foster the addressing of key scientific gaps in knowledge, methods, or tools needed to make regulatory assessments of USFDA-regulated products that either contain nanomaterial or otherwise involve the application of nanotechnology. Led by the USFDA Nanotechnology Task Force formed in 2006, issues such as the development of measurement tools for the determination of physicochemical properties of engineered nanomaterials and the development of appropriate in vitro and in vivo assays and models to predict human responses are scheduled to be addressed in partnership with other government agencies, academia, and industry (USFDA 2013).

The EMA equally takes action to ensure that nanomedicines safely and timely enter the clinic, and the European Commission has developed several initiatives to stimulate research and facilitate commercialization of the technologies, including a consultation on nanotoxicology and nano-ecotoxicology and a round table promoted by the European Group of Ethics in Science and New Technologies (CHMP 2006). In 2009, the EMA's Committee for Medicinal Products for Human Use (CHMP) established an ad hoc expert group on nanomedicines that comprises selected experts from academia and the European regulatory network to provide specialist input and assist with the review of respective guidelines. Furthermore, the CHMP began to develop a series of four reflection papers on nanotechnology in 2011 to offer guidance to sponsors, covering the development of both new nanomedicines and nanosimilars (CHMP 2009; Ehmann et al. 2013).

7.3.6 CURRENT SITUATION AND FUTURE OF THE FIELD

In recent studies, the potential of nanomedicines to treat asthma could clearly be demonstrated. It was shown that the effects of steroids against airway inflammation are prolonged and enhanced when packaging them within nanoparticles (Matsuo et al. 2009), and that particles with smaller sizes induce bronchodilation more effectively (Usmani et al. 2005). The advantages of nanosizing respective substances were revealed by incorporating several drugs routinely used in asthma treatment in nanosystems. Salbutamol nanoparticles were shown to interact more intensively with the lung membrane due to augmented peripheral deposition and mucociliary movement to the tracheobronchial region, resulting in a higher and sustained drug concentration in the desired region (Bhavna et al. 2009; Da Silva et al. 2013). Budesonide showed preferable properties in nanocluster formulations for efficient drug delivery (Pornputtapitak et al. 2014) and was successfully manufactured in nanosuspensions by the microfluidizer method (Zhang and Zhang 2014), as well as in freeze-dried soy phophatidylcholine–cholesterol liposome vehicles, in combination with salbutamol (Elhissi et al. 2010), with all mentioned approaches aiming for pulmonary delivery. Solid lipid nanoparticles and nanostructured lipid carriers containing beclomethasone were produced by high-shear homogenization and efficiently nebulized as aerosols with a suitable particle size for deep lung delivery (Jaafar-Maalej et al. 2011). Furthermore, indomethacin was incorporated in nanostructured lipid carriers showing a controlled drug release (Castelli et al. 2005). Curcumin, an anti-inflammatory substance with favorable pharmacological effects but poor bioavailability and rapid metabolization, was formulated in solid lipid nanoparticles and tested in an OVA-induced allergic asthma model in rats. The obtained release profile showed an initial burst, followed by sustained release, which resulted in significantly higher plasma concentrations than those after applying curcumin alone. Besides that, the particles were able to effectively diminish AHR and inflammatory cell influx, as well as the expression of T helper 2–type cytokines in BALF (Gaur et al. 2012).

A different approach was developed by the 2006 founded biotechnology company Revalesio. Its lead product, RNS60, is based on charge-stabilized nanostructure technology, created by a patented device that produces stable nanostructures in an aqueous suspension by generating rotational forces, cavitation, and high-energy fluid dynamics. The obtained particles are smaller than 100 nm and consist of a stabilized oxygen bubble core. Other than traditional therapies, RNS60 does not target single cellular proteins, but aims to alter cellular inflammatory signaling in order to prevent cell and tissue damage by modulating the PI3K-Akt pathway, among others responsible for cellular protection from apoptosis and reduction of inflammation. In cell-based in vitro assays, it was shown to change the responsiveness of epithelial, smooth muscle, and immune cells to inflammatory stimuli that can trigger asthma attacks. Moreover, in vivo in an OVA challenge model in rats, it achieved a significant improvement in tidal volume, as well as downregulation of inflammatory cytokines and chemokines in blood and BALF. Two clinical safety studies of RNS60 administered via nebulizer have already been completed, showing no concerns regarding safety or tolerability in either healthy volunteers or asthma patients, but significant improvements in PEF and QoL in diseased individuals. A clinical trial studying the effects of RNS60 on the late-phase asthmatic response to allergen challenge was just completed, and final data collection is being conducted. A further study determining the impact on regional inflammation and allergen-induced bronchoconstriction is soon to be opened for participant recruitment (Dunnhaupt et al. 2015; Revalesio 2016).

Additionally, despite their promising potential, most of the new asthma therapies in preclinical investigations have failed to show convincing effects in clinical studies thus far. Besides the challenge that asthma is a very patient-specific disease, a crucial factor leading to this discrepancy is the absence of relevant models that are really able to accurately mimic the actual conditions in human asthmatic airway tissue.

One auspicious progression that could help reduce this lack is the recent development of a novel human airway musculature on a chip that is able to simulate the contraction of smooth muscles, and hence imitate bronchoconstriction and bronchodilation in vitro under asthmatic as well as healthy conditions. In this model, asthmatic inflammation was imitated by exposition of the muscle tissue on a thin elastomeric film to IL-13, a native protein that often is hypersecreted in asthmatic airways. When consecutively subjecting it to acetylcholine, the airway muscle responds with hypercontractility, just as it is observed in both asthmatic patients and animal tissue studies. The reverse reaction, a relaxation, can be achieved by using a muscarinic antagonist and a β agonist that are both utilized against constricted airways in the clinic. On the cellular level, the chip's reaction coincides with the *in vivo* conditions, too: the known phenomena of atrophying smooth muscle cells in the presence of IL-13 and the increased alignment of actin fibers in asthmatic airways were also observed *in vitro*. As a first example, HA1077, a RhoA-targeting inhibitor, was tested. It resulted in decreased basal tone, which prevented hypercontraction and improved relaxation, and therefore made the muscle tissue less sensitive to the asthma-triggering IL-13. Consequently, this tool potentially enables an innovative approach to evaluating the efficacy and safety of novel asthma treatments (Nesmith et al. 2014).

Another hurdle to overcome on the way to clinical translation is to find the ideal biocompatible nanocarrier with respective characteristics. Although several studies have been conducted on the safety of nano-based systems and materials, extensive biocompatibility investigations are rather scarce, yet it is essential to scrutinize all possible levels of toxicity in a responsible way. Equally important is the securing of *in vivo* stability concerning both the nanovector and the active compound. Biodegradable nanoparticles indeed can be applied for sustained-release effects; however, the exact drug release mechanisms still have to be clarified. Degradation rates of the polymer, as well as release profiles and bioavailability of the incorporated drug, demand comprising investigations. To successfully commercialize inhalable nanoparticles for asthma treatment, moreover, the development of suitable inhalation devices is of great importance. Apart from the appropriate device type and, if applicable, respective excipients and attached equipment, inhaler performance factors such as flow rate, administered volume, and dose reproducibility need to be addressed.

Concerning the lung as a target organ, the largest gap of knowledge is the uptake and clearance of particles in the cells, as the exact transport mechanisms across the pulmonary epithelium, especially under realistic disease-specific conditions, are still not understood in a satisfactory way (Paranjpe and Muller-Goymann 2014).

GLOSSARY

airway hyperresponsiveness (AHR): an exaggerated response of the airways to nonspecific stimuli that results in airway obstruction and can be assessed with bronchial challenge tests, for example, with methacholine or histamine.

area under the curve (AUC): a pharmacokinetic parameter that indicates the area under the curve in a plot of plasma drug concentration against time.

early asthmatic response (EAR): an allergic reaction that occurs shortly after allergen exposure and results in bronchoconstriction.

forced expiratory volume in 1 second (FEV$_1$): the volume of air that is forcefully exhaled within the first second.

forced vital capacity (FVC): the greatest volume of air that can be expelled when performing a rapid, forced exhalation; includes the FEV$_1$. An FEV$_1$/FVC ratio of <80% is an indicator for an obstructive defect.

hypothalamic–pituitary–adrenocortical (HPA) axis: a complex set of influences and feedback interactions among the hypothalamus, the pituitary gland, and the adrenal glands; an important part of the neuroendocrine system that controls several body processes, such as digestion, immune system, emotions, and energy storage.

late asthmatic response (LAR): after allergen exposure of 3–4 h, a second phase of decreased lung function, which may persist up to 24 h, occurs due to released cytokines and recruited leukocytes (e.g., eosinophils and Th2 cells) during EAR.

mass median aerodynamic diameter (MMAD): median of the distribution of airborne particle mass with respect to the aerodynamic diameter, which is based on the particles' inertial and gravitational motion in air.

maximal concentration (C$_{max}$): a pharmacokinetic parameter that indicates the maximum drug concentration in plasma.

MTT assay: colorimetric assay for assessing the metabolic activity and therefore viability of cells. NAD(P)H-dependent oxidoreductase enzymes, reflecting the number of viable cells, can reduce the tetrazolium dye MTT to insoluble purple-colored formazan.

PC$_{20}$: concentration of inhaled methacholine at 20% FEV$_1$ decrease.

peak expiratory flow (PEF): the maximum flow generated during expiration performed with maximal force and started after a full inspiration.

residual volume (RV): residual air; the volume of air still remaining in the lungs after the most forcible expiration possible.

vital capacity (VC): inspiratory capacity; the greatest volume of gas that can be expelled during a complete, slow, forced exhalation, following maximum inhalation.

REFERENCES

Abraham, W. M., A. Gill, A. Ahmed, M. W. Sielczak, I. T. Lauredo, Y. Botinnikova, K. C. Lin, B. Pepinsky, D. R. Leone, R. R. Lobb, and S. P. Adams. 2000. A small-molecule, tight-binding inhibitor of the integrin alpha(4)beta(1) blocks antigen-induced airway responses and inflammation in experimental asthma in sheep. *Am J Respir Crit Care Med* 162 (2 Pt 1):603–11.

Al-Qadi, S., A. Grenha, D. Carrión-Recio, B. Seijo, and C. Remuñán-López. 2012. Microencapsulated chitosan nanoparticles for pulmonary protein delivery: In vivo evaluation of insulin-loaded formulations. *J Control Release* 157 (3):383–90.

Anderson, P. J., J. D. Wilson, and F. C. Hiller. 1990. Respiratory tract deposition of ultrafine particles in subjects with obstructive or restrictive lung disease. *Chest* 97 (5):1115–20.

Arulmuthu, E. R., D. J. Williams, H. Baldascini, H. K. Versteeg, and M. Hoare. 2007. Studies on aerosol delivery of plasmid DNA using a mesh nebulizer. *Biotechnol Bioeng* 98 (5):939–55.

ATS (American Thoracic Society). 2015. Asthma Quality of Life Questionnaire (AQLQ). http://www.thoracic.org/members/assemblies/assemblies/srn/questionaires/aqlq.php (accessed December 5, 2015).

Barnes, N., I. Pavord, A. Chuchalin, J. Bell, M. Hunter, T. Lewis, D. Parker, M. Payton, L. Pearce Collins, and R. Pettipher. 2012. A randomized, double-blind, placebo-controlled study of the CRTH2 antagonist OC000459 in moderate persistent asthma. *Clin Exp Allergy* 42 (1):38–48.

Barnes, P. J. 2010. New therapies for asthma: Is there any progress? *Trends Pharmacol Sci* 31 (7):335–43.

BCC Research. 2012. Global market for asthma and COPD drugs. Wellesley, MA: BCC Research.

Beck-Broichsitter, M., M.-C. Knuedeler, T. Schmehl, and W. Seeger. 2013. Following the concentration of polymeric nanoparticles during nebulization. *Pharm Res* 30 (1):16–24.

Beck-Broichsitter, M., C. Schweiger, T. Schmehl, T. Gessler, W. Seeger, and T. Kissel. 2012. Characterization of novel spray-dried polymeric particles for controlled pulmonary drug delivery. *J Control Release* 158 (2):329–35.

Becker, A. B., J. Hershkovich, F. E. Simons, K. J. Simons, M. K. Lilley, and M. W. Kepron. 1989. Development of chronic airway hyperresponsiveness in ragweed-sensitized dogs. *J Appl Physiol* 66 (6):2691–7.

Belvisi, M. G., D. S. Bundschuh, M. Stoeck, S. Wicks, S. Underwood, C. H. Battram, E.-B. Haddad, S. E. Webber, and M. L. Foster. 2005. Preclinical profile of ciclesonide, a novel corticosteroid for the treatment of asthma. *J Pharmacol Exp Ther* 314 (2):568–74.

Beyerle, A., A. Braun, O. Merkel, F. Koch, T. Kissel, and T. Stoeger. 2011. Comparative in vivo study of poly (ethylene imine)/siRNA complexes for pulmonary delivery in mice. *J Control Release* 151 (1):51–6.

Bhaskar, S., F. Tian, T. Stoeger, W. Kreyling, J. M. de la Fuente, V. Grazu, P. Borm, G. Estrada, V. Ntziachristos, and D. Razansky. 2010. Multifunctional nanocarriers for diagnostics, drug delivery and targeted treatment across blood-brain barrier: Perspectives on tracking and neuroimaging. *Part Fibre Toxicol* 7:3.

Bhavna, F., J. Ahmad, G. Mittal, G. K. Jain, G. Malhotra, R. K. Khar, and A. Bhatnagar. 2009. Nano-salbutamol dry powder inhalation: A new approach for treating broncho-constrictive conditions. *Eur J Pharm Biopharm* 71 (2):282–91.

Birchall, J. C., I. W. Kellaway, and M. Gumbleton. 2000. Physical stability and in-vitro gene expression efficiency of nebulised lipid–peptide–DNA complexes. *Int J Pharm* 197 (1):221–31.

Bischof, R. J., K. Snibson, R. Shaw, and E. N. T. Meeusen. 2003. Induction of allergic inflammation in the lungs of sensitized sheep after local challenge with house dust mite. *Clin Exp Allergy* 33 (3):367–75.

Boswell-Smith, V., D. Spina, A. W. Oxford, M. B. Comer, E. A. Seeds, and C. P. Page. 2006. The pharmacology of two novel long-acting phosphodiesterase 3/4 inhibitors, RPL554[9,10-dimethoxy-2(2,4,6-trimethylphenylimino)-

3-(N-carbamoyl-2-aminoethyl)-3,4,6,7-tetrahydro-2H-pyrimido[6,1-a]isoquinolin-4-one] and RPL565 [6,7-dihydro-2-(2,6-diisopropylphenoxy)-9,10-dimethoxy-4H-pyrimido[6,1-a]isoquinolin-4-one]. *J Pharmacol Exp Ther* 318 (2):840–8.

Bur, M., A. Henning, S. Hein, M. Schneider, and C. M. Lehr. 2009. Inhalative nanomedicine—Opportunities and challenges. *Inhal Toxicol* 21 (Suppl 1):137–43.

Card, J. W., D. C. Zeldin, J. C. Bonner, and E. R. Nestmann. 2008. Pulmonary applications and toxicity of engineered nanoparticles. *Am J Physiol Lung Cell Mol Physiol* 295 (3):L400–11.

Castelli, F., C. Puglia, M. G. Sarpietro, L. Rizza, and F. Bonina. 2005. Characterization of indomethacin-loaded lipid nanoparticles by differential scanning calorimetry. *Int J Pharm* 304 (1–2):231–8.

Chalupa, D. C., P. E. Morrow, G. Oberdorster, M. J. Utell, and M. W. Frampton. 2004. Ultrafine particle deposition in subjects with asthma. *Environ Health Perspect* 112 (8):879–82.

Chen, X., W. Huang, B. C. Wong, L. Yin, Y. F. Wong, M. Xu, and Z. Yang. 2012. Liposomes prolong the therapeutic effect of anti-asthmatic medication via pulmonary delivery. *Int J Nanomed* 7:1139.

CHMP (Committee for Medicinal Products for Human Use). 2006. Reflection paper on nanotechnology-based medicinal products for human use. http://www.ema.europa.eu/docs/en_GB/document_library/Regulatory_and_procedural_guideline/2010/01/WC500069728.pdf (accessed December 3, 2015).

CHMP (Committee for Medicinal Products for Human Use). 2009. Guideline on the requirements for clinical documentation for orally inhaled products (OIP). http://www.ema.europa.eu/docs/en_GB/document_library/Scientific_guideline/2009/09/WC500003504.pdf (accessed December 5, 2015).

Cloutier, M. M., M. Schatz, M. Castro, N. Clark, H. W. Kelly, R. Mangione-Smith, J. Sheller, C. Sorkness, S. Stoloff, and P. Gergen. 2012. Asthma outcomes: Composite scores of asthma control. *J Allergy Clin Immunol* 129 (3 Suppl):S24–33.

Conti, D. S., D. Brewer, J. Grashik, S. Avasarala, and S. R. P. da Rocha. 2014. Poly (amidoamine) dendrimer nanocarriers and their aerosol formulations for siRNA delivery to the lung epithelium. *Mol Pharm* 11 (6):1808–22.

Daigle, C. C., D. C. Chalupa, F. R. Gibb, P. E. Morrow, G. Oberdorster, M. J. Utell, and M. W. Frampton. 2003. Ultrafine particle deposition in humans during rest and exercise. *Inhal Toxicol* 15 (6):539–52.

Dailey, L. A., N. Jekel, L. Fink, T. Gessler, T. Schmehl, M. Wittmar, T. Kissel, and W. Seeger. 2006. Investigation of the proinflammatory potential of biodegradable nanoparticle drug delivery systems in the lung. *Toxicol Appl Pharmacol* 215 (1):100–8.

Dailey, L. A., T. Schmehl, T. Gessler, M. Wittmar, F. Grimminger, W. Seeger, and T. Kissel. 2003. Nebulization of biodegradable nanoparticles: Impact of nebulizer technology and nanoparticle characteristics on aerosol features. *J Control Release* 86 (1):131–44.

Damgé, C., P. Maincent, and N. Ubrich. 2007. Oral delivery of insulin associated to polymeric nanoparticles in diabetic rats. *J Control Release* 117 (2):163–70.

Da Silva, A. L., R. S. Santos, D. G. Xisto, V. Alonso Sdel, M. M. Morales, and P. R. Rocco. 2013. Nanoparticle-based therapy for respiratory diseases. *An Acad Bras Cienc* 85 (1):137–46.

da Silva, A. L., S. V. Martini, S. C. Abreu, C. dos S. Samary, B. L. Diaz, S. Fernezlian, V. Karen de Sá, V. L. Capelozzi, N. J. Boylan, and R. G. Goya. 2014. DNA nanoparticle-mediated thymulin gene therapy prevents airway remodeling in experimental allergic asthma. *J Control Release* 180:125–33.

Davies, L. A., G. McLachlan, S. G. Sumner-Jones, D. Ferguson, A. Baker, P. Tennant, C. Gordon, C. Vrettou, E. Baker, and J. Zhu. 2008. Enhanced lung gene expression after aerosol delivery of concentrated pDNA/PEI complexes. *Mol Ther* 16 (7):1283–90.

Davis, M. S., S. McCulloch, T. Myers, and A. N. Freed. 2002. Eicosanoids modulate hyperpnea-induced late phase airway obstruction and hyperreactivity in dogs. *Respir Physiol* 129 (3):357–65.

DeYoung, M. B., L. MacConell, V. Sarin, M. Trautmann, and P. Herbert. 2011. Encapsulation of exenatide in poly-(D,L-lactide-co-glycolide) microspheres produced an investigational long-acting once-weekly formulation for type 2 diabetes. *Diabetes Technol Ther* 13 (11):1145–54.

Diamant, Z., M. C. Timmersa, H. van der Veena, B. S. Friedman, M. De Smet, M. Depré, D. Hilliard, E. H. Bel, and P. J. Sterk. 1995. The effect of MK-0591, a novel 5-lipoxygenase activating protein inhibitor, on leukotriene biosynthesis and allergen-induced airway responses in asthmatic subjects in vivo. *J Allergy Clin Immunol* 95 (1):42–51.

Dicke, T., I. Pali-Schöll, A. Kaufmann, S. Bauer, H. Renz, and H. Garn. 2012. Absence of unspecific innate immune cell activation by GATA-3-specific DNAzymes. *Nucleic Acid Ther* 22 (2):117–26.

Dunnhaupt, S., O. Kammona, C. Waldner, C. Kiparissides, and A. Bernkop-Schnurch. 2015. Nano-carrier systems: Strategies to overcome the mucus gel barrier. *Eur J Pharm Biopharm* 96:447–53.

EFPIA (European Federation of Pharmaceutical Industries and Associations). 2014. The right prevention and treatment for the right patient at the right time—Strategic Research Agenda for Innovative Medicines Initiative 2. Brussels: EFPIA.

Ehmann, F., K. Sakai-Kato, R. Duncan, D. Hernan Perez de la Ossa, R. Pita, J. M. Vidal, A. Kohli et al. 2013. Next-generation nanomedicines and nanosimilars: EU regulators' initiatives relating to the development and evaluation of nanomedicines. *Nanomedicine (Lond)* 8 (5):849–56.

Elhissi, A. M. A., M. A. Islam, B. Arafat, M. Taylor, and W. Ahmed. 2010. Development and characterisation of freeze-dried liposomes containing two anti-asthma drugs. *Micro Nano Lett* 5 (3):184–8.

Ensign, L. M., R. Cone, and J. Hanes. 2012. Oral drug delivery with polymeric nanoparticles: The gastrointestinal mucus barriers. *Adv Drug Deliv Rev* 64 (6):557–70.

European Medicines Agency. 2009. Withdrawal assessment report for Bosatria. London: European Medicines Agency.

Evans, R. L., A. T. Nials, R. G. Knowles, E. J. Kidd, W. R. Ford, and K. J. Broadley. 2012. A comparison of antiasthma drugs between acute and chronic ovalbumin-challenged guinea-pig models of asthma. *Pulm Pharmacol Ther* 25 (6):453–64.

Ewe, A., and A. Aigner. 2014. Nebulization of liposome–polyethylenimine complexes (lipopolyplexes) for DNA or siRNA delivery: Physicochemical properties and biological activity. *Eur J Lipid Sci Technol* 116 (9):1195–204.

Farkas, A., I. Balashazy, and K. Szocs. 2006. Characterization of regional and local deposition of inhaled aerosol drugs in the respiratory system by computational fluid and particle dynamics methods. *J Aerosol Med* 19 (3):329–43.

Flood-Page, P., C. Swenson, I. Faiferman, J. Matthews, M. Williams, L. Brannick, D. Robinson, S. Wenzel, W. Busse, and T. T. Hansel. 2007. A study to evaluate safety and efficacy of mepolizumab in patients with moderate persistent asthma. *Am J Respir Crit Care Med* 176 (11):1062–71.

Franciosi, L. G., Z. Diamant, K. H. Banner, R. Zuiker, N. Morelli, I. M. C. Kamerling, M. L. de Kam, J. Burggraaf, A. F. Cohen, and M. Cazzola. 2013. Efficacy and safety of RPL554, a dual PDE3 and PDE4 inhibitor, in healthy volunteers and in patients with asthma or chronic obstructive pulmonary disease: Findings from four clinical trials. *Lancet Respir Med* 1 (9):714–27.

Freed, A. N., S. McCulloch, and Y. Wang. 2000. Eicosanoid and muscarinic receptor blockade abolishes hyperventilation-induced bronchoconstriction. *J Appl Physiol* 89 (5):1949–55.

Fuhst, R., F. Runge, J. Buschmann, H. Ernst, C. Praechter, T. Hansen, J. von Erichsen, A. Turowska, H.-G. Hoymann, and M. Müller. 2013. Toxicity profile of the GATA-3-specific DNAzyme hgd40 after inhalation exposure. *Pulm Pharmacol Ther* 26 (2):281–9.

Gamble, J., M. Stevenson, E. McClean, and L. G. Heaney. 2009. The prevalence of nonadherence in difficult asthma. *Am J Respir Crit Care Med* 180 (9):817–22.

Gaur, S., L. Chen, T. Yen, Y. Wang, B. Zhou, M. Davis, and Y. Yen. 2012. Preclinical study of the cyclodextrin-polymer conjugate of camptothecin CRLX101 for the treatment of gastric cancer. *Nanomedicine* 8 (5):721–30.

Gautam, A., C. L. Densmore, B. Xu, and J. C. Waldrep. 2000. Enhanced gene expression in mouse lung after PEI–DNA aerosol delivery. *Mol Ther* 2 (1):63–70.

GINA (Global Initiative for Asthma). 2015. Global strategy for asthma management and prevention. Global Initiative for Asthma.

Global Asthma Network. 2014. The global asthma report. Auckland: Global Asthma Network.

Grenha, A., B. Seijo, and C. Remunán-López. 2005. Microencapsulated chitosan nanoparticles for lung protein delivery. *Eur J Pharm Sci* 25 (4):427–37.

Haldar, P., C. E. Brightling, B. Hargadon, S. Gupta, W. Monteiro, A. Sousa, R. P. Marshall, P. Bradding, R. H. Green, and A. J. Wardlaw. 2009. Mepolizumab and exacerbations of refractory eosinophilic asthma. *N Engl J Med* 360 (10):973–84.

Harrington, K. J., G. Rowlinson-Busza, K. N. Syrigos, P. S. Uster, R. M. Abra, and J. S. Stewart. 2000. Biodistribution and pharmacokinetics of 111In-DTPA-labelled pegylated liposomes in a human tumour xenograft model: Implications for novel targeting strategies. *Br J Cancer* 83 (2):232–8.

Hart, T. K., R. M. Cook, P. Zia-Amirhosseini, E. Minthorn, T. S. Sellers, B. E. Maleeff, S. Eustis, L. W. Schwartz, P. Tsui, and E. R. Appelbaum. 2001. Preclinical efficacy and safety of mepolizumab (SB-240563), a humanized monoclonal antibody to IL-5, in cynomolgus monkeys. *J Allergy Clin Immunol* 108 (2):250–7.

Hertel, S., T. Pohl, W. Friess, and G. Winter. 2014. That's cool! Nebulization of thermolabile proteins with a cooled vibrating mesh nebulizer. *Eur J Pharm Biopharm* 87 (2):357–65.

Hertel, S. P., G. Winter, and W. Friess. 2015. Protein stability in pulmonary drug delivery via nebulization. *Adv Drug Deliv Rev* 93:79–94.

Hirai, H., K. Tanaka, O. Yoshie, K. Ogawa, K. Kenmotsu, Y. Takamori, M. Ichimasa, K. Sugamura, M. Nakamura, and S. Takano. 2001. Prostaglandin D2 selectively induces chemotaxis in T helper type 2 cells, eosinophils, and basophils via seven-transmembrane receptor CRTH2. *J Exp Med* 193 (2):255–62.

Homburg, U., H. Renz, W. Timmer, J. M. Hohlfeld, F. Seitz, K. Lüer, A. Mayer, A. Wacker, O. Schmidt, and J. Kuhlmann. 2015. Safety and tolerability of a novel inhaled GATA3 mRNA targeting DNAzyme in patients with TH2-driven asthma. *J Allergy Clin Immunol* 136 (3):797.

House, S. L., K. Matsuda, G. O'Brien, M. Makhay, Y. Iwaki, I. Ferguson, L. M. Lovato, and L. M. Lewis. 2015. Efficacy of a new intravenous β 2-adrenergic agonist (bedoradrine, MN-221) for patients with an acute exacerbation of asthma. *Respi Med* 109 (10):1268–73.

Huang, F., E. Watson, C. Dempsey, and J. Suh. 2013. Real-time particle tracking for studying intracellular trafficking of pharmaceutical nanocarriers. *Methods Mol Biol* 991:211–23.

Huang, Y.-Y., and C.-H. Wang. 2006. Pulmonary delivery of insulin by liposomal carriers. *J Control Release* 113 (1):9–14.

Hutson, P. A., S. T. Holgate, and M. K. Church. 1988. The effect of cromolyn sodium and albuterol on early and late phase bronchoconstriction and airway leukocyte infiltration after allergen challenge of nonanesthetized guinea pigs. *Am Rev Respir Dis* 138 (5):1157–63.

Ibrahim, M., R. Verma, and L. Garcia-Contreras. 2015. Inhalation drug delivery devices: Technology update. *Med Devices (Auckland, NZ)* 8:131.

ICH (International Conference on Harmonisation of Technical Requirements for Registration of Pharmaceuticals for Human Use). 1993. Studies in support of special populations: Geriatrics. http://www.ema.europa.eu/docs/en_GB/document_library/Scientific_guideline/2009/09/WC500002875.pdf (accessed December 28, 2015).

ICH (International Conference on Harmonisation of Technical Requirements for Registration of Pharmaceuticals for Human Use). 2000. Clinical investigation of medicinal products in the pediatric population. http://www.ich.org/fileadmin/Public_Web_Site/ICH_Products/Guidelines/Efficacy/E11/Step4/E11_Guideline.pdf (accessed December 28, 2015).

Inchaurraga, L., N. Martin-Arbella, V. Zabaleta, G. Quincoces, I. Penuelas, and J. M. Irache. 2015. In vivo study of the mucus-permeating properties of PEG-coated nanoparticles following oral administration. *Eur J Pharm Biopharm* 97 (Pt A):280–9.

Jaafar-Maalej, C., V. Andrieu, A. Elaissari, and H. Fessi. 2011. Beclomethasone-loaded lipidic nanocarriers for pulmonary drug delivery: Preparation, characterization and in vitro drug release. *J Nanosci Nanotechnol* 11 (3):1841–51.

Jensen, D. K., L. B. Jensen, S. Koocheki, L. Bengtson, D. Cun, H. M. Nielsen, and C. Foged. 2012. Design of an inhalable dry powder formulation of DOTAP-modified PLGA nanoparticles loaded with siRNA. *J Control Release* 157 (1):141–8.

Kashyap, U. N., V. Gupta, and H. V. Raghunandan. 2013. Comparison of drug approval process in United States and Europe. *J Pharm Sci Res* 5 (6):131–6.

Krug, N., J. M. Hohlfeld, A.-M. Kirsten, O. Kornmann, K. M. Beeh, D. Kappeler, S. Korn, S. Ignatenko, W. Timmer, and C. Rogon. 2015. Allergen-induced asthmatic responses modified by a GATA3-specific DNAzyme. *N Engl J Med* 372:1987–95.

Kumar, M., X. Kong, A. K. Behera, G. R. Hellermann, R. F. Lockey, and S. S. Mohapatra. 2003. Chitosan IFN-γ-pDNA nanoparticle (CIN) therapy for allergic asthma. *Genet Vaccines Ther* 1 (1):1.

Leckie, M. J., A. ten Brinke, J. Khan, Z. Diamant, B. J. O'Connor, C. M. Walls, A. K. Mathur, H. C. Cowley, K. F. Chung, and R. Djukanovic. 2000. Effects of an interleukin-5 blocking monoclonal antibody on eosinophils, airway hyper-responsiveness, and the late asthmatic response. *Lancet* 356 (9248):2144–8.

Li, G., Z. Liu, N. Zhong, B. Liao, and Y. Xiong. 2006. Therapeutic effects of DNA vaccine on allergen-induced allergic airway inflammation in mouse model. *Cell Mol Immunol* 3 (5):379–84.

Lipworth, B. J. 2005. Phosphodiesterase-4 inhibitors for asthma and chronic obstructive pulmonary disease. *Lancet* 365 (9454):167–75.

Liu, S., K. Chihara, and K. Maeyama. 2005. The contribution of mast cells to the late-phase of allergic asthma in rats. *Inflamm Res* 54 (5):221–8.

Lu, D., and A. J. Hickey. 2005. Liposomal dry powders as aerosols for pulmonary delivery of proteins. *AAPS PharmSciTech* 6 (4):E641–8.

Lührmann, A., T. Tschernig, H. von der Leyen, M. Hecker, R. Pabst, and A. H. Wagner. 2010. Decoy oligodeoxynucleotide against STAT transcription factors decreases allergic inflammation in a rat asthma model. *Exp Lung Res* 36 (2):85–93.

Mansour, H. M., Y.-S. Rhee, and X. Wu. 2009. Nanomedicine in pulmonary delivery. *Int J Nanomedicine* 4:299–319.

Matsuda, K., M. Makhay, K. Johnson, and Y. Iwaki. 2012. Evaluation of bedoradrine sulfate (MN-221), a novel, highly selective beta2-adrenergic receptor agonist for the treatment of asthma via intravenous infusion. *J Asthma* 49 (10):1071–8.

Matsuo, Y., T. Ishihara, J. Ishizaki, K. Miyamoto, M. Higaki, and N. Yamashita. 2009. Effect of betamethasone phosphate loaded polymeric nanoparticles on a murine asthma model. *Cell Immunol* 260 (1):33–8.

Merkel, O. M., and T. Kissel. 2012. Nonviral pulmonary delivery of siRNA. *Acc Chem Res* 45 (7):961–70.

Merkel, O. M., I. Rubinstein, and T. Kissel. 2014. siRNA delivery to the lung: What's new? *Adv Drug Deliv Rev* 75:112–28.

Mills, N. L., N. Amin, S. D. Robinson, A. Anand, J. Davies, D. Patel, J. M. de la Fuente et al. 2006. Do inhaled carbon nanoparticles translocate directly into the circulation in humans? *Am J Respir Crit Care Med* 173 (4):426–31.

Misbah, S., M. H. Sturzenegger, M. Borte, R. S. Shapiro, R. L. Wasserman, M. Berger, and H. D. Ochs. 2009. Subcutaneous immunoglobulin: Opportunities and outlook. *Clin Exp Immunol* 158 (Suppl 1):51–9.

Moller, W., K. Felten, K. Sommerer, G. Scheuch, G. Meyer, P. Meyer, K. Haussinger, and W. G. Kreyling. 2008. Deposition, retention, and translocation of ultrafine particles from the central airways and lung periphery. *Am J Respir Crit Care Med* 177 (4):426–32.

Morimoto, Y., H. Izumi, Y. Yoshiura, T. Tomonaga, B. W. Lee, T. Okada, T. Oyabu et al. 2015. Comparison of pulmonary inflammatory responses following intratracheal instillation and inhalation of nanoparticles. *Nanotoxicology* 10 (5):607–18.

Nair, P., M. M. M. Pizzichini, M. Kjarsgaard, M. D. Inman, A. Efthimiadis, E. Pizzichini, F. E. Hargreave, and P. M. O'Byrne. 2009. Mepolizumab for prednisone-dependent asthma with sputum eosinophilia. *N Engl J Med* 360 (10):985–93.

Nesmith, A. P., A. Agarwal, M. L. McCain, and K. K. Parker. 2014. Human airway musculature on a chip: An in vitro model of allergic asthmatic bronchoconstriction and bronchodilation. *Lab Chip* 14 (20):3925–36.

Nielsen, E. J. B., J. M. Nielsen, D. Becker, A. Karlas, H. Prakash, S. Z. Glud, J. Merrison, F. Besenbacher, T. F. Meyer, and J. Kjems. 2010. Pulmonary gene silencing in transgenic EGFP mice using aerosolised chitosan/siRNA nanoparticles. *Pharm Res* 27 (12):2520–7.

Nishitsuji, M., M. Fujimura, Y. Oribe, and S. Nakao. 2008. Effect of montelukast in a guinea pig model of cough variant asthma. *Pulm Pharmacol Ther* 21 (1):142–5.

Oh, Y. J., J. Lee, J. Y. Seo, T. Rhim, S.-H. Kim, H. J. Yoon, and K. Y. Lee. 2011. Preparation of budesonide-loaded porous PLGA microparticles and their therapeutic efficacy in a murine asthma model. *J Control Release* 150 (1):56–62.

Ortega, H. G., M. C. Liu, I. D. Pavord, G. G. Brusselle, J. M. FitzGerald, A. Chetta, M. Humbert, L. E. Katz, O. N. Keene, and S. W. Yancey. 2014. Mepolizumab treatment in patients with severe eosinophilic asthma. *N Engl J Med* 371 (13):1198–207.

Oyarzun-Ampuero, F. A., J. Brea, M. I. Loza, D. Torres, and M. J. Alonso. 2009. Chitosan–hyaluronic acid nanoparticles loaded with heparin for the treatment of asthma. *Int J Pharm* 381 (2):122–9.

Palmans, E. L. S., J. C. Kips, and R. A. Pauwels. 2000. Prolonged allergen exposure induces structural airway changes in sensitized rats. *Am J Respir Crit Care Med* 161 (2):627–35.

Paranjpe, M., and C. C. Muller-Goymann. 2014. Nanoparticle-mediated pulmonary drug delivery: A review. *Int J Mol Sci* 15 (4):5852–73.

Park, J., S. H. Wrzesinski, E. Stern, M. Look, J. Criscione, R. Ragheb, S. M. Jay, S. L. Demento, A. Agawu, and P. L. Limon. 2012. Combination delivery of TGF-β inhibitor and IL-2 by nanoscale liposomal polymeric gels enhances tumour immunotherapy. *Nat Mater* 11 (10):895–905.

Patel, B., N. Gupta, and F. Ahsan. 2014. Low-molecular-weight heparin (LMWH)–loaded large porous PEG-PLGA particles for the treatment of asthma. *J Aerosol Med Pulm Drug Deliv* 27 (1):12–20.

Patton, J. S., J. D. Brain, L. A. Davies, J. Fiegel, M. Gumbleton, K. J. Kim, M. Sakagami, R. Vanbever, and C. Ehrhardt. 2010. The particle has landed—Characterizing the fate of inhaled pharmaceuticals. *J Aerosol Med Pulm Drug Deliv* 23 (Suppl 2):S71–87.

Pettipher, R., M. G. Hunter, C. M. Perkins, L. P. Collins, T. Lewis, M. Baillet, J. Steiner, J. Bell, and M. A. Payton. 2014. Heightened response of eosinophilic asthmatic patients to the CRTH2 antagonist OC000459. *Allergy* 69 (9):1223–32.

Pettipher, R., S. L. Vinall, L. Xue, G. Speight, E. R. Townsend, L. Gazi, C. J. Whelan, R. E. Armer, M. A. Payton, and M. G. Hunter. 2012. Pharmacologic profile of OC000459, a potent, selective, and orally active D prostanoid receptor 2 antagonist that inhibits mast cell-dependent activation of T helper 2 lymphocytes and eosinophils. *J Pharmacol Exp Ther* 340 (2):473–82.

Pettipher, R., and M. Whittaker. 2012. Update on the development of antagonists of chemoattractant receptor-homologous molecule expressed on Th2 cells (CRTH2). From lead optimization to clinical proof-of-concept in asthma and allergic rhinitis. *J Med Chem* 55 (7):2915–31.

Pietropaoli, A. P., M. W. Frampton, R. W. Hyde, P. E. Morrow, G. Oberdorster, C. Cox, D. M. Speers, L. M. Frasier, D. C. Chalupa, L. S. Huang, and M. J. Utell. 2004. Pulmonary function, diffusing capacity, and inflammation in healthy and asthmatic subjects exposed to ultrafine particles. *Inhal Toxicol* 16 (Suppl 1):59–72.

Pillai, G. 2014. Nanomedicines for cancer therapy: An update of FDA approved and those under various stages of development. *SOJ Pharm Pharm Sci* 1 (2):13.

Popescu, F.-D., and F. Popescu. 2007. A review of antisense therapeutic interventions for molecular biological targets in asthma. *Biologics* 1 (3):271.

Pornputtapitak, W., N. El-Gendy, J. Mermis, A. O'Brien-Ladner, and C. Berkland. 2014. Nanocluster budesonide formulations enable efficient drug delivery driven by mechanical ventilation. *Int J Pharm* 462 (1–2):19–28.

Revalesio. 2016. Our technology/programs-pipeline: Asthma. http://revalesio.com/about-our-technology/ (accessed January 1, 2015).

Rudolph, C., R. H. Müller, and J. Rosenecker. 2002. Jet nebulization of PEI/DNA polyplexes: Physical stability and in vitro gene delivery efficiency. *J Gene Med* 4 (1):66–74.

Salmon, M., S. L. Tannheimer, T. T. Gentzler, Z.-H. Cui, E. A. Sorensen, K. C. Hartsough, M. Kim, L. J. Purvis, E. G. Barrett, and J. D. McDonald. 2014. The in vivo efficacy and side effect pharmacology of GS-5759, a novel bifunctional phosphodiesterase 4 inhibitor and long-acting β2-adrenoceptor agonist in preclinical animal species. *Pharmacol Res Perspect* 2 (4).

Santoro, S. W., and Joyce, G. F. 1997. A general purpose RNA-cleaving DNA enzyme. *Proc Natl Head Sci USA* 94:4262–6.

Sel, S., M. Wegmann, T. Dicke, S. Sel, W. Henke, A. Ö. Yildirim, H. Renz, and H. Garn. 2008. Effective prevention and therapy of experimental allergic asthma using a GATA-3–specific DNAzyme. *J Allergy Clin Immunol* 121 (4):910–6.e5.

Singh, D., P. Cadden, M. Hunter, L. P. Collins, M. Perkins, R. Pettipher, E. Townsend, S. Vinall, and B. O'Connor. 2013. Inhibition of the asthmatic allergen challenge response by the CRTH2 antagonist OC000459. *Eur Respir J* 41 (1):46–52.

Singh, D., D. Richards, R. G. Knowles, S. Schwartz, A. Woodcock, S. Langley, and B. J. O'Connor. 2007. Selective inducible nitric oxide synthase inhibition has no effect on allergen challenge in asthma. *Am J Respir Crit Care Med* 176 (10):988–93.

Singh, P., M. Daniels, D. W. Winsett, J. Richards, D. Doerfler, G. Hatch, K. B. Adler, and M. I. Gilmour. 2003. Phenotypic comparison of allergic airway responses to house dust mite in three rat strains. *Am J Physiol* 284 (4):L588–98.

Snibson, K. J., R. J. Bischof, R. F. Slocombe, and E. N. Meeusen. 2005. Airway remodelling and inflammation in sheep lungs after chronic airway challenge with house dust mite. *Clin Exp Allergy* 35 (2):146–52.

Stevens, W. H., C. G. Lane, M. J. Woolley, R. Ellis, P. Tagari, C. Black, and A. Ford-Hutchinson. 1994. Effect of FLAP antagonist MK-0591 on leukotriene production and ozone-induced airway responses in dogs. *J Appl Physiol* 76 (4):1583–8.

Tang, B. C., M. Dawson, S. K. Lai, Y. Y. Wang, J. S. Suk, M. Yang, P. Zeitlin, M. P. Boyle, J. Fu, and J. Hanes. 2009. Biodegradable polymer nanoparticles that rapidly penetrate the human mucus barrier. *Proc Natl Acad Sci USA* 106 (46):19268–73.

Taylor, D. L., E. S. Woo, and K. A. Giuliano. 2001. Real-time molecular and cellular analysis: The new frontier of drug discovery. *Curr Opin Biotechnol* 12 (1):75–81.

Tsukamoto, N., N. Takahashi, H. Itoh, and I. Pouliquen. 2015. Pharmacokinetics and pharmacodynamics of mepolizumab, an anti-interleukin 5 monoclonal antibody, in healthy Japanese male subjects. *Clin Pharmacol Drug Dev* 5 (2):102–8.

Turowska, A., D. Librizzi, N. Baumgartl, J. Kuhlmann, T. Dicke, O. Merkel, U. Homburg, H. Höffken, H. Renz, and H. Garn. 2013. Biodistribution of the GATA-3-specific DNAzyme hgd40 after inhalative exposure in mice, rats and dogs. *Toxicol Appl Pharmacol* 272 (2):365–72.

Ungaro, F., I. d'Angelo, C. Coletta, R. d'Emmanuele di Villa Bianca, R. Sorrentino, B. Perfetto, M. A. Tufano, A. Miro, M. I. La Rotonda, and F. Quaglia. 2012. Dry powders based on PLGA nanoparticles for pulmonary delivery of antibiotics: Modulation of encapsulation efficiency, release rate and lung deposition pattern by hydrophilic polymers. *J Control Release* 157 (1):149–59.

USFDA (U.S. Food and Drug Administration). 2004. Innovation/stagnation: Challenge and opportunity on the critical path to new medical products. Silver Spring, MD: USFDA.

USFDA (U.S. Food and Drug Administration). 2013. 2013 nanotechnology regulatory science. http://www.fda.gov/ScienceResearch/SpecialTopics/Nanotechnology/ucm273325.htm (accessed December 3, 2015).

USFDA (U.S. Food and Drug Administration). 2015. New drug application. http://www.fda.gov/Drugs/DevelopmentApprovalProcess/HowDrugsareDevelopedandApproved/ApprovalApplications/NewDrugApplicationNDA/ (accessed December 3, 2015).

Usmani, O. S., M. F. Biddiscombe, and P. J. Barnes. 2005. Regional lung deposition and bronchodilator response as a function of beta2-agonist particle size. *Am J Respir Crit Care Med* 172 (12):1497–504.

Van Der Velden, J., G. Sum, D. Barker, E. Koumoundouros, G. Barcham, H. Wulff, N. Castle, P. Bradding, and K. Snibson. 2013. K Ca 3.1 channel-blockade attenuates airway pathophysiology in a sheep model of chronic asthma. *PLoS One* 8 (6):e66886.

Wang, Y., K. Kho, W. S. Cheow, and K. Hadinoto. 2012. A comparison between spray drying and spray freeze drying for dry powder inhaler formulation of drug-loaded lipid–polymer hybrid nanoparticles. *Int J Pharm* 424 (1):98–106.

Watt, A. P., I. I. Morrison, and D. C. Evans. 2000. Approaches to higher-throughput pharmacokinetics (HTPK) in drug discovery. *Drug Discov Today* 5 (1):17–24.

Webster, T. J. 2006. Nanomedicine: What's in a definition? *Int J Nanomed* 1 (2):115.

Weersink, E. J. M., D. S. Postma, R. Aalbers, and J. G. R. De Monchy. 1994. Early and late asthmatic reaction after allergen challenge. *Respir Med* 88 (2):103–14.

Wei-hong, T., G. Min-chang, X. Zhen, and S. Jie. 2014. Pharmacological and pharmacokinetic studies with vitamin D-loaded nanoemulsions in asthma model. *Inflammation* 37 (3):723–8.

Wiebert, P., A. Sanchez-Crespo, R. Falk, K. Philipson, A. Lundin, S. Larsson, W. Moller, W. G. Kreyling, and M. Svartengren. 2006. No significant translocation of inhaled 35-nm carbon particles to the circulation in humans. *Inhal Toxicol* 18 (10):741–7.

Woolley, M. J., J. Wattie, R. Ellis, C. G. Lane, W. H. Stevens, K. L. Woolley, M. Dahlback, and P. M. O'Byrne. 1994. Effect of an inhaled corticosteroid on airway eosinophils and allergen-induced airway hyperresponsiveness in dogs. *J Appl Physiol* 77 (3):1303–8.

Wulff, H., and N. A. Castle. 2010. Therapeutic potential of KCa3. 1 blockers: Recent advances and promising trends. *Expert Rev Clin Pharmacol* 3 (3):385–96.

Xie, Y., and O. M. Merkel. 2015. Pulmonary delivery of siRNA via polymeric vectors as therapies of asthma. *Arch Pharm* 348 (10):681–8.

Yang, W., J. I. Peters, and R. O. Williams 3rd. 2008. Inhaled nanoparticles—A current review. *Int J Pharm* 356 (1–2):239–47.

Young, N. A., M. S. Bruss, M. Gardner, W. L. Willis, X. Mo, G. R. Valiente, Y. Cao, Z. Liu, W. N. Jarjour, and L.-C. Wu. 2014. Oral administration of nano-emulsion curcumin in mice suppresses inflammatory-induced NFκB signaling and macrophage migration. *PLoS One* 9 (11):e111559.

Zhang, Y., and J. Zhang. 2014. Preparation of budesonide nanosuspensions for pulmonary delivery: Characterization, in vitro release and in vivo lung distribution studies. *Artif Cells Nanomed Biotechnol* 44:285–9.

Zia-Amirhosseini, P., E. Minthorn, L. J. Benincosa, T. K. Hart, C. S. Hottenstein, L. A. P. Tobia, and C. B. Davis. 1999. Pharmacokinetics and pharmacodynamics of SB-240563, a humanized monoclonal antibody directed to human interleukin-5, in monkeys. *J Pharmacol Exp Ther* 291 (3):1060–7.

PART THREE

Introduction

THE EMERGING ROLE OF INFLAMMATION
IN COMMON DISEASES

We are living in an exciting age of scientific progress in molecular and cellular biology. Although it is not a new concept that inflammation plays a role in common diseases such as neurodegenerative disease, cancer, and diabetes, the concept is just now becoming more widespread and being investigated in the larger context of disease pathology. The simple connection here is that these diseases are all a distribution of homeostasis, and as Ruslan Medzhitov so eloquently depicts in his commentaries on inflammation, inflammation is a tissue stress response: cells are at first in homeostasis and there is a cellular stress response, followed by a tissue stress response (a stage referred to as parainflammation) and, finally, an inflammatory response (Medzhitov 2010). The parainflammatory phase is mediated by macrophages and mast cells (Medzhitov 2010).

Macrophages have demonstrated a role in the pathology of diabetes, cancer, and neurodegenerative disease. Under conditions of obesity, macrophages actually infiltrate adipose cells and result in insulin resistance (Schenk et al. 2008). This recruitment is in response to hypoxia inducible factor-1 (HIF-1) activation (by local hypoxia and excess nutrients) (Schenk et al. 2008; Donath 2014). Pro-inflammatory islet macrophages also impair β-cell secretory ability through cytokine signaling (Donath 2014). Microglia are the tissue-specific macrophages of the central nervous system (CNS); M1 activated microglia are the pro-inflammatory phenotype that can result in chronic brain inflammation, neuronal damage, and dysfunction (Heneka et al. 2014). Neuroinflammation is a hallmark of neurodegenerative disease (Heneka et al. 2014). Dysfunction of the cerebrovascular and blood–brain barrier contribute to neurodegenerative disease, and pericyte dysfunction may

contribute to local ischemia and hypoxia in the brain (Nelson et al. 2016). Macrophages in cancer are actually so supportive of tumor growth and survival that the subset of enabling macrophages in the tumor microenvironment have been called tumor-associated macrophages (TAMs) (Engblom et al. 2016). TAMs promote tumor survival and metastasis in many ways, including remodeling of the extracellular matrix through secretion of matrix metalloproteinases and proliferative activation through the secretion of growth factors, such as epidermal growth factor (Engblom et al. 2016). Hypoxia also contributes to tumor survival as hypoxia (through perpetually destroyed and created vasculature), and cell stress can activate HIF and the target genes of HIF include many enabling proteins, such as drug efflux pumps that contribute to multidrug resistance and growth factor receptors that promote proliferation (Milane et al. 2011).

Is hypoxia the cell stress trigger for macrophage activation in all of these diseases? Is there an ultimate common cell stress trigger? Assuredly, the role of macrophage dysfunction is just one common thread in the chronic inflammatory component of these diseases. Mitochondrial dysfunction is central to cancer (apoptotic resistant) and neurodegeneration (apoptotic hypersensitization), and diabetes is characterized by decreased oxidative phosphorylation. Mitochondria are involved in pro-inflammatory signaling (López-Armada et al. 2013). Is mitochondrial dysfunction an overarching relationship between primary (Part 2) and secondary (Part 3) inflammatory disease? There is an autophagy–mitochondria–inflammation–cell death axis (Green et al. 2011). In this axis, decreased autophagy (and mitophagy) allows pro-inflammatory mitochondrial signaling and increased cell death; autophagy is required for macrophage clearance of apoptotic cells. This balance is clearly tipped in opposite directions for cancer and neurodegeneration. Is this axis the key to understanding the overarching role of inflammation in primary and secondary inflammatory disease? On a more basic note, how are the transformed fibroblasts in rheumatoid arthritis and other primary inflammatory diseases similar to cancer-associated fibroblasts? As we expand our understanding of the inflammatory components of these diseases, assuredly, new treatment strategies will emerge, including translational nanomedicine therapies. Although there are current Food and Drug Administration (FDA)–approved nanomedicines for the treatment of cancer, translation of nanomedicines for neurodegenerative disease and for diabetes is not yet at the level of market approval. Part 3 discusses Alzheimer's and neurodegenerative disease, cancer, and diabetes in the context of inflammation and translational nanomedicine.

REFERENCES

Donath, M. Y. 2014. Targeting Inflammation in the Treatment of Type 2 Diabetes: Time to Start. *Nature Reviews: Drug Discovery* 13 (6): 465–76.

Engblom, C., C. Pfirschke, and M. J. Pittet. 2016. The Role of Myeloid Cells in Cancer Therapies. *Nature Reviews: Cancer* 16 (7): 447–62.

Green, D. R., L. Galluzzi, and G. Kroemer. 2011. Mitochondria and the Autophagy–Inflammation–Cell Death Axis in Organismal Aging. *Science* 333 (6046): 1109–12.

Heneka, M. T., M. P. Kummer, and E. Latz. 2014. Innate Immune Activation in Neurodegenerative Disease. *Nature Reviews: Immunology* 14 (7): 463–77.

López-Armada, M. J., R. R. Riveiro-Naveira, C. Vaamonde-García, and M. N. Valcárcel-Ares. 2013. Mitochondrial Dysfunction and the Inflammatory Response. *Mitochondrion* 13 (2): 106–18.

Medzhitov, R. 2010. Inflammation 2010: New Adventures of an Old Flame. *Cell* 140 (6): 771–6.

Milane, L., Z. Duan, and M. Amiji. 2011. Role of Hypoxia and Glycolysis in the Development of Multi-Drug Resistance in Human Tumor Cells and the Establishment of an Orthotopic Multi-Drug Resistant Tumor Model in Nude Mice Using Hypoxic Pre-Conditioning. *Cancer Cell International* 11: 3.

Nelson, A. R., M. D. Sweeney, A. P. Sagare, and B. V. Zlokovic. 2016. Neurovascular Dysfunction and Neurodegeneration in Dementia and Alzheimer's Disease. *Biochimica et Biophysica Acta* 1862 (5): 887–900.

Schenk, S., M. Saberi, and J. M. Olefsky. 2008. Insulin Sensitivity: Modulation by Nutrients and Inflammation. *Journal of Clinical Investigation* 118 (9): 2992–3002.

CHAPTER EIGHT

Neurodegenerative Disease

Neha N. Parayath, Grishma Pawar, Charul Avachat,
Marcel Menon Miyake, Benjamin Bleier, and Mansoor M. Amiji

CONTENTS

8.1 INTRODUCTION

Neurodegenerative disorders are characterized by progressive loss of neuronal function and structure, subsequently leading to neuronal cell death. The incidences of neurodegenerative disease are predicted to reach more than 70 million in 2030 and 106 million in 2050 according to the demographics of dementia published by the World Health Organization (WHO) (Amor et al. 2014). The extensive rise in neurodegenerative disease cases is primarily attributed to the increase in the aging population, the population that is a result of medical advances that have prolonged the life span of individuals (Akagi et al. 2015). Thus, it becomes essential to develop effective treatments and preventive interventions to reduce the financial and emotional strain of these aging-related disorders. Neurodegenerative diseases are complex disorders since multiple events constitute the development of these diseases. The molecular events accountable for neurodegeneration include protein oligomerization and aggregation, oxidative stress, neuroinflammation, axonal transport deficits, neuron–glial interactions, DNA damage, aberrant RNA processing, calcium deregulation, and mitochondrial dysfunction (Forman et al. 2004). Major neurodegenerative diseases include Parkinson's disease; Alzheimer's disease; Huntington's disease, along with other forms, such as amyotrophic lateral sclerosis (ALS); Creutzfeldt–Jakob disease; corticobasal degeneration; frontotemporal dementia; Batten disease; and vascular dementia. Thus, extensive research is focused on understanding the pathogenesis of neurodegenerative disorders and developing effective treatment options.

Most of the current available therapies for neurodegenerative diseases are based on alleviating the symptoms without any disease-modifying effects. Furthermore, the high doses and multiple drugs necessary for chronic treatment may result in major side effects. The first part of this chapter discusses the pathophysiology of the major neurodegenerative diseases and the therapeutic strategies that are extensively researched for these diseases. The newer approaches, such as stem cell therapies, monoclonal antibodies (mAbs), and epigenetic molecules, expose an array of therapeutics that focus on having disease-modifying effects rather than just treating the symptoms. However, the major hurdle toward clinical translation of these therapeutic agents is their inability to traverse the central nervous system (CNS) barriers, which are deliberated in the later part of this chapter. These innate barriers, mainly the blood–brain barrier (BBB) and the blood–cerebrospinal fluid barrier (BCSFB), have an important role in protecting the CNS against toxic and infectious agents and maintaining an ionic environment. Conversely, these barriers also form an obstacle for effective systemic drug delivery to the CNS. Varied approaches for efficient CNS delivery that have been researched, broadly classified as invasive and noninvasive, are discussed in detail in the final part of this chapter.

8.2 PATHOPHYSIOLOGY OF NEURODEGENERATIVE DISORDERS

8.2.1 Alzheimer's Disease

Alzheimer's is a progressive disorder associated with memory and cognitive dysfunction. The pathology of the disease is characterized by senile plaques that are composed of β-amyloid (cleavage product of amyloid precursor protein [APP]) and neurofibrillary tangles (NFTs), which are associated with microtubule-associated protein tau (Lotharius and Brundin 2002). The functions of APP are not well identified; however, the metabolism of APP by a group of enzymes called secretases, which generate β-amyloid, has been studied extensively (Forman et al. 2004). There are three secretases characterized for the metabolism of APP. The enzyme α secretase cleaves APP in the middle of the APP to produce nonamyloidogenic fragments of APP. Conversely, sequential cleavage by β secretase and γ secretase produces an amyloidogenic fragment (Forman et al. 2004).

In cases of familial Alzheimer's disease (FAD), the mutations disrupt this proteolytic cleavage process, thereby generating longer and larger numbers of β-amyloid products (especially β-amyloid$_{42}$) or promoting β-amyloid fibrillization. In presenilin mutations, the C-terminal of γ secretase is altered, thereby generating greater numbers of β-amyloid species (Veugelen et al. 2016). Various animal models for Alzheimer's are generated based on these mutations (Table 8.1). Another mechanism that facilitates accumulation of β-amyloid species is mediated through the apolipoprotein E4 (apoE) haplotype. In transgenic mice, apoE modulated production of β-amyloid species, although this mechanism is not yet completely understood. However, this effect is associated with the ability of apoE to bind to β-amyloid species, thereby influencing its clearance and aggregation (Forman et al. 2004). Since apoE plays a role in cholesterol packaging and transport, this correlates the increased cholesterol levels with incidence of Alzheimer's disease (Puglielli et al. 2003). Furthermore, cholesterol is shown to be involved in APP processing and increased levels of cholesterol promote β-amyloid species accumulation. This has been supported by epidemiological studies that demonstrate a correlation between elevated cholesterol levels and Alzheimer's disease (Puglielli et al. 2003). Additionally, epidemiological studies also show lower incidences of Alzheimer's disease in individuals who are on nonsteroidal anti-inflammatory drugs. In vitro and in vivo studies in mice also demonstrate reduction of β-amyloid species upon treatment with nonsteroidal anti-inflammatory drugs (In'T Veld et al. 2001; Weggen et al. 2001). Thus, treatment strategies for Alzheimer's are targeted at prevention of

TABLE 8.1

Commonly used animal models for neurodegenerative diseases.

Disease	Model	Type	Advantages	Disadvantages	References
Parkinson's	6-Hydroxydopamine	Neurotoxin model	• Multisystemic lesions • Dorsoventral gradient of striatal denervation	• Rodents • No α-synuclein aggregation • No Lewy bodies	Dauer and Przedborski 2003; Bezard et al. 2013; Schneider et al. 2014
Parkinson's	1-Methyl-4-phenyl-1,2,3,6-tetrahydropyridine	Neurotoxin model	• Dorsoventral gradient of striatal denervation • Rodents and nonhuman primates • α-Synuclein aggregation	• Specific toxicity to dopaminergic neurons • No Lewy bodies	Dauer and Przedborski 2003; Bezard et al. 2013; Schneider et al. 2014
Parkinson's	α-Synuclein mutant	Transgenic model	• Lack of neuronal cell death • Presence of α-synuclein aggregation • Mimics early stages of Parkinson's disease	• Inconsistent transgene expression in substantia nigra based on promoter	Dauer and Przedborski 2003; Bezard et al. 2013; Schneider et al. 2014
Parkinson's	Leucine-rich repeat kinase-2 mutant	Transgenic model	• Impairment of striatal DA transmission • Abnormal phospho-tau levels • Mimics early stages of Parkinson's	• No neurodegeneration or α-synuclein accumulation	Dauer and Przedborski 2003; Bezard et al. 2013; Schneider et al. 2014
Parkinson's	Viral vector–based models (adeno-associated virus and lentivirus)	Transgenic model	• Ability to use wild-type animal • Unilateral transduction • Transgene expression in targeted cell population	• Variable degree of transgene expression based on surgical intervention	Dauer and Przedborski 2003; Bezard et al. 2013; Schneider et al. 2014
Alzheimer's	APP mutant	Transgenic model	• Aβ42 accumulation • Plaque deposition • Cognitive impairment	• May or may not show NFTs	LaFerla and Green 2012; Lee and Han 2013
Alzheimer's	β-Amyloid species	Transgenic model	• Early deposition of β-amyloid species at 3 months • Plaque deposition	• Do not show NFTs	LaFerla and Green 2012; Lee and Han 2013

(Continued)

TABLE 8.1 (CONTINUED)

Commonly used animal models for neurodegenerative diseases.

Disease	Model	Type	Advantages	Disadvantages	References
Alzheimer's	Presenilin mutant	Transgenic model	• β-Amyloid species accumulation	• No plaque deposition	LaFerla and Green 2012; Lee and Han 2013
Alzheimer's	APP + presenilin Mutant	Transgenic model	• β-Amyloid species accumulation • Plaque deposition	• Do not show NFTs	LaFerla and Green 2012; Lee and Han 2013
Alzheimer's	Tau mutant	Transgenic model	• NFTs	• Do not show β-amyloid species accumulation	LaFerla and Green 2012; Lee and Han 2013
Alzheimer's	APP + presenilin + tau mutant	Transgenic model	• β-Amyloid species accumulation • Plaque deposition • NFTs	—	LaFerla and Green 2012; Lee and Han 2013
Huntington's	QA	Neurotoxin model	• Applicable to rodents and large animals • Cognitive defects • Increase in GABAA receptor	• Nonprogressive cell death	Ramaswamy et al. 2007; Pouladi et al. 2013; Menalled and Brunner 2014
Huntington's	3-Nitropropionic acid	Metabolic model	• Impaired glucose metabolism in neuronal cells	• Nonprogressive cell death	Ramaswamy et al. 2007; Pouladi et al. 2013; Menalled and Brunner 2014
Huntington's	R6/2	Transgenic model	• Mimic juvenile Huntington's symptoms • Motor and cognitive symptoms	• Aggressive behavioral phenotype	Ramaswamy et al. 2007; Pouladi et al. 2013; Menalled and Brunner 2014
Huntington's	R6/1	Transgenic model	• Mild phenotype • Motor and cognitive symptoms	• Multiple copies of huntingtin causes additional pathology not associated with Huntington's	Ramaswamy et al. 2007; Pouladi et al. 2013; Menalled and Brunner 2014
Huntington's	YAC	Transgenic model	• Motor and cognitive symptoms	• Multiple copies of human htt causes additional pathology not associated with Huntington's	Ramaswamy et al. 2007; Pouladi et al. 2013
Huntington's	N171–82Q	Transgenic model	• Mimic adult Huntington's symptoms • Motor and cognitive symptoms	• Mutant expressed only on neurons and not glial cells	Ramaswamy et al. 2007; Pouladi et al. 2013; Menalled and Brunner 2014
Huntington's	Knock-in models	Transgenic model	• Mimic protein pathology of Huntington's	• Do not show cognitive symptoms • No reports of neuronal loss	Ramaswamy et al. 2007; Pouladi et al. 2013; Menalled and Brunner 2014

β-amyloid species production and accumulation or degradation of β-amyloid species, or a combination of both (Forman et al. 2004).

8.2.2 Parkinson's Disease

Parkinson's disease is the second most common neurodegenerative disorder, with a mean onset age of 60 (Dexter and Jenner 2013). Parkinson's is more prevalent in men than women, with a ratio from 1.1:1 to 3:1, which has been attributed to protective effects of estrogen in women (Schrag et al. 2000). Parkinson's is associated with impaired motor and nonmotor functions with symptoms such as bradykinesia, rigidity, and tremors (Jankovic 2008). The impaired motor functions are a result of neuronal loss in the substantia nigra pars compacta and the loss of striatal dopaminergic neurons. Nondopaminergic nuclei such as the locus coeruleus, reticular formation of the brain stem, raphe nucleus, dorsal motor nucleus of the vagus, basal nucleus of the Meynert, amygdala, and hippocampus are also known to be affected in Parkinson's disease (Dexter and Jenner 2013). The loss of all of these neurons is associated with the presence of filamentous Lewy bodies, which is a pathological characteristic of Parkinson's disease. However, Lewy bodies are associated with non-Parkinson's pathologies, such as dementia and a subset of Alzheimer's disease (Goedert 2001). Autosomal dominant Parkinson's disease involves point mutation in the SNCA gene, encoding synuclein (Polymeropoulos et al. 1997), while familial Parkinson's disease is associated with two different point mutations in the SNCA gene (Lotharius and Brundin 2002; Zarranz et al. 2004), as well as triplication of the SNCA gene (Singleton et al. 2003).

The synuclein protein is the principal component of the Lewy bodies (Polymeropoulos et al. 1997). The function of this presynaptic synuclein protein is not yet well understood; however, this protein undergoes biophysical conformational changes to form a β-pleated sheet structure, which facilitates polymerization of α-synuclein to form filamentous Lewy bodies (Goedert 2001). Braak et al. (2003) proposed staging of Parkinson's disease based on accumulation of synuclein in different parts of the brain. Stage I shows the deposition in the dorsal motor nucleus of the vagus. Stage II leads to the accumulation in the upward direction via the pons eventually reaching the midbrain (stage III). Stage IV represents accumulation of synuclein in the basal prosencaphalon and mesocortex, while stages V and VI are represented by accumulation in the neocortex. Even though this pattern of synuclein accumulation is not true for all Parkinson's cases, the spreading pathology of synuclein has been validated in Parkinson's disease. However, the exact mechanism of this propagation from one neuron to another is not yet completely understood. Other than the genetic mutation, environmental factors such as exposure of rotenone may trigger formation of Lewy bodies, as shown in rats treated with rotenone (Betarbet et al. 2000), and oxidative injury is shown to either promote Lewy body formation or stabilize the formed Lewy bodies (Ischiropoulos 2003).

Apart from the SNCA gene, mutations in other genes, such as PARK2, PARK7, UHCL1, PINK1, and MAPT, have been involved in the progression of Parkinson's disease. The PARK2 gene encodes a putative E3 ligase protein called parkin (Kitada et al. 1998). Autosomal recessive mutations in this gene are seen in juvenile and early-onset parkinsonism (Kitada et al. 1998). However, autopsy of these patients does not show the presence of Lewy body pathology (Giasson and Lee 2003), leaving it unclear if Parkinson's progresses through some other mechanisms in these patients. Furthermore, other proteins that are also involved in ubiquitination, such as ubiquitin carboxyl terminal hydrolase-1 (encoded by UHCL1) (Leroy et al. 1998) and PARK7 (encoding protein DJ-1), are linked to Parkinson's disease (Bonifati et al. 2003). Additionally, autosomal recessive mutations in PINK1 (encoding PTEN-induced kinase-1) are indicative of mitochondrial dysfunction in Parkinson's disease (Valente et al. 2004). Moreover, frontotemporal dementia with parkinsonism-17 is associated with mutation in the MAPT gene on chromosome 17, which encodes proteins involved in tau synthesis (Lee et al. 2001). Animal models mimicking these disease scenarios are generated through the action of neurotoxins or genetic mutations (Table 8.1).

8.2.3 Huntington's Disease

Huntington's disease is an autosomal dominant disorder associated with atrophy of the caudate nucleus and putamen in the striatum, as well as the cerebral cortex region, resulting in involuntary movements and dementia (Padowski et al. 2014). Huntington's disease is characterized by expansion of CAG repeat nucleotides of the Huntington gene, located on the short arm of chromosome 4, leading to abnormal expansion of the polyglutamine tract at the N-terminus of the Huntington protein (Rubinsztein 2003). The number of CAG repeats is correlated to the onset of the symptoms of Huntington's disease, with higher repeats associated with early onset of the disease (Duyao et al. 1993). The function of the Huntington protein is not well understood; however, this protein is normally located in the cytoplasm (Lipinski and Yuan 2004). Studies indicate that nuclear localization of the abnormal Huntington protein (with expanded polyglutamine tract) leads to toxicity, causing cell death (Lipinski and Yuan 2004). Table 8.1 describes various animal models developed to mimic Huntington's disease. Notably, expansion of the polyglutamine tract is associated with at least nine inherited neurodegenerative diseases, such as Huntington's disease, spinobulbar muscular atrophy, dentatorubro-pallidoluysian atrophy, and six forms of spinocerebellar ataxia (Gemayel et al. 2015).

8.3 THERAPEUTICS FOR NEURODEGENERATIVE DISEASES

8.3.1 Neurotrophic Factors

Neurotrophic factors (neurotrophins) are a class of growth factors that regulate growth and differentiation of neurons. More than 30 different neurotrophins, such as the nerve growth factor (NGF), brain-derived nerve growth factor (BDNF), and glial cell–derived neurotrophic factor (GDNF), have been identified (Bickel et al. 2001). Neurotrophins bind to tyrosine receptor kinase, in order to elicit a neuronal survival response (Chao 2003). Furthermore, these neurotrophins bind to a 75 kD neurotrophin receptor, to respond through the apoptotic pathway. Thus, the balance between neuronal survival and death depends on the level of neurotrophin expressed and its binding either to tyrosine kinase receptors or neurotrophin receptors (Chao 2003). In the case of the development and progression of neurodegenerative diseases, neurotrophins play a crucial factor.

In Parkinson's disease, both protein and mRNA levels of BDNF are decreased in dopaminergic neurons of the substantia nigra (Mogi et al. 1999).

Additionally, external administration of BDNF results in protection of cultured dopaminergic neurons from neurotoxins such as 6-OHDA or MPP (Emerich et al. 1994). Under *in vivo* conditions, administration of BDNF results in attenuation of the degenerative process of dopaminergic neurons both in the MPPT and in the 6-OHDA animal models of Parkinson's disease (Chen and Gage 1995; Smith et al. 1999; Conner et al. 2001; Mendez et al. 2005). Furthermore, administration of levodopa, which is a prodrug of BDNF (discussed in Section 8.4), shows increased levels of BDNF messenger RNA (mRNA) in the striatum of healthy mice (Bankiewicz et al. 2006) and increases BDNF expression in a rat model of Parkinson's disease (Hadaczek et al. 2010). MANF and CDNF belong to the neurotrophic protein family. MANF is a 20 kDa secreted human protein that was described in 2003 as a survival-promoting factor for embryonic midbrain dopaminergic neurons *in vitro* (Petrova et al. 2003). Striatal administration of CDNF shows both neuroprotective and neurorestorative properties for dopaminergic cells in the MPTP mice model and 6-OHDA rat model (Lindholm et al. 2007; Voutilainen et al. 2011; Bäck et al. 2013). The striatal injection of MANF protects and restores dopaminergic cells in the 6-OHDA rat model (Voutilainen et al. 2011). However, the mechanisms of action of the putative receptors of MANF and CDNF are unknown. Similarly, in Huntington's disease, BDNF transport from cortical to striatal neurons is deficient, contributing to selective loss of striatal neurons and voluntary muscle movements in Huntington's patients (Gauthier et al. 2004; Strand et al. 2007).

8.3.2 Monoclonal Antibodies for Therapeutics for Neurodegenerative Diseases

The development of antibodies for neurodegenerative diseases has been a controversial field due to the debate about the ability of antibodies to cross the BBB (discussed in Section 8.4). However, recent evidence in Alzheimer's therapeutics that showed reduced plaque following active immunization against β-amyloid species in APP transgenic mice (Schenk et al. 1999) has led to further investigations related to immunotherapy for neurodegenerative diseases. Furthermore, clinical trials have shown β-amyloid species immunotherapy leading to a decrease in plaque formation and increased levels of peripheral β-amyloid species in patients (Rinne et al. 2010; Farlow et al. 2012; Ostrowitzki et al. 2012). Additionally, unlike other small-molecule therapeutics, such as inhibitors of cathepsin D, which are associated with specificity-related risks (Koike et al. 2000; Follo et al. 2011), mAbs offer high specificity. Hitt et al. (2012) showed that complete inhibition of small-molecule therapeutics such as BACE1 (enzyme cleaving APP) leads to a deleterious effect. Thus, the specificity of targeting offered by the antibodies makes them extremely suitable as therapeutics.

The efficacy of antibodies against β-amyloid species in the Alzheimer's model led to two theories being proposed to explain the mechanism of action of the antibody. The first theory was based on the "peripheral sink" mechanism. This theory hypothesized that β-amyloid species targeted by the antibody would shift the equilibrium and pull β-amyloid species from the brain into the blood to maintain equilibrium, and is based on the assumption that antibodies are unable to cross the BBB, and that β-amyloid species maintain a passive equilibrium between the blood and the brain (DeMattos et al. 2002). However, this theory was negated by the fact that the peripheral sink effect was not true for other molecules, such as secretase inhibitors, which are unable to cross the BBB (Ghosh et al. 2012). Furthermore, the antibody against BACE1 did not show a proportional reduction of β-amyloid species in the brain and the peripheral organs, and conversely, the reduction of β-amyloid species was proportional to the anti-BACE1, which crosses the BBB (Atwal et al. 2011). Additionally, an increase in peripheral administration of anti-β-amyloid species antibodies leads to slower clearance of β-amyloid species–antibody complexes from the brain, further contradicting the hypothesis (Yamada et al. 2009). The second theory, "direct action," hypothesizes that the antibodies cross the BBB and facilitate the clearance of β-amyloid species by promoting microglial engulfment or clearance as a complex via CSF–interstitial fluid bulk flow (Yu and Watts 2013). This CSF–interstitial fluid bulk flow could explain the increase in peripheral β-amyloid species after peripheral antibody administration (DeMattos et al. 2002; Yamada et al. 2009). However, as the antibody employs the microglia for clearance, this also results in disruption of the BBB, causing edema (amyloid-related imaging abnormality edema), which was the major dose-limiting factor for treatment with bapineuzumab (Rinne et al. 2010; Sperling et al. 2011). This edema was confirmed to be an effect of bapineuzumab treatment, since bapineuzumab showed dose-related edema and is on the immunoglobulin (Ig) G1 backbone, which has maximum affinity for Fc gamma receptors for microglia activation (Yu and Watts 2013). Thus, different antibodies were designed to minimize the dose-limiting effect using various epitomes, such as polyglutamate N-terminally truncated β-amyloid species p53-42 and crenezumab (Adolfsson et al. 2012; DeMattos et al. 2012).

Furthermore, apart from antibodies against β-amyloid species, other antibodies against tau aggregates and α-synuclein for Parkinson's have been researched (Yu and Watts 2013). However, unlike β-amyloid species, these aggregates are intracellular, which further complicates the targeting process and could lead to neuronal toxicity (Yu and Watts 2013). Thus, further studies are needed to analyze the cellular processes responsible for antibodies to target intracellular aggregates for developing a therapeutic approach for these aggregates. However, the major issue of antibody-based therapeutics for neurodegenerative diseases is that there is a steady-state relationship between antibody levels in the brain and blood at multiple doses, which is approximately 1:1000, as reported previously (Felgenhauer 1974; Poduslo et al. 1994), for antibody against BACE1, making it necessary to deliver high doses such that a sufficient dose reaches the brain on crossing the BBB.

8.3.3 Stem Cell–Based Therapies

The ability of stem cells to differentiate into mature cells of a particular lineage can be harnessed as a therapeutic option for neurodegenerative diseases. A stem cell–based therapeutic approach could be used for either replacing the lost neurons and glial cells by transplantation of *in vitro* predifferentiated stem cells or implanting endogenous stem cells in the CNS to form new neurons and glial cells. Apart from replacing the neurons, stem cell–based therapies induce functional improvement by releasing therapeutic molecules against neurodegeneration. Even though some scientifically founded clinical trials using stem cells to treat neurodegenerative disorders are in process, a stem cell–based approach has not yet been proven beneficial for neurodegenerative conditions.

8.3.3.1 Stem Cell–Based Therapies for Parkinson's Disease

The human embryonic mesencephalic tissue is rich in postmitotic dopamine neuroblasts (Lindvall and Björklund 2004). These dopamine neurons from the tissue reinervate the denervated striatum and become functionally integrated, releasing dopamine (Lindvall and Björklund 2004). This has been shown to help symptomatic relief in patients. Intrastriatal transplantation of this mesencephalic tissue demonstrated that neuronal replacement could be an effective therapeutic strategy in Parkinson's patients (Lindvall and Björklund 2004). An important factor for the effectiveness of stem cell therapy relies on the ability of the transplanted dopaminergic neurons' stem cell to undergo maturation and exhibit properties of substantia nigra neurons (Isacson et al. 2003; Mendez et al. 2005). The dopaminergic neurons have been generated in vitro from stem cells derived from various sources, such as embryonic stem cells, therapeutically cloned embryonic stem cells, neural stem cells and progenitors of embryonic ventral mesencephalon, adult neural stem cells from the subventricular zone, bone marrow stem cells, and fibroblast-derived induced pluripotent cells (Lindvall et al. 2012). Human stem cell–derived dopaminergic precursors, which are required for clinical application, have been shown to survive and mature in animal models of Parkinson's and exert a functional effect (Sánchez-Pernaute et al. 2001; Dezawa et al. 2004; Roy et al. 2006; Cho et al. 2008). However, the ability of human stem cells to reinnervate the striatum and restore dopamine levels in vivo has not yet been proven (Lindvall and Kokaia 2010). Furthermore, for long-term functional restoration, combining dopamine cell replacement with neuroprotective therapy would be an effective strategy (Olanow et al. 2008). For this, genetically modified stem cells can be generated to secrete neurotropic factors such as GDNF or BDNF, which could help restore the neuronal population in the striatum and substantia nigra (Behrstock et al. 2006).

8.3.3.2 Stem Cell–Based Therapies for Alzheimer's Disease

The pathological features of Alzheimer's disease include loss of neurons in different parts, such as the basal forebrain cholinergic system, amygdala, hippocampus, and cortical tissue (Lindvall and Kokaia 2010). Thus, functional restoration using stem cell therapy in Alzheimer's disease is complex, as the stem cells would have to be predifferentiated in vitro into many types of neuroblasts for implantation in different brain areas (Lindvall and Kokaia 2010).

Furthermore, for long-term effectiveness, cholinergic neuron replacement would require intact host neurons that the new cholinergic neurons can act on, which are in most Alzheimer's cases damaged by then (Lindvall and Kokaia 2010). Another approach to generate an effective treatment strategy for Alzheimer's disease is to deliver factors modifying the course of Alzheimer's. Preclinical study in vivo demonstrated that a basal forebrain graft of fibroblasts producing NGF counteracts cholinergic neuronal cell death, stimulates cell function, and improves memory (Tuszynski et al. 2005). The basal forebrain implantation of encapsulated retinal pigment epithelial cells releasing NGF is presently under clinical trial.

8.3.3.3 Stem Cell Therapy for Huntington's Disease

Huntington's disease is characterized by loss of GABAergic neurons in the striatum, which affects GABA containing medium-sized spiny neurons and their function in the damaged circuit (Ferrante et al. 1985). Bosch et al. (2004) showed that retinoic acid and potassium chloride depolarization can help a homogenous population of functional GABAergic neurons from a neural stem cell line. Additionally, when transplanted in the quinolinic acid (QA) model, these cells survived and improved functional deficits in vivo (Bosch et al. 2004).

Transplantation of F3 human neural stem cells secreting BDNF showed both behavioral and anatomical recovery (Ryu et al. 2004), as well as an increase of neural growth factor production (Kordower et al. 1997) in the Huntington's disease model. Furthermore, the decreased striatal atrophy suggested that neural stem cell transplantation protects the host brain from further destruction (Lee et al. 2005). However, Lescaudron et al. (2003) showed that autologous adult bone marrow mesenchymal stem cell transplantation in the QA rat model resulted in an improved behavioral function, but only a few cells expressed the neural phenotype. This suggested that the release of the growth factors by the grafted cells helped the survival of host cells. Additionally, autologous mesenchymal stem cells carry the mutant huntingtin gene, which is not ideal for human clinical study (Kim et al. 2008).

8.3.4 Epigenetic Therapies in Neurodegenerative Diseases

Epigenetic processes such as DNA methylation and histone modifications impact the gene expression without changing the primary DNA sequence and play a major role in embryonic development (Martin-Subero 2011). In neurons, epigenetic processes are dynamically regulated by response to stimulus to regulate the expression of memory-related genes (Lattal and Wood 2013). Thus, it was expected that the onset and progression of neurodegenerative diseases would involve epigenetic changes (Coppedè 2013). Thus, studies have focused on the use of compounds that exert epigenetic properties, as a therapeutic approach for neurodegenerative diseases.

Alzheimer's disease in 95%–98% of cases is a late-onset (>65 years) sporadic condition, resulting from interactions between environmental, genetic, and stochastic factors superimposed on the decline of cognitive functions (Coppedè 2013). Epigenetic mechanisms form the basis of gene environment interactions and are thought to play a role in Alzheimer's disease (Migliore and Coppedè 2009). Studies have shown reduced levels of 5-mC and 5-hmC (Mastroeni et al. 2010; Chouliaras et al. 2013), as well as histone tail modifications, such as decreased levels of H3 acetylation (Zhang et al. 2012) and increased levels of HDAC6 (related to increased tau phosphorylation) and HDAC2 (required for learning and memory) in human postmortem Alzheimer's patient's brains (Ding et al. 2008; Gräff et al. 2012). The vitamin B group is known to play a role in one-carbon metabolism, which is the metabolic pathway necessary for S-adenosylmethionime (SAM) production. In Alzheimer's patients, there is an indication of dysfunctional one-carbon metabolism and reduced DNA methylation (Coppedè 2010). Furthermore, transgenic mice and human neuroblastoma cells, when maintained under vitamin B deficiency conditions, showed PSEN1 promoter demethylation, followed by increased presenilin 1, BACE1, and APP, as well as β-amyloid species deposition in mice brain. Conversely, supplementation with SAM maintained PSEN1 methylation levels and reduced the progression of Alzheimer's-like features, indicating the potential of these epigenetic drugs in Alzheimer's (Fuso et al. 2008, 2005, 2012). Furthermore, administration of sodium butyrate (HDAC inhibitor) for 4 weeks was able to reinstate learning and memory in mice with severe Alzheimer's pathology (Fischer et al. 2007). Sodium butyrate is also correlated with reduced tau phosphorylation and restoration of dendritic spine density in hippocampal neurons in a transgenic mouse model of Alzheimer's disease (Ricobaraza et al. 2012).

In other neurodegenerative diseases, such as Parkinson's and Huntington's, the role of epigenetic modifications has been studied as well. Reduced methylation levels of α-synuclein have been observed in several brain regions, especially the substantia nigra of sporadic Parkinson's patients (Jowaed et al. 2010; Matsumoto et al. 2010). In Huntington's disease, the mutant Huntington protein interacts with HAT proteins, resulting in altered histone acetylation (Steffan et al. 2000; Jiang et al. 2006). Studies show that treatment with HDAC inhibitors arrests the progression of neuronal degeneration in fly and mouse models of Huntington's disease (Gray 2010). The major concern with using epigenetic drugs as therapeutics for neurodegenerative diseases is the multitargeted, multicellular effects exerted by drugs, leading to the possibility of a wide range of side effects, and thus extensive research is mandatory before the epigenetic therapeutics reach the clinic.

8.4 CHALLENGES IN CNS DRUG DELIVERY

The reason that the majority of drug candidates are unable to reach the brain is the presence of the BBB and BCSFB. These physical barriers are important for maintaining the extracellular fluid environment for the neurons and the glial cells in the CNS (Begley 2004). They function by protecting the brain against varied infectious agents and bacteria (Chen and Liu 2012). The BBB and BCSFB pose a great challenge to developing effective therapies to treat CNS disorders. Many drugs, in spite of being effective at their targeted site of action, have failed to reach the market, as most of them could not be clinically developed due to inefficient delivery to the CNS (Begley 2004).

The BBB is composed of endothelial cells lining the blood microvessels surrounded by other cell types, such as astrocytes end feet, pericytes, and microglia. The endothelial cells in the BBB differ from the rest of the cells in the body due to the presence of very few fenestrations and reduced pinocytic vesicles that impede the vesicular transport across the BBB (Ballabh et al. 2004). The adjacent endothelial cells are connected to each other with tight junctions. These components, together with neurons, form the "neurovascular unit" (Abbott 2013). The intact BBB, with its low passive permeability, limited paracellular transport, and presence of various receptors and transporters, limits the entry of many drug molecules, and therefore is the main reason for many CNS diseases still being untreated. Hence, extensive efforts are being taken to overcome the BBB and deliver drugs effectively to the brain (Chen and Liu 2012).

The tight junctions are formed by the interaction of several transmembrane proteins, such as occludins and claudins, and the complexes between these transmembrane proteins and cytoplasmic accessory proteins, such as zonula occludins-1 and -2, cingulin, and AF-6. The interactions of these proteins block the aqueous pores and prevent the transport of polar solutes from the blood into the brain. The tight junctions are further strengthened by the interactions of astrocytes and pericytes with the brain endothelial cells. The endothelial cells are surrounded by an extracellular collagen matrix and the pericytes that are located in the basement membrane. The outer surface of this basement membrane is covered by astrocytes. The astrocytes and pericytes are responsible for the proliferation and differentiation of the endothelial cells and increase the formation of tight junction complexes between endothelial cells, strengthening the BBB (Mankowski et al. 1999). Furthermore, as the tight junctions prevent many polar solutes from entering the brain, the presence of transporter proteins on the endothelial cells allows the entry of nutrient molecules, such as glucose and other essential amino acids (Begley 2004; Chen and Liu 2012).

Small essential lipid-soluble molecules such as CO_2 and O_2 can pass through the BBB through a concentration gradient, and large lipid molecules such as transferrin and insulin are taken up by other processes, such as receptor-mediated transcytosis (Ballabh et al. 2004). Various transporters, such as GLUT1, LAT1, transferrin, insulin receptors, and lipoprotein receptors, are present on the BBB. Additional efflux transporters are also present on BBB, such as the ATP family of transporters, for example, P-glycoprotein (p-gp) and MDR1 transporters. The MDR1 is mainly responsible for the resistance developed by many anti-cancer agents. Hence, several drugs blocking these efflux transporters are being developed to reduce efflux across the barrier. Similarly, agents blocking p-gp have been developed to increase drug delivery to the brain (Chen and Liu 2012).

Furthermore, the diffusion of drug molecules across the BBB depends on the molecular weight of the drug, log P values (lipophilicity), hydrogen bond donors, and hydrogen bond acceptors. A low-molecular-weight drug, of about 500 kDa, shows a higher log P value (highly lipophilic). However, most of the drugs do not satisfy all these properties, and in such cases, the drugs could be modified chemically or biologically to increase the brain permeability (Salameh et al. 2015). Hence, it can be concluded that the BBB is the biggest hurdle in developing effective therapies for treating CNS diseases. All the efforts made in the past until now by researchers and industries have resulted in the development of various strategies for better permeation of drugs into the brain by either disrupting the BBB (invasive strategies), chemically or biologically modifying the existing drug candidates (noninvasive strategies), or bypassing the BBB by taking advantage of the olfactory pathways (intranasal administration).

Another barrier for delivery of drugs to the CNS apart from the BBB and BCSFB is the systemic distribution and clearance of the drugs administered in local systemic circulation (Barchet and Amiji 2009). Lipophilic drugs or prodrugs having lipophilic analogues after entering the systemic circulation can get distributed to other nontarget organs, such as the liver or spleen, and can get metabolized and eliminated. Hence, the total drug entering the brain turns out to be further reduced. This necessitates administration of higher doses systemically so that therapeutic concentrations of drug can reach the brain (Barchet and Amiji 2009).

8.5 STRATEGIES TO ENHANCE CNS DELIVERY

The CNS barriers maintain a tight regulation of substances reaching the CNS. Thus, various strategies are explored to enhance drug delivery to the CNS (Figure 8.1).

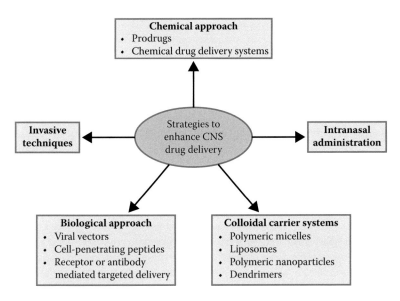

Figure 8.1. Strategies to enhance CNS delivery. These are broadly classified as invasive approaches, chemical-based approaches, biological approaches, colloidal carrier–based approaches, and intranasal administration.

8.5.1 Invasive Strategies for CNS Delivery

Invasive techniques to enable drug delivery to the CNS aim to physically disrupt the BBB and inject the drug directly into the CSF space or parenchyma of the brain. Lumbar puncture and intraventricular and intracerebral injections are the routes that have been most extensively used. Since the drug is administered directly in the CSF fluid, invasive approaches avoid the drug manipulation needed in the noninvasive approaches, thereby minimizing costs and regulatory processes.

Intrathecal drug delivery has shown efficacy in a variety of studies. The control of malignancy or neuropathic pain with an implantable intrathecal system has strong and moderate evidence for short- and long-term improvement, respectively (Boswell et al. 2007). Several orthopedic disorders can also be treated with analgesics administered through invasive techniques (Manchikanti et al. 2013). Other drugs that can be directly injected in the CSF and are currently in use are chemotherapies for CNS cancers and carcinomatosis (Siegal 1998; Calias et al. 2014; Hoang-Xuan et al. 2015), antibiotics for meningitis (Tunkel et al. 2004; Nau et al. 2010), and antispastics (Hasnat and Rice 2015).

However, several adverse effects are associated with implanting a catheter into the CNS. Catheter manipulation and removal are frequently required due to malposition and clogging (Gill et al. 2003). The direct trauma to the brain during implantation and injection of the drug can cause vasogenic edema around the catheter tip (Miyake and Bleier 2015a). Infections are also a problem, especially in patients that require a long-term implanted catheter. Besides that, even being administered directly in the CNS, drugs may not reach the targeted area with a sufficient pharmacological concentration due to the high turnover rate of the CSF, the efflux system, and the low coefficient of diffusion into the brain (Miyake and Bleier 2015b).

8.5.2 Noninvasive Strategies for CNS Delivery

To overcome the BBB and improve the drug delivery to the CNS, there are various noninvasive strategies. These mainly include chemical or biological approaches. These approaches have the advantage of not being invasive and prevent any kind of injury to the brain. These techniques take advantage of the various receptors and transporters present on the BBB to increase the therapeutic accumulation of drugs in the brain. This kind of delivery relies on chemical alterations to the currently used drugs or new drug candidates, such as prodrugs, lipophilic analogues, or carrier or receptor-mediated analogues. Intranasal is one of the recent noninvasive approaches that take advantage of the olfactory and trigeminal neuronal pathways.

8.5.2.1 Prodrug Approach

The chemical approach for CNS drug delivery mainly includes modification of the chemical structure of the drug to enhance its solubility and lipophilicity (permeability through the membrane) (Akagi et al. 2015). This in turn modifies the drugs' functionalities and improves their delivery. Such chemical modification to improve the deficient physiochemical properties of drugs is called a prodrug approach. This approach mainly involves adding particular analogues to a drug to make it more lipophilic or hydrophilic (Pathan et al. 2009).

Prodrugs are biologically inactive compounds that undergo a chemical transformation by enzymes through a metabolic process in vivo to release the active compound (Rautio et al. 2008). The prodrug approach has gained increased importance, as the biopharmaceutical and pharmacokinetic properties of drug molecules can be modified to make them more therapeutically active (Rautio et al. 2008). During the process of drug development, novel drug candidates are selected employing various techniques, such as high-throughput screening and combinatorial chemistry. In this process, the biopharmaceutical, pharmacokinetic, and physiochemical properties of drug molecules are neglected, and thus these novel drug candidates face problems in later stages of development (Rautio et al. 2008). Hence, under such conditions, if a prodrug approach is applied in the early stages of development, it can overcome the problems of novel drug candidates being rejected at later stages of drug development and clinical trials (Salameh et al. 2015).

However, the disadvantage of this prodrug approach is the lack of selectivity due to activation of nonspecific sites, which can lead to off-target toxicity. This can be overcome by a targeted prodrug approach in which the drug molecule can be covalently attached to a moiety that is specific for a particular transporter or a receptor present at that particular site of administration, in this case, the BBB (Salameh et al. 2015). In order to achieve this, a detailed knowledge of the transporter and receptor system, including the structure and functionalities of the receptor and transporter system, is required (Salameh et al. 2015).

Also, this approach of adding lipophilic analogues to polar drugs increases the volume of distribution due to excessive plasma protein binding. This affects all the pharmacokinetic parameters, thus affecting the bioavailability of the drug molecules. The other pharmacokinetic parameter affected is the oxidative metabolism, which tends to be higher for the lipophilic drugs. Due to an increased volume of distribution, the lipophilic drugs tend to accumulate in nontarget organs, which can lead to toxicity problems (Akagi et al. 2015). Apart from these shortcomings, this approach has proven to be a success, as increased lipophilicity by this approach has helped in delivering drugs across the BBB (Salameh et al. 2015).

Levodopa is an example of the prodrug approach. Parkinson's disease is characterized by deficiency of dopamine (Tolosa et al. 1998). Dopamine, being highly polar, is not able to permeate through the BBB, and hence the lipid-modified analogue of dopamine, levodopa, came into the picture. The development of levodopa and the application of the prodrug approach date to as early as the 1960s. After it was discovered, by 1960, that dopamine deficiency was responsible for Parkinson's, there was a need for a dopamine analogue or a dopamine agonist to treat Parkinson's (Tolosa et al. 1998). By 1961, levodopa was administered to patients, and by 1967, its effectiveness in treating Parkinson's was proven. However, side effects of levodopa, such as dyskinesias and motor fluctuations, came into the picture, and by 1970, this was resolved by administering a dopamine decarboxylase inhibitor, carbidopa. By 1975, the combination therapy of levodopa and carbidopa became available. Levodopa, which is a substrate for the neutral amino acid transporter LAT1 expressed at the BBB (Rautio et al. 2008) on permeation through the BBB, is converted into dopamine by dopamine decarboxylase. This approach enhances the drug delivery into the brain, and the active component dopamine, being polar, cannot leave the brain once it is inside, which increases the bioavailability and therapeutic concentrations of the drug in the brain.

Opioids have been used for a long time as potent analgesics. They are mainly prescribed for pain relief. Several clinical uses of opioids include pain relief after surgery, injury, or trauma and from diseases and cancer. Other uses include anesthesia and cough suppression (Hug 1992). The most commonly used opioid for all the uses mentioned above is morphine. It is an alkaloid naturally present in poppy plants. Morphine is soluble in aqueous solutions, but it cannot dissolve in lipids. Hence, it cannot permeate lipophilic barriers, such as the BBB, and thus cannot reach therapeutic concentrations in the brain after intravenous administration. Therefore, there are several morphine prodrugs that, on administration *in vivo*, get converted to active morphine (Salameh et al. 2015).

One such example is codeine, which undergoes oxidative demethylation *in vivo* to release the active parent compound morphine (Salameh et al. 2015). Also, two N-alkyl esters of morphine, morphine propionate and morphine ethanate, were synthesized by Wang et al. (2007) for efficient transdermal delivery of morphine. These esters had better permeability and an enhanced therapeutic effect than morphine due to the lipophilic analogue attached to them. Heroin is also an example of a morphine prodrug; it was synthesized as early as 1874 by adding two acetyl groups to the morphine, which makes it more lipophilic. This enhanced the BBB permeability, and thus improved the therapeutic effect. Gabapentin, an analogue of GABA, has better BBB permeability than GABA, another opioid (Salameh et al. 2015). Thus, the prodrug approach has been used for a long time and has been proven useful in administering drugs into the brain by overcoming one of the biggest barriers to CNS drug delivery, the BBB.

8.5.2.2 Chemical Drug Delivery System

A chemical drug delivery system is a modified prodrug approach that utilizes the enhanced lipophilicity, as well as sequential bioactivation steps, for conversion of the active drug to an intermediate that can be easily trapped inside the brain. The term was coined by Bodor and Farag (1983) and is used to distinguish this approach from a prodrug approach that requires only a single activation step (Rautio et al. 2008).

The chemical drug delivery system takes advantage of lipidization of the drug molecule and, in addition, exploits specific properties of the BBB to accumulate the drug in the brain so that the therapeutic effect can be enhanced. For example, one of the most studied lipophilic target moieties is dihydrotrigonelline. This lipophilic analogue attached to the drug molecule results in better penetration of the drug into the brain. Once in the brain, the

trigonelline is further oxidized to its cationic moiety, which helps in elimination of the hydrophilic intermediate and also locks it inside the brain, resulting in a slow and sustained release of the active parent drug in the brain. The target can be efficiently removed by active processes. Esterodex (chemical drug delivery system of estradiol) is one such example that is being investigated for its improved effects and is currently being investigated in clinical trials (Brewster et al. 1996).

Apart from these other targeted prodrugs utilizing the phosphonates, phosphates as ionic trapped intermediates are being investigated. One such example is testosterone. The target moiety attached to testosterone in this example is the (acyloxy)-alkyl phosphonate. The testosterone anionic chemical delivery system (T-aCDS) enters the brain readily due to its lipophilic nature. After entering the brain, the hydrolysis by esterases cleaves the CDS into a negatively charged intermediate, which now gets accumulated in the brain due to its polar nature and can release the drug slowly for a sustained effect. Additionally, being hydrophilic (polar), the intermediate is readily eliminated (Somogyi et al. 2002).

Other examples of prodrugs include carrier-mediated prodrugs. Such prodrugs are linked to amino acid conjugates and can be made specific to a particular transporter or receptor present on the BBB. One example is dopamine attached to a sugar moiety, which increases its uptake through the GLUT-1 transporter (Salameh et al. 2015). As can be seen, such chemical approaches of modifying drugs for enhanced permeability through the BBB and sustained drug release in the brain have contributed toward providing better alternatives for currently existing drug molecules that in their natural form prove to be ineffective.

8.5.2.3 Biological-Based Approach

The biological approach for enhancing CNS drug delivery mainly depends on the physiological and anatomical properties of the BBB. These include the various receptors (transferrin) and transporters present on the BBB (Pathan et al. 2009). This approach mainly includes conjugating the drug with an antibody or a ligand, such as sugars or lectins that can directly bind to the various receptors and transporters present on the BBB. Once such a conjugate reaches the BBB, the ligand attached to the drug (peptide or protein) can directly recognize the specific receptor on the endothelial cells, and this can trigger endocytosis of the drug conjugate (Pathan et al. 2009). This approach is also called a targeted drug delivery, as the ligands conjugated to the drug molecule directly target their specific receptors and not the other receptors present and enhance the drug delivery by receptor-mediated or absorption-mediated endocytosis. After endocytosis, the drug is then released in the interstitial space of the brain (Akagi et al. 2015). An enzymatic reaction can break down the bond between the drug and the ligand conjugate, and thus release the active drug molecule. In the case of the antibody-conjugated drug, they are targeted to the various antigens present on the targeted tissue. The antibodies that can recognize the extracellular epitopes of various BBB receptors that are present are employed. Nevertheless, it is challenging to find a targeted antigen for the BBB receptors (Akagi et al. 2015; Salameh et al. 2015).

8.5.2.3.1 VIRAL VECTORS

Viruses are one of the most widely used viral vectors for gene delivery. They have been used widely, as they have a high encapsulation, as well as transfection efficiency. However, the use of viral vectors has several shortcomings, such as eliciting immune responses, endogenous recombination and mutations leading to oncogenesis, and the inability to transfect certain types of cells

(Akagi et al. 2015). Examples of widely used viral vectors are herpes simplex virus type 1 (HSV-1) (100 nm particles), lentiviral vectors (100 nm particles), and adenoviral vectors (100 nm particles) (Gray et al. 2010; Akagi et al. 2015). A wide variety of proteins, peptides, and genes have been delivered using the viral vectors directly in the brain ventricles, which have been extensively discussed in the review by Akagi et al. (2015).

Apart from utilization of a single type of viral vector, a combination of two viral vectors can be used to enhance the drug delivery to the brain. Such vectors are also called hybrid vectors. The examples are HSV-AAV (adeno-associated virus), amplicon hybrid vectors and HSV-EBV (Epstein–Barr virus) hybrid vectors (Back et al. 2013). As shown in Table 8.2, varied viral vectors have been used to enhance the delivery of different proteins and nucleic acids for treating neurodegenerative disorders, and have been proven effective owing to their good transfection efficiencies. Thus, a particular viral vector–based drug delivery strategy should be selected based on the transfection agent and the desired outcome (Back et al. 2015).

8.5.2.3.2 CELL-PENETRATING PEPTIDES

Almost 20 years ago, in the late eighties, certain proteins were identified that could easily penetrate the cell membrane, and it was demonstrated that TAT peptide derived from human immunodeficiency

TABLE 8.2
Viral vectors for enhancing CNS drug delivery.

Viral vectors	Cargo	Applications	Reference
HSV-1	Fibroblast growth factor-2	HSV-1-based amplicon vectors to increase the delivery into the CNS for amplification and differentiation of embryonic stem cells to neurons	Vicario and Schimmang 2003
HSV-1	Interleukin-2 (IL-2) and IL-4	HSV-1 carrying IL-2-induced demyelination and optic neuropathy	Mott et al. 2013
Adeno-associated viral vector	Insulin-like growth factor-1 (IGF-1)	The viral vector encoding IGF-1 was administered to the cerebellar nucleus of a type III spinal muscular atrophy mouse model and showed antiapoptotic effects on motor neurons	Tsai et al. 2012
Adeno-associated viral vector	Striatal DOPA	Vector encoding DOPA delivery to dopaminergic neurons in the midbrain in a rodent model helped in restoring sensorimotor functions and preventing dyskinesias in Parkinson's	Björklund et al. 2010
Lentiviral vector 24	Dominant negative tumor necrosis factor	Direct delivery to the rat substantia niagra could attenuate dopaminergic neuron loss	McCoy et al. 2008

virus could be easily taken up by the cells from the surrounding media (Bechara and Sagan 2013). Along with TAT, another protein, known as Antp, derived from the third helix of the antennapedia transcription factor of *Drosophilia melanogaster*, is considered the first group of cell-penetrating peptides (CPPs) to be derived from natural proteins (Zou et al. 2013). The second group consists of chimeric molecules, such as the transportan derived from wasp venom mastoparan, and the third group includes synthetic peptides, of which polyarginines are extensively studied (Zou et al. 2013).

Most of these peptides are used to deliver drugs across the BBB. These peptides can be covalently bonded to a number of drug molecules, such as proteins, peptides, and oligonucleotides, and this can enhance the drug delivery to the targeted cell. These can also be conjugated to liposomes, and thus serve as an attractive approach for enhancing CNS drug delivery (Bechara and Sagan 2013). These peptides can enter almost any cell owing to their highly cationic nature. These peptides are made up of lysine and arginine amino acids. There are two mechanisms by which CPPs enhance the drug delivery, namely, direct translocation by interaction of positively charged groups of the CPPs with negatively charged groups on lipid membranes, and endocytosis of the cargo molecules by large vesicles formed at the cell membrane (Akagi et al. 2015).

Apart from CPP–protein conjugates, nucleic acid–CPP conjugates have also been developed. Transfecting cells with nucleic acids for therapeutic effects using viral vectors is difficult owing to low transfection efficiencies and difficulties in working with the viral vectors (Zou et al. 2013). The viral vectors also lead to immune responses *in vivo*, which makes them undesirable. The nonviral vectors are the next choice; however, the transfection efficiencies of these are very low. Hence, nucleic acid–CPP conjugates have been developed to overcome all these problems (Zou et al. 2013). The nucleic acids can be conjugated to the CPPs through a labile linker, such as a disulfide linker, or due to strong interaction between positively charged CPP and negatively charged nucleic acid (Zou et al. 2013).

Many SiRNAs and nucleic acids have been tested for delivery efficiency after conjugation with CPP, and there has been an increase in the transfection efficiencies. Most of these CPP–nucleic acid conjugates can be delivered using a colloidal drug delivery system to enhance the drug delivery (Zou et al. 2013). Another study by Kanazawa et al. (2013) showed that the MPEG-PCL-Tat micelles (PED-PCL copolymers conjugated with CPP) could deliver the encapsulated SiRNA better than intravenous delivery without using the micelles. Furthermore, CPP conjugates have also been used to enhance the delivery of small molecules, such as chemotherapeutic drugs (Zou et al. 2013).

8.5.2.3.3 RECEPTOR OR ANTIBODY-MEDIATED TARGETED DELIVERY

As explained in the previous sections, the BBB has several receptors, such as the transferrin receptors, lipoprotein receptors, and insulin receptors, expressed on the endothelial cells, along with various transporters and efflux pumps, such as the glucose transporter and p-gp efflux pumps (Chen and Liu 2012). A delivery strategy of conjugating drug molecules, such as protein, peptides, and small molecules to the ligands of these specific receptors, can help in directing the drug cargo to the particular receptor and can enhance the uptake into the brain via receptor-mediated or absorption-mediated endocytosis (Pathan et al. 2009).

Also, certain antibodies to the various receptors present can be conjugated to the drug cargo for targeted therapy and better uptake into the brain. The drug conjugated to the antibody is directed toward the antigen present on the target tissue. For example, OX26, 8D3 mAb, and CD71 are all antibodies to transferrin receptors expressed on the BBB. Drugs conjugated to these receptors can undergo endocytosis easily and can be taken up in the BBB (Pathan et al. 2009). Several growth factors, such as BDNF, GDNF, FGF-2, and NGF, can be successfully delivered to the CNS by conjugating them to mAbs to specific receptors (Pathan et al. 2009). A study conducted by Pardridge (2003) utilized BDNF chimeric peptide, which was formed by conjugating BDNF to a mAb for the transferrin receptor, and resulted in better neuroprotective effects in rat brain. Another study demonstrated that the conjugate of BDNF-PEG-biotin and OX26–streptavidin had better neuroprotective effects and increased hippocampal CA1 neuronal density than unconjugated BDNF (Wu and Pardridge 1999).

8.5.2.3.4 NANOMEDICINE AND COLLOIDAL DRUG DELIVERY SYSTEMS

Colloidal drug delivery systems are particulate or vesicular delivery systems dispersed uniformly, usually in the size range of 100–1000 nm (Akagi et al. 2015). The colloidal drug systems mainly include nanoparticle delivery systems such as liposomes, nanoemulsion, micelles, and dendrimers. With most of these being lipophilic, they can be easily transported across the BBB via absorption or receptor-mediated endocytosis. The factors that affect the efficacy of these systems are the particle size, charge on the surface, encapsulation efficiency, and circulation time. These systems can be targeted by binding them to specific ligands that are targeted to the specific receptors present on the endothelial cells on BBB. This enhances the targeted delivery of the required drug molecule (Akagi et al. 2015).

The main advantages of using these systems are enhanced permeability due to their lipophilic nature, increased protection of the encapsulated drug molecule from enzymatic cleavage, improved bioavailability due to enhanced permeation through biological membranes, and the ease of modifying them to target them to a particular site of interest by attaching certain ligands to their functional groups. One disadvantage of these systems is that after entering the blood circulation following intravenous administration, the particles get coated with plasma components called opsonins (Pathan et al. 2009). These opsonins are recognized by macrophages of the liver and spleen, as the macrophages have specific receptors to recognize them. The particles are thus removed from the circulation, which decreases the therapeutic effect. However, PEGylation increases the hydrophilicity of the nanoparticles and prevents this process and increases the blood circulation time (Pathan et al. 2009).

The colloidal nanoparticle systems can also be made specific to a target by attaching ligands to the polyethylene glycol (PEG) molecules so that the PEG does not mask the ligands. This decreases nontarget toxicity. Also, cell-penetrating peptides such as TAT can be covalently linked to these systems to enhance their permeation through BBB (Pathan et al. 2009). Table 8.3 explicitly describes the various colloidal-based nanocarriers that have been currently explored.

8.5.2.3.4.1 MICELLAR FORMULATIONS Polymeric micelles are one of the colloidal carrier systems being investigated for CNS drug delivery. The polymeric micelles have a hydrophobic polymer core and a hydrophilic outer shell made of polymer blocks such as PEG, resulting in kinetic stability in aqueous media (Akagi et al. 2015). Different polymers can be used to enhance the size distribution, stability, pharmacokinetic parameters, tissue distribution, and blood circulation time, as well as inhibit the p-gp efflux pumps and enhance CNS drug delivery (Pathan et al. 2009).

8.5.2.3.4.2 LIPOSOMAL FORMULATIONS Liposomes are vesicular particle systems in the size range of 150–250 nm made of a hydrophilic core and have an outer shell of single or multiple lipid bilayers. The hydrophilic drugs can be encapsulated in the hydrophilic core, and the hydrophobic drugs can be solubilized in the hydrophobic outer shell (Akagi et al. 2015). The liposomes in blood circulation can undergo elimination by the reticuloendothelial system due to their recognition by macrophages through opsonization (Akagi et al. 2015).

Liposomes can also be used to encapsulate oligonucleotides such as plasmid DNA and SiRNA. The use of cationic lipids is beneficial for encapsulating nucleic acids since they can help in forming an electrostatic interaction between the negatively charged nucleic acids (MacLachlan 2007). The liposomes can also be targeted to a particular receptor or transporter present on the BBB by incorporating the PEG molecules on the surface of liposomes for conjugation to various ligands and antibodies. Additionally, peptides such as TAT can be attached to the liposomes to increase penetration through the BBB (Akagi et al. 2015).

8.5.2.3.4.3 POLYMERIC NANOPARTICLES Polymeric nanoparticles are made of biodegradable polymers that release the drug from their core slowly by diffusion of the drug from the core as the polymer degrades. Biodegradable polymers are widely used for synthesis of polymeric nanoparticles since they do not induce any kind of inflammatory response and degrade in biological fluids to form lactic acid and glycolic acid, which are eventually converted to carbon dioxide and water, which is used in Krebs cycle (Akagi et al. 2015). The drugs encapsulated in the matric core of these nanoparticles can be released over a longer period of time, giving a sustained release of the drugs (Akagi et al. 2015). However, even though these particles can permeate through

TABLE 8.3

Nanocarrier systems for enhancing CNS drug delivery.

Nanocarrier system	Therapeutic agent	Disorder	Efficacy	Reference
Polymeric micelles	Flurbiprofen-encapsulated FB4 aptamer functionalized polyethylene glycol–polylactic acid (PEG-PLA) micelles	Alzheimer's disease	The micelles significantly enhanced the intracellular uptake by 1.41 times when compared with only drug in phosphate-buffered saline (PBS) and unmodified micelles loaded with the drug.	Mu et al. 2013
Liposomes	NGF-encapsulated lactoferrin-modified liposomes	Alzheimer's disease	The formulations were tested on neuron-like SK-N-MC cells with deposited Aβ peptide. The lactoferrin-modified NGF-encapsulated-only liposomes showed better cell viability than only NGF and NGF liposomes.	Kuo and Wang 2014
Liposomes	Dopamine-encapsulated OX26 mAb (antitransferrin receptor mAb)-modified liposomes	Parkinson's disease	The antibody-conjugated liposomes showed 8-fold uptake in Parkinson's disease model (transection of medial forebrain bundle) rat brains compared with only dopamine and about 3-fold uptake compared with only PEGylated liposomes.	Kang et al. 2016
Liposomes	Levodopa-encapsulated chlorotoxin-modified liposomes	Parkinson's disease	In the MPTP rat model of Parkinson's disease, the rotarod test showed that levodopa-encapsulated chlorotoxin-modified liposomes showed an increase in the time on the rod compared with only levodopa and unmodified liposomes.	Xiang et al. 2012
Polymeric nanoparticles	PLA-polysorbate 80 (PS80) nanoparticles loaded with resveratrol	Parkinson's disease	PLA nanoparticles significantly reduced the rate of lipid peroxidation and tyrosine hydroxylase reduction induced by MPTP in the MPTP rat model of Parkinson's.	da Rocha Lindner et al. 2015
Polymeric nanoparticles	Poly-(D,L)-lactide-co-glycolide (PLGA) nanoparticles with bacoside A (neuroprotective drug) coated with PS80	Alzheimer's disease	In male Wistar rats, the brain targeting ratio for nanoparticles was significantly higher than that of drug in solution.	Jose et al. 2014

(Continued)

Nanocarrier system	Therapeutic agent	Disorder	Efficacy	Reference
Polymeric nanoparticles	Polyethlene glycol–polyethyleneimine nanoparticles loaded with α-synuclein siRNA (siSNCA)	Parkinson's disease	Nanoparticles showed increased transfection in PC12 cells compared with naked siRNA. The nanoparticles also suppressed the SNCA gene expression in the cells and prevented the cells from apoptosis.	Liu et al. 2014
Solid lipid nanoparticle	Solid lipid nanoparticle composed of compritol as lipid and Tween 80 as surfactant was loaded with drug quercitin	Alzheimer's disease	Quercetin nanoparticles showed significant improvement in the elevated plus maze test and spatial navigation test compared with only quercetin in the Alzheimer's model (aluminum chloride).	Dhawan et al. 2011
Solid lipid nanoparticles	Solid lipid nanoparticles loaded with thymoquinone	Huntington's disease (HD)	Nanoparticles caused significant improvements in movement and memory and reduction in oxidative stress markers when compared with thymoquinone in suspension only in a 3-nitropropionic acid model of HD.	Surekha and Sumathi 2016
Dendrimers	Pyridylphenylene cationic dendrimers	Prion disease	Cationic dendrimers could destroy the ovine prion aggregates by complexing with them and not allowing them to form aggregates. An advantage of this dendrimer was the constant positive charge at any pH, which could help to interact with proteins with different isoelectric points.	Sorokina et al. 2016
Dendrimers	Neutral maltose-modified poly(propyleneimine) dendrimers (generation 5) mPPIg5	Prion disease	Dendrimer could efficiently remove the PrP^{SC} from an *in vitro* model of prion disease.	McCarthy et al. 2013b

BBB, they cannot efficiently be transported across the BBB that reduces the drug delivery to the CNS (Akagi et al. 2015). Hence, the efficiency can be improved by conjugating these nanoparticles to a peptide, ligand, or antibody specific to a particular receptor or transporter present on the endothelial cells of the BBB that can enhance its permeation through BBB and enhance the release of the drug in the CNS. Furthermore, surface modification, such as PEGylation, can enhance the circulation time and in turn increase drug delivery to the CNS.

These polymers can also adsorb apoE and apoB, and thus mimic the low-density lipoproteins, and can enhance the uptake of drugs in the brain by receptor-mediated endocytosis by the LDL receptor present on the endothelial cells (ECs). Transport across the ECs can be reduced due to the accumulation of the drugs as large aggregates in the ECs (Pathan et al. 2009). This is addressed by using the shuttle-mediated transport strategy. This is a new approach, as not much attention has been paid to that happens after the drug conjugates are taken up by endocytosis. The drugs might just get stuck in the ECs. Hence, attention to such minute details could help improve drug delivery to the CNS (Akagi et al. 2015).

8.5.2.3.4.4 DENDRIMERS Prion disease, also called transmissible spongiform encephalopathies (TSEs), is a rare type of neurodegenerative disorder that is caused by agents called prions. The prions are the infectious agents that can fold in multiple ways and can cause normal proteins to fold in an abnormal manner. The main component of these prion infectious agents is the PrPsc, which is the abnormally folded isoform of PrPC (McCarthy et al. 2013a). These abnormally folded proteins are called prions, and they lead to brain damage, further causing neurodegeneration. Disruption of these prions is one of the ways to treat this disease. It has been challenging to treat this disease, as it is difficult to identify compounds that can affect the prions (McCarthy et al. 2013a; Sorokina et al. 2016).

Dendrimers are the synthetic three-dimensional molecules that have developed to be promising candidates for treating prion disease. The dendrimers have a central core from which monomers branch out, and these can be easily modified by attaching ligands to them. The generations of dendrimers can be determined based on the branching of the functional groups on the surface, determining the charge. The antiprion activity of dendrimers is mainly dependent on the generation number and is time dependent. Different examples of dendrimers that have been developed include phosphorous dendrimers, poly(propyleneimine) dendrimers, and maltose-based dendrimers. Until now, these are the only ones found to be effective to treat prion disease (McCarthy et al. 2013a). The only limitation of these systems is the way they release the encapsulated drug and the short-term release of the drug. Dendrimers have a tendency to release the drug even before the nanoparticles reach the target site of action. Table 8.3 describes

examples of colloidal carriers for treating neurodegenerative diseases.

8.5.2.4 Intranasal Delivery

Intranasal drug delivery is a technique that takes advantage of the connections between nasal mucosa and the brain, and thus bypasses the BBB. The intranasal method of administering drugs to the CNS for neurodegeneration was invented by William H. Frey (William 1997). However, Faber showed for the first time in 1937 that the olfactory nerve pathway existed and nonviral agents could be administered to the brain using this pathway. He administered materials such as potassium ferrocyanide to animals intranasally (Faber 1937). In 1985, Shipley first confirmed that neuronal pathways did exist and that drugs could be administered to the brain through the nose. He carried out the study using gelfoam implants soaked in 1% wheat germ agglutinin–horseradish peroxidase. The study showed transneuronal labeling and confirmed that drugs could be transported to the brain via the nasal neuronal pathways (Shipley 1985). Also, most of the delivery strategies used include intracerebroventricular injections, intracranial injections that injure the brain and require special surgical expertise. Systemic administration of therapeutics has not proven to be of any advantage, as the amount of drug crossing the BBB and reaching the brain is minimal. Hence, the intranasal route, being noninvasive, has gained importance recently and has been explored by researchers (Dhuria et al. 2010; Abbott 2013). The main advantages of this route are the ability to bypass first-pass metabolism by the liver, patient compatibility, high bioavailability, and rapid onset of action (Minn et al. 2002). This route of administration has been explored for evaluating the treatment efficacy of numerous CNS drug candidates, as shown in Table 8.4.

8.5.2.4.1 TRIGEMINAL PATHWAY AND OLFACTORY NERVE PATHWAY The nasal cavity has an intimate connection with the nervous system through the olfactory and trigeminal nerves. Since the subarachnoid space extends into the nasal cavity through these nerves, intranasal administration of therapeutics could potentially deliver biologics to the CNS (Betarbet et al. 2000). The olfactory nerve pathway is a major pathway involved in intranasal

TABLE 8.4
Intranasally administered therapeutic agents for neurodegenerative diseases.

Neurodegenerative disorder	Therapeutic agent	Efficacy	Reference
Parkinson's disease	DNSP-11 (neuroactive peptide)	Repeated administration of DNSP-11 intranasally to normal rats increased the dopamine turnover in both striatum and substantia niagra. In the 6-OHDA lesion model, the intranasal administration showed a decrease in rotations induced by amphetamine, and there was an increase in TH-positive neurons.	Stenslik et al. 2015
Alzheimer's disease	Deferoxime (iron chelator)	Intranasal administration of deferoxime changed the processing of APP and also reversed the behavioral changes in transgenic mice with APP and presenilin 1 administered at high doses of iron.	Guo et al. 2013
Huntington's disease	Insulin-like growth factor-1 (IGF-1)	Intranasal administration of IGF-1 increased cortical levels of IGF-1 in wild-type mice as well as YAC128. The intranasal administration reduced the motor deficits and promoted the signaling pathways in the striatum.	Lopes et al. 2014
Alzheimer's disease	Insulin	Part of the clinical trial. Intranasal administration of insulin to patients suffering from Alzheimer's disease showed improvements in memory retention and also increases in CSF markers.	Hölscher 2014

administration. The olfactory pathways responsible for intranasal transport are located in the upper parts of the olfactory region, where olfactory receptor neurons are present. These neurons are dispersed among different cell populations, such as supporting cells, basal cells, and microvillar cells (Dhuria et al. 2010). These neurons are bipolar, and the dendrites of these extend into the olfactory epithelium and the axons of these extend into the cribriform plate (Figure 8.2). This plate separates the nasal and cranial cavities (Dhuria et al. 2010). The axons, after entering the cribriform plate, pass through the subarachnoid space and end up in olfactory bulbs

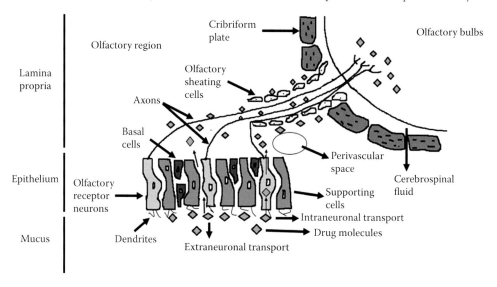

Figure 8.2.   Intranasal delivery through the olfactory nerve. The dendrites of the olfactory nerve extend into the olfactory epithelium in the nasal cavity, and the axons extend into the olfactory bulbs in the CNS, which facilitates transport of molecules from the nasal cavity to the CNS.

on mitral cells, and from there, these axons enter various brain regions, such as the hypothalamus and amygdala (Figure 8.2) (Dhuria et al. 2010). This feature of olfactory receptor neurons is mainly responsible for intranasal delivery, and these neurons regenerate every 3–4 weeks as they come in contact with the external environment (Dhuria et al. 2010).

The trigeminal pathway consists of the trigeminal nerve that innervates the respiratory and olfactory epithelium of the nasal passage and enters into the CNS, and some of these nerves also end up in the olfactory bulbs. The trigeminal nerve helps in communicating the signals from the nasal cavity to the CNS (Dhuria et al. 2010). The trigeminal nerve carries this out through the ophthalmic division (V1), maxillary division (V2), or mandibular division. These three branches of the trigeminal nerve meet at the trigeminal ganglion and extend into the brain and terminate in the brain stem. The trigeminal nerve enters the brain at two different sites from the respiratory epithelium, through the anterior foramen near the pons and the cribriform plate located near the olfactory bulbs. This creates entry points at both the cauda and rostral parts of the brain after intranasal administration. There is no evidence for the presence of the unsheathing cells on the trigeminal nerve (Dhuria et al. 2010). In one study, it was successfully demonstrated for the first time that drugs on intranasal administration enter the brain via two pathways, namely, the olfactory pathway and the other trigeminal pathway. The researchers demonstrated this using insulin, such as growth factor-1 (Thorne et al. 2004). Several studies were conducted that resulted in the conclusion that trigeminal pathways are involved in transporting drugs to the brain via intranasal drug delivery. One such study was conducted where interferon β 1b was administered to monkeys, and it was concluded that the trigeminal pathway was involved (Thorne et al. 2008).

A new perspective to bypass the BBB and deliver drugs to the CNS through the nose after a surgical procedure is being developed. Based on the well-established technique of skull base reconstruction with a nasal septal flap after an endonasal procedure (Bernal-Sprekelsen et al. 2005; Bleier et al. 2011), researchers hypothesized whether this could be a reliable pathway to transport drugs from the nasal cavity to the CNS. Previous findings demonstrated that nasal mucosa is permeable to high-molecular-weight polar molecules (Fisher et al. 1992; Zhao et al. 2004). In order to access the skull base through the nose, removal of bone, dura, and arachnoid is required and the reconstruction is made only with an autologous mucosal flap that becomes the only barrier between the nasal cavity and subarachnoid space. To test this theory, a murine model was developed (Bleier et al. 2013). A mucosal septal flap was implanted in a craniotomy, and its viability and capacity of absorbing different types of molecules were tested. Results showed that even large and polar molecules were capable of diffusing through the mucosal flap, raising a new potential route for drugs in the treatment of patients with chronic neurologic diseases. Besides being noninvasive, another advantage of this method is that it reduces the adverse effects of systemic administration of drugs (Miyake and Bleier 2015a). Animal experiments in mice showed promising results (Dahlin et al. 2000), but clinical findings are controversial (Born et al. 2002; van den Berg et al. 2004). One possible explanation is that while in rodents the olfactory epithelium occupies up to 50% of the nose, the human olfactory epithelium represents only 3% of the total nasal surface and is located in a region with little access for topical agents (Morrison and Costanzo 1992).

8.6 CONCLUSION

With the inherent complexity of neurodegenerative disorders, the molecular mechanisms leading to pathophysiology of the disease are still under extensive study. Thus, newer therapeutic targets are being identified continually. The efficiency of these therapeutic agents differs from the projected theoretical efficiency due to the CNS barriers. The invasive strategies to enhance the efficacy involve surgical risks and seem unsuitable for chronic neurodegenerative diseases. The intranasal route of administration offers a noninvasive method for the administration of multiple doses. However, the hindering factors include limited volume and permeability issues across the nasal mucosa. The use of nanoformulations demonstrates a pronounced potential in enhancing the therapeutic efficacy. Evaluating the efficacy of these therapeutics in appropriate disease models is an important aspect of the study design. Additionally, the safety of the nanoformulations requires a critical evaluation for further clinical translation.

REFERENCES

Abbott, N. J. 2013. Blood–brain barrier structure and function and the challenges for CNS drug delivery. *Journal of Inherited Metabolic Disease* 36 (3):437–449.

Adolfsson, O., M. Pihlgren, N. Toni, Y. Varisco, A. L. Buccarello, K. Antoniello, S. Lohmann, K. Piorkowska, V. Gafner, and J. K. Atwal. 2012. An effector-reduced anti-β-amyloid (Aβ) antibody with unique Aβ binding properties promotes neuroprotection and glial engulfment of Aβ. *Journal of Neuroscience* 32 (28):9677–9689.

Akagi, M., N. Matsui, H. Akae, N. Hirashima, N. Fukuishi, Y. Fukuyama, and R. Akagi. 2015. Nonpeptide neurotrophic agents useful in the treatment of neurodegenerative diseases such as Alzheimer's disease. *Journal of Pharmacological Sciences* 127 (2):155–163.

Amor, S., L. A. N. Peferoen, D. Y. S. Vogel, M. Breur, P. van der Valk, D. Baker, and J. M. van Noort. 2014. Inflammation in neurodegenerative diseases—An update. *Immunology* 142 (2):151–166.

Atwal, J. K., Y. Chen, C. Chiu, D. L. Mortensen, W. J. Meilandt, Y. Liu, C. E. Heise, K. Hoyte, W. Luk, and Y. Lu. 2011. A therapeutic antibody targeting BACE1 inhibits amyloid-β production in vivo. *Science Translational Medicine* 3 (84):84ra43.

Bäck, S., J. Peränen, E. Galli, P. Pulkkila, L. Lonka-Nevalaita, T. Tamminen, M. H. Voutilainen, A. Raasmaja, M. Saarma, and P. T. Männistö. 2013. Gene therapy with AAV2-CDNF provides functional benefits in a rat model of Parkinson's disease. *Brain and Behavior* 3 (2):75–88.

Ballabh, P., A. Braun, and M. Nedergaard. 2004. The blood–brain barrier: An overview: Structure, regulation, and clinical implications. *Neurobiology of Disease* 16 (1):1–13.

Bankiewicz, K. S., J. Forsayeth, J. L. Eberling, R. Sanchez-Pernaute, P. Pivirotto, John Bringas, P. Herscovitch, R. E. Carson, W. Eckelman, and B. Reutter. 2006. Long-term clinical improvement in MPTP-lesioned primates after gene therapy with AAV-hAADC. *Molecular Therapy* 14 (4):564–570.

Barchet, T. M., and M. M. Amiji. 2009. Challenges and opportunities in CNS delivery of therapeutics for neurodegenerative diseases. *Expert Opinion on Drug Delivery* 6 (3):211–225.

Bechara, C., and S. Sagan. 2013. Cell-penetrating peptides: 20 years later, where do we stand? *FEBS Letters* 587 (12):1693–1702.

Begley, D. J. 2004. Delivery of therapeutic agents to the central nervous system: The problems and the possibilities. *Pharmacology & Therapeutics* 104 (1):29–45.

Behrstock, S., A. Ebert, J. McHugh, S. Vosberg, J. Moore, B. Schneider, E. Capowski, D. Hei, J. Kordower, and P. Aebischer. 2006. Human neural progenitors deliver glial cell line-derived neurotrophic factor to parkinsonian rodents and aged primates. *Gene Therapy* 13 (5):379–388.

Bernal-Sprekelsen, M., I. Alobid, J. Mullol, F. Trobat, and M. Tomás-Barberán. 2005. Closure of cerebrospinal fluid leaks prevents ascending bacterial meningitis. *Rhinology* 43 (4):277.

Betarbet, R., T. B. Sherer, G. MacKenzie, M. Garcia-Osuna, A. V. Panov, and J. Timothy Greenamyre. 2000. Chronic systemic pesticide exposure reproduces features of Parkinson's disease. *Nature Neuroscience* 3 (12):1301–1306.

Bezard, E., Z. Yue, D. Kirik, and M. G. Spillantini. 2013. Animal models of Parkinson's disease: Limits and relevance to neuroprotection studies. *Movement Disorders* 28 (1):61–70.

Bickel, U., T. Yoshikawa, and W. M. Pardridge. 2001. Delivery of peptides and proteins through the blood–brain barrier. *Advanced Drug Delivery Reviews* 46 (1):247–279.

Björklund, T., T. Carlsson, E. A. Cederfjäll, M. Carta, and D. Kirik. 2010. Optimized adeno-associated viral vector-mediated striatal DOPA delivery restores sensorimotor function and prevents dyskinesias in a model of advanced Parkinson's disease. *Brain* 133 (2):496–511.

Bleier, B. S., R. E. Kohman, R. E. Feldman, S. Ramanlal, and X. Han. 2013. Permeabilization of the blood-brain barrier via mucosal engrafting: Implications for drug delivery to the brain. *PloS One* 8 (4):e61694.

Bleier, B. S., E. W. Wang, W. Alex Vandergrift, and R. J. Schlosser. 2011. Mucocele rate after endoscopic skull base reconstruction using vascularized pedicled flaps. *American Journal of Rhinology & Allergy* 25 (3):186–187.

Bodor, N., and H. H. Farag. 1983. Improved delivery through biological membranes. 11. A redox chemical drug-delivery system and its use for brain-specific delivery of phenylethylamine. *Journal of Medicinal Chemistry* 26 (3):313–318.

Bonifati, V., P. Rizzu, M. J. van Baren, O. Schaap, G. J. Breedveld, E. Krieger, M. C. J. Dekker, F. Squitieri, P. Ibanez, and M. Joosse. 2003. Mutations in the DJ-1 gene associated with autosomal recessive early-onset parkinsonism. *Science* 299 (5604):256–259.

Born, J., T. Lange, W. Kern, G. P. McGregor, U. Bickel, and H. L. Fehm. 2002. Sniffing neuropeptides: A transnasal approach to the human brain. *Nature Neuroscience* 5 (6):514–516.

Bosch, M., J. R. Pineda, C. Suñol, J. Petriz, E. Cattaneo, J. Alberch, and J. M. Canals. 2004. Induction of GABAergic phenotype in a neural stem cell line for transplantation in an excitotoxic model of Huntington's disease. *Experimental Neurology* 190 (1):42–58.

Boswell, M. V., A. M. Trescot, S. Datta, D. M. Schultz, H. C. Hansen, S. Abdi, N. Sehgal, R. V. Shah, V. Singh, and R. M. Benyamin. 2007. Interventional techniques: Evidence-based practice guidelines in the management of chronic spinal pain. *Pain Physician* 10 (1):7–111.

Braak, H., K. Del Tredici, U. Rüb, R. A. I. de Vos, E. N. H. Jansen Steur, and E. Braak. 2003. Staging of brain pathology related to sporadic Parkinson's disease. *Neurobiology of Aging* 24 (2):197–211.

Brewster, M. E., T. Murakami, W. R. Anderson, N. Bodor, and E. Pop. 1996. Improved oral bioavailability of the brain-targeting estrogen, E2-CDS, through the use of carboxymethylethyl-β-cyclodextrin. Presented at Proceedings of the Eighth International Symposium on Cyclodextrins, Budapest.

Calias, P., W. A. Banks, D. Begley, M. Scarpa, and P. Dickson. 2014. Intrathecal delivery of protein therapeutics to the brain: A critical reassessment. *Pharmacology & Therapeutics* 144 (2):114–122.

Chao, M. V. 2003. Neurotrophins and their receptors: A convergence point for many signalling pathways. *Nature Reviews: Neuroscience* 4 (4):299–309.

Chen, K. S., and F, H. Gage. 1995. Somatic gene transfer of NGF to the aged brain: Behavioral and morphological amelioration. *Journal of Neuroscience* 15 (4):2819–2825.

Chen, Y., and L. Liu. 2012. Modern methods for delivery of drugs across the blood–brain barrier. *Advanced Drug Delivery Reviews* 64 (7):640–665.

Cho, M. S., Y.-E. Lee, J. Y. Kim, S. Chung, Y. H. Cho, D.-S. Kim, S.-M. Kang, H. Lee, M.-H. Kim, and J.-H. Kim. 2008. Highly efficient and large-scale generation of functional dopamine neurons from human embryonic stem cells. *Proceedings of the National Academy of Sciences of the United States of America* 105 (9):3392–3397.

Chouliaras, L., D. Mastroeni, E. Delvaux, A. Grover, G. Kenis, P. R. Hof, H. W. M. Steinbusch, P. D. Coleman, B. P. F. Rutten, and D. L. A. van den Hove. 2013. Consistent decrease in global DNA methylation and hydroxymethylation in the hippocampus of Alzheimer's disease patients. *Neurobiology of Aging* 34 (9):2091–2099.

Conner, J. M., M. A. Darracq, Jeff Roberts, and M. H. Tuszynski. 2001. Nontropic actions of neurotrophins: Subcortical nerve growth factor gene delivery reverses age-related degeneration of primate cortical cholinergic innervation. *Proceedings of the National Academy of Sciences of the United States of America* 98 (4):1941–1946.

Coppedè, F. 2010. One-carbon metabolism and Alzheimer's disease: Focus on epigenetics. *Current Genomics* 11 (4):246–260.

Coppedè, F. 2013. Advances in the genetics and epigenetics of neurodegenerative diseases. *Epigenetics of Degenerative Diseases* 1 (1).

Dahlin, M., U. Bergman, B. Jansson, E. Björk, and E. Brittebo. 2000. Transfer of dopamine in the olfactory pathway following nasal administration in mice. *Pharmaceutical Research* 17 (6):737–742.

da Rocha Lindner, G., D. B. Santos, D. Colle, E. L. G. Moreira, R. D. Prediger, M. Farina, N. M. Khalil, and R. M. Mainardes. 2015. Improved neuroprotective effects of resveratrol-loaded polysorbate 80-coated poly (lactide) nanoparticles in MPTP-induced Parkinsonism. *Nanomedicine* 10 (7):1127–1138.

Dauer, W., and S. Przedborski. 2003. Parkinson's disease: Mechanisms and models. *Neuron* 39 (6):889–909.

DeMattos, R. B., K. R. Bales, D. J. Cummins, S. M. Paul, and D. M. Holtzman. 2002. Brain to plasma amyloid-β efflux: A measure of brain amyloid burden in a mouse model of Alzheimer's disease. *Science* 295 (5563):2264–2267.

DeMattos, R. B., J. Lu, Y. Tang, M. M. Racke, C. A. DeLong, J. A. Tzaferis, J. T. Hole, B. M. Forster, P. C. McDonnell, and F. Liu. 2012. A plaque-specific antibody clears existing β-amyloid plaques in Alzheimer's disease mice. *Neuron* 76 (5):908–920.

Dexter, D. T., and P. Jenner. 2013. Parkinson disease: From pathology to molecular disease mechanisms. *Free Radical Biology and Medicine* 62:132–144.

Dezawa, M., H. Kanno, M. Hoshino, H. Cho, N. Matsumoto, Y. Itokazu, N. Tajima, H. Yamada, H. Sawada, and H. Ishikawa. 2004. Specific induction of neuronal cells from bone marrow stromal cells and application for autologous transplantation. *Journal of Clinical Investigation* 113 (12):1701–1710.

Dhawan, S., R. Kapil, and B. Singh. 2011. Formulation development and systematic optimization of solid lipid nanoparticles of quercetin for improved brain delivery. *Journal of Pharmacy and Pharmacology* 63 (3):342–351.

Dhuria, S. V., L. R. Hanson, and W. H. Frey. 2010. Intranasal delivery to the central nervous system: Mechanisms and experimental considerations. *Journal of Pharmaceutical Sciences* 99 (4):1654–1673.

Ding, H., P. J. Dolan, and G. V. W. Johnson. 2008. Histone deacetylase 6 interacts with the microtubule-associated protein tau. *Journal of Neurochemistry* 106 (5):2119–2130.

Duyao, M., C. Ambrose, R. Myers, A. Novelletto, F. Persichetti, M. Frontali, S. Folstein, C. Ross, M. Franz, and M. Abbott. 1993. Trinucleotide repeat length instability and age of onset in Huntington's disease. *Nature Genetics* 4 (4):387–392.

Emerich, D. F., S. R. Winn, J. Harper, J. P. Hammang, E. E. Baetge, and J. H. Kordower. 1994. Implants of polymer-encapsulated human NGF-secreting cells in the nonhuman primate: Rescue and sprouting of degenerating cholinergic basal forebrain neurons. *Journal of Comparative Neurology* 349 (1):148–164.

Faber, W. M. 1937. The nasal mucosa and the subarachnoid space. *American Journal of Anatomy* 62 (1):121–148.

Farlow, M., S. E. Arnold, C. H. Van Dyck, P. S. Aisen, B. J. Snider, A. P. Porsteinsson, S. Friedrich, R. A. Dean, C. Gonzales, and G. Sethuraman. 2012. Safety and biomarker effects of solanezumab in patients with Alzheimer's disease. *Alzheimer's & Dementia* 8 (4):261–271.

Felgenhauer, K. 1974. Protein size and cerebrospinal fluid composition. *Klinische Wochenschrift* 52 (24):1158–1164.

Ferrante, R. J., N. W. Kowall, M. F. Beal, E. P. Richardson, E. D. Bird, and J. B. Martin. 1985. Selective sparing of a class of striatal neurons in Huntington's disease. *Science* 230 (4725):561–563.

Fischer, A., F. Sananbenesi, X. Wang, M. Dobbin, and L.-H. Tsai. 2007. Recovery of learning and memory is associated with chromatin remodelling. *Nature* 447 (7141):178–182.

Fisher, A. N., L. Illum, S. S. Davis, and E. H. Schacht. 1992. Di-iodo-L-tyrosine-labelled dextrans as molecular size markers of nasal absorption in the rat. *Journal of Pharmacy and Pharmacology* 44 (7):550–554.

Follo, C., M. Ozzano, V. Mugoni, R. Castino, M. Santoro, and C. Isidoro. 2011. Knock-down of cathepsin D affects the retinal pigment epithelium, impairs swim-bladder ontogenesis and causes premature death in zebrafish. *PLoS One* 6 (7):e21908.

Forman, M. S., J. Q. Trojanowski, and V. M. Y. Lee. 2004. Neurodegenerative diseases: A decade of discoveries paves the way for therapeutic breakthroughs. *Nature Medicine* 10 (10):1055–1063.

Fuso, A., V. Nicolia, R. A. Cavallaro, L. Ricceri, F. D'Anselmi, P. Coluccia, G. Calamandrei, and S. Scarpa. 2008. B-vitamin deprivation induces hyperhomocysteinemia and brain S-adenosylhomocysteine, depletes brain S-adenosylmethionine, and enhances PS1 and BACE expression and amyloid-β deposition in mice. *Molecular and Cellular Neuroscience* 37 (4):731–746.

Fuso, A., V. Nicolia, L. Ricceri, R. A. Cavallaro, E. Isopi, F. Mangia, M. T. Fiorenza, and S. Scarpa. 2012. S-Adenosylmethionine reduces the progress of the Alzheimer-like features induced by B-vitamin deficiency in mice. *Neurobiology of Aging* 33 (7):1482-e1.

Fuso, A., L. Seminara, R. A. Cavallaro, F. D'Anselmi, and S. Scarpa. 2005. S-adenosylmethionine/homocysteine cycle alterations modify DNA methylation status with consequent deregulation of PS1 and BACE and beta-amyloid production. *Molecular and Cellular Neuroscience* 28 (1):195–204.

Gauthier, L. R., B. C. Charrin, M. Borrell-Pagès, J. P. Dompierre, H. Rangone, F. P. Cordelières, J. De Mey, M. E. MacDonald, V. Leßmann, and S. Humbert. 2004. Huntingtin controls neurotrophic support and survival of neurons by enhancing BDNF vesicular transport along microtubules. *Cell* 118 (1):127–138.

Gemayel, R., S. Chavali, K. Pougach, M. Legendre, B. Zhu, S. Boeynaems, E. van der Zande, K. Gevaert, F. Rousseau, and J. Schymkowitz. 2015. Variable glutamine-rich repeats modulate transcription factor activity. *Molecular Cell* 59 (4):615–627.

Ghosh, A. K., M. Brindisi, and J. Tang. 2012. Developing β-secretase inhibitors for treatment of Alzheimer's disease. *Journal of Neurochemistry* 120 (s1):71–83.

Giasson, B. I., and V. M. Y. Lee. 2003. Are ubiquitination pathways central to Parkinson's disease? *Cell* 114 (1):1–8.

Gill, S. S., N. K. Patel, G. R. Hotton, K. O'Sullivan, R. McCarter, M. Bunnage, D. J. Brooks, C. N. Svendsen, and P. Heywood. 2003. Direct brain infusion of glial cell line–derived neurotrophic factor in Parkinson disease. *Nature Medicine* 9 (5):589–595.

Goedert, M. 2001. Alpha-synuclein and neurodegenerative diseases. *Nature Reviews: Neuroscience* 2 (7):492–501.

Gräff, J., D. Rei, J.-S. Guan, W.-Y. Wang, J. Seo, K. M. Hennig, T. J. F. Nieland, D. M. Fass, P. F. Kao, and M. Kahn. 2012. An epigenetic blockade of cognitive functions in the neurodegenerating brain. *Nature* 483 (7388):222–226.

Gray, S. G. 2010. Targeting histone deacetylases for the treatment of Huntington's disease. *CNS Neuroscience & Therapeutics* 16 (6):348–361.

Gray, S. J., K. T. Woodard, and R. J. Samulski. 2010. Viral vectors and delivery strategies for CNS gene therapy. *Therapeutic Delivery* 1 (4):517–534.

Hadaczek, P., J. L. Eberling, P. Pivirotto, J. Bringas, J. Forsayeth, and K. S. Bankiewicz. 2010. Eight years of clinical improvement in MPTP-lesioned primates after gene therapy with AAV2-hAADC. *Molecular Therapy* 18 (8):1458–1461.

Hasnat, M. J., and J. E. Rice. 2015. Intrathecal baclofen for treating spasticity in children with cerebral palsy. *Cochrane Database of Systematic Reviews* 2004 (1): CD004552.

Hitt, B., S. M. Riordan, L. Kukreja, W. A. Eimer, T. W. Rajapaksha, and R. Vassar. 2012. β-Site amyloid precursor protein (APP)-cleaving enzyme 1 (BACE1)-deficient mice exhibit a close homolog of L1 (CHL1) loss-of-function phenotype involving axon guidance defects. *Journal of Biological Chemistry* 287 (46):38408–38425.

Hoang-Xuan, K., E. Bessell, J. Bromberg, A. F. Hottinger, M. Preusser, R. Rudà, U. Schlegel, T. Siegal, C. Soussain, and U. Abacioglu. 2015. Diagnosis and treatment of primary CNS lymphoma in immunocompetent patients: Guidelines from the European Association for Neuro-Oncology. *Lancet Oncology* 16 (7):e322.

Hug, C. C. 1992. Opioids: Clinical use as anesthetic agents. *Journal of Pain and Symptom Management* 7 (6):350–355.

In'T Veld, B. A., A. Ruitenberg, A. Hofman, L. J. Launer, C. M. van Duijn, T. Stijnen, M. M. B. Breteler, and B. H. C. Stricker. 2001. Nonsteroidal antiinflammatory drugs and the risk of Alzheimer's disease. *New England Journal of Medicine* 345 (21):1515–1521.

Isacson, O., L. M. Bjorklund, and J. M. Schumacher. 2003. Toward full restoration of synaptic and terminal function of the dopaminergic system in Parkinson's disease by stem cells. *Annals of Neurology* 53 (S3):S135–S148.

Ischiropoulos, H. 2003. Oxidative modifications of α-synuclein. *Annals of the New York Academy of Sciences* 991 (1):93–100.

Jankovic, J. 2008. Parkinson's disease: Clinical features and diagnosis. *Journal of Neurology, Neurosurgery & Psychiatry* 79 (4):368–376.

Jiang, H., M. A. Poirier, Y. Liang, Z. Pei, C. E. Weiskittel, W. W. Smith, D. B. DeFranco, and C. A. Ross. 2006. Depletion of CBP is directly linked with cellular toxicity caused by mutant huntingtin. *Neurobiology of Disease* 23 (3):543–551.

Jose, S., S. Sowmya, T. A. Cinu, N. A. Aleykutty, S. Thomas, and E. B. Souto. 2014. Surface modified PLGA nanoparticles for brain targeting of bacoside-A. *European Journal of Pharmaceutical Sciences* 63:29–35.

Jowaed, A., I. Schmitt, O. Kaut, and U. Wüllner. 2010. Methylation regulates alpha-synuclein expression and is decreased in Parkinson's disease patients' brains. *Journal of Neuroscience* 30 (18):6355–6359.

Kanazawa, T., F. Akiyama, S. Kakizaki, Y. Takashima, and Y. Seta. 2013. Delivery of siRNA to the brain using a combination of nose-to-brain delivery and cell-penetrating peptide-modified nano-micelles. *Biomaterials* 34 (36):9220–9226.

Kang, Y.-S., H.-J. Jung, J.-S. Oh, and D.-Y. Song. 2016. Use of PEGylated immunoliposomes to deliver dopamine across the blood–brain barrier in a rat model of Parkinson's disease. *CNS Neuroscience & Therapeutics* 22 (10):817–823.

Kim, M., S.-T. Lee, K. Chu, and S. U. Kim. 2008. Stem cell-based cell therapy for Huntington disease: A review. *Neuropathology* 28 (1):1–9.

Kitada, T., S. Asakawa, N. Hattori, H. Matsumine, Y. Yamamura, S. Minoshima, M. Yokochi, Y. Mizuno, and N. Shimizu. 1998. Mutations in the parkin gene cause autosomal recessive juvenile parkinsonism. *Nature* 392 (6676):605–608.

Koike, M., H. Nakanishi, P. Saftig, J. Ezaki, K. Isahara, Y. Ohsawa, W. Schulz-Schaeffer, T. Watanabe, S. Waguri, and S. Kametaka. 2000. Cathepsin D deficiency induces lysosomal storage with ceroid lipofuscin in mouse CNS neurons. *Journal of Neuroscience* 20 (18):6898–6906.

Kordower, J. H., E.-Y. Chen, C. Winkler, R. Fricker, V. Charles, A. Messing, E. J. Mufson et al. 1997. Grafts of EGF-responsive neural stem cells derived from GFAP-hNGF transgenic mice: Trophic and tropic effects in a rodent model of Huntington's disease. *Journal of Comparative Neurology* 387 (1):96–113.

Kuo, Y.-C., and C.-T. Wang. 2014. Protection of SK-N-MC cells against β-amyloid peptide-induced degeneration using neuron growth factor-loaded liposomes with surface lactoferrin. *Biomaterials* 35 (22):5954–5964.

LaFerla, F. M., and K. N. Green. 2012. Animal models of Alzheimer disease. *Cold Spring Harbor Perspectives in Medicine* 2 (11):a006320.

Lattal, K. M., and M. A. Wood. 2013. Epigenetics and persistent memory: Implications for reconsolidation and silent extinction beyond the zero. *Nature Neuroscience* 16 (2):124–129.

Lee, J.-E., and P.-L. Han. 2013. An update of animal models of Alzheimer disease with a reevaluation of plaque depositions. *Experimental Neurobiology* 22 (2):84–95.

Lee, S.-T., K. Chu, J.-E. Park, K. Lee, L. Kang, S. U. Kim, and M. Kim. 2005. Intravenous administration of human neural stem cells induces functional recovery in Huntington's disease rat model. *Neuroscience Research* 52 (3):243–249.

Lee, V. M. Y., M. Goedert, and J. Q. Trojanowski. 2001. Neurodegenerative tauopathies. *Annual Review of Neuroscience* 24 (1):1121–1159.

Leroy, E., R. Boyer, G. Auburger, B. Leube, G. Ulm, E. Mezey, G. Harta, M. J. Brownstein, S. Jonnalagada, and T. Chernova. 1998. The ubiquitin pathway in Parkinson's disease. *Nature* 395 (6701):451–452.

Lescaudron, L., D. Unni, and G. L. Dunbar. 2003. Autologous adult bone marrow stem cell transplantation in an animal model of Huntington's disease: Behavioral and morphological outcomes. *International Journal of Neuroscience* 113 (7):945–956.

Lindholm, P., M. H. Voutilainen, J. Laurén, J. Peränen, V.-M. Leppänen, J.-O. Andressoo, M. Lindahl, S. Janhunen, N. Kalkkinen, and T. Timmusk. 2007. Novel neurotrophic factor CDNF protects and rescues midbrain dopamine neurons in vivo. *Nature* 448 (7149):73–77.

Lindvall, O., R. A. Barker, O. Brüstle, O. Isacson, and C. N. Svendsen. 2012. Clinical translation of stem cells in neurodegenerative disorders. *Cell Stem Cell* 10 (2):151–155.

Lindvall, O., and A. Björklund. 2004. Cell therapy in Parkinson's disease. *NeuroRx* 1 (4):382–393.

Lindvall, O., and Z. Kokaia. 2010. Stem cells in human neurodegenerative disorders—Time for clinical translation? *Journal of Clinical Investigation* 120 (1):29–40.

Lipinski, M. M., and J. Yuan. 2004. Mechanisms of cell death in polyglutamine expansion diseases. *Current Opinion in Pharmacology* 4 (1):85–90.

Liu, Y.-Y., X.-Y. Yang, Z. Li, Z.-L. Liu, D. Cheng, Y. Wang, X.-J. Wen, J.-Y. Hu, J. Liu, and L.-M. Wang. 2014. Characterization of polyethylene glycol-polyethyleneimine as a vector for alpha-synuclein siRNA delivery to PC12 cells for Parkinson's disease. *CNS Neuroscience & Therapeutics* 20 (1):76–85.

Lotharius, J., and P. Brundin. 2002. Pathogenesis of Parkinson's disease: Dopamine, vesicles and α-synuclein. *Nature Reviews: Neuroscience* 3 (12):932–942.

MacLachlan, I. 2007. Liposomal formulations for nucleic acid delivery. *Antisense Drug Technology: Principles, Strategies, and Applications* 2:237–270.

Manchikanti, L., F. J. Falco, V. Singh, R. M. Benyamin, G. B. Racz, S. Helm 2nd, D. L. Caraway, A. K. Calodney, L. T. Snook, and H. S. Smith. 2013. An update of comprehensive evidence-based guidelines for interventional techniques in chronic spinal pain. Part I. Introduction and general considerations. *Pain Physician* 16 (2 Suppl):S1–S48.

Mankowski, J. L., S. E. Queen, L. M. Kirstein, J. P. Spelman, J. Laterra, I. A. Simpson, R. J. Adams, J. E. Clements, and M. Christine Zink. 1999. Alterations in blood-brain barrier glucose transport in SIV-infected macaques. *Journal of Neurovirology* 5 (6):695–702.

Martin-Subero, J. I. 2011. How epigenomics brings phenotype into being. *Pediatric Endocrinology Reviews: PER* 9:506–510.

Mastroeni, D., A. Grover, E. Delvaux, C. Whiteside, P. D. Coleman, and J. Rogers. 2010. Epigenetic changes in Alzheimer's disease: Decrements in DNA methylation. *Neurobiology of Aging* 31 (12):2025–2037.

Matsumoto, L., H. Takuma, A. Tamaoka, H. Kurisaki, H. Date, S. Tsuji, and A. Iwata. 2010. CpG demethylation enhances alpha-synuclein expression and affects the pathogenesis of Parkinson's disease. *PLoS One* 5 (11):e15522.

McCarthy, J. M., D. Appelhans, J. Tatzelt, and M. S. Rogers. 2013a. Nanomedicine for prion disease treatment: New insights into the role of dendrimers. *Prion* 7 (3):198–202.

McCarthy, J. M., M. Franke, U. K. Resenberger, S. Waldron, J. C. Simpson, J. Tatzelt, D. Appelhans, and M. S. Rogers. 2013b. Anti-prion drug mPPIg5 inhibits PrP C conversion to PrP Sc. *PloS One* 8 (1):e55282.

McCoy, M. K., K. A. Ruhn, T. N. Martinez, F. E. McAlpine, A. Blesch, and M. G. Tansey. 2008. Intranigral lentiviral delivery of dominant-negative TNF attenuates neurodegeneration and behavioral deficits in hemiparkinsonian rats. *Molecular Therapy* 16 (9):1572–1579.

Menalled, L., and D. Brunner. 2014. Animal models of Huntington's disease for translation to the clinic: Best practices. *Movement Disorders* 29 (11):1375–1390.

Mendez, I., R. Sanchez-Pernaute, O. Cooper, A. Viñuela, D. Ferrari, L. Björklund, A. Dagher, and O. Isacson. 2005. Cell type analysis of functional fetal dopamine cell suspension transplants in the striatum and substantia nigra of patients with Parkinson's disease. *Brain* 128 (7):1498–1510.

Migliore, L., and F. Coppedè. 2009. Genetics, environmental factors and the emerging role of epigenetics in neurodegenerative diseases. *Mutation Research/Fundamental and Molecular Mechanisms of Mutagenesis* 667 (1):82–97.

Minn, A., S. Leclerc, J.-M. Heydel, A.-L. Minn, C. Denizot, M. Cattarelli, P. Netter, and D. Gradinaru. 2002. Drug transport into the mammalian brain: The nasal pathway and its specific metabolic barrier. *Journal of Drug Targeting* 10 (4):285–296.

Miyake, M. M., and B. S. Bleier. 2015a. Bypassing the blood–brian barrier using established skull base reconstruction techniques. *World Journal of Otorhinolaryngology—Head and Neck Surgery* 1 (1):11–16.

Miyake, M. M., and B. S. Bleier. 2015b. The blood-brain barrier and nasal drug delivery to the central nervous system. *American Journal of Rhinology & Allergy* 29 (2):124–127.

Mogi, M., A. Togari, T. Kondo, Y. Mizuno, O. Komure, S. Kuno, H. Ichinose, and T. Nagatsu. 1999. Brain-derived growth factor and nerve growth factor concentrations are decreased in the substantia nigra in Parkinson's disease. *Neuroscience Letters* 270 (1):45–48.

Morrison, E. E., and R. M. Costanzo. 1992. Morphology of olfactory epithelium in humans and other vertebrates. *Microscopy Research and Technique* 23 (1):49–61.

Mott, K. R., M. Zandian, S. J. Allen, and H. Ghiasi. 2013. Role of interleukin-2 and herpes simplex virus 1 in central nervous system demyelination in mice. *Journal of Virology* 87 (22):12102–12109.

Mu, C., N. Dave, J. Hu, P. Desai, G. Pauletti, S. Bai, and J. Hao. 2013. Solubilization of flurbiprofen into aptamer-modified PEG–PLA micelles for targeted delivery to brain-derived endothelial cells in vitro. *Journal of Microencapsulation* 30 (7):701–708.

Nau, R., F. Sörgel, and H. Eiffert. 2010. Penetration of drugs through the blood-cerebrospinal fluid/blood-brain barrier for treatment of central nervous system infections. *Clinical Microbiology Reviews* 23 (4):858–883.

Olanow, C. W., K. Kieburtz, and A. H. V. Schapira. 2008. Why have we failed to achieve neuroprotection in Parkinson's disease? *Annals of Neurology* 64 (S2):S101–S110.

Ostrowitzki, S., D. Deptula, L. Thurfjell, F. Barkhof, B. Bohrmann, D. J. Brooks, W. E. Klunk, E. Ashford, K. Yoo, and Z.-X. Xu. 2012. Mechanism of amyloid removal in patients with Alzheimer disease treated with gantenerumab. *Archives of Neurology* 69 (2):198–207.

Padowski, J. M., K. E. Weaver, T. L. Richards, M. Y. Laurino, A. Samii, E. H. Aylward, and K. E. Conley. 2014. Neurochemical correlates of caudate atrophy in Huntington's disease. *Movement Disorders* 29 (3):327–335.

Pardridge, W. M. 2003. Blood-brain barrier drug targeting enables neuroprotection in brain ischemia following delayed intravenous administration of neurotrophins. In *Molecular and Cellular Biology of Neuroprotection in the CNS*, ed. C. Alzheimer, 397–430. Berlin: Springer.

Pathan, S. A., Z. Iqbal, S. Zaidi, S. Talegaonkar, D. Vohra, G. K. Jain, A. Azeem, N. Jain, J. R. Lalani, and R. K. Khar. 2009. CNS drug delivery systems: Novel approaches. *Recent Patents on Drug Delivery & Formulation* 3 (1):71–89.

Petrova, P. S., A. Raibekas, J. Pevsner, N. Vigo, M. Anafi, M. K. Moore, A. E. Peaire, V. Shridhar, D. I. Smith, and J. Kelly. 2003. MANF. *Journal of Molecular Neuroscience* 20 (2):173–187.

Poduslo, J. F., G. L. Curran, and C. T. Berg. 1994. Macromolecular permeability across the blood-nerve and blood-brain barriers. *Proceedings of the National Academy of Sciences of the United States of America* 91 (12):5705–5709.

Polymeropoulos, M. H., C. Lavedan, E. Leroy, S. E. Ide, A. Dehejia, A. Dutra, B. Pike, H. Root, J. Rubenstein, and Re. Boyer. 1997. Mutation in the α-synuclein gene identified in families with Parkinson's disease. *Science* 276 (5321):2045–2047.

Pouladi, M. A., A. J. Morton, and M. R. Hayden. 2013. Choosing an animal model for the study of Huntington's disease. *Nature Reviews: Neuroscience* 14 (10):708–721.

Puglielli, L., R. E. Tanzi, and D. M. Kovacs. 2003. Alzheimer's disease: The cholesterol connection. *Nature Neuroscience* 6 (4):345–351.

Ramaswamy, S., J. L. McBride, and J. H. Kordower. 2007. Animal models of Huntington's disease. *ILAR Journal* 48 (4):356–373.

Rautio, J., H. Kumpulainen, T. Heimbach, R. Oliyai, D. Oh, T. Järvinen, and J. Savolainen. 2008. Prodrugs: Design and clinical applications. *Nature Reviews: Drug Discovery* 7 (3):255–270.

Ricobaraza, A., M. Cuadrado-Tejedor, S. Marco, I. Pérez-Otaño, and A. García-Osta. 2012. Phenylbutyrate rescues dendritic spine loss associated with memory deficits in a mouse model of Alzheimer disease. *Hippocampus* 22 (5):1040–1050.

Rinne, J. O., D. J. Brooks, M. N. Rossor, N. C. Fox, R. Bullock, W. E. Klunk, C. A. Mathis, K. Blennow, J. Barakos, and A. A. Okello. 2010. 11 C-PiB PET assessment of change in fibrillar amyloid-β load in patients with Alzheimer's disease treated with bapineuzumab: A phase 2, double-blind, placebo-controlled, ascending-dose study. *Lancet Neurology* 9 (4):363–372.

Roy, N. S., C. Cleren, S. K. Singh, L. Yang, M. Flint Beal, and S. A. Goldman. 2006. Functional engraftment of human ES cell–derived dopaminergic neurons enriched by coculture with telomerase-immortalized midbrain astrocytes. *Nature Medicine* 12 (11):1259–1268.

Rubinsztein, D. C. 2003. The molecular pathology of Huntington's disease (HD). *Current Medicinal Chemistry—Immunology, Endocrine & Metabolic Agents* 3 (4):329–340.

Ryu, J. K., J. Kim, S. J. Cho, K. Hatori, A. Nagai, H. B. Choi, M. C. Lee, J. G. McLarnon, and S. U. Kim. 2004. Proactive transplantation of human neural stem cells prevents degeneration of striatal neurons in a rat model of Huntington disease. *Neurobiology of Disease* 16 (1):68–77.

Salameh, F., D. Karaman, G. Mecca, L. Scrano, S. A. Bufo, and R. Karaman. 2015. Prodrugs targeting the central nervous system (CNS). *World Journal of Pharmacy and Pharmaceutical Sciences* 4 (8):208–237.

Sánchez-Pernaute, R., L. Studer, K. S. Bankiewicz, E. O. Major, and R. D. G. McKay. 2001. In vitro generation and transplantation of precursor-derived human dopamine neurons. *Journal of Neuroscience Research* 65 (4):284–288.

Schenk, D., R. Barbour, W. Dunn, G. Gordon, H. Grajeda, T. Guido, K. Hu, J. Huang, K. Johnson-Wood, and K. Khan. 1999. Immunization with amyloid-β attenuates Alzheimer-disease-like pathology in the PDAPP mouse. *Nature* 400 (6740):173–177.

Schneider, L. S., F. Mangialasche, N. Andreasen, H. Feldman, E. Giacobini, R. Jones, V. Mantua, P. Mecocci, L. Pani, B. Winblad, and M. Kivipelto. 2014. Clinical trials and late-stage drug development for Alzheimer's disease: An appraisal from 1984 to 2014. *Journal of Internal Medicine* 275 (3):251–283.

Schrag, A., Y. Ben-Shlomo, and N. P. Quinn. 2000. Cross sectional prevalence survey of idiopathic Parkinson's disease and parkinsonism in London. *BMJ* 321 (7252):21–22.

Shipley, M. T. 1985. Transport of molecules from nose to brain: Transneuronal anterograde and retrograde labeling in the rat olfactory system by wheat germ agglutinin-horseradish peroxidase applied to the nasal epithelium. *Brain Research Bulletin* 15 (2):129–142.

Siegal, T. 1998. Leptomeningeal metastases: Rationale for systemic chemotherapy or what is the role of intra-CSF-chemotherapy? *Journal of Neuro-Oncology* 38 (2–3):151–157.

Singleton, A. B., M. Farrer, J. Johnson, A. Singleton, S. Hague, J. Kachergus, M. Hulihan, T. Peuralinna, A. Dutra, and R. Nussbaum. 2003. α-Synuclein locus triplication causes Parkinson's disease. *Science* 302 (5646):841–841.

Smith, D. E., J. Roberts, F. H. Gage, and M. H. Tuszynski. 1999. Age-associated neuronal atrophy occurs in the primate brain and is reversible by growth factor gene therapy. *Proceedings of the National Academy of Sciences of the United States of America* 96 (19):10893–10898.

Somogyi, G., P. Buchwald, and N. Bodor. 2002. Targeted drug delivery to the central nervous system via phosphonate derivatives (anionic delivery system for testosterone). *Die Pharmazie* 57 (2):135–137.

Sorokina, S. A., Y. Y. Stroylova, Z. B. Shifrina, and V. I. Muronetz. 2016. Disruption of amyloid prion protein aggregates by cationic pyridylphenylene dendrimers. *Macromolecular Bioscience* 16 (2):266–275.

Sperling, R. A., C. R. Jack, S. E. Black, M. P. Frosch, S. M. Greenberg, B. T. Hyman, P. Scheltens, M. C. Carrillo, W. Thies, and M. M. Bednar. 2011. Amyloid-related imaging abnormalities in amyloid-modifying therapeutic trials: Recommendations from the Alzheimer's Association Research Roundtable Workgroup. *Alzheimer's & Dementia* 7 (4):367–385.

Steffan, J. S., A. Kazantsev, O. Spasic-Boskovic, M. Greenwald, Y.-Z. Zhu, H. Gohler, E. E. Wanker, G. P. Bates, D. E. Housman, and L. M. Thompson. 2000. The Huntington's disease protein interacts with p53 and CREB-binding protein and represses transcription. *Proceedings of the National Academy of Sciences of the United States of America* 97 (12):6763–6768.

Strand, A. D., Z. C. Baquet, A. K. Aragaki, P. Holmans, L. Yang, C. Cleren, M. F. Beal, L. Jones, C. Kooperberg, and J. M. Olson. 2007. Expression profiling of Huntington's disease models suggests that brain-derived neurotrophic factor depletion plays a major role in striatal degeneration. *Journal of Neuroscience* 27 (43):11758–11768.

Surekha, R., and T. Sumathi. 2016. A novel therapeutic application of solid lipid nanoparticles encapsulated thymoquinone (TQ-SLNs) on 3-nitroproponic acid induced Huntington's disease-like symptoms in Wistar rats. *Chemico-biological Interactions* 256:25–36.

Thorne, R. G., L. R. Hanson, T. M. Ross, D. Tung, and W. H. Frey. 2008. Delivery of interferon-β to the monkey nervous system following intranasal administration. *Neuroscience* 152 (3):785–797.

Thorne, R. G., G. J. Pronk, V. Padmanabhan, and W. H. Frey 2nd. 2004. Delivery of insulin-like growth factor-I to the rat brain and spinal cord along olfactory and trigeminal pathways following intranasal administration. *Neuroscience* 127 (2):481–496.

Tolosa, E., M. J. Martí, F. Valldeoriola, and J. L. Molinuevo. 1998. History of levodopa and dopamine agonists in Parkinson's disease treatment. *Neurology* 50 (6 Suppl 6):S2–S10.

Tsai, L.-K., Y.-C. Chen, W.-C. Cheng, C.-H. Ting, J. C. Dodge, W.-L. Hwu, S. H. Cheng, and M. A. Passini. 2012. IGF-1 delivery to CNS attenuates motor neuron cell death but does not improve motor function in type III SMA mice. *Neurobiology of Disease* 45 (1):272–279.

Tunkel, A. R., B. J. Hartman, S. L. Kaplan, B. A. Kaufman, K. L. Roos, W. M. Scheld, and R. J. Whitley. 2004. Practice guidelines for the management of bacterial meningitis. *Clinical Infectious Diseases* 39 (9):1267–1284.

Tuszynski, M. H., L. Thal, M. Pay, D. P. Salmon, R. Bakay, P. Patel, A. Blesch, H. L. Vahlsing, G. Ho, and G. Tong. 2005. A phase 1 clinical trial of nerve growth factor gene therapy for Alzheimer disease. *Nature Medicine* 11 (5):551–555.

Valente, E. M., P. M. Abou-Sleiman, V. Caputo, M. M. K. Muqit, K. Harvey, S. Gispert, Z. Ali, D. Del Turco, A. R. Bentivoglio, and D. G. Healy. 2004. Hereditary early-onset Parkinson's disease caused by mutations in PINK1. *Science* 304 (5674):1158–1160.

van den Berg, M. P., P. Merkus, S. G. Romeijn, J. Coos Verhoef, and F. W. H. M. Merkus. 2004. Uptake of melatonin into the cerebrospinal fluid after nasal and intravenous delivery: Studies in rats and comparison with a human study. *Pharmaceutical Research* 21 (5):799–802.

Veugelen, S., T. Saito, T. C. Saido, L. Chávez-Gutiérrez, and B. De Strooper. 2016. Familial Alzheimer's disease mutations in presenilin generate amyloidogenic Aβ peptide seeds. *Neuron* 90 (2):410–416.

Vicario, I., and T. Schimmang. 2003. Transfer of FGF-2 via HSV-1-based amplicon vectors promotes efficient formation of neurons from embryonic stem cells. *Journal of Neuroscience Methods* 123 (1):55–60.

Voutilainen, M. H., S. Bäck, J. Peränen, P. Lindholm, A. Raasmaja, P. T. Männistö, M. Saarma, and R. K. Tuominen. 2011. Chronic infusion of CDNF prevents 6-OHDA-induced deficits in a rat model of Parkinson's disease. *Experimental Neurology* 228 (1):99–108.

Wang, J.-J., K. C. Sung, J.-F. Huang, C.-H. Yeh, and J.-Y. Fang. 2007. Ester prodrugs of morphine improve transdermal drug delivery: A mechanistic study. *Journal of Pharmacy and Pharmacology* 59 (7):917–925.

Weggen, S., J. L. Eriksen, P. Das, S. A. Sagi, R. Wang, C. U. Pietrzik, K. A. Findlay, T. E. Smith, M. P. Murphy, and T. Bulter. 2001. A subset of NSAIDs lower amyloidogenic Aβ42 independently of cyclooxygenase activity. *Nature* 414 (6860):212–216.

William II, H. F. 1997. Method for administering neurologic agents to the brain. Google Patents.

Wu, D., and W. M. Pardridge. 1999. Neuroprotection with noninvasive neurotrophin delivery to the brain. *Proceedings of the National Academy of Sciences of the United States of America* 96 (1):254–259.

Xiang, Y., Q. Wu, L. Liang, X. Wang, J. Wang, X. Zhang, X. Pu, and Q. Zhang. 2012. Chlorotoxin-modified stealth liposomes encapsulating levodopa for the targeting delivery against the Parkinson's disease in the MPTP-induced mice model. *Journal of Drug Targeting* 20 (1):67–75.

Yamada, K., C. Yabuki, P. Seubert, D. Schenk, Y. Hori, S. Ohtsuki, T. Terasaki, T. Hashimoto, and T. Iwatsubo. 2009. Aβ immunotherapy: Intracerebral sequestration of Aβ by an anti-Aβ monoclonal antibody 266 with high affinity to soluble Aβ. *Journal of Neuroscience* 29 (36):11393–11398.

Yu, Y. J., and R. J. Watts. 2013. Developing therapeutic antibodies for neurodegenerative disease. *Neurotherapeutics: The Journal of the American Society for Experimental NeuroTherapeutics* 10 (3):459–472.

Zarranz, J. J., J. Alegre, J. C. Gómez-Esteban, E. Lezcano, R. Ros, I. Ampuero, L. Vidal, J. Hoenicka, O. Rodriguez, and B. Atarés. 2004. The new mutation, E46K, of α-synuclein causes Parkinson and Lewy body dementia. *Annals of Neurology* 55 (2):164–173.

Zhang, K., M. Schrag, A. Crofton, R. Trivedi, H. Vinters, and W. Kirsch. 2012. Targeted proteomics for quantification of histone acetylation in Alzheimer's disease. *Proteomics* 12 (8):1261–1268.

Zhao, H. M., X. F. Liu, X. W. Mao, and C. F. Chen. 2004. Intranasal delivery of nerve growth factor to protect the central nervous system against acute cerebral infarction. *Chinese Medical Sciences Journal* 19 (4):257–261.

Zou, L.-L., J.-L. Ma, T. Wang, T.-B. Yang, and C.-B. Liu. 2013. Cell-penetrating peptide-mediated therapeutic molecule delivery into the central nervous system. *Current Neuropharmacology* 11 (2):197–208.

CHAPTER NINE

Cancer

Lara Scheherazade Milane

CONTENTS

9.1 BIOLOGY AND CLINICAL TREATMENT OF CANCER

9.1.1 Hallmarks of Cancer: 10 + 3

The best way to understand cancer as a disease is in terms of Darwinian evolution; cancer is survival of the fittest at its finest. Each cancer cell is the epitome of Darwinian evolution at the molecular level; through natural selection of variations in genotypes and phenotypes, each cancer cell optimizes its ability to compete, survive, and reproduce. Darwin himself would be astounded at the miraculous speed with which these molecular adaptations can occur. A cancer cell, most simply, is a cell that will do and can do anything it has to do in order to survive. With survival as a primary goal, it is easy to conceive how it is essential for a cancer cell to have distinct behaviors that distinguish it from a normal cell. In 2000, Douglas Hanahan and Robert Weinberg wrote a seminal paper, "Hallmarks of Cancer," classifying the characteristics of cancer into six categories (Hanahan and Weinberg 2000). In 2011, Hanahan and Weinberg deepened their contribution to cancer biology when they expanded their classifications in a second seminal review, "Hallmarks of Cancer: The Next Generation" (Hanahan and Weinberg 2011). The six initial hallmarks identified by Hanahan and Weinberg are

1. Resisting cell death
2. Sustaining proliferative signaling
3. Evading growth suppressors
4. Activating invasion and metastasis

5. Enabling replicative immortality

6. Inducing angiogenesis

The two emerging hallmarks that Hanahan and Weinberg identified in 2011 are

1. Deregulating cellular energetics

2. Avoiding immune destruction

The two enabling characteristics that Hanahan and Weinberg identified in 2011 are

1. Genome instability and mutation

2. Tumor-promoting inflammation

The 2016 additions that I add to Hanahan and Weinberg's hallmarks of cancer are

1. Cellular plasticity (+1)

2. Tumor heterogeneity (+2)

3. Quiescence and stemness (+3)

This addition is the basis for the title of this section, "Hallmarks of Cancer: 10 + 3"—Hanahan and Weinberg's 10 defined hallmarks, plus 3 of my own defining hallmarks of cancer. These hallmarks are depicted in Figure 9.1 (Hanahan and Weinberg 2011).

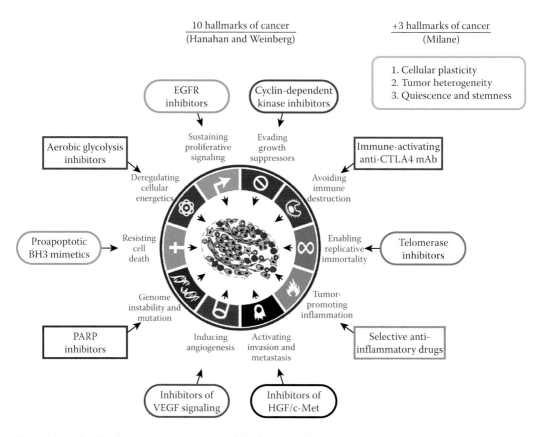

Figure 9.1. Hallmarks of cancer. Cancer is survival of the fittest at its finest. A cancer cell will change and adapt in almost any way that it needs to achieve the one main goal: survival. There are drugs and therapies that target each hallmark (see the text for a full description of each hallmark). Cellular plasticity represents this adaptability; cancer cells undergo perpetual change. This is one reason why clinical dosing presents an advantage for the creation of acquired MDR. Tumor heterogeneity is why the standard of care is to treat with more than one drug. A tumor is not one population of cells; a tumor is a polyclonal mass of cells, with many different phenotypes. Quiescence and stemness represent the ability of the cells to enter and exist G_0; these persister cells are hard to reach, and they are a clinical challenge, as their low basal stage facilitates MDR and metabolic protection. (Adapted from Hanahan, D., and Weinberg, R. A., Cell, 144(5), 646–674, 2011. With permission.)

9.1.1.1 Resisting Cell Death

Returning to the concept that a cancer cell is survival of the fittest at its finest, a simplification of this is that a cancer cell is a cell that is resistant to death. This category includes a multitude of cellular adaptations that ensure that the extrinsic (death receptor) and intrinsic apoptotic pathways do not result in cell death.

The extrinsic death receptor apoptotic pathway is initiated through engagement of the death receptor at the cellular membrane level. Death receptor proteins that have a recognized death domain include TNF-R1, CD95 (APO1/FAS), TRAIL-R1 (DR4), TRAIL-R2 (APO-2/TRICK/DR5/KILLER), DR3 (TRAMP/APO-3), and DR6 (Walczak 2013). The ligands for the death receptors include tumor necrosis factor (TNF), CD95L (FasL), TL1A, and TRAIL (Walczak 2013). Upon ligand binding, the intracellular death domain of the receptor associates with the TNF receptor–associated death domain (TRADD) of the Fas-associated death domain (FADD). FADD then interacts with pro-caspase 8 to form the death-inducing signaling complex (DISC); caspase 8 is then activated, beginning the caspase cascade of the extrinsic pathway (subsequent activation of effector caspases 3, 6, and 7), which allows for eventual apoptosis (Walczak 2013). There is intersection between the extrinsic and intrinsic pathways at the level of caspase 8; caspase 8 can activate Bid, which in turn activates Bax or Bak, leading to mitochondrial outer membrane permeability, cytochrome c release, and eventual apoptosis (Tait and Green 2010). The intrinsic pathway, on the other hand, does not involve activation of a cell membrane receptor. The intrinsic pathway begins when DNA damage or cellular stress activates BH3-only proteins, which in turn activate Bax or Bak to induce mitochondrial outer membrane permeability, which subsequently leads to the release of cytochrome c from mitochondria (Tait and Green 2010). Cytochrome c complexes with apoptotic protease-activating factor 1 (APAF1), which initiates the formation of the apoptosome, incorporating the initiator caspase 9 (Tait and Green 2010). Caspase 9 then begins the caspase cascade of the intrinsic pathway, cleaving and activating the effector caspases 3 and 7, which allows for eventual apoptosis (Tait and Green 2010). To further promote apoptosis through the intrinsic pathway, mitochondria release SMAC/

DIABLO (second mitochondria-derived activator of caspase) and OMI (HTRA2); SMAC/DIABLO and OMI bind to the X-linked inhibitor of apoptosis protein (XIAP) and prevent XIAP inhibition of caspases 3, 7, and 9 (Tait and Green 2010). From this abbreviated description of the intrinsic and extrinsic apoptotic pathways, it is easy to conceptualize that there is an apoptotic threshold that must be overcome for apoptosis to occur. It is also easy to conceive with such a sequential and cumulative series of events that the apoptotic process could be diverted even after it has begun. There are many opportunities for intersection and for cancer cells to evade apoptosis. Many cancers are resistant to death receptor activation (such as by TNF-related apoptosis-inducing ligand [TRAIL]) through the extrinsic pathway (Carr et al. 2016). Some cancer cells upregulate antiapoptotic proteins such as Bcl-2, Bcl-XL, Mcl-1, and IAPs, conferring resistance to the intrinsic apoptotic pathway (Indran et al. 2011).

Resisting apoptosis also includes resistance to cell death through exogenous factors (drugs), the acquisition of multidrug resistance (MDR), and the acquisition of stemness, as stemness promotes both resistance to cell death and immortalization. A significant mechanism by which cancer cells resist apoptosis is through mutations or losses of p53. p53 is a tumor suppressor that functions as a master regulator of gene expression; p53 is a transcription factor that governs many processes, including the cell cycle, metabolism, apoptosis, senescence, and immune regulation. This master tumor suppressor is inactive in more than half of all sporadic cancers, and these p53-deficient cancers are more malignant than cancers with retained p53 (Bieging et al. 2014). Hypoxia, DNA damage, hyperproliferative signaling, oxidative stress, nutrient depletion, ribonucleotide depletion, DNA replicative stress, and telomere depletion can trigger p53 activation (release from negative suppression) (Bieging et al. 2014). The completion of the Human Genome Project in 2003 and the subsequent burst of genomic research produced an overwhelming amount of data, but we are just now perfecting the tools for processing this data. A recently developed analytical tool is expression quantitative trait loci (eQTL) mapping, which can process the results of genome-wide association studies, provide a phenotype, and categorize single-nucleotide polymorphisms (Stracquadanio et al. 2016). Recent genomic analysis using *e*QTL and other analytical tools has revealed that single-nucleotide polymorphisms of the p53 pathway are similar to cancer-causing somatic mutations of the pathway, and variations in the p53 pathway dominate the incidence of more cancers than any other molecular pathway studies to date (Stracquadanio et al. 2016). p53 is truly the master tumor suppressor; although p53 has various mechanisms of antitumor activity, such as permeabilization of mitochondria to induce apoptosis, the primary function of p53 in tumor suppression is its role as a transcription factor. Analysis of p53 mutations has revealed a cancer hot spot zone on the gene; this zone lies in the DNA binding domain of p53, and more than 80% of p53 mutations are associated with this region (Olivier et al. 2010; Bieging et al. 2014).

These mutations disrupt the ability of p53 to bind to some target genes and regulate transcription (Olivier et al. 2010; Bieging et al. 2014). The target genes of p53 regulate cell cycle arrest (*Cdkn1a*, Reprimo, mir-34a, mir-34 b/c, and Gadd45a), apoptosis (Apaf1, Bax, mir-34, Noxa, Puma, and Fas), senesence (Pml, Pai1, and *Cdkn1a*), DNA repair (Mgmt, Ddb2, Fancc, and Gadd45a), metabolism (Parkin, Pten, Aldh4, and Sco1), autophagy (Puma, Foxo3, Atg, and Vamp4), tumor microenvironment regulation (Icam1, Bai1, Tlr1-10, and Tsp1), invasion and metastasis regulation (Cdkn1a, mir-200c, and mir-34a), and stem cell biology (Cdkn1a, Notch1, mir-34a, mir-34 b/c, and mir-145) (Bieging et al. 2014). Interestingly, activation of p53 resists the Warburg effect (increased cytoplasmic aerobic glycolysis and decreased mitochondrial oxidative phosphylation) through the activation of *Tigar* (TP53-induced glycolysis and apoptosis regulator) and inhibition of glucose transporters (1 and 4) (Bensaad et al. 2006; Bieging et al. 2014; Schwartzenberg-Bar-Yoseph et al. 2004). Recent molecular studies have identified that p53 prevents stemness through suppression of pluripotency (Choi et al. 2011; Lin et al. 2012; Bieging et al. 2014). Pluripotency is inhibited through p53 activation of mir-145 and mir-34a-c (Choi et al. 2011; Lin et al. 2012; Bieging et al. 2014).

9.1.1.2 Sustaining Proliferative Signaling

A second hallmark of cancer is the ability to sustain proliferative signaling (Hanahan and Weinberg 2011). Chronic, deregulated growth defines cancer's ability to maintain itself, perpetuate, and propagate. A central mechanism for achieving this is the upregulation of growth factor receptors, constitutive signaling of growth pathways (such as the Raf MAP kinase pathway and PI3-kinase), paracrine signaling to increase the production of growth factors in the tumor microenvironment, and dissolution of negative feedback loops (Ras oncoprotein; mTOR pathway) (Hanahan and Weinberg 2011).

Although Hanahan and Weinberg discuss senescence as a response to repeated cell divisions (consequence of proliferative signaling) and telomere shortening, senescence of cancer cells is more than a response to repeated cell divisions; it can be a mechanism of survival. Senescence is classically an irreversible stage, but as cancer cells defy normal cellular processes, they also defy senescence and, as discussed shortly, appear to be able to transition from senescence to quiescence and subsequent rescue. In all regards, when cancer cells enter (and exit) senescence, it is a protective mechanism to sustain existence even in the presence of excessive proliferative signaling that exceeds cellular capacity.

9.1.1.3 Evading Growth Suppressors

A third hallmark of cancer is evasion of growth suppression (Hanahan and Weinberg 2011). One mechanism of growth suppression evasion is through loss of tumor suppressor function, including p53 and RB (retinoblastoma-associated protein; RB protein regulates cell cycle progression) (Hanahan and Weinberg 2011). Other mechanisms include deregulation of transforming growth factor (TGF) β and loss of contact inhibition (Hanahan and Weinberg 2011). Loss of contact inhibition actually promotes aggressive phenotypes; constant growth, overgrowth, destruction, and remodeling of the tumor microenvironment are due in part to the lack of contact inhibition of cancer cells. Although tumors induce angiogenesis, the vasculature is constantly destroyed as cancer cells crush the disorganized vessels and new vessels are formed. This phenomenon of constant tumor microenvironment remodeling creates a situation where cancer cells do not have consistent oxygen and nutrients from the vasculature. This, of course, depends on the rate of growth of the cancer cells, and can be distinct within different regions of a tumor. However, this occurrence actually creates a scenario where cancer cells are fighting against their own microenvironmental conditions to survive as cells with access to newly formed vasculature may lose it and have to adapt to conditions of hypoxia (or anoxia) and nutrient deprivation. The disorganized growth also effects the lymphatics of the solid tumor. Tumors have poor lymphatic drainage, and this, combined with the interstitial fluid pressure, is the basis of the enhanced permeability and retention (EPR) effect, yet the magnitude of this can also change as the tumor microenvironment is remodeled. This is an important consideration in nanomedicine design, as passive targeting of tumors with nanomedicine relies on exploitation of the EPR effect.

9.1.1.4 Activating Invasion and Metastasis

A fourth hallmark of cancer is the ability to activate invasion and metastasis (Hanahan and Weinberg 2011). For this to occur, cancer cells lose adhesion molecules and then undergo a mobilization process known as epithelial-to-mesenchymal transition

(EMT), followed by establishment of a secondary tumor site in the process of mesenchymal-to-epithelial transition (Milane et al. 2015). Exosomes are endogenous cellular nanoparticles released as a subsequent product of the endocytotic pathway and during cellular migration. Exosomes actually play a significant role in cancer cell invasion and metastasis. Figure 9.2 portrays the exosomal pathway in cancer as a 10-step process. The process begins with altered membrane dynamics (different lipid compositions and altered receptor expression) and increased plasma membrane remodeling through endocytosis (steps 1 and 2). Exosomes are eventually formed within the endosome, and specific proteins and miRNAs are trafficked to the endosome for incorporation into preexosomes (steps 3–6). Exosomal content is selected through the endosomal sorting complex required for transcript (ESCRT) machinery, through lipid-mediated interactions, and through receptor–ligand interactions (step 7). Multivesicular bodies (MVBs) are trafficked to the plasma membrane by Rab GTPases (step 8). MVBs fuse with the plasma membrane via SNARE-dependent and SNARE-independent processes

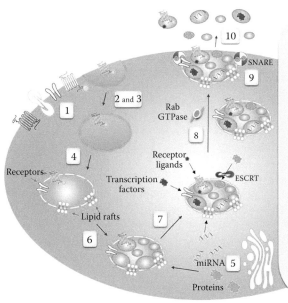

1. Fundamental abnormalities in cancer (such as HIF activity and p53 oncogene activity), result in subsequent plasma membrane alterations such as altered lipid composition and altered receptor expression (increased expression of drug efflux pumps, growth factors, and nutrient importers).
2. Abnormal plasma membrane constituents (altered receptor expression) and constant phenotypic flux requires increased plasma membrane remodeling through endocytosis.
3. Increased endocytosis results in an eventual increase in exosome formation.
4. The endosome is a signaling entity (likened to a transient organelle).
5. Synthesis of microRNAs and molecules for exosomal export are trafficked towards endosomes.
6. The endosome MVB interacts with cytosolic components; ILVs (pre-exosomes) begin to form.
7. Content is selected through ESCRT machinery, through receptor mediated internalization, and through lipid mediated interactions such as microRNA engagement with ceramide lipid rafts. ESCRT machinery and/or lipids function to complete exosome formation.
8. Rab GTPases traffic MVBs to plasma membrane.
9. SNARE dependent or SNARE independent fusion of MVB with plasma membrane and exosome secretion.
10. Exosomal function in local invasion of ECM and enhancing drug resistance of tumor microenvironment, or as cellular derived nanoparticles that can engage specific receptors and transform target cells.

Figure 9.2. Exosomal biology in cancer. Cancer exosomes are a key mechanism of immune cell and cancer cell communication within the tumor microenvironment. Exosomes are released as a by-product of endocytosis, as a form of intracellular signaling, during cellular migration (movement requires the leading edge of the cell to secrete exosomes) and invasion (exosomes can contain MMPs that remodel the extracellular matrix). (ESCRT machinery = endosomal sorting complex required for transcript; MVB = multivesicular bodies; ILV = intraluminal vesicles; SNARE = soluble NSF attachment protein receptors; ECM = extracellular matrix). (Reprinted from Milane, L. et al., *J. Control. Release*, 219, 278–294, 2015. Illustrated by Lara Milane. With permission.)

and exosomes are secreted (step 9). Exosomes can function in digestion of the extracellular matrix, migration, invasion, MDR, and the establishment of premetastatic niches at distal sites. Exosome release occurs prior to any cellular movement, and cellular migration is associated with a higher level of exosome release in the polarized direction of movement (Milane et al. 2015). Exosomes are actually associated with invadopdia, the actin-rich extensions of cancer cells that begin the degradation of the extracellular matrix for invasion (Hoshino et al. 2013). Exosomes extracted from active invadopodia regions have the ability the transform noninvading cells into cells with active invadopodia (Hoshino et al. 2013). Exosomes have also been demonstrated to prime and establish metastatic nodes, and exosomes from metastatic cells can transform nonmetastatic cells into metastatic cells (governed by the transfer of miR200s) (Milane et al. 2015). Exosomes are also a mechanism of cell signaling and communication within the tumor microenvironment. Cancer-associated fibroblasts (CAFs) and tumor-associated macrophages (TAMs) also play a role in invasion and metastasis, as these cells can supply growth factors to cancer cells, and these cells are diverted (without subsequent immune clearance of tumor) during invasion.

9.1.1.5 Enabling Replicative Immortality and Inducing Angiogenesis

A fifth hallmark of cancer is the ability to enable replicative immortality (Hanahan and Weinberg 2011). Telomeres are recognized as the cellular timer for replication; telomeres protect the ends of DNA and successively shorten with each cell replication. Once telomeres are depleted, DNA is not protected from fusion and chromosomes can become destabilized (Hanahan and Weinberg 2011). Telomerase adds telomeres to the ends of DNA, and this polymerase is overexpressed in cancer cells (Hanahan and Weinberg 2011). In this sense, the biological clock of a cancer cell does not tick forward. A sixth hallmark of cancer is the ability to induce angiogenesis (Hanahan and Weinberg 2011). Vascular endothelial growth factor A (VEGF-A) is a central mediator of neovasculature induction (Hanahan and Weinberg 2011). Although tumors can induce the creation of their own blood supply, the vasculature of a solid tumor is very disorganized

and "leaky." The leaky vasculature is exploited in the EPR effect of passive nanomedicine tumor targeting.

9.1.1.6 Emerging Hallmarks: Deregulating Cellular Energetics and Avoiding Immune Destruction

A seventh characteristic of cancer cells categorized by Hanahan and Weinberg as an emerging hallmark is deregulated cellular energetics. In the 1930s, Otto Warburg first identified that cancer cells resort to aerobic glycolysis for energy production (glycolysis even in the presence of oxygen), whereas normal cells resort to glycolysis under conditions of oxygen depletion (anaerobic glycolysis) (Milane et al. 2011a). Aerobic glycolysis in cancer cells is termed the Warburg effect. In normal cells, the presence of oxygen inhibits glycolysis (this is the Pasteur effect). Even though some cancer cells resort to the Warburg effect, oxidative phosphorylation still occurs. The impetus for aerobic glycolysis is multifold. One reason is that glycolysis is safer than oxidative phosphorylation. Oxidative phosphorylation produces reactive oxygen species (ROS) that can jeopardize the survival of a cancer cell, whereas glycolysis is not associated with free radical production. Also, due to the dynamic microenvironment of a tumor, it is a survival benefit for cancer cells to decrease reliance on oxygen (required for oxidative phosphorylation). Another reason is that glycolysis is a faster process (occurs in cytoplasm vs. mitochondrial oxidative phosphorylation). Of note, even though glycolysis is faster, the reason it is not employed as a primary mechanism of energy production is that it is very inefficient compared with oxidative phosphorylation (2 ATP vs. 36). Glycolysis actually provides biomolecules required for proliferation, and the accumulation of lactic acid reduces the pH, which may enhance invasion. An eighth characteristic of cancer also classified by Hanahan and Weinberg as an emerging hallmark is the ability of cancer cells to avoid immune destruction (Hanahan and Weinberg 2011). This hallmark is biologically intuitive; the function of the immune system is to protect tissues, maintain homeostasis, and eliminate damaged cells. In this sense, cancer is a failure of the immune system.

9.1.1.7 Enabling Characteristics: Genome Instability and Mutation and Tumor-Promoting Inflammation

Hanahan and Weinberg categorize two characteristics as enabling, meaning that these characteristics foster the maintenance of the other hallmarks (Hanahan and Weinberg 2011). As discussed so far in the context of the other hallmarks, genetic mutations and genome instability are central to cancer survival; mutations in p53 and RB protein are two prime examples already discussed. Cells of the innate and adaptive immune system infiltrate the tumor microenvironment and actually aid tumor survival by providing signaling molecules (growth factors, angiogenic factors, and extracellular matrix–modifying enzymes) that support the propagation of a tumor (Hanahan and Weinberg 2011). Elaboration of cancer as an inflammatory disease is discussed shortly.

9.1.1.8 +3 Hallmarks: Cellular Plasticity, Tumor Heterogeneity, Quiescence, and Stemness

Hanahan and Weinberg's last review of cancer hallmarks was published in 2011; our understanding of cancer has progressed since then. Three additions that I contribute to Hanahan and Weinberg's classical hallmarks of cancer are cellular plasticity, tumor heterogeneity, and quiescence and stemness. Although Hanahan and Weinberg's seminal paper discusses these topics briefly, they are not categorized by them as hallmarks. In light of the research conducted in the past few years, these categories are entitled to being unique hallmarks, not merely consequences. Cellular plasticity is a primary characteristic of cancer cells; cancer cells have the ability to adapt to maintain survival—and they continually do this. As indicated by biomarker signatures, many cancer cells will retain some of their molecular signatures, but an important occurrence is that many cells within a solid tumor undergo perpetual flux in order to maintain survival. The microenvironment of a solid tumor is also continually changing, so each cancer cell must have the ability to adapt survival mechanisms to maintain itself under a state of perpetual flux. Cancer cell plasticity can be thought of as the ability of a cancer cell to change and adapt its phenotype to maintain survival. A prime example of cancer cell plasticity

is the relationship between the hypoxia-inducible factor (HIF) and the emergence of the MDR phenotype. In the presence of cell stress or hypoxia, HIFα (there are multiple isoforms, but 1α and 2α are the most common) translocates from the cytoplasm to the nucleus, where it forms an active transcription factor complex with HIFβ. The HIF complex now functions as an active transcription factor and is able to activate target genes, including P-glycoprotein and MDR protein 1 (two drug efflux pumps that are central to the MDR phenotype), epidermal growth factor receptor (EGFR), and GLUT1 glucose receptor. This plasticity promotes survival; in the case of hypoxia, HIF upregulates growth factor receptors and nutrient importers to make the cell hypersensitive to available factors in a deprived tumor microenvironment and also upregulates drug efflux pumps as a protective mechanism.

Likewise, tumor heterogeneity is a hallmark of cancer. This means that cancers are not 100% clonal. Tumors are composed of a diverse population of cancer cells, immune cells, and stromal cells. The significance of tumor heterogeneity is demonstrated by the clinical management of the disease; combination therapy is the standard. Why is combination therapy the standard? Because the tumor mass is composed of a diverse population of cells—if tumors were truly 100% clonal and each cancer cell in a tumor had the same phenotype and molecular signature, then treatment with one drug would be effective (or the same) for a population of tumor cells. However, this is not the case. Combination therapy is the standard of care in order to target and molecularly treat a polyclonal mass of cells (by addressing more than one phenotype). Tumor heterogeneity is a consequence of the "messy" and disorganized tumor microenvironment; the tumor microenvironment is not the same throughout the mass—there are differences in extracellular matrix components, stromal cells, immune cells, lymphatics, and vasculature that in turn result in different phenotypes from the localized cells, as each is adapting to survival within its region of the tumor. Recognizing that a tumor is not neat and structured as a normal organ aids in understanding tumor heterogeneity; it parallels the constant remodeling of the tumor microenvironment that is necessary to maintain cancer cell survival.

A third hallmark that is an addition to Hanahan and Weinberg's classical hallmarks is the ability to enter quiescence and exhibit stemness. As demonstrated in Figure 9.3, cancer cells are on the opposite side of the cellular scale to a normal cell. A normal cell begins from a stem cell and becomes differentiated, whereas a cancer cell is a differentiated cell that becomes dysregulated and loses its molecular signatures that distinguish it as a particular cell type. This loss of character can result in stemness, which is a precursor to entering quiescence for many cancer-associated stem cells.

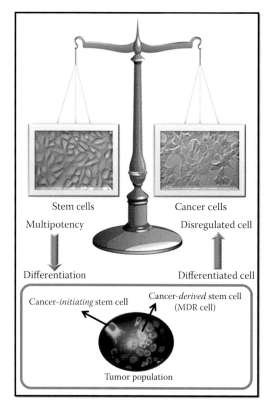

Figure 9.3. Cancer stem cell theory. Cancer cells and normal (healthy) cells are on opposite sides of the cellular scale; a normal healthy cell is a cell that begins from a stem cell linage and becomes differentiated, whereas a cancer cell is a differentiated cell that becomes disregulated and loses its molecular characteristics. In this sense, stem cells and cancer cells seem to balance each other on the cellular scale (equal but opposite). Stem cells are multipotent and can differentiate and specialize, whereas cancer cells are differentiated cells that lose their distinguishing properties and gain stemness. Cancer cells that gain stemness and become quiescent are referred to as cancer-derived stem cells; progression to this phenotype often leads to MDR before stemness is attained. Cancer-initiating stem cells, on the other hand, can give rise to a tumor. (Adapted from Milane, L. et al., *Journal of Controlled Release: Official Journal of the Controlled Release Society*, 155(2), 237–47, 2011b. With permission.)

There also appears to be distinct populations of cancer stem cells: cancer-initiating stem cells that can give rise to tumor cells and cancer-derived stem cells that result from molecular "stripping" of cell-type signatures. Along the progression to stemness, a subpopulation of these transforming cells often exhibit multi-drug-resistant properties. There is a subpopulation of cancer cells that have stemness and multi-drug-resistant phenotypes, as well as distinct stem cell subpopulations and distinct multi-drug-resistant subpopulations. Quiescence is entered when cancer stem cells have a high degree of stem character and enter G_0. When quiescent cancer stem cells are "rescued" from G_0, they can be even more aggressive and resistant; these are the persister cells that often result in relapsing disease that is unresponsive to therapy.

9.1.2 Cancer as an Inflammatory Disease

9.1.2.1 Cancer versus Wound Healing

The concept of cancer as an inflammatory disease originated in 1863 when the scientist Virchow rationalized that regions of chronic inflammation are the sites where cancer originates (Balkwill and Mantovani 2001). Solid tumors have further been described as wounds that do not heal (Coussens and Werb 2002). The similarities of cancer formation to wound healing are demonstrated in Figure 9.4 (Coussens and Werb 2002). Upon normal tissue injury, leukocytes are activated to migrate to the region from the vasculature (Coussens and Werb 2002). The extracellular matrix is a critical supportive environment for repair (Coussens and Werb 2002). Neutrophils are recruited through activation of selectin adhesion molecules that permit "rolling" along vasculature, cytokine-mediated upregulation of leukocyte integrins, localization of neutrophils to the vascular endothelium via integrin adhesion to adhesion molecules such as vascular cell adhesion molecule-1 (VCAM-1), and migration through the endothelium to the injury (Coussens and Werb 2002). As demonstrated in Figure 9.4a, wound healing is a very organized and prescribed process with the basement membrane functioning as a clear segregation between the epithelial cells and vascularized stromal region (Coussens and Werb 2002). Activated platelets at the site of a wound mediate the formation of a hemostatic plug and fibrin clot from the recruitment of fibrinogen

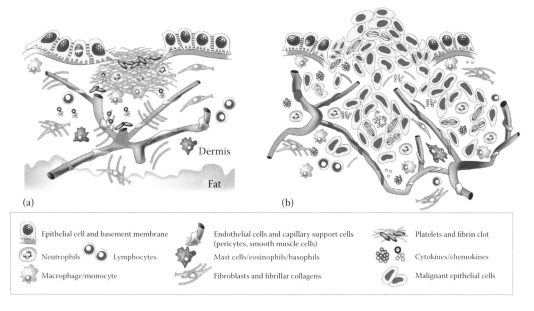

Epithelial cell and basement membrane	Endothelial cells and capillary support cells (pericytes, smooth muscle cells)	Platelets and fibrin clot
Neutrophils Lymphocytes	Mast cells/eosinophils/basophils	Cytokines/chemokines
Macrophage/monocyte	Fibroblasts and fibrillar collagens	Malignant epithelial cells

Figure 9.4. Cancer as abnormal wound healing. (a) Organized process of natural wound healing. (b) Haphazard and dysfunctional wound healing process of the tumor microenvironment. See the text for a more detailed description. (Reprinted from Coussens, L. M., and Werb, Z., *Nature*, 420(6917), 860–867, 2002. With permission.)

(Coussens and Werb 2002). The activated platelets also secrete chemotactic factors that activate fibroblasts and granulocytes and mediate extracellular matrix remodeling (Coussens and Werb 2002). A dynamic interplay of cytokine and chemokine signaling between epithelial cells (and the epithelial region; platelets, fibrin clot, neutrophils, and macrophages) and the stromal cell population (macrophages, lymphocytes, mast cells, eosinophils, and fibroblasts) continues until complete wound healing is achieved (Coussens and Werb 2002). On the other hand, Figure 9.4b demonstrates the haphazard disarray of a tumor. Invasion involves destruction of the basement membrane and loss of epithelial integrity (transformed cells). The vascular reconstruction is disorganized, and the cancer cells occupy the stromal region that, in the normal wound healing process, is dominated by immune cells in a supportive microenvironment (Coussens and Werb 2002). The tumor microenvironment is also supportive, but instead of supporting healing, due to the dynamic and direct interaction of immune cells with the transformed cancer cells, the tumor microenvironment is supportive of tumor sustention and growth. The same players are involved, but they function as directed by the cancer cells; that is why subpopulations of fibroblasts are referred to as CAFs and subpopulations

of macrophages are referred to as TAMs. These cells support the growth and maintenance of a tumor in the same way that they promote wound healing. It is also easy to conceptualize from Figure 9.4 how cancer progression is from a failed immune response.

Dendritic cells play an initial role in cancer-associated inflammation as they encounter early cancer cells and present antigens to T and B lymphocytes in lymphoid organs, initiating an adaptive immune response (de Visser et al. 2006). Pro-inflammatory cytokines and B cell activation lead to a chronic innate immune response that enables tumor sustention and growth, angiogenesis, growth of a supportive extracellular matrix, activated fibroblasts, inhibition of apoptosis, genomic instability, and malignant conversion (de Visser et al. 2006).

9.1.2.2 Cancer-Associated Fibroblasts and Tumor-Associated Macrophages

Cancer cells do not exist and propagate in isolation; an enabling microenvironment supports their growth, and fibroblasts are an enabling constituent of this tumor stroma. Fibroblasts in the tumor microenvironment are referred to as TAFs, activated fibroblasts, or CAFs (Kalluri 2016).

Fibroblasts are normally involved in wound healing and remain quiescent until activated; fibroblasts are an obvious contributor to cancer as an inflammatory disease (Kalluri 2016). Once a quiescent fibroblast is activated for wound healing, the shape of the cell changes (broadens); the fibroblasts express vimentin and α-smooth muscle actin, are more contractile, remodel the extracellular matrix, and secrete factors such as VEGF-A, TNF, interferon (IFN) γ, interleukin (IL) 6, TGF-β, and EGF (Kalluri 2016). These normal activated fibroblasts are capable of further transformation; under conditions of sustained injury-related stimuli or epigenetic activation, normal activated fibroblasts can take on the phenotype of CAFs (Kalluri 2016). CAFs can be highly proliferative, mediate immune responses, extensively remodel the extracellular matrix, and secrete more (and modified) proteins and factors that promote vascularization, growth, and immune resistance (Kalluri 2016). CAFs are undoubtedly supportive cells in the tumor microenvironment, especially since they have demonstrated the ability to become induced pluripotent stem cells, adipocytes, chondrocyte-like cells, or endothelial cells (Kalluri 2016).

Macrophages within the tumor microenvironment are often referred to as Tumor Associated Macrophages (TAMs) or Cancer Associated Macrophages (CAMs). Macrophages have two states of polarization: an M1 state and an M2 state. M1 macrophages are pro-inflammatory (releasing IL-1, IL-6, TNF, and ROS), whereas M2 macrophages are anti-inflammatory (secreting IL-4, IL-10, IL-13, and TGF-β) (Fridman et al. 2012). Chronic inflammation and M2 macrophages support tumor growth (Fridman et al. 2012). M2 polarization within the tumor microenvironment has been linked to HIF1α-mediated action of lactic acid (Colegio et al. 2014). The promotion of tumor growth by M2 macrophages may be linked to the higher levels of VEGF and arginase 1 (Arg1); VEGF contributes to angiogenesis, whereas Arg1 is a substrate for cancer cell growth and proliferation (Colegio et al. 2014).

9.1.2.3 Exosomes

Immune cells rely on cytokine and chemokine signaling, as well as communication through exosomes and tunneling nanotubes; exosomes are endogenous nanocarriers released through endocytosis, as a means of signaling, and through cellular migration, whereas tunneling nanotubes are physical extensions that connect neighboring cells and are capable of content transfer (including large organelles such as mitochondria) between cells (McCoy-Simandle et al. 2016). The role of exosomes and tunneling nanotubes in immune cell signaling has recently been reviewed (McCoy-Simandle et al. 2016); these processes are critical communication pathways within the tumor microenvironment. Exosomes from cancer cells have distinct content from normal cells (microRNA that are not present in normal cells, matrix metalloproteases [MMPs] that digest extracellular matrix and enable invasion); exosomal communication between the immune cells and cancer cells in a solid tumor undoubtedly contributes to progression of the disease, evasion of immune clearance, drug resistance, and creating a microenvironment supportive of tumor growth (Milane et al. 2015).

9.1.3 Nanomedicine for Cancer: The Translational Precedent

According to the U.S. National Cancer Institute (NCI), cancer is a leading cause of death worldwide, and 2012 was marked with 8.2 million deaths globally (NCI 2016). Due to the cellular plasticity of cancer cells (adaptability) and tumor heterogeneity (more than one phenotype within a solid tumor), innate and acquired MDR continues to present a clinical challenge to the treatment of cancer. As such, new therapies are in great demand. Immunotherapies such as cancer vaccines and CAR–T cell therapy (chimeric antigen receptors) have been studied extensively in recent years as an avenue for using the body's immune system to attack a tumor. CAR–T cell therapy is a form of adoptive cell transfer where a patient's own immune cells are extracted and engineered to produce artificial receptors that have a high affinity for tumor cell antigens. Despite three deaths due to cerebral edema in a clinical trial of a CAR–T cell therapy ("Rocket," for refractory B cell acute lymphoblastic leukemia), research and development of improved CAR–T cell therapies has continued. Although cancer vaccines have yet to produce remarkable results, CAR–T cell therapy has transformed the prognosis for many patients in clinical trials.

Although nanotechnology has received less media attention, nanomedicine offers an alternative approach to immunotherapy that may be safer than adoptive cell transfer techniques. A nanocarrier system was recently developed as an adjuvant to PD-L1 antibody therapy for the treatment of advanced colon cancer (He et al. 2016). PD-L1 antibody therapy functions as a checkpoint inhibitor that allows progressive activation of T cells; however, the response rate to PD-L1 antibody is low (He et al. 2016). Researchers designed an elaborate nanoscale coordination polymer core–shell nanocarrier system (NCP@pyrolipid) with oxaliplatin in the core and pyrolipid in the shell (pyrolipid is a photosensitizer pyropheophorbide–lipid conjugate) (He et al. 2016). Combination therapy with the nanocarrier system and with the PD-L1 antibody resulted in regression of the primary tumor and metastatic regions (He et al. 2016). However, this system has not progressed past preclinical *in vivo* testing.

To address the challenge of clinical translation, a group of scientists, including researchers from academia, hospitals, and the NCI's Nanotechnology Characterization Laboratory, formed a list of six tenets to follow for designing targeted nanomedicines for cancer (Goldberg et al. 2013). These tenets are critical, as nanomedicine for cancer is the most fluent area of nanomedicine research, yet this activity has not correlated to clinical translation. The first tenet is "sights on the target"; it includes selection parameters for the target, such as selecting a target with high upregulation or exclusive presentation on cancer cells; kinetic parameters of the receptor, such as saturation; and internalization of receptor ligands (Goldberg et al. 2013). The second tenet is "leaping biological hurdles"; it includes designing a treatment plan that addresses tumor heterogeneity and differences in the rate and amount of drug absorption (Goldberg et al. 2013). The third tenet is "these are not tablets"; nanomedicine formulations have unique material properties, including particle composition, shape, size, charge, and surface modification (Goldberg et al. 2013). The fourth tenet is "from one to millions"; it includes the procedures and complications associated with scale-up manufacturing (Goldberg et al. 2013). The fifth tenet is "technology meets reality"; it addresses the concept that nanomedicines, even if they use previously approved materials, are new drug entities, and validating approval requires that the nanomedicine has a good safety profile and improved efficacy over the standard of care (Goldberg et al. 2013). The sixth tenet is "at what cost?"; the manufacture of nanomedicines is expensive compared with traditional medicines, and this translates to a substantial cost increase that must be justified and in line with the patient's needs (Goldberg et al. 2013). These tenets are important for designing translational nanomedicines for cancer. To assist with translation, the NCI's Alliance for Nanotechnology in Cancer formed Translation of Nanotechnology in Cancer (TONIC). TONIC is a consortium of government, public, and private entities working together to facilitate the evaluation and translation of promising nanomedicines for cancer therapy.

Cancer nanomedicine truly is the archetype of translational nanomedicine. There are nine Food and Drug Administration (FDA)–approved nanomedicines for cancer in the United States.

1. Doxil™ was the first FDA-approved nanomedicine, in 1995. It is a liposomal form of doxorubicin marketed by Janssen for the treatment of Kaposi's sarcoma, second-line ovarian cancer, and multiple myeloma.

2. Myocet™ is a liposomal doxorubicin formulation marketed by Teva UK for the treatment of metastatic breast cancer.

3. Marqibo™ is a liposomal vincristine formulation marketed by Spectrum Pharmaceuticals for the treatment of acute lymphoblastic leukemia.

4. DaunoXome™ is approved for the treatment of HIV-related Kaposi's sarcoma; it is a daunorubicin liposome formulation.

5. Pacira Pharmaceuticals markets Depocyt™, cytarabine lipsosomes, for the treatment of lymphomatous meningitis.

6. Onivyde™ is the most recent approval, irinotecan liposomes marketed by Merrimack Pharmaceuticals as second-line therapy for metastatic pancreatic cancer.

7. Baxalta markets Oncaspar™, polyethylene glycol–asparaginase polymeric conjugates for treating acute lymphoblastic leukemia.

8. Samyang Biopharmaceuticals markets paclitaxel polymeric micelles, Genexol-PM™, for treating breast cancer, non-small-cell lung cancer, and ovarian cancer.

9. Abraxane™ is an albumin-bound paclitaxel formulation marketed by Celgene for treating advanced breast cancer, advanced non-small-cell lung cancer, and advanced pancreatic cancer.

Translational nanomedicine for the treatment of cancer truly is leading the field. Although there are nine approved nanomedicine therapies for the treatment of cancer, characterization remains a challenge. For years, researchers such as Rakesh Jain have been exploring how to improve the delivery of nanomedicine to the site of a tumor by normalizing the tumor blood vessels (Chauhan et al. 2012). Jain and colleagues have found that inhibiting VEGF-2 in tumor vessels impedes the delivery of larger nanoparticles (more than 100 nm) and aids the delivery of smaller particles (diameter 12 nm) (Chauhan et al. 2012). Expanding characterization studies such as this to include a wide range of tumors could aid in nanomedicine design for clinical translation, as size can easily be optimized. However, current data is not broad and extensive. There are a plethora of new cancer nanomedicines under experimental development, such as the NCP@ pyrolipid system discussed earlier. The arena of cancer nanomedicines covers a wide range of materials (from polymers to metals), designs (from liposomes to dendrimers), payloads, surface modifications, and surface properties. Yet, until protocols for evaluating new nanomedicines and characterization studies solidify the biological behavior of nanomedicines, clinical translation will remain a challenge, even for cancer nanomedicines.

REFERENCES

Balkwill, F., and A. Mantovani. 2001. Inflammation and Cancer: Back to Virchow? *Lancet* 357 (9255): 539–45.

Bensaad, K., A. Tsuruta, M. A. Selak, M. N. C. Vidal, K. Nakano, R. Bartrons, E. Gottlieb, and K. H. Vousden. 2006. TIGAR, a p53-Inducible Regulator of Glycolysis and Apoptosis. *Cell* 126 (1): 107–20.

Bieging, K. T., S. S. Mello, and L. D. Attardi. 2014. Unravelling Mechanisms of p53-Mediated Tumour Suppression. *Nature Reviews: Cancer* 14 (5): 359–70.

Carr, R. M., G. Qiao, J. Qin, S. Jayaraman, B. S. Prabhakar, and A. V. Maker. 2016. Targeting the Metabolic Pathway of Human Colon Cancer Overcomes Resistance to TRAIL-Induced Apoptosis. *Cell Death Discovery* 2: 16067.

Chauhan, V. P., T. Stylianopoulos, J. D. Martin, Z. Popović, O. Chen, W. S. Kamoun, M. G. Bawendi, D. Fukumura, and R. K. Jain. 2012. Normalization of Tumour Blood Vessels Improves the Delivery of Nanomedicines in a Size-Dependent Manner. *Nature Nanotechnology* 7 (6): 383–88.

Choi, Y. J., C.-P. Lin, J. J. Ho, X. He, N. Okada, P. Bu, Y. Zhong et al. 2011. miR-34 miRNAs Provide a Barrier for Somatic Cell Reprogramming. *Nature Cell Biology* 13 (11): 1353–60.

Colegio, O. R., N.-Q. Chu, A. L. Szabo, T. Chu, A. M. Rhebergen, V. Jairam, N. Cyrus et al. 2014. Functional Polarization of Tumour-Associated Macrophages by Tumour-Derived Lactic Acid. *Nature* 513 (7519): 559–63.

Coussens, L. M., and Z. Werb. 2002. Inflammation and Cancer. *Nature* 420 (6917): 860–67.

de Visser, K. E., A. Eichten, and L. M. Coussens. 2006. Paradoxical Roles of the Immune System during Cancer Development. *Nature Reviews: Cancer* 6 (1): 24–37.

Fridman, W. H., F. Pagès, C. Sautès-Fridman, and J. Galon. 2012. The Immune Contexture in Human Tumours: Impact on Clinical Outcome. *Nature Reviews: Cancer* 12 (4): 298–306.

Goldberg, M. S., S. S. Hook, A. Z. Wang, J. W. M. Bulte, A. K. Patri, F. M. Uckun, V. L. Cryns et al. 2013. Biotargeted Nanomedicines for Cancer: Six Tenets before You Begin. *Nanomedicine (London, England)* 8 (2): 299–308.

Hanahan, D., and R. A. Weinberg. 2000. The hallmarks of cancer. *Cell* 100 (1): 57–70.

Hanahan, D., and R. A. Weinberg. 2011. Hallmarks of Cancer: The Next Generation. *Cell* 144 (5): 646–74.

He, C., X. Duan, N. Guo, C. Chan, C. Poon, R. R. Weichselbaum, and W. Lin. 2016. Core-Shell Nanoscale Coordination Polymers Combine Chemotherapy and Photodynamic Therapy to Potentiate Checkpoint Blockade Cancer Immunotherapy. *Nature Communications* 7: 12499.

Hoshino, D., K. C. Kirkbride, K. Costello, E. S. Clark, S. Sinha, N. Grega-Larson, M. J. Tyska, and A. M. Weaver. 2013. Exosome Secretion Is Enhanced by Invadopodia and Drives Invasive Behavior. *Cell Reports* 5 (5): 1159–68.

Indran, I. R., G. Tufo, S. Pervaiz, and C. Brenner. 2011. Recent Advances in Apoptosis, Mitochondria and Drug Resistance in Cancer Cells. *Biochimica et Biophysica Acta (BBA)—Bioenergetics of Cancer* 1807 (6): 735–45.

Kalluri, R. 2016. The Biology and Function of Fibroblasts in Cancer. *Nature Reviews: Cancer* 16 (9): 582–98.

Lin, C.-P., Y. J. Choi, G. G. Hicks, and L. He. 2012. The Emerging Functions of the p53-miRNA Network in Stem Cell Biology. *Cell Cycle (Georgetown, Tex.)* 11 (11): 2063–72.

McCoy-Simandle, K., S. J. Hanna, and D. Cox. 2016. Exosomes and Nanotubes: Control of Immune Cell Communication. *International Journal of Biochemistry & Cell Biology* 71: 44–54.

Milane, L., Z. Duan, and M. Amiji. 2011a. Role of Hypoxia and Glycolysis in the Development of Multi-Drug Resistance in Human Tumor Cells and the Establishment of an Orthotopic Multi-Drug Resistant Tumor Model in Nude Mice Using Hypoxic Pre-Conditioning. *Cancer Cell International* 11: 3.

Milane, L., S. Ganesh, S. Shah, Z.-F. Duan, and M. Amiji. 2011b. Multi-Modal Strategies for Overcoming Tumor Drug Resistance: Hypoxia, the Warburg Effect, Stem Cells, and Multifunctional Nanotechnology. *Journal of Controlled Release: Official Journal of the Controlled Release Society* 155 (2): 237–47.

Milane, L., A. Singh, G. Mattheolabakis, M. Suresh, and M. M. Amiji. 2015. Exosome Mediated Communication within the Tumor Microenvironment. *Journal of Controlled Release: Official Journal of the Controlled Release Society* 219: 278–94.

NCI (National Cancer Institute). 2016. Cancer Statistics. Bethesda, MD: National Cancer Institute. https://www.cancer.gov/about-cancer/understanding/statistics (accessed October 5).

Olivier, M., M. Hollstein, and P. Hainaut. 2010. TP53 Mutations in Human Cancers: Origins, Consequences, and Clinical Use. *Cold Spring Harbor Perspectives in Biology* 2 (1): a001008.

Schwartzenberg-Bar-Yoseph, F., M. Armoni, and E. Karnieli. 2004. The Tumor Suppressor p53 Down-Regulates Glucose Transporters GLUT1 and GLUT4 Gene Expression. *Cancer Research* 64 (7): 2627–33.

Stracquadanio, G., X. Wang, M. D. Wallace, A. M. Grawenda, P. Zhang, J. Hewitt, J. Zeron-Medina et al. 2016. The Importance of p53 Pathway Genetics in Inherited and Somatic Cancer Genomes. *Nature Reviews: Cancer* 16 (4): 251–65.

Tait, S. W. G., and D. R. Green. 2010. Mitochondria and Cell Death: Outer Membrane Permeabilization and Beyond. *Nature Reviews: Molecular Cell Biology* 11 (9): 621–32.

Walczak, H. 2013. Death Receptor–Ligand Systems in Cancer, Cell Death, and Inflammation. *Cold Spring Harbor Perspectives in Biology* 5 (5): a008698.

CHAPTER TEN

Diabetes

Antonio J. Ribeiro, Marlene Lopes, Raquel Monteiro,
Gaia Cilloni, Francisco Veiga, and P. Arnaud

CONTENTS

10.1 INTRODUCTION

Nanotechnology-based approaches toward oral delivery of macromolecules such as insulin are increasing therapeutic approaches toward the prevention or treatment of diabetes. Nanoencapsulation of insulin increases its protection against enzymatic degradation, and facilitates its absorption through intestinal membranes.

However, the intestinal absorption of insulin nanoparticles (NPs) may be related to a stimulatory reaction induced by the local mucosa immune system. Intestinal epithelium is an immune-privileged organ capable of mediating immune reactions by either playing a local protective role or triggering an inflammatory response to the presence of NPs.

Compromising the safety of insulin delivered by NPs, by inadequacy or insufficiency of studies, may lead to exacerbation of the inflammatory pathways conducive to unwanted local and other severe adverse effects. Therefore, it is imperative to include comprehensive safety assays in early preclinical studies in order to increase the representativeness of the results and strengthen the potential of oral delivery of insulin by means of NPs.

Herein, focus will be put on recent reports in the oral delivery of insulin NPs for diabetes, as well as a critical analysis of the safety studies supporting their preclinical development. The improvement of early safety assessment by transitioning to quantitative, NP composition–immune performance relationship studies in representative models will also be elucidated. Thereby, the role and importance of rational optimization in the development of "safe by design" insulin NPs will be contextualized in the field of diabetes.

10.1.1 Background

Diabetes is a chronic, progressive, medically incurable disease, poorly controlled in a vast majority, in spite of tremendous advancements in pharmacotherapy (Sanyal 2013).

It is estimated that the worldwide prevalence of diabetes in adults was 6.6% in 2010 and will increase to 7.8% by the year 2030 (Card and Magnuson 2011). Today, according to the World Health Organization, 422 million people have diabetes (WHO 2016). Intensive insulin in Type 1 diabetes (T1D) patients is likely to reduce the risks of nephropathy, neuropathy, and retinopathy (Reichard et al. 1993), and initiating insulin early in the management of Type 2 diabetes (T2D) will prevent the development of most complications associated with diabetes.

Subcutaneous insulin therapy is a common mode of diabetes treatment, which significantly reduces morbidity and mortality, but can be burdened by complications (hypoglycemia and edema) and low patient compliance owing to the use of needles and the complexity of the insulin-based treatments and fear of unwanted hypoglycemia events and weight gain. To overcome such hurdles, considerable effort has been put into the development of alternative formulations that can provide bioactive insulin in a noninvasive manner posing minimal patient risk (Ribeiro et al. 2013). The oral route remains the preferred choice because of its noninvasive nature. However, proteins such as insulin have poor oral bioavailability, mainly due to their low permeability through the intestinal epithelium. The development of a delivery system intended to provide insulin orally requires a proper understanding of the mucosal microenvironment and intestinal physiology. Such efforts are important, as the identification of factors that influence the immune responses underlying the pathogenesis of T1D could provide an opportunity for therapeutic measures to prevent and/or treat the disease, as well as developing improved biomarkers for predicting future cases of diabetes.

The term *gut microbiota* represents a complex microbial community within the body, one capable of affecting health by contributing to nutrition and the prevention of colonization of the host by pathogens, and through influencing the development and maintenance of the immune system. Experimental evidence suggests that alterations in the gut microbiota are associated with the development of a number of disorders attributed to an overly activated immune system or autoimmunity (e.g., ulcerative colitis and Crohn's disease),

including an influence on T1D (Atkinson and Chervonsky 2012).

Although human clinical studies are still required, especially for long-term applications, the investigation has experienced difficulty in the development of an efficient and safe oral insulin delivery system (Lopes et al. 2014).

Potential protein therapies that have failed so far outnumber the successes, mainly due to a number of challenges that are faced in the development and use of protein therapeutics, among which route of administration is a relevant factor in any therapeutic intervention that governs both the biodistribution and efficacy of the protein (Muheem et al. 2014). The Food and Drug Administration (FDA) laid down the foundation for the popularity of protein therapeutics with the regulatory approval of recombinant insulin in 1982.

This chapter focuses on recent reports in oral delivery of insulin for diabetes, including diabetes-related inflammation diseases, as well as the suitability of the safety studies supporting its preclinical development. The improvement of early safety assessment by transitioning to quantitative, NP composition–safety performance relationship studies in representative models is also discussed. Thereby, the role and importance of rational optimization in the development of safety by design oral delivery of insulin NPs are contextualized in the field of diabetes.

10.2 ORAL ADMINISTRATION: THE PROMISING ROUTE

The route of administration is a relevant factor that governs both the pharmacokinetics and efficacy of insulin. The high molecular weight and hydrophilicity of insulin hinder its intestinal absorption, leading to poor oral bioavailability, low plasma levels, and variability. Various challenges, such as overcoming enzymatic degradation, solving the problem of poor absorption in the intestinal tract, and preserving its biological activity during the formulation process (Fonte et al. 2013; Muheem et al. 2014; Ruiz et al. 2014), have to be overcome. These obstacles have led researchers to develop more effective delivery systems toward the oral delivery of insulin.

The oral route is the most intensively investigated alternative route for insulin delivered parenterally. Despite several barriers to proteins' drug delivery that exist through the gastrointestinal tract (GIT), the oral route presents advantages, such as patient acceptability and ease of administration (Gamboa and Leong 2013; Ribeiro et al. 2013). Moreover, there are several cost savings because oral formulations do not require high-cost sterile manufacturing facilities or the direct intervention of healthcare professionals (Muheem et al. 2014).

Even though the oral delivery of insulin is an attractive option, there are still some challenges to be met before reaching its true potential. Oral delivery of proteins has long been considered the "holy grail" of drug development by showing high potential, but also presenting several problems to overcome (Shen 2003).

The main barrier to the development of a clinically usable oral insulin formulation is the low permeability of proteins across the intestinal wall coupled with high susceptibility to acid denaturation in the stomach and enzymatic degradation throughout the gut.

10.2.1 Oral Insulin Delivery Systems

Insulin is released from pancreatic β-cells into the hepatic portal vein and then into the liver, which is the primary site of action, whereas parenteral route and other delivery systems (buccal, pulmonary, and nasal) deliver the drug directly into the systemic circulation. In this delivery system, the drug reaches the systemic circulation, bypassing the first-pass metabolism, but in the case of oral delivery, insulin first reaches the liver (20% of the dose is available in the liver) and then the peripheral tissue. Thus, the oral route of administration is closer to the natural physiological route of insulin (Lopes et al. 2015).

An ideal delivery system for oral administration of insulin should prolong its intestinal residence time, reversibly increase the permeability of the mucosal epithelium to enhance the absorption of insulin, and provide intact insulin to the systemic circulation. Additionally, this delivery system must be safe after oral administration (Sonaje et al. 2009).

NPs are able to permeate the intestine by different pathways (Figure 10.1). In the case of insulin, it is adsorbed on the apical membrane and is internalized by specific types of endocytosis processes.

NPs hardly diffuse through the paracellular route, although some of their components, such as chitosan, may affect tight junctions (TJs) (Werle et al. 2009; Plapied et al. 2011). NPs can be uptaken by M-cells, which reveal higher transport activity than enterocytes. The type of target cells and the physicochemical properties of the NPs are two main factors that determine absorption pathways that may be utilized: phagocytosis or endocytic pathways (i.e., clathrin- and caveolae-mediated endocytosis) (Lopes et al. 2014). The main assembly unit is clathrin, a cytosolic coat protein. Clathrin-mediated endocytosis (CME) occurs via a specific receptor. A systematic evaluation of intestinal absorption studies for insulin NP (Figure 10.1), based on characterized pathways, was previously described (Lopes et al. 2014).

NPs can protect insulin from enzymes and acidic medium, as well as provide controlled release through enzyme degradation of the carrier,

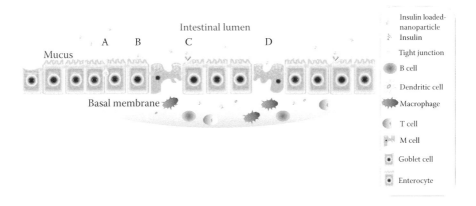

Figure 10.1. Pathways for insulin NP translocation through the intestinal epithelium. (From Lopes, M. A. et al., *Nanomedicine*, 10(6), 1139–1151, 2014. Copyright 2014. Reprinted with permission from Elsevier B.V.)

which varies according to the enzymes present at different parts in the GIT (Glangchai et al. 2008). Hence, pH-sensitive polymers such as acrylates or anionic polymers can also result in controlled release of insulin (Ramesan and Sharma 2009). Commonly used pH-sensitive polymers include Eudragit® (L100-55, L100, and S100), hydroxypropyl methylcellulose phthalate, and hydroxypropyl methylcellulose acetate succinate, which dissolve at pH values of 5.5, 6.0, 7.0, 5.5, and 5.5, respectively (Gamboa and Leong 2013). When applied in vivo, these pH-sensitive polymers show release profiles sensitive to a certain pH (Wu et al. 2010).

The nanoencapsulation of protein drugs can also offer targeting ability through the inclusion of ligands (Tahara et al. 2011), for example, the prohibitin-homing peptide ligand via a polyethylene glycol (PEG) linker to target adipose endothelial cells (Hossen et al. 2012) and the Fc portion of an immunoglobulin G (IgG) to target the Fc receptor to facilitate uptake and transport across epithelial cells (Foss et al. 2016). In addition, NPs are also engineered to provide sustained release of proteins (Tahara et al. 2011; Van Hove et al. 2014), which can be particularly beneficial for the treatment of chronic diseases such as diabetes. Tuning of NP properties for increased oral availability can be achieved by modulating size, charge, hydrophilicity, or hydrogen bonding capabilities, which mask the encapsulation characteristics and therefore increase transport of the NPs, together with their loaded proteins, through both the cell membrane and possibly the epithelial layer of the gut to provide systemic circulation (Gamboa and Leong 2013).

10.3 DIABETES AND INTESTINAL IMMUNE SYSTEM

The human intestinal microbiome is a symbiotic community that influences human health and development, including the development and maintenance of the human immune system. The intestinal mucosa is a common entry site for pathogens, even though there are a significant proportion of cells of the immune system. An intact mucosa provides the first line of defense against pathogens and exogenous matter (Atkinson and Chervonsky 2012).

With the importance of the gut microbiota for autoimmunity established, the mechanisms by which exogenous matter, such as NPs, specifically influence the pathogenesis of diabetes are of utmost importance.

Accumulating data indicate that dysregulation of the gut immune system may seem to have some relevance in the development of β-cell autoimmunity and T1D (Vaarala 2002). Following exposure to oral antigens, the dual nature of the gut's immune system may lead to tolerance and/or immunization.

In diabetic subjects, there is a potential risk of metabolic anomalies, such as autoimmune and inflammatory diseases. Multifactorial organ white adipose tissue secretes adipose-derived factors that have been collectively termed "adipokines" (Andrade-Oliveira et al. 2015). Adipokines may exert specific effects on a variety of biological processes, including inflammation, for example, interleukin (IL)-1β, IL-6, IL-8, and IL-10 and tumor necrosis factor α (TNFα) (Fasshauer and Blueher 2015). Adipokines can be an interesting drug target to treat autoimmune diseases, obesity, insulin resistance, and adipose tissue inflammation (Andrade-Oliveira et al. 2015).

10.3.1 Intestinal Immune System

The intestine forms a barrier not only against invading agents but also against dietary antigens and commensal bacteria. The mucosal barrier (Figure 10.2) consists of an extrinsic barrier with nonspecific defense mechanisms, that is, low pH, digestive enzymes, mucus, and peristalsis, and an immunological barrier with secretory IgA and IgM, whereas the intrinsic barrier is based on the integrity of the intestine itself (Paronen 2001).

The complexity of the immune system is the result of multiple, complementary, and sometimes conflicting functions that are fundamental to maintain homeostasis in organisms. Among these functions are the defense against invasive pathogens, the capacity to tolerate innocuous particles to which the intestine is often exposed, and self-components of the organism.

Mucosal surfaces, such as the intestine, represent a barrier between the human body and the external environment (Figure 10.3) (Sorini and Falcone 2013). Specialized epithelial cells constitute barrier surfaces that separate mammalian hosts from the external environment. The GIT is the largest of these barriers and is specially adapted to colonization by commensal bacteria that aid in digestion and markedly influence the development

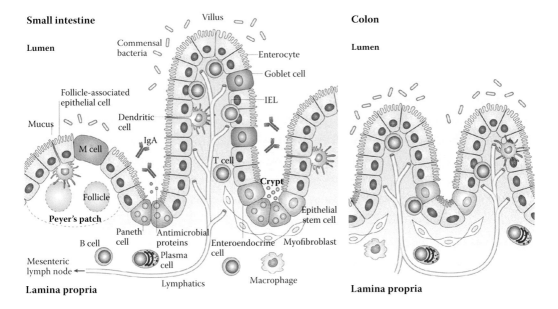

Figure 10.2. Anatomy of intestinal immune system. A single layer of IECs provides a physical barrier that separates the trillions of commensal bacteria in the intestinal lumen from the underlying lamina propria. (From Abreu, M. T., *Nat. Rev. Immunol.*, 10(2), 131–144, 2010. Copyright 2010. Reprinted with permission from Macmillan Publishers Ltd.)

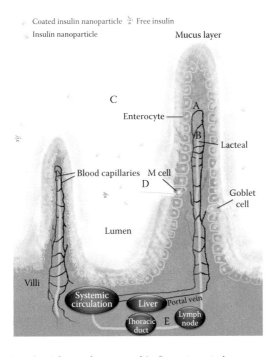

Figure 10.3. Schematic illustrations (particles not drawn to scale) of some intestinal structures where the flow and transport of NPs can occur. This can be in the direction of the (A) blood and (B) lymphatic flow. Following oral administration, the insulin NP may (C) flow into the GIT, (D) access and adhere to M-cells of the Peyer's patches or enterocytes, and (E) flow into the lymph vessels and become entrapped in the lymphatic pathway. (From Lopes, M.A. et al. *Nanomedicine: Nanotechnology, Biology and Medicine*, 10(6); 1139–51, 2014. Copyright 2014. Reprinted with permission from Elsevier B.V.)

and function of the mucosal immune system. However, microbial colonization carries with it the risk of infection and inflammation if epithelial or immune cell homeostasis is disrupted. The capacity to maintain the segregation between host and microorganism is essential to keeping the coexistence of commensal microbial communities and mucosal immune cells. The intestinal epithelium accomplishes this by forming a physical and biochemical barrier to commensal and pathogenic microorganisms. Intestinal epithelial cells (IECs) can sense and respond to microbial stimuli to reinforce their barrier function and to participate in the coordination of appropriate immune responses, ranging from tolerance to antipathogen immunity. This way, IECs maintain a fundamental immunoregulatory function that influences the development and homeostasis of mucosal immune cells (Peterson and Artis 2014).

Enterocytes and goblet cells (mucus secreting) cover the villi, which are interspersed with follicle-associated epithelium (FAE). These lymphoid regions, Peyer's patches, are covered with M-cells specialized in antigen sampling. M-cells are significant for insulin delivery, since they are relatively less protected by mucus and have a high transcytotic capacity (Ensign et al. 2012).

Foreign exogenous particles that enter the body throughout the intestinal mucosa must be rapidly removed by a protective immune response. On the other hand, the intestine is commonly home to a high number of commensal bacteria and is constantly exposed to large amounts of food proteins (Sorini and Falcone 2013). Thus, the intestinal immune system needs to discriminate between invasive particles and harmless antigens, activating immunological tolerance in relation to the latter. The anatomical organization and particular cell subsets that are present in the gut mucosa reveal the complexity of its functions (Sorini and Falcone 2013).

Insulin NPs comprise multiplexed formulations with a wide property range. Their physicochemical and biological properties (such as size, shape, and composition) can modulate the individual cell immune response (Dwivedi et al. 2011), which at this point may represent a "black box" in mucosal tissues (Cario 2012). Comprehensive data on the immunomodulatory behavior of NPs in mucosal cells are lacking. Establishing standardized and objective in vitro immunotoxicological assays for NPs is challenging. Most foreign substances at mucosal surfaces should be considered with caution, and some NPs regarded as biologically inert can be recognized as "nonself."

Although the evidence for the role of the gut immune cells in the pathogenesis of β-cell autoimmunity is indirect (Table 10.1), this possibility

TABLE 10.1
Evidence for the role of the gut immune system in the pathogenesis of Type 1 diabetes.

Animal models of autoimmune diabetes	Human T1D
Diet modifies the incidence of diabetes in nonobese diabetic (NOD) mice and in biobreeding (BB) rats	Cow milk exposure implies a risk of diabetes
Lymphocytes expressing mucosal adhesion molecules infiltrate the islets in NOD mice	Enhanced immune responsiveness to food antigens occurs in the patients with T1D
Monoclonal antibodies to MadCAM-1 and b 7-integrin prevent autoimmune diabetes in NOD mice and in the adoptive transfer model	GAD-reactive T-cells express gut-associated homing receptor
Diets with low diabetogenecity induce a Th2-type cytokine profile in the islet-infiltrating T-cells	Mucosal homing of lymphocytes derived from human diabetic pancreas
Mesenterial lymphocytes from young NOD mice transfer diabetes to healthy recipients	Increased permeability of the gut
Feeding autoantigen may prevent or accelerate autoimmune diabetes	Association of celiac disease with T1D
Dietary manipulations cause changes in the cytokine profile of the islets infiltrating lymphocytes in BB rats	Immunohistology of intestine shows markers of inflammation in T1D

SOURCE: Vaarala, O., *Diabetes Metab. Res. Rev.*, 15(5), 353–361, 1999; Vaarala, O., *Ann. NY Acad. Sci.*, 958, 39–46, 2002.

is fascinating since it may lead to the development of novel strategies for diabetes prevention (Vaarala 1999).

10.3.2 Humoral Immune Response to Insulin

Insulin is one of the major autoantigens of T1D and has some unique features, compared with other autoantigens. It is the major product of pancreatic islet β-cells, which are the specific target of autoimmune destruction. Insulin is the only T1D-associated autoantigen that is exclusively expressed in the β-cells, with the exception of self-antigen-expressing cells in lymphoid tissues such as the thymus, where insulin is expressed at low levels without hormonal importance. The other autoantigens are expressed in other islet cells and in other tissues, in addition to the β-cells. Insulin is secreted into the bloodstream and is a ubiquitous antigen in this sense (Ruiz et al. 2014).

In humans, insulin was the first autoantigen identified to which autoantibodies were proven to exist (Tiittanen 2006; Ruiz et al. 2014).

Environment is a critical factor in shaping the immune system, under physiologic and pathologic conditions, and it modulates the pathogenesis of autoimmune diseases such as T1D (Sorini and Falcone 2013). The gut microbiota can have a fundamental role as an intermediary between the variety of environmental triggers likely to alter autoimmune processes and the immune cells, possibly including autoimmune T-cells that patrol our mucosal surfaces (Sorini and Falcone 2013). Moreover, a direct relationship between microbiota changes and autoimmunity has been disclosed. The discovery that specific nutrients and dietary supplements can selectively affect the colonization capacity of either beneficial or detrimental species of the microbiota opens up the possibility of their therapeutic exploitation in the prevention or treatment of autoimmune diseases (Vaarala 1999; Burcelin 2012; Sorini and Falcone 2013). All these factors, together with certain genetic factors, may influence the regulation of immune responses in the gut immune system and possibly influence the development of β-cell autoimmunity. Prevention strategies of T1D based on manipulation of the gut immune system may provide new realism to the search for a cure to autoimmune diabetes (Vaarala 1999).

Many of the nanoparticulate-based formulations are capable of eliciting both cellular and humoral immune responses. While NPs may present high potential for oral delivery of insulin, it is also worth noting their potential drawbacks, particularly those associated with cytotoxicity. Since NPs have a relatively short history in medicine, they do not have a long-standing safety profile in human use. It is therefore essential that further research be carried out in NP toxicity to fully address these questions if they are to be accepted as an alternative method for the delivery of insulin and are licensed more widely for human use (Gregory et al. 2013).

One of the ways in which NPs are able to elicit different immune responses is through their size, moving into cells via nonclassical pathways and then being processed as such. Delivering antigens in different ways also has a profound effect on the resulting immune response, whether the antigen is decorated on the NP surface for presentation to antigen-presenting cells or encapsulated for slow release and prolonged exposure to the immune system. NPs are also versatile and can be modified with immunostimulatory compounds to enhance the intensity of the immune response or with molecules to increase their stability *in vivo* (PEG).

There are concerns about unwanted absorption of antigens or toxic substances from the gut lumen during oral delivery of nanoencapsulated insulin. In the case of chitosan insulin NPs, absorption enhancement was specific for the loaded insulin only and did not promote the intestinal absorption of the endotoxin lipopolysaccharide. On the basis of these results, it was concluded that chitosan NPs can be used as a safe carrier for the oral delivery of insulin (Sonaje et al. 2011).

Other therapeutic concerns are related to sustained immune responses to oral insulin by the production of neutralizing antibodies that can compromise its efficacy or safety (Barbosa and Celis 2007). This requires special attention since the role of M-cells in NP transport was elucidated, because these specialized epithelial cells are responsible for antigen sampling at the interface of mucosal surfaces (Lopes et al. 2014).

Once insulin has a relatively narrow therapeutic window, factors such as age, genomic factors, pathophysiological conditions, and other individual variations must be thoroughly evaluated, since they could affect GI transport.

Moreover, since insulin is a mitogen implicated in the increased risk of colorectal cancer (Giovannucci 1995; Argiles and Lopez-Soriano 2001), the risk of its oral administration also needs to be considered, although there are already studies showing that in fact insulin does not contribute to this disease process (Bao et al. 2010).

The diabetic condition is one example where pathophysiological-induced changes occur in the absorptive capacity of the GI mucosa for particulates (McMinn et al. 1996). Thus, insulin NP absorption studies should take this fact into account, since it can influence the results, namely, the absorption of NPs through the GIT and their transit to secondary organs.

10.4 INFLAMMATION ASSOCIATED WITH DIABETES AND NANOPARTICLES

Insulin resistance can precede the clinical onset of both T1D and T2D (Prentki and Nolan 2006; Razavi et al. 2006), and early in these two diseases, nondiabetic subjects can adapt to insulin resistance by increasing β-cell mass and function toward maintenance of euglycemic values (Weir and Bonner-Weir 2004; Terauchi et al. 2007). Since in T1D inflammation of pancreatic islets is a major factor in β-cell death, the activation of the acute-phase response and systemic inflammation plays a fundamental role for the pathoetiology of T2D and metabolic syndrome (Weir and Bonner-Weir 2004; Terauchi et al. 2007).

There are convincing data that support a role for inflammation in the pathogenesis of T2D, and anti-inflammatory drugs can improve glycemia without the danger of inducing hypoglycemia (Donath 2014). Treatments addressing inflammation can be used to prevent the progressive decrease in insulin secretion and effectiveness (Barzilay et al. 2001; Varvarovska et al. 2003).

NPs are capable of delivering inflammation-resolving drugs to sites of tissue injury. Several NP formulations that have potential for the treatment of a wide array of diseases characterized by excessive inflammation, such as diabetes, are displayed in Table 10.2.

From Table 10.2, it is evident that the results have been obtained through both *in vitro* (Chen et al. 2012; Ghosh et al. 2012; Ganguly et al. 2016; Kasiewicz and Whitehead 2016) and *in vivo* tests, mainly on induced-diabetes rodent models (Leuschner et al. 2011; Chen et al. 2012; Joshi et al. 2013; Karthick et al. 2014).

According to these findings, the possible connection between antidiabetic and anti-inflammatory effects has been confirmed first of all at a molecular and cellular level, through the modulation of the expression of different inflammation-related molecules (Chen et al. 2012; Leuschner et al. 2011).

However, this relationship has been demonstrated considering the long-term consequences of this disease as well, such as diabetic ulcers (Chen et al. 2012) or cardiovascular disease (Ganguly et al. 2016).

Moreover, it is clear that the main purpose of the findings highlighted in Table 10.2 is to prove that NPs have the fundamental benefit of allowing drug targeting, which consequently leads to less systemic adverse effects and to a reduction of the effective drug concentration.

In particular, the targeting has been developed with the use of siRNAs in several studies (Kasiewicz and Whitehead 2016) and always focusing primarily on molecules and genes connected to the inflammation process, such as ILs and TNFα (Karthick et al. 2014), CCR2, chemokine receptor (Leuschner et al. 2011), or the receptor for advanced glycation end products (RAGE) (Chen et al. 2012).

Furthermore, it has ultimately been revealed that NPs are important not only as delivery systems but also as an intrinsic therapeutic interest (Ghosh et al. 2012), a conclusion that increases interest in them even more.

10.5 TOXICOLOGICAL AND SAFETY ISSUES

Toxicity is a critical factor to be considered when evaluating the potential of insulin-loaded NPs. Given that NPs are designed to interact with cells, it is important to ensure that they do not cause any adverse effects or even damage the intestinal epithelium. The important issue is that, whether uncoated or coated, NPs will undergo biodegradation in the cellular environment and may affect cellular responses. For instance, biodegraded NPs can accumulate inside the cells and lead to intracellular changes, such as disruption of organelle integrity or gene alterations, which cause severe toxicity (Fonte et al. 2013).

Cytotoxicity may not be the only adverse effect, because cell immunological response may also be affected. Furthermore, molecules delivered to unnatural sites in unnatural quantities are likely to behave in unexpected ways, so from a toxicological perspective, oral delivery of macromolecules such as insulin may be questionable. If insulin is entrapped and not released

TABLE 10.2

Composition and study models of nanoparticle formulations used for inflammation associated with diabetes.

NP composition	Purpose	Study model	Output	Reference
Amphiphilic hyaluronic acid conjugates	To investigate the antiatherogenic potential of trivalent chromium, loaded cowpea mosaic virus (CPMV) NPs under hyperglycemic conditions	Human aortic smooth muscle cells (HASMCs)	Cellular evidence for an atheroprotective effect of CPMV-Cr in vascular smooth muscle cells (VSMCs)	Ganguly et al. 2016
Lipidoid NPs	Potential of RNA interference therapy to reduce the inappropriately high levels of TNFα in the wound bed	In vitro macrophage–fibroblast coculture model	Single lipidoid NP dose of 100 nM siTNFα downregulated TNFα and MCP-1 by 64% and 32%, respectively	Kasiewicz and Whitehead 2016
Gold nanoparticles (AuNPs) synthesized using the antidiabetic potent plant *Gymnema sylvestre* R. Br.	Anti-inflammatory effect by estimating the serum levels of TNFα, IL-6, and highly sensitive C-reactive protein (CRP)	Diabetes-induced Wistar albino rats	AuNPs decreased serum levels of TNFα, IL-6, and CRP to normal compared with those of the diabetic group and standard drug (glibenclamide), indicating a suppressing effect on inflammation	Karthick et al. 2014
Self-nano-emulsifying drug delivery system (SNEDDS) curcumin formulation	Efficacy in experimental diabetic neuropathy	Male Sprague Dawley rats	Greater neuroprotective action of SNEDDS curcumin when compared with naïve curcumin	Joshi et al. 2013
Mixture of AuNP, epigallocatechin gallate (EGCG), and α-lipoic acid (ALA) (AuEA) for topical treatment	To verify the capability of AuEA in changing the RAGE and being helpful in diabetic wound (in vivo study)	In vitro human foreskin fibroblasts (Hs68); in vivo male BALB/c mice	AuEA significantly accelerated diabetic wound healing through anti-inflammation and angiogenesis modulation	Chen et al. 2012
Pancreatic islet microvessels targeting cyclic peptide (CHVLWSTRKC) conjugated to the amphiphilic PLGA-b-PEG-COOH block copolymer	To investigate the potential of a new islet-targeted immunomodulatory approach	Mouse islet capillary endothelium (CE) compared with mouse skin CE to demonstrate the islet-homing capability of the NPs	These nanomaterials exhibit selective Islet-targeting capability and offer a tremendous advantage over systemic drug delivery, as they obtain a similar immunosuppressive response with a 200-fold lower drug concentration	Ghosh et al. 2012
NP prepared using phospholipids, cholesterol, polyethylene glycol-dimyristoglycerol (PEG-DMG) and siRNA with specific sequence against CCR2 (chemokine receptor) (siCCR2)	To investigate the capability of intravenous injection of NP-encapsulated siCCR2 in prolonging the normoglycemic period and, by association, islet graft function	Streptozotocin-induced-diabetes mice	Efficient degradation of CCR2 mRNA in monocytes prevents their accumulation in sites of inflammation; a prolonged normoglycemia after pancreatic islet transplantation	Leuschner et al. 2011

NOTE: PEG-DMG, polyethylene glycol-dimyristolglycerol.

from carrier systems until it reaches the systemic circulation, then this may not be an issue, but this approach is questionable, because insulin may cause gastroparesis (Fonte et al. 2013). The use of absorption enhancers may lead to a long-term toxicity, and surfactants can also damage intestinal epithelium. Indeed, absorption enhancers, when administered in a continuous manner, may also promote permeation of pathogens and toxins. Moreover, mucoadhesive systems may affect mucus turnover and consequently alter the physiology of the intestinal membrane (Fonte et al. 2013).

10.6 CURRENT SAFETY AND TOXICOLOGY STUDIES

It is known that the natural pH environment in the GIT varies from acidic in the stomach to slightly alkaline in the small intestine. Studies reported that mucoadhesive properties of NPs were found to be affected by the pH conditions in the small intestine, and also that with increasing pH, the amount of insulin transported decreased significantly (Sonaje et al. 2009). Results of *in vivo* toxicity studies demonstrated that there was no apparent toxicity observed for the animals treated with empty NPs, even with a dose 18 times higher than that used in the pharmacokinetic study (Sonaje et al. 2009).

Other studies reported no increase in hemolysis in the presence of the NPs, suggestive of suitable blood compatibility epithelial integrity, and cellular TJs in the ileum remained intact in the presence of the NPs. No statistical differences from the control values were noted in any of the measured outcomes, and no evidence of gross or histological abnormalities was reported, suggesting that the NPs were well tolerated over the course of a 14-day repeat-dose regimen (Sonaje et al. 2009).

A large number of clinical trials were retrieved from the clinical trials registry of the U.S. National Library of Medicine (www.clinicaltrials .gov; accessed July 5, 2014) when a search was conducted using the terms "oral AND insulin." Some studies on the safety and efficacy of the administration of oral insulin are completed, others are waiting for recruitment, and others were suspended (Table 10.3).

No published data are available yet, and it is unclear how many of these pertain to NP-based delivery systems.

Oramed Pharmaceuticals is developing an oral insulin product that consists of unmodified recombinant human insulin combined with adjuvants that protect it from enzymatic degradation in the GIT and promote its absorption from the gut. The aim of this study was to determine the optimal adjuvant-to-insulin ratio that can provide for the best pharmacodynamic profile, while maintaining the safety of the product. A decreased risk of hypoglycemia has been observed in numerous studies where insulin was administered either directly to the portal vein or indirectly by way of peritoneal insulin administration or peritoneal dialysates (Eldor et al. 2010).

A pill formulation of insulin has met safety and pharmacokinetic endpoints in a phase IIa trial (Fiore 2014). The compound, ORMD-0801, met its primary endpoint of safety and tolerability, as well as secondary endpoints of pharmacodynamics and pharmacokinetics. Oramed did not release numerical data, but said the full results would be presented at a scientific conference in the near future.

In April 2014, Oramed admitted that in a 6-week trial in 30 patients, a "formulation issue resulted in diminished and inconsistent release of study drug." So a third of the patients—those designed to receive the higher doses of the drug at 24 mg—were compromised as far as the study was concerned, receiving only 8 mg of the drug. Despite this obvious setback, the study did highlight some of the drug's potential. For the uncompromised group who received the 16 mg dose, the patients showed a mean reduction in nighttime glucose levels of about 23 mg/dl for the week compared with a placebo. And the fasting session from 5:00 a.m. to 7:00 a.m. provided a greater reduction of more than 30 mg/dl on average. Three patients in the group reported adverse events, which Oramed states were not related to the drug (Gibney 2014).

However, limited information is available, so no conclusions can be made regarding the toxicity profiles of the different formulations and their components. However, many of the components of the various NP formulations are included in approved oral drug products, as indicated by their listing in the *Physician's Desk Reference* and the U.S. FDA Inactive Ingredient Database.

At the present time, information available in the published literature on the oral safety of food-related nanomaterials is lacking in terms of both quantity and quality.

TABLE 10.3
Oral insulin delivery systems undergoing clinical trials.

Product name	Company	Technology	Status
Capsulin	Diabetology (Jersey, UK)	Axcess™; enteric-coated capsule filled with a mixture of insulin, an absorption enhancer, and a solubilizer	Phase IIa in T1D and phase II in T2D completed; agreement with USV Limited (Mumbai, India) to complete the development and commercialize for Indian market
ORMD-0801	Oramed (Jerusalem, Israel)	Enteric-coated capsule containing insulin and adjuvants to protect the protein and promote its intestinal uptake	Phase IIa in T1D and phase IIb in T2D
ORA2	BOWS Pharmaceuticals AG (Zug, Switzerland)	Capsule containing insulin in dextran matrix	Phase II in T2D; agreement with Orin Pharmaceuticals AG (Stockholm) for the development
–	Emisphere Technologies (Cedar Knolls, NJ)	Eligen®; capsule containing insulin and an absorption enhancer that facilitates the passive transcellular transport	Phase II in T2D suspended
NN1952	Novo Nordisk (Bagsvaerd, Denmark)	GIPET® from Merrion Pharmaceuticals (Dublin); capsule or tablet containing absorption enhancers that activate micelle formation, facilitating transport of insulin	Canceled after phase II
NN1953; NN1954	Novo Nordisk (Bagsvaerd, Denmark)	Tablet of long-acting insulin analog	Phase I in T1D and T2D
IN-105	Biocon (Bangalore)	Insulin modified with a small PEG	Phase II: Searching for other company to pursue development
HDV-I	Diasome (Conshohocken, PA)	Liposomal insulin, which is hepatic-directed vesicle insulin (HDV-I), in orally administered forms	Phase III
–	Biolaxy (Shanghai)	NOD Technology; insulin-loaded bioadhesive NPs	Phase I
–	Access Pharmaceuticals (Dallas, TX)	CobaCyte™; NP or polymer containing insulin, coated with vitamin B_{12} for targeted delivery	Phase I

SOURCE: Lopes, M. et al., *Ther. Deliv.*, 6(8), 973–987, 2015.

It is clear that assessment of the safety aspects of oral insulin dosing via NP-mediated systems has not received as much attention as the assessment of efficacy. This is not surprising, given that formal toxicology testing is not expected to be initiated until a suitable NP formulation for the oral delivery of insulin has been identified and demonstrated to be effective in relevant animal models (Card and Magnuson 2011).

A lack of adequate physicochemical characterization places a limit on the value and significance of the results of a given study and makes it difficult, if not impossible, to compare studies and identify parameters that might influence efficacy and/or safety.

10.7 DISCUSSION

Oral delivery of insulin is the most physiological way to replace the invasive parenteral route, as well as a very promising area for research. The strategy for the development of oral insulin has always been a challenge for researchers due to its high molecular weight, chemical or enzymatic degradation susceptibility, and low permeability through the intestinal mucosa. The high molecular weight of this class of drugs, coupled with their hydrophilic nature, restricts their transcellular permeation, perhaps the most difficult hurdle to overcome. Nanotechnology offers various efficient carriers for the delivery of proteins, namely, solid lipid NPs, nanostructured lipid carriers, liposomes, niosomes, cubosomes, and polymeric NPs.

NPs could be identified as foreign substances by the immune system, causing the cells to react against their surface and the contents. This reaction can result in an inflammatory response by the body.

However, with all the benefits provided by nanotechnology, one has to look at the safety and toxicity of the NPs that are being inserted into the bloodstream.

Oral delivery of insulin is often limited because of its long-term efficacy, and safety concerns need to be demonstrated through adequately powered studies in different patient populations across the diabetes spectrum. Furthermore, a reproducible absorption of insulin and an understanding of meal-related absorption are also important goals for developing drug delivery systems that need lifelong administration.

To obtain more information about the permeation of the GI barriers and the subsequent biological effects, physiologically relevant *in vitro* models should be used, which enable controlled variation of the most important parameters involved. Particle properties should be recorded in mucus of different pH values, and the extent of binding to proteins and other macromolecules should be studied. Physiologically relevant *in vitro* (coculture) models, including mucus, should be established to also investigate the effect of changed mucus structure, inflammation, and pH changes. It is obvious that *in vivo* experiments are also needed, but without good knowledge of the influence of GI variations on particle parameters and penetration *in vivo*, data may be difficult to interpret.

Clinical studies need to clearly demonstrate the superiority of insulin NPs over parenteral insulin formulations and oral hypoglycemic agents, including an improved antihyperglycemic profile, reduced weight gain, and better disease progression outcome in long-term studies. The toxicological profile of the developed delivery systems must be also properly assessed.

10.8 CONCLUSION

Recent advances have highlighted the great promise of NP-based insulin delivery for the prevention or treatment of diabetes. But it remains unclear whether these nanotechnology-based delivery systems can reach the clinic in the near future. Nanomedicine still requires proof-of-concept studies, opening a multitude of exciting research opportunities for insulin delivery. Many research studies have focused on the development of oral insulin delivery systems, able to circumvent the obstacles presented by the GIT enabling suitable insulin.

Microbiota influence the immune system through their ability to affect immune responses to pathogens and commensals, and most likely also autoimmunity response, since the microbial symbiotic colonization of the GIT may present a risk if epithelial or immune homeostasis is disturbed. Since insulin is a product that is subject to autoimmune destruction, intestinal microbiome alterations have a particular role in autoimmune disease development. Prevention strategies based on the manipulation of the gut immune system may provide new paths for a cure to autoimmune diabetes. Mucuspenetrating particles have the capability to improve oral drug delivery by penetrating into zones with a decreased mucus barrier, being retained longer in the firmly adherent layer and increasing distribution over the epithelium, leading to a more effective treatment.

Some problems have not yet been overcome, and extensive clinical trials are still needed. Many promising studies have been completed with various drugs. The vast array of *in vitro* systems and animal models that have been used has produced discordant results regarding the optimum characteristics for efficient NP-based delivery of insulin in the GIT. Fewer data are available concerning the safety of these systems once they interfere with the physicochemical and pharmacodynamic properties of the GI mucosa.

Successful NPs for the oral delivery of insulin must be able to increase insulin bioavailability to therapeutic levels, with minimal interindividual variability and lower doses achieving the desired antihyperglycemic effect.

REFERENCES

Abreu, M. T. 2010. Toll-like receptor signalling in the intestinal epithelium: How bacterial recognition shapes intestinal function. *Nature Reviews: Immunology* 10 (2):131–144.

Andrade-Oliveira, V., N. O. S. Camara, and P. M. Moraes-Vieira. 2015. Adipokines as drug targets in diabetes and underlying disturbances. *Journal of Diabetes Research* 2015:681612.

Argiles, J. M., and F. J. Lopez-Soriano. 2001. Insulin and cancer [review]. *International Journal of Oncology* 18 (4):683–687.

Atkinson, M. A., and A. Chervonsky. 2012. Does the gut microbiota have a role in type 1 diabetes? Early evidence from humans and animal models of the disease. *Diabetologia* 55 (11):2868–2877.

Bao, Y., K. Nimptsch, J. A. Meyerhardt, A. T. Chan, K. Ng, D. S. Michaud, J. C. Brand-Miller, W. C. Willett, E. Giovannucci, and C. S. Fuchs. 2010. Dietary insulin load, dietary insulin index, and colorectal cancer. *Cancer Epidemiology Biomarkers & Prevention* 19 (12):3020–3026.

Barbosa, M. D. F. S., and E. Celis. 2007. Immunogenicity of protein therapeutics and the interplay between tolerance and antibody responses. *Drug Discovery Today* 12 (15–16):674–681.

Barzilay, J. I., L. Abraham, S. R. Heckbert, M. Cushman, L. H. Kuller, H. E. Resnick, and R. P. Tracy. 2001. The relation of markers of inflammation to the development of glucose disorders in the elderly—The cardiovascular health study. *Diabetes* 50 (10):2384–2389.

Burcelin, R. 2012. Regulation of metabolism: A cross talk between gut microbiota and its human host. *Physiology (Bethesda)* 27 (5):300–307.

Card, J. W., and B. A. Magnuson. 2011. A review of the efficacy and safety of nanoparticle-based oral insulin delivery systems. *American Journal of Physiology: Gastrointestinal and Liver Physiology* 301 (6):G956–G967.

Cario, E. 2012. Nanotechnology-based drug delivery in mucosal immune diseases: Hype or hope? *Mucosal Immunology* 5 (1):2–3.

Chen, S.-A., H.-M. Chen, Y.-D. Yao, C.-F. Hung, C.-S. Tu, and Y.-J. Liang. 2012. Topical treatment with anti-oxidants and Au nanoparticles promote healing of diabetic wound through receptor for advance glycation end-products. *European Journal of Pharmaceutical Sciences* 47 (5):875–883.

Donath, M. Y. 2014. Targeting inflammation in the treatment of type 2 diabetes: Time to start. *Nature Reviews: Drug Discovery* 13 (6):465–476.

Dwivedi, P. D., A. Tripathi, K. M. Ansari, R. Shanker, and M. Das. 2011. Impact of nanoparticles on the immune system. *Journal of Biomedical Nanotechnology* 7 (1):193–194.

Eldor, R., M. Kidron, and E. Arbit. 2010. Open-label study to assess the safety and pharmacodynamics of five oral insulin formulations in healthy subjects. *Diabetes Obes Metab* 12 (3):219–223.

Ensign, L. M., R. Cone, and J. Hanes. 2012. Oral drug delivery with polymeric nanoparticles: The gastrointestinal mucus barriers. *Advanced Drug Delivery Reviews* 64 (6):557–570.

Fasshauer, M., and M. Blueher. 2015. Adipokines in health and disease. *Trends in Pharmacological Sciences* 36 (7):461–470.

Fiore, K. 2014. Insulin pill passes safety test. http://www.medpagetoday.com/Endocrinology/Diabetes/44072 (accessed July 5).

Fonte, P., F. Araujo, S. Reis, and B. Sarmento. 2013. Oral insulin delivery: How far are we? *Journal of Diabetes Science and Technology* 7 (2):520–531.

Foss, S., A. Grevys, K. M. K. Sand, M. Bern, P. Blundell, T. E. Michaelsen, R. J. Pleass, I. Sandlie, and J. T. Andersen. 2016. Enhanced FcRn-dependent transepithelial delivery of IgG by Fc-engineering and polymerization. *Journal of Controlled Release* 223:42–52.

Gamboa, J. M., and K. W. Leong. 2013. In vitro and in vivo models for the study of oral delivery of nanoparticles. *Advanced Drug Delivery Reviews* 65 (6):800–810.

Ganguly, R., A. M. Wen, A. B. Myer, T. Czech, S. Sahu, N. F. Steinmetz, and P. Raman. 2016. Anti-atherogenic effect of trivalent chromium-loaded CPMV nanoparticles in human aortic smooth muscle cells under hyperglycemic conditions in vitro. *Nanoscale* 8 (12):6542–6554.

Ghosh, K., M. Kanapathipillai, N. Korin, J. R. McCarthy, and D. E. Ingber. 2012. Polymeric nanomaterials for islet targeting and immunotherapeutic delivery. *Nano Letters* 12 (1):203–208.

Gibney, M. 2014. Oramed's oral insulin PhIIa details show a compromised study, and shares continue to fall. April 30. http://www.fiercepharma.com/regulatory/oramed-s-oral-insulin-phiia-details-show-a-compromised-study-and-shares-continue-to-fall (accessed July 10).

Giovannucci, E. 1995. Insulin and colon-cancer. *Cancer Causes & Control* 6 (2):164–179.

Glangchai, L. C., M. Caldorera-Moore, L. Shi, and K. Roy. 2008. Nanoimprint lithography based fabrication of shape-specific, enzymatically-triggered smart nanoparticles. *Journal of Controlled Release* 125 (3):263–272.

Gregory, A. E., R. Titball, and D. Williamson. 2013. Vaccine delivery using nanoparticles. *Frontiers in Cellular and Infection Microbiology* 3:13.

Hossen, Md. N., K. Kajimoto, H. Akita, M. Hyodo, and H. Harashima. 2012. Vascular-targeted nanotherapy for obesity: Unexpected passive targeting mechanism to obese fat for the enhancement of active drug delivery. *Journal of Controlled Release* 163 (2):101–110.

Joshi, R. P., G. Negi, A. Kumar, Y. B. Pawar, B. Munjal, A. K. Bansal, and S. S. Sharma. 2013. SNEDDS curcumin formulation leads to enhanced protection from pain and functional deficits associated with diabetic neuropathy: An insight into its mechanism for neuroprotection. *Nanomedicine: Nanotechnology, Biology, and Medicine* 9 (6):776–785.

Karthick, V., V. Ganesh Kumar, T. Stalin Dhas, G. Singaravelu, A. Mohamed Sadiq, and K. Govindaraju. 2014. Effect of biologically synthesized gold nanoparticles on alloxan-induced diabetic rats—An in vivo approach. *Colloids and Surfaces B—Biointerfaces* 122:505–511.

Kasiewicz, L. N., and K. A. Whitehead. 2016. Silencing TNF alpha with lipidoid nanoparticles downregulates both TNF alpha and MCP-1 in an in vitro co-culture model of diabetic foot ulcers. *Acta Biomaterialia* 32:120–128.

Leuschner, F., P. Dutta, R. Gorbatov, T. I. Novobrantseva, J. S. Donahoe, G. Courties, K. M. Lee et al. 2011. Therapeutic siRNA silencing in inflammatory monocytes in mice. *Nature Biotechnology* 29 (11):1005-U73.

Lopes, M., S. Simoes, F. Veiga, R. Seica, and A. Ribeiro. 2015. Why most oral insulin formulations do not reach clinical trials. *Therapeutic Delivery* 6 (8):973–987.

Lopes, M. A., B. A. Abrahim, L. M. Cabral, C. R. Rodrigues, R. M. F. Seiça, F. José de Baptista Veiga, and A. J. Ribeiro. 2014. Intestinal absorption of insulin nanoparticles: Contribution of M cells. *Nanomedicine: Nanotechnology, Biology and Medicine* 10 (6):1139–1151.

McMinn, L. H., G. M. Hodges, and K. E. Carr. 1996. Gastrointestinal uptake and translocation of microparticles in the streptozotocin-diabetic rat. *Journal of Anatomy* 189:553–559.

Muheem, A., F. Shakeel, M. A. Jahangir, M. Anwar, N. Mallick, G. K. Jain, M. H. Warsi, and F. J. Ahmad. 2014. A review on the strategies for oral delivery of protein and peptides and their clinical perspectives. *Saudi Pharmaceutical Journal* 25 (4).

Paronen, J. 2001. Dietary insulin and the gut immune system in type 1 diabetes. Molecular Medicine, University of Helsinki.

Peterson, L. W., and D. Artis. 2014. Intestinal epithelial cells: Regulators of barrier function and immune homeostasis. *Nature Reviews: Immunology* 14 (3):141–153.

Plapied, L., N. Duhem, A. des Rieux, and V. Préat. 2011. Fate of polymeric nanocarriers for oral drug delivery. *Current Opinion in Colloid & Interface Science* 16 (3):228–237.

Prentki, M., and C. J. Nolan. 2006. Islet beta cell failure in type 2 diabetes. *Journal of Clinical Investigation* 116 (7):1802–1812.

Ramesan, R. M., and C. P. Sharma. 2009. Challenges and advances in nanoparticle-based oral insulin delivery. *Expert Review of Medical Devices* 6 (6):665–676.

Razavi, R., Y. Chan, F. N. Afifiyan, X. J. Liu, X. Wan, J. Yantha, H. Tsui et al. 2006. TRPV1(+) sensory neurons control beta cell stress and islet inflammation in autoimmune diabetes. *Cell* 127 (6):1123–1135.

Reichard, P., B. Y. Nilsson, and U. Rosenqvist. 1993. The effect of long-term intensified insulin-treatment on the development of microvascular complications of diabetes-mellitus. *New England Journal of Medicine* 329 (5):304–309.

Ribeiro, A. J., R. Seiça, and F. Veiga. 2013. Nanoparticles for oral delivery of insulin. In *Drug Delivery Systems: Advanced Technologies Potentially Applicable in Personalised Treatment*, ed. J. Coelho, 109–125. Amsterdam: Springer.

Ruiz, E. C., M. G. Hierro, and M. Torres. 2014. Oral insulin delivery systems. *European Industrial Pharmacy* (14).

Sanyal, D. 2013. Diabetes is predominantly an intestinal disease. *Indian Journal of Endocrinology and Metabolism* 17 (Suppl 1):S64–S67.

Shen, W. C. 2003. Oral peptide and proteindelivery: Unfulfilled promises? *Drug Discovery Today* 8 (14):607–608.

Sonaje, K., K. J. Lin, M. T. Tseng, S. P. Wey, F. Y. Su, E. Y. Chuang, C. W. Hsu, C. T. Chen, and H. W. Sung. 2011. Effects of chitosan-nanoparticle-mediated tight junction opening on the oral absorption of endotoxins. *Biomaterials* 32 (33):8712–8721.

Sonaje, K., Y.-H. Lin, J.-H. Juang, S.-P. Wey, C.-T. Chen, and H.-W. Sung. 2009. In vivo evaluation of safety and efficacy of self-assembled nanoparticles for oral insulin delivery. *Biomaterials* 30 (12):2329–2339.

Sorini, C., and M. Falcone. 2013. Shaping the (auto) immune response in the gut: The role of intestinal immune regulation in the prevention of type 1 diabetes. *American Journal of Clinical and Experimental Immunology* 2 (2):156–171.

Tahara, K., S. Samura, K. Tsuji, H. Yamamoto, Y. Tsukada, Y. Bando, H. Tsujimoto, R. Morishita, and Y. Kawashima. 2011. Oral nuclear factor-kappa B decoy oligonucleotides delivery system with chitosan modified poly(D,L-lactide-co-glycolide) nanospheres for inflammatory bowel disease. *Biomaterials* 32 (3):870–878.

Terauchi, Y., I. Takamoto, N. Kubota, J. Matsui, R. Suzuki, K. Komeda, A. Hara et al. 2007. Glucokinase and IRS-2 are required for compensatory beta cell hyperplasia in response to high-fat diet-induced insulin resistance. *Journal of Clinical Investigation* 117 (1):246–257.

Tiittanen, M. 2006. Immune response to insulin and changes in the gut immune system in children with or at risk for type 1 diabetes. Viral Deseases and Immunology, University of Helsinki.

Vaarala, O. 1999. Gut and the induction of immune tolerance in type 1 diabetes. *Diabetes/Metabolism Research and Reviews* 15 (5):353–361.

Vaarala, O. 2002. The gut immune system and type 1 diabetes. *Annals of the New York Academy of Sciences* 958:39–46.

Van Hove, A. H., M.-J. G. Beltejar, and D. S. W. Benoit. 2014. Development and in vitro assessment of enzymatically-responsive poly(ethylene glycol) hydrogels for the delivery of therapeutic peptides. *Biomaterials* 35 (36):9719–9730.

Varvarovska, J., J. Racek, F. Stozicky, J. Soucek, L. Trefil, and R. Pomahacova. 2003. Parameters of oxidative stress in children with type 1 diabetes mellitus and their relatives. *Journal of Diabetes and Its Complications* 17 (1):7–10.

Weir, G. C., and S. Bonner-Weir. 2004. Five stages of evolving beta-cell dysfunction during progression to diabetes. *Diabetes* 53:S16–S21.

Werle, M., H. Takeuchi, and A. Bernkop-Schnuerch. 2009. Modified chitosans for oral drug delivery. *Journal of Pharmaceutical Sciences* 98 (5):1643–1656.

WHO (World Health Organization). 2016. Global report on diabetes. Geneva: WHO.

Wu, C.-S., X.-Q. Wang, M. Meng, M.-G. Li, H. Zhang, X. Zhang, J.-C. Wang, T. Wu, W.-H. Nie, and Q. Zhang. 2010. Effects of pH-sensitive nanoparticles prepared with different polymers on the distribution, adhesion and transition of rhodamine 6G in the gut of rats. *Journal of Microencapsulation* 27 (3):205–217.

CHAPTER ELEVEN

Concluding Remarks

Inflammation is an important protective, immune response. However, chronic inflammation and dysfunction in the process can result in primary inflammatory disease or secondary inflammatory disease. The role of inflammation in such a wide array of diseases is astounding—from inflammatory bowel disease, to asthma, to multiple sclerosis, to cancer, neurodegenerative disease, and diabetes. These are a mere handful of examples. The understanding of inflammation as an essential component to many diseases suggests that perhaps we should be asking, which diseases do not have an inflammatory component?

Treating inflammatory disease with nanomedicine is not an outrageous concept; the benefits of nanomedicine, such as reduced residual toxicity and the opportunity for molecularly targeted design, could enhance treatment options for primary and secondary inflammatory disease. The challenge for nanomedicine for inflammatory disease is not in identifying the clear benefits, but in the clinical translation of nanomedicine therapies from the bench to the bedside.

Despite the challenge of translation, and encouraged by the great demand for standardized characterization techniques for nanomedicine, the National Cancer Institute's Nanotechnology Characterization Laboratory (NCL) has established a precedent for successful nanomedicine translation. In concert with the National Institute of Standards and Technology, the NCL is developing protocols for the standardized characterization of nanomedicines for cancer. This is exactly what translational nanomedicine need for all therapeutic applications. The NCL is also a resource for translation, assisting researchers through the process from the bench into clinical trials. An NCL of the National Institutes of Health would accelerate the translational process and create a scripted pathway for translation. Such a program has the potential to revolutionize the translation of nanomedicines and, subsequently, result in the pharmaceutical "boom" that has been expected since the completion of the Human Genome Project.

Although there are challenges, the future is bright for translational nanomedicine for inflammatory disease. The question is, which nanomedicine formulation and which disease application will be the next success story to attain market approval?

INDEX

Page numbers followed by f and t indicate figures and tables, respectively.